Birkhäuser

Statistics for Industry and Technology

For additional information on this series, go to
www.springer.com/series/4982

Mathematical and Statistical Models and Methods in Reliability

Applications to Medicine, Finance, and Quality Control

V.V. Rykov
N. Balakrishnan
M.S. Nikulin
Editors

Birkhäuser

Editors
V.V. Rykov
Department of Applied Mathematics
Gubkin Russian State University of Oil and Gas
Leninskii Prospekt 65
117917 Moskva
Russia
vladimir_rykov@mail.ru

N. Balakrishnan
Department of Mathematics and Statistics
McMaster University
1280 Main Street West
Hamilton, Ontario L8S 4K1
Canada
bala@univmail.cis.mcmaster.ca

M.S. Nikulin
l'Institut de Mathématiques de Bordeaux, IMB
Université Victor Segalen Bordeaux 2
33076 Bordeaux Cedex
France
nikou@sm.u-bordeaux2.fr

ISBN 978-0-8176-4970-8 e-ISBN 978-0-8176-4971-5
DOI 10.1007/978-0-8176-4971-5
Springer New York Dordrecht Heidelberg London

Library of Congress Control Number: 2010938856

Mathematics Subject Classification (2010): 60K10, 62Mxx, 90B25

Printed on acid-free paper

www.birkhauser-science.com

Contents

Part III Applications

Preface

The contributions in this volume were all presented as invited papers at the Sixth International Conference on Mathematical Methods in Reliability: Theory, Methods, Applications (MMR 2009), which was held at Gubkin Russian State University of Oil and Gas (Gubkin University, Moscow, Russia) during June 22–26, 2009. The International Organizing Committee of this conference included organizers of the previous conferences, namely, Professors Nikolaos Limnios (France), Mikhail Nikulin (France, Russia), Bo Lindqwist (Norway), Sally McNulty (USA), Tim Bedford (UK), and Vladimir Rykov (Russia). In addition to Gubkin University, the Peoples Friendship University of Russia (PFUR), and the University of Bordeaux-2 (France) participated in the meeting's organization.

Reliability theory is a multidisciplinary science aiming to provide complex technical, computer, and informational systems and processes that are resistant to failure. Catastrophic events of the recent past, such as the explosion of the 4th block of Chernobyl's nuclear power station in April 1986, the failure of the blocking system that switched out 21 United States electrical stations in August 2003, and the 2009 breakdown of a turbine in the Sayano-Shushenskaya electrical station, show the necessity for the scientific community to pay more serious attention to reliability problems. Although human error played an important role in most of these events, mathematical modeling and careful investigation into causes of failure are nevertheless very important.

During the early stages of research on reliability theory, the primary focus was on developing mathematical terminology and formalism. These rudiments were established in works such as B. V. Gnedenko, Yu. K. Belyaev, A. D. Solov'ev's *Mathematical Methods in Reliability Theory* and R. E. Barlow F. Proschan's *Mathematical Theory of Reliability*. More modern developments and practical problems began to receive attention at the end of the last century, when the First International Conference on Mathematical Methods in Reliability Theory (MMR 1997) was organized in Bucharest in 1997. Since then, six more conferences have been undertaken as part of the MMR series:

the second, in Bordeaux (France, 2000);
the third, in Trondheim (Norway, 2002);
the fourth, in Stanta Fe (New Mexico, USA, 2004);
the fifth, in Glasgow (Scotland, UK, 2007);

the sixth, in Moscow (Russia, 2009, based on which this volume is being prepared); and

the seventh, in Beijing (China, planned for 2011).

More than 200 people from 35 countries participated in the sixth conference and presented a total of 167 talks. Ten plenary talks (1 hour) were also presented; most appear in this volume. All the talks given at the conference were broadly classified according to the following topics: "Mathematical models and methods in reliability theory" (22 sessions), "Statistical methods in reliability theory" (10 sessions), "Computer tools and support of reliability problems solution" (3 sessions), "Applications of reliability theory in industry, medicine, power stations, transport and other spheres" (9 sessions). Accordingly, these topics are well represented in the present collection.

It is worth noting that the conference also included a "round table discussion" devoted to the memory of one of the field's greatest pioneers, B. V. Gnedenko. Professors V. Korolyuk, Yu. Belyaev, I. Ushakov, I. Kovalenko, and D. B. Gnedenko (B.V.'s son) all discussed their memories and experiences with this remarkable scientist and man. Further information on this and other conference events can be found at http://mmr.gubkin.ru.

We would like to thank the Russian Foundation of Fundamental Investigation and the open joint-stock company "Gasprom. Promgas" for their financial support of the conference. Our final thanks go to Mr. Tom Grasso (Editor, Birkhäuser, Boston) for his support and encouragement in producing this book, and to Mrs. Debbie Iscoe for her fine work on the entire manuscript.

Moscow, Russia
Hamilton, Canada
Bordeaux Cedex, France
June 2010

V.V. Rykov
N. Balakrishnan
M.S. Nikulin

List of Contributors

Ahmadi Reza
City University, London, UK
ahmadi_stat@yahoo.com

Akushkina Kseniya A.
Novosibirsk State Technical University,
Novosibirsk, Russia
ksushco@ngs.ru

Andronov Alexander
Riga Technical University Riga Latvia, 1
Kalku Str., LV-1658
Aleksandrs.Andronovs@rtu.lv

Antonov Alexander
Obninsk state technical university for
nuclear power engeneering
(INPE), Obninsk, Russia
antonov@iate.obninsk.ru

Bagdonavičius Vilijandas
Vilnius University, Vilnius, Lithuania
Vilijandas.Bagdonavicius@mif.vu.lt

Barreau Mihaela
LASQUO, University of Angers, Angers,
France

Belova Kristina
INPE, Obninsk, Russia
kiivanova@gmail.com

Belyaev Yuri K.
Dept. of Mathematics and Mathematical
Statistics,
Umeå University, Umeå;
and
Dept. of Forest Economics,
Swedish University of Agricultural
Sciences, Umeå,
Sweden
yuri.belyaev@math.umu.se;
yuri.belyaev@sekon.se

Bousseboua Moussedek
University Mentouri, Constantine.
moussedek1@yahoo.fr

Cha Ji Hwan
Department of Statistics, Ewha
Womans University, Seoul, 120-750,
Korea
jhcha@ewha.ac.kr

Chepurko Valeriy
INPE, Obninsk, Russia
chepurko@iate.obninsk.ru

Chimitova Ekaterina V.
Novosibirsk State Technical University,
Novosibirsk, Russia
chim@mail.ru

Cloupet Sylvain
LASQUO, University of Angers, Angers,
France

Demri Amel
LASQUO, University of Angers, Angers,
France

Dimitrov Boyan
Kettering University, Flint, Michigan, USA
bdimitro@kettering.edu

Yi Ding
School of EEE Nanyang Technological University Singapore
dingyi@ntu.edu.sg

Döring Maik
Institut für Angewandte Mathematik und Statistik,
Universität Hohenheim, 70599 Stuttgart, Germany,
maik.doering@uni-hohenheim.de

Finkelstein Maxim
Department of Mathematical Statistics, University of the Free State, 339 Bloemfontein 9300, South Africa/Max Planck
Institute for Demographic Research, Germany
FinkelM@ufs.ac.za

Foschi Rachele
Università degli Studi di Roma "La Sapienza", Roma, Italy

Guérin Fabrice
LASQUO, University of Angers, Angers, France
fabrice.guerin@univ-angers.fr

Hambli Ridha
Polytech Orléans, Orléans, France
ridha.hambli@univ-orleans.fr

Harlamov B.P.
Institute of Problems of Mechanical Engineering
Russian Academy of Science. St.Petersburg
b.p.harlamov@gmail.com

Hersant Julien
LASQUO, University of Angers, Angers, France

Hong Yili
Department of Statistics,
Virginia Tech, Blacksburg, VA, 24060
yilihong@vt.edu

Huber Catherine
Université Paris Descartes, 45 rue des Saints-Pères, 75 270 Paris Cedex 06, and U1018 INSERM, 94 800, Villejuif.
catherine.huber@parisdescartes.fr,
http://www.biomedicale.univ-paris5.fr/catherine.huber/

Jensen Uwe
Institut für Angewandte Mathematik und Statistik,
Universität Hohenheim, 70599 Stuttgart, Germany,
jensen@uni-hohenheim.de

Kahle Waltraud
Institute of Mathematical Stochastics, Otto-von-Guericke-University, D-39016 Magdeburg, Germany,
waltraud.kahle@ovgu.de

Kaminskiy M.P.
Center of Technology and Systems Management, University of Maryland, College Park, USA
mkaminsk@umd.edu

Karagrigoriou Alex
Department of Mathematics and Statistics,
University of Cyprus, Cyprus
alex@ucy.ac.cy

Koroliuk V. S.
Institute of Mathematics,
Ukrainian National Academy of Science, Kiev, Ukraine
korol@imath.kiev.ua

Kozyrev Dmitry
Dept. of Probability Theory
and Mathematical Statistics,
Peoples' Friendship University of Russia, Ordzhonikidze st., 3, Moscow 117198, Russia
kozyrevdv@gmail.com

Krivtsov V.V.
Office of Technical Fellow in Quality Engineering
Ford Motor Company, Dearborn, MI USA
vkrivtso@ford.com

Kumazaki Chiharu
Department of Systems Engineering,
Univeristy of Electro-Communications,
Tokyo, Japan

Läuter Henning
University of Potsdam,
Institute of Mathematics
laeuter@uni-potsdam.de

Lee Mei-Ling Ting
University of Maryland, College Park, USA
mltlee@umd.edu

Lemeshko Boris Yu.
Novosibirsk State Technical University,
Novosibirsk, Russia
Lemeshko@fpm.ami.nstu.ru

Lemeshko Stanislav B.
Novosibirsk State Technical University,
Novosibirsk, Russia
skyer@mail.ru

Liero Hannelore
University of Potsdam,
Institute of Mathematics
liero@uni-potsdam.de

Limnios N.
Laboratoire de Mathématiques Appliquées,
Université de Technologie de Compiègne, France
nikolaos.limnios@utc.fr

Lindqvist Bo Henry
Department of Mathematical Sciences,
Norwegian University of Science and Technology,
N-7491 Trondheim, Norway
bo@math.ntnu.no

Anatoly Lisnianski
Israel Electric Corporation Ltd, Israel
anatoly-l@iec.co.il

Lundgren Robin
Mälardalen University, Västerås, Sweden
robin.lundgren@mdh.se

Masiulaitytė Inga
Vilnius University, Vilnius, Lithuania
imasiulaityte@gmail.com

Mattheou Kyriacos
Department of Mathematics and Statistics,
University of Cyprus, Cyprus

Meeker William Q.
Department of Statistics, Iowa State University, Ames, IA, 50011
wqmeeker@iastate.edu

Mesbah Mounir
University Pierre et Marie Curie, Paris 6
mounir.mesbah@upmc.fr

Musal R. Muzaffer
Texas State University, San Marcos,
TX 78666, USA
rm84@txstate.edu

Newby Martin
City University, London, UK
n.j.newby@city.ac.uk

Nikulin Mikhail
IMB, Victor Segalen University,
Bordeaux, France
mikhail.nikouline@u-bordeaux2.fr

Panayiotou Panayiotis
Department of Mathematics and Statistics,
University of Cyprus, Cyprus

Postovalov Sergey N.
Novosibirsk State Technical University,
Novosibirsk, Russia
Postovalov@ngs.ru

Ramadan Ayad
University of Potsdam,
Institute of Mathematics
ayad_math@yahoo.com

Rogozhnikov Andrey P.
Novosibirsk State Technical University,
Novosibirsk, Russia
rogozhnikov.andrey@gmail.com

Rosner Bernard
Harvard Medical School, Boston, MA,
USA
Stbar@channing.harvard.edu

Rykov Vladimir
Dept. of Probability Theory
and Mathematical Statistics,
Peoples' Friendship University of Russia,
Ordzhonikidze st., 3, Moscow 117198,
Russia
vladimir_rykov@mail.ru

Saaidia Noureddine
Université Victor Segalen
Bordeaux 2, France, et
Université Badji Mokhtar, Annaba,
Algérie
saaidianoureddine@yahoo.fr

Sherman E.D.
Kazan State Univercity, 420008,
Kremlevskaya-st., 18, Kazan, Russia
Evgenyi.Sherman@ksu.ru

Silvestrov Dmitrii S.
Stockholm University, Stockholm,
Sweden
silvestrov@math.su.se

Solev Valentin N.
St.Petersburg Department of V.A.Steklov
Institute of Mathematics of the Russian
Academy of Sciences, Russia
solev@pdmi.ras.ru

Soyer Refik
George Washington University,
Washington, DC 20052, USA
soyer@gwu.edu

Fabio Spizzichino
Università degli Studi di Roma
"La Sapienza", Roma, Italy
Fabio.Spizzichino@uniroma1.it

Suzuki Kazuyuki
Department of Systems Engineering,
Univeristy of Electro-Communications,
Tokyo, Japan
suzuki@se.uec.ac.jp

Volodin I.N.
Kazan State Univercity, 420008,
Kremlevskaya-st., 18, Kazan, Russia
Igor.Volodin@ksu.ru

Vonta Filia
Department of Mathematics, National
Technical University of Athens, Greece,
vonta@math.ntua.gr

Whitmore G. A.
McGill University, Montreal, Canada
George.Whitmore@mcgill.ca

Yamamoto Watalu
Department of Systems Engineering,
Univeristy of Electro-Communications,
Tokyo, Japan
watalu@se.uec.ac.jp

List of Tables

List of Figures

Part I

Reliability Models, Methods, and Optimization

1

Reliability of Semi-Markov Systems with Asymptotic Merging Phase Space

V.S. Koroliuk[1] and N. Limnios[2]*

[1] Institute of Mathematics, Ukrainian National Academy of Science, Kiev, Ukraine
[2] Laboratoire de Mathématiques Appliquées, Université de Technologie de Compiègne, France

Abstract: The aim of this chapter is to present, under an unified framework, asymptotic merging of phase state space of complex systems in reliability. Such simplification methods are important in reliability since the most system have very large phase spaces and it is almost impossible to handle them by usual analytical methods. Results presented here are of averaging type and obtained by weak convergence techniques. Nevertheless, a result of diffusion approximation is also given.

Keywords and phrases: Reliability, Semi-Markov process, Renewal process, Markov process, Merging, Random evolution, Weak convergence, Singular perturbation

1.1 Introduction

Modeling reliability of real systems is often hard, or impossible to handle by direct analytical-numerical methods, due essentially to the very large number of components. For example, a system of twenty binary components (yet a small system for real problems) gives a phase (state) space of more than one million of states which is almost impossible to handle by the usual analytical-numerical methods.

A common point of methods developed in order to handle such systems is the reduction of the number of states called aggregation or merging of phase space methods. Of course, in such methods the merging system has to keep the essential characteristics of the original system in regards of reliability.

The exact aggregation is a natural candidate method to this end. This method, in the case of a system described by a Markov process, $x(t), t \geq 0$, with (a large) phase space, E, is to consider another Markov process, $\widehat{x}(t), t \geq 0$, with (a much smaller) state space, $\widehat{E}$. This means that there exists a (merging) function $v : E \to \widehat{E}$ such that

$$Q(f \circ v) = (\widehat{Q} f) \circ v,$$

for any bounded and measurable function f on $\widehat{E}$; where Q and $\widehat{Q}$ are the generators of the Markov processes x and $\widehat{x}$, respectively.

V.V. Rykov et al. (eds.), *Mathematical and Statistical Models and Methods in Reliability: Applications to Medicine, Finance, and Quality Control*, Statistics for Industry and Technology, DOI 10.1007/978-0-8176-4971-5_1, © Springer Science+Business Media, LLC 2010

Unfortunately, the exact aggregation method usually does not work for reliability problems, essentially, due to the fact that failure rate and repair rates values are of different magnitudes. So, there is need to provide approximation methods for aggregation. The method that seems to give interesting results for merging the phase space is the method of (functional) asymptotic merging by means of weak convergence techniques, (see, e.g., [KL05a,b, KT93, KS95, A08, SV79, SK89, SHS02, S04, LS]). The most known are average, diffusion and Poisson approximation methods (see, e.g., [KL05a,b, K90, LO01]).

We are proposed here to present some kind of asymptotic methods in the case where the temporal behavior of the system is described by a semi-Markov process which is the most general process encountered in the literature of reliability modeling. The problem here can be formulated as follows. Given a family of semi-Markov processes in series scheme, that is, for $\varepsilon > 0$, the process $x^\varepsilon(t), t \geq 0$, is a semi-Markov process, we have

$$v(x^\varepsilon(t)) \Longrightarrow \widehat{x}(t), \quad \varepsilon \to 0,$$

where the limit process $\widehat{x}(t)$ is a Markov process. So, in this way, we not only reduce the initial phase space to a simpler one, but also we get a Markov process instead of the initial semi-Markov one.

The asymptotic considered here is of functional type in series scheme where the considered systems are indexed by a series parameter $\varepsilon > 0$ in a weak convergence framework. Average and diffusion approximations are considered.

This article is a continuation of our work presented in the MMR2000 [KL00, KL04] where we presented asymptotic methods in the particular case of Markov switching stochastic systems. We present here, asymptotic results for semi-Markov systems, under a unified framework for reliability problems.

The particular systems studied here are: the semi-Markov process in merging phase space, the integral functional, the dynamical system, and an heuristic principle for superposed renewal processes.

This chapter is organized as follows. Section 1.2 presents a short review of some asymptotic results. Section 1.3 presents an asymptotic merging result for the supporting semi-Markov process and some additional results especially for the failure time. Section 1.4 presents reward functional asymptotic results. Section 1.5 presents dynamical system asymptotic merging results, and fluctuations of such functionals in Sect. 1.6. Section 1.7 presents a heuristic method for superposition of two renewal processes. Finally, Sect. 1.8 presents results of the stationary phase merging

---

1.2 Reliability of the Renewal System

Let us review here some results on repairable systems with two identical components where they are solved by a renewal process approach and a generalization to semi-Markov process approach (Gnedenko [G64a,b, GS74], Soloviev [SL64, SL71], and Korolyuk [K89]). In both cases, their solution is based on the solution of a singular perturbation problem for reducible-invertible operators.

1. Renewal duplicated system. Let us consider a two component cold standby system. The lifetimes of which are iid, with common distribution F; and their repair (or replacement) times are also iid with common distribution function G and moreover they are independent of the lifetimes. Denote by α the lifetime and by β the repair time. Thus, we have $F(t) = \mathbf{P}(\alpha \le t)$ and $G(t) = \mathbf{P}(\beta \le t)$.

The system fails at time τ when the working component fails while the repaired component is still under repair. So, the reliability problem consists to find the distribution function $\Phi(t) = \mathbf{P}(\tau \le t)$. The solution of this problem is given by Gnedenko [G64a] in terms of Laplace transforms as follows

$$\varphi(s) = \frac{\psi(s)}{1 - g(s)}, \tag{1.1}$$

where:

$$\begin{aligned} \varphi(s) &:= \mathbf{E}\mathrm{e}^{-s\tau} = \int_0^\infty \mathrm{e}^{-st} \mathrm{d}\Phi(t), \\ \psi(s) &:= \int_0^\infty \mathrm{e}^{-st} \overline{G}(t) \mathrm{d}F(t), \\ g(s) &:= \int_0^\infty \mathrm{e}^{-st} G(t) \mathrm{d}F(t), \end{aligned}$$

with $\overline{G}(t) := 1 - G(t)$.

We denote by $q := \psi(0) = \int_0^\infty \overline{G}(t)\mathrm{d}F(t) = \mathbf{P}(\beta > \alpha)$ which is the probability of failure of the system on every working interval. This is also called the terminating probability in Feller [F66] in the case of a terminating renewal process, where q represents the defect of the distribution in the renewal process.

An obvious view of solution (1.1) gives possibility to get a limit result as $q \to 0$ (Soloviev [SL64]), that is

$$\lim_{q \to 0} \mathbf{P}(q\tau > t) = \mathrm{e}^{-t/a}, \quad a = \mathbf{E}\alpha. \tag{1.2}$$

The proof of the limit result (1.2) is based on the asymptotic representation of functions ψ and g in the formula (1.1), as $q \to 0$, namely:

$$\psi(qs) = q + o(q), \quad 1 - g(qs) = q(1 + as) + o(q). \tag{1.3}$$

The problem of reliability of the duplicated system is generalized in the case of different distributions of working times and renewal times (Gnedenko [G64a]).

Now working times α_k and renewal times β_k, $k = 1, 2$ have different distribution functions:

$$F_k(t) = \mathbf{P}(\alpha_k \le t), \quad G_k(t) = \mathbf{P}(\beta_k \le t), \quad k = 1, 2.$$

Working times to the first failure of the system τ_k, $k = 1, 2$ also depend on the number of initial working components. For the Laplace transform functions of working times up to failure, that is,

$$\varphi_k(s) = \mathbf{E}\mathrm{e}^{-s\tau_k} = \int_0^\infty \mathrm{e}^{-st} \mathrm{d}\Phi_k(t), \quad k = 1, 2$$

may be obtained a system of algebraic equations

$$Q(s)\varphi(s) = \psi(s), \tag{1.4}$$

where $\varphi(s) = (\varphi_1(s), \varphi_2(s)), \psi(s) = (\psi_1(s), \psi_2(s))$, and the matrix Q is

$$Q(s) = \begin{bmatrix} 1 & -g_1(s) \\ -g_2(s) & 1 \end{bmatrix} \tag{1.5}$$

with

$$g_1(s) = \int_0^\infty \mathrm{e}^{-st} G_2(t) \mathrm{d}F_1(t), \quad g_2(s) = \int_0^\infty \mathrm{e}^{-st} G_1(t) \mathrm{d}F_2(t).$$

System (1.4) is obtained from stochastic relations between unknown working times $\tau_k, k = 1, 2$:

$$\begin{aligned} \tau_1 &\overset{d}{=} \alpha_1 + I(\beta_2 \le \alpha_1)\tau_2, \\ \tau_2 &\overset{d}{=} \alpha_2 + I(\beta_1 \le \alpha_2)\tau_1, \end{aligned} \tag{1.6}$$

where equality $\xi \overset{d}{=} \zeta$ means equality of the distribution functions of random variables ξ and ζ; and the indicator function of event A, that is, $I(A) = 1$ if A is realized and $I(A) = 0$ otherwise.

2. The singular perturbation problem. We set a small parameter of series $\varepsilon \to 0$ ($\varepsilon > 0$) so that the following conditions are true:
C1: $\psi_k^\varepsilon(s) = \int_0^\infty \mathrm{e}^{-st} \overline{G}_{k'}^\varepsilon(t) \mathrm{d}F_k(t) = \varepsilon q_k + o(\varepsilon)$, $k = 1, 2$, $k' = 2, 1$;
C2: $1 - f_k^\varepsilon(s) = s \int_0^\infty \mathrm{e}^{-st} \overline{F}_k^\varepsilon(t) \mathrm{d}t = \varepsilon s a_k + o(\varepsilon)$, $k = 1, 2$.

Then the matrix of the system (1.4) in a series scheme has asymptotic representation

$$Q^\varepsilon(s) = Q_0 + \varepsilon Q_1(s) + o(\varepsilon), \tag{1.7}$$

where

$$Q_0 = \begin{bmatrix} 1 & -1 \\ -1 & 1 \end{bmatrix}, \quad Q_1 = \begin{bmatrix} 0 & q_1 + s a_1 \\ q_2 + s a_2 & 0 \end{bmatrix}. \tag{1.8}$$

The singularity of matrix Q_0 (that is, $\det Q_0 = 0$) means singularity of the system (1.4) in the series scheme:

$$Q^\varepsilon(s) \varphi^\varepsilon(s) = \psi^\varepsilon(s). \tag{1.9}$$

3. Renewal systems in series scheme. A two component system with two different workings components may be described by a semi-Markov process (Korolyuk [Chap. 2, K89]) in a phase space $E = \{1, 2, 0\}$, in which states 1 and 2 are working, and state 0 is absorbing. Times of staying in the working states are given by working times of devices α_1 and α_2. Embedded Markov chain $x_n, n \ge 0$ is given by transition probability matrix

$$P = \begin{bmatrix} 0 & 1 - q_1 & q_1 \\ 1 - q_2 & 0 & q_2 \\ 0 & 0 & 1 \end{bmatrix},$$

here $q_k = P(\beta_{k'} > \alpha_k) = \int_0^\infty \overline{G}_{k'}(t) \mathrm{d}F_k(t)$, $k = 1, 2$, $k' = 2, 1$.

Now the system lifetime is determined by the time of staying of semi-Markov process in the subset of the working states $E_0 = \{1, 2\}$ before absorption in state 0.

We now set the small series parameter $\varepsilon \to 0$ $(\varepsilon > 0)$ to transition probability matrix of embedded Markov chain:

$$P^{\varepsilon} = \begin{bmatrix} 0 & 1-\varepsilon q_1 \\ 1-\varepsilon q_2 & 0 \end{bmatrix} = P - \varepsilon P_1, \tag{1.10}$$

where

$$P = \begin{bmatrix} 0 & 1 \\ 1 & 0 \end{bmatrix}, \quad \text{and} \quad P_1 = \begin{bmatrix} 0 & q_1 \\ q_1 & 0 \end{bmatrix}.$$

It is worth noticing that the transition probability matrix is restricted only on the working states 1 and 2.

Let us rewrite the equality (1.10) with the generator of embedded Markov chain $Q^{\varepsilon} := I - P^{\varepsilon}$:

$$Q^{\varepsilon} = Q_0 + \varepsilon P_1, \quad Q_0 = \begin{bmatrix} 1 & -1 \\ -1 & 1 \end{bmatrix}. \tag{1.11}$$

Representation (1.11) is analog of (1.7)–(1.8).

In the asymptotic problem of calculation of holding (sojourn) time of semi-Markov process in the subset of the states until absorption appears a problem of singular perturbation of the operator Q_0, that defines a supporting Markov chain (Koroliuk [K65, K69]). Naturally that now the phase space of "working" states may be arbitrary it is only required that the generator Q_0 defines ergodic (supporting) Markov chain (Koroliuk [Chap. 3, [K89]]).

We will notice that in the series scheme, given by formula (11), rather difficult analytical problem of asymptotic analysis of characteristics under conditions **C1–C2** disappeared. Substantial results in this problem are obtained by Gnedenko [G64a] and Soloviev [SL71, SL64].

Thus, from one side representation (1.11) generalizes the problem of reliability of renewal systems, and on the other side it simplifies analytical problem of research of concrete characteristics of renewal systems.

Generalization of the problem about reliability of renewal system resulted in creation of a new direction of reliability theory of stochastic systems, based on the method of phase merging of the states of Markov and semi-Markov processes with the use of solution of singular perturbation problem for reducible-invertible operators (Koroliuk [KK99, KT82, K89]).

1.3 Absorbing Time of Semi-Markov Process

The semi-Markov process $x(t), t \geq 0$ on the standard (Polish) phase space $(E, \mathcal{E})$ is given by the semi-Markov kernel

$$Q(x, B, t) = \mathbf{P}(x_{n+1} \in B, \theta_{n+1} \leq t | x_n = x) = P(x, B) F_x(t), \tag{1.12}$$

where $\quad x \in E, B \in \mathcal{E}, t \geq 0.$

The stochastic kernel $P(x, B)$ defines the transition probabilities of embedded Markov chain $x_n, n \geq 0$:

$$P(x, B) = \mathbf{P}(x_{n+1} \in B | x_n = x). \tag{1.13}$$

The family of distribution functions $F_x(t), t \geq 0, x \in E$ gives the sojourn times $\theta_x, x \in E$:

$$F_x(t) = \mathbf{P}(\theta_{n+1} \leq t | x_n = x) =: \mathbf{P}(\theta_x \leq t). \tag{1.14}$$

The jump times of the semi-Markov process are

$$\tau_{n+1} = \tau_n + \theta_{n+1}, \quad n \geq 0, \quad \text{and} \quad \tau_0 = 0. \tag{1.15}$$

The counting process which counts the jumps of the semi-Markov process in the time interval $(0, t]$, is

$$\nu(t) = \max\{n : \tau_n \leq t\}, \quad t \geq 0.$$

The semi-Markov process is connected to the embedded Markov chain by the following relation

$$x(t) = x_{\nu(t)}, \quad t \geq 0.$$

Let us now consider a family of semi-Markov processes $x^\varepsilon(t), t \geq 0$, indexed by the small parameter $\varepsilon > 0, \varepsilon \to 0$, and semi-Markov kernels Q^ε. *Absorbing time of semi-Markov process* $x^\varepsilon(t), t \geq 0$ is considered in the series scheme with the parameter series $\varepsilon \to 0 (\varepsilon > 0)$ on the split phase space

$$E^0 = E \cup \{0\}, \quad E = \bigcup_{k=1}^{N} E_k, \quad E_k \cap E_{k'} = \emptyset, \quad k \neq k' \tag{1.16}$$

with absorbing state 0 (see Fig. 1.1).

The main assumptions are:

MA1: The semi-Markov kernel is dependent on the series parameter ε as follows:

$$\begin{aligned} Q^\varepsilon(x, B, t) &= P^\varepsilon(x, B) F_x(t), \\ P^\varepsilon(x, B) &= P(x, B) + \varepsilon P_1(x, B). \end{aligned} \tag{1.17}$$

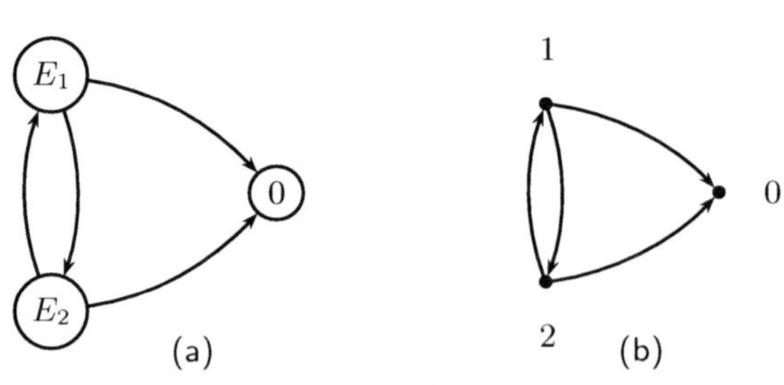

Figure 1.1. A merged system with $N = 2$: (**a**) initial system and (**b**) merged system

The stochastic kernel $P(x, B)$ is coordinated with the split phase space (1.16) as follows:

$$P(x, E_k) = \begin{cases} 1, & x \in E_k, \\ 0, & x \notin E_k, \end{cases}, \quad k \in \widehat{E} := \{1, 2, ..., N\} \tag{1.18}$$

and defines the supporting Markov chain $x_n^0, n \geq 0$ on E, which is uniformly ergodic in every class E_k with the stationary distributions $\rho_k(B), k \in \widehat{E}$.

MA2: The perturbing kernel $P_1(x, B)$ satisfies the following absorption condition:

$$\max_{k \in \widehat{E}} p_{k0} > 0, \quad p_{k0} := -\int_{E_k} \rho_k(\mathrm{d}x) P_1(x, E), \quad k \in \widehat{E}. \tag{1.19}$$

Note that, according to (1.17)–(1.18),

$$\varepsilon P_1(x, E) = P^\varepsilon(x, E) - 1 = -P^\varepsilon(x, 0)$$

are the absorbing probabilities.

Theorem 1. *([KL05a,b], Theorem 4.2) Under Assumptions **MA1–MA2** the weak convergence*

$$v(x^\varepsilon(t/\varepsilon)) \Rightarrow \widehat{x}(t), \quad \varepsilon \to 0$$

takes place ($v(x) = k, x \in E_k$).

The limit Markov process $\widehat{x}(t), t \geq 0$ on the merging phase space $\widehat{E}^0 = \widehat{E} \cup \{0\}$ is defined by the generating matrix

$$\widehat{Q} = \left(\widehat{q}_{kr}; k, r \in \widehat{E}^0\right), \tag{1.20}$$

$$\widehat{q}_{kr} = q_k p_{kr}, \quad p_{kr} = \int_{E_k} \rho_k(\mathrm{d}x) P_1(x, E_r), \quad k, r \in \widehat{E}, \tag{1.21}$$

$$q_k = 1/m_k, \quad m_k = \int_{E_k} \rho_k(\mathrm{d}x) m(x), \quad m(x) := \mathbf{E}\theta_x = \int_0^\infty \overline{F}_x(t)\mathrm{d}t. \tag{1.22}$$

Corollary 1. *The transition intensities $\widehat{q}_{kr}$ are represented as follows:*

$$\widehat{q}_{kr} = \widehat{q}_k \widehat{p}_{kr},$$

where the intensities of sojourn times are given by

$$\widehat{q}_k = q_k p_k, \quad p_k = -\int_{E_k} \rho_k(\mathrm{d}x) P_1(x, E_k),$$

the transition probabilities are given by

$$\widehat{p}_{kr} = p_{kr}/p_k, \quad k, r \in \widehat{E},$$

and the absorption probabilities by

$$\widehat{p}_{k0} = p_{k0}/p_k, \quad p_{k0} := -\int_{E_k} \rho_k(\mathrm{d}x) P_1(x, E).$$

The following relations are obvious

$$\sum_{r\neq k} p_{kr} = \int_{E_k} \rho_k(\mathrm{d}x)P_1(x, E\setminus E_k) = \int_{E_k} \rho_k(\mathrm{d}x)P_1(x, E) - \int_{E_k} \rho_k(\mathrm{d}x)P_1(x, E_k).$$

Hence

$$\sum_{r\in\widehat{E}^0} \widehat{p}_{kr} = 1.$$

Remark 1. The intensities of sojourn times $\widehat{q}_k = q_k p_k, k \in \widehat{E}$ are constructed according to the stoppage principle for a renewal process: the intensity of stoppage time $\widehat{q}$ is a product of the intensity of renewal time q and the stoppage probability p.

Corollary 2. *In the main applicable case in reliability of $N = 1$ (with one class of working states), the limit Markov process with absorbing state $\widehat{E}^0 = \{1, 0\}$ is given by the renewal process with the exponentially distributed renewal time θ with the intensity $q = 1/m, m = \int_E \rho(\mathrm{d}x)m(x)$ and with the stoppage probability $p = 1-\int_E \rho(\mathrm{d}x)P_1(x, E)$. That is, the series parameter ε.*

The common working time is exponentially distributed random variable ζ:

$$P(\zeta > t) = \exp(-\Lambda t),$$

where:

$$\Lambda = qp, \quad q = \int_E \pi(\mathrm{d}x)q(x), \quad q(x) = 1/m(x).$$

Remark 2. Theorem 1 is used in analysis of reliability of the stochastic system with some classes of working states used during long time almost separately. The simplified stochastic system is described by the Markov process with the finite state space and with absorbing state. The common working time is defined by the Laplace transform of the distribution function satisfies the linear equation. So, the working time has the Erlang distribution. By Theorem 1 the distribution of working time of the semi-Markov stochastic system can be approximated by the Erlang distribution.

Remark 3. The simplification formulae in Theorem 1 are natural from heuristic point of view. Indeed, having the stationary distributions $\rho_k(B), k \in \widehat{E}$ of the support embedded Markov chains $x_n^{(k)}, n \geq 0, k \in \widehat{E}$ the merging transition probabilities are given as follows $(k \neq r)$:

$$\widehat{p}_{kr} = \int_{E_k} \rho_k(\mathrm{d}x)P_1(x, E_r)/\left[1 - \int_{E_k} \rho_k(\mathrm{d}x)P_1(x, E_k)\right]$$

and the merging absorbing probabilities are

$$\widehat{p}_{k0} = \left[1 - \int_{E_k} \rho_k(\mathrm{d}x)P_1(x, E)\right] / \left[1 - \int_{E_k} \rho_k(\mathrm{d}x)P_1(x, E_k)\right].$$

The main problem is to get the stationary probabilities $\rho_k(B), k \in \widehat{E}$ of the ideal stochastic system without absorption.

The parameter series ε disappeared in the last two formulae, but the real value of ε is given by the formula:

$$p_k = 1 - \int_{E_k} \rho_k(\mathrm{d}x) P_1(x, E_k), \quad k \in \widehat{E}$$

that are the probabilities of leaving the class $E_k, k \in \widehat{E}$.

Remark 4. The phase merging algorithm provides an estimation of its application effectiveness. That is the methodological principle of the applied mathematics.

1.4 Reward Functional in the Semi-Markov Environment

Integral functional with semi-Markov switching is represented by

$$A^{\varepsilon}(t) = \int_0^t a(x^{\varepsilon}(s/\varepsilon))\mathrm{d}s, \quad t \geq 0,$$

with given bounded real-valued function $a(x), x \in E^0$, $E^0 = E \cup \{0\}$, and $a(x) \in \mathbb{R}^d, d \geq 1$.

The semi-Markov switching process $x^{\varepsilon}(t), t \geq 0, \varepsilon > 0$ in the series scheme with the series parameter $\varepsilon \to 0 (\varepsilon > 0)$ is given on the phase space E^0 with absorbing state $\{0\}$ and set of working states E by the semi-Markov kernel (1.17) and (1.18).

Theorem 2. *([KL05a,b], Corollary 4.3) Under Assumptions **MA1-MA2** the weak convergence*

$$A^{\varepsilon}(t) \Rightarrow A^0(t), \quad \varepsilon \to 0$$

takes place.

The limit functional

$$A^0(t) = \int_0^{t \wedge \zeta} \widehat{a}(\widehat{x}(s))\mathrm{d}s, \quad t \geq 0$$

is defined by the average rate of reward

$$\widehat{a}(k) = q_k \int_{E_k} \rho_k(\mathrm{d}x) a(x), \quad k \in \widehat{E}.$$

The limit merged Markov process $\widehat{x}(t), t \geq 0$, defined on the merged phase space $\widehat{E}^0 = \widehat{E} \cup \{0\}$, $\widehat{E} = \{1, 2, ..., N\}$, with absorbing state $\{0\}$, is given by the Markov matrix (1.20).

Corollary 3. *In the particular case of $N = 1$, the limit reward functional is represented as follows:*

$$A^0(t) = \widehat{a}(t \wedge \widehat{\zeta}), \quad \widehat{a} = q \int_E \rho(\mathrm{d}x) a(x).$$

The stoppage time $\widehat{\zeta}$ has the exponential distribution:

$$\mathbf{P}(\widehat{\zeta} > t) = \mathrm{e}^{-\widehat{\Lambda} t}, \quad \widehat{\Lambda} = qp, \quad p = 1 - \int_E \rho(\mathrm{d}x) P_1(x, E)$$

$$q = \int_E \pi(\mathrm{d}x) q(x), \quad q(x) = 1/m(x), \quad m(x) = \int_0^\infty \overline{F}_x(t)\mathrm{d}t.$$

Remark 5. The phase merging scheme gives the most simplified model for the reward functional in a semi-Markov environment.

Remark 6. Of particular interest is the case when $a(x) = \mathbf{1}_A(x)$, $x \in E$ where $A \subset E$. In that case $A^\varepsilon(t)$ is the holding time of process in A in the time interval $[0, t]$ (see [KKL]). If A is the subset of working states of E, then $\mathbf{E}A(t)/t$ is the mean availability of system in the time interval $[0, t]$ which is equal to reliability at time t.

1.5 Dynamic Reward Functional

The dynamic reward functional in the semi-Markov medium is defined by a solution of the evolutionary equation

$$\frac{\mathrm{d}}{\mathrm{d}t} u^\varepsilon(t) = C(u^\varepsilon(t); x^\varepsilon(t/\varepsilon)), \quad u^\varepsilon(0) = u \in \mathbb{R}^d.$$

As in previous section, the semi-Markov switching process $x^\varepsilon(t), t \geq 0, \varepsilon > 0$ in the series scheme is considered on the split phase space $E^0 = E \cup \{0\}$ with absorbing state $\{0\}$ and set of working states E given by the semi-Markov kernel (1.17)–(1.18).

Theorem 3. *([KL05a,b], Theorem 4.5) Under the merging conditions* ***MA1–MA2*** *the weak convergence*

$$u^\varepsilon(t) \Rightarrow u^0(t), \quad \varepsilon \to 0$$

takes place.

The limit functional $u^0(t), t \geq 0$ defined by a solution of the average evolutionary equation

$$\frac{\mathrm{d}}{\mathrm{d}t} u^0(t) = \widehat{C}(u^0(t); \widehat{x}(t)), \quad u^0(0) = u^0$$

on the time interval $0 \leq t \leq \zeta$ with the stoppage time ζ of the merged phase space $\widehat{E}^0 = \widehat{E} \cup \{0\}, \widehat{E} = \{1, 2, ..., N\}$.

The average velocity in state k is

$$\widehat{C}(u; k) = \int_{E_k} \pi_k(\mathrm{d}x) C(u; x), \quad k \in \widehat{E}.$$

The stationary distributions $\pi_k(\mathrm{d}x)$, $1 \leq k \leq N$ are defined by the relations:

$$\pi_k(\mathrm{d}x) q(x) = q\rho_k(\mathrm{d}x), \quad 1 \leq k \leq N.$$

1.6 Fluctuations of the Reward Functional

In this section we present a diffusion approximation result on the fluctuation of reward functional. The method represented in our book [KL05a,b] can be applied to investigation of the fluctuation

$$B^\varepsilon(t) = [A^\varepsilon(t) - A^0(t)]/\varepsilon, \quad t \geq 0.$$

Here, for simplicity $N = 1$. The integral functional is defined now as follows

$$A^\varepsilon(t) = \int_0^t a(x^\varepsilon(s/\varepsilon^2))\mathrm{d}s, \quad t \geq 0,$$

where the time is scaled by ε^{-2}.

Theorem 4. *Under Assumptions **MA1-MA2** the weak convergence*

$$B^\varepsilon(t) \Rightarrow B^0(t \wedge \widehat{\zeta}), \quad \varepsilon \to 0$$

takes place.

The limit diffusion process $B^0(t), t \geq 0$ is the Wiener process with the variance coefficient

$$\sigma^2 = 2\int_E \pi(\mathrm{d}x)b(x)R_0 b(x) + \sigma_\mu, \quad \sigma_\mu := \int_E \pi(\mathrm{d}x)\mu(x)b^2(x),$$

$$\mu(x) := [m_2(x) - 2m^2(x)]/m(x), \quad m_2(x) := \mathbf{E}\theta_x^2, \quad b(x) := a(x) - \widehat{a}.$$

Remark 7. The fluctuation is considered as follows

$$B^\varepsilon(t) = \varepsilon^{-1}\int_0^t b(x^\varepsilon(s/\varepsilon^2))\mathrm{d}s.$$

Note that the function $b(x)$ satisfies *the balance condition*

$$\Pi b(x) = \int \pi(\mathrm{d}x)b(x) = 0.$$

According to Theorem 3.3, [KL05a,b] we get $\sigma^2 = \Pi b R_0 b \Pi + \sigma_\mu$.

1.7 Heuristic Phase Merging

The phase merging principles can be formulated as an *heuristic* phase merging based on the renewal theorem (see, e.g., [Chap. 5, KL05a,b]).

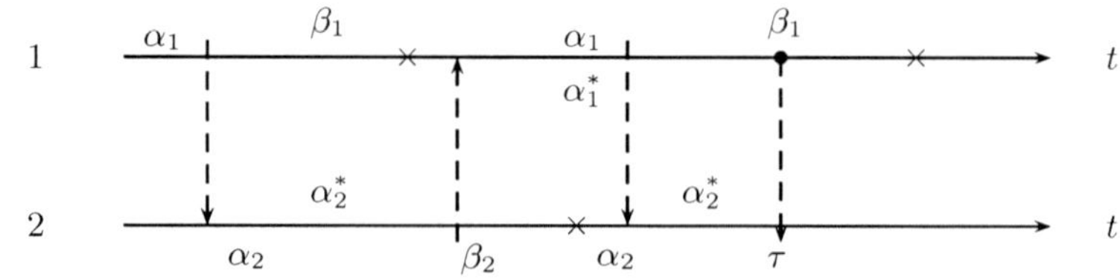

Figure 1.2. Double renewal system

The limit distribution of *remaining waiting time* α^* for the renewal process given by the distribution function $F(t) = P\{\alpha \leq t\}, t \geq 0$ is defined by the formula

$$\mathbf{P}(\alpha^* \leq t) = \int_0^t \overline{F}(s)\mathrm{d}s/a, \quad a = \mathbf{E}\alpha. \tag{1.23}$$

The double renewal system (Fig. 1.2) is defined by the superposition of two alternating renewal processes given by the distribution functions $F_k(t) = \mathbf{P}(\alpha_k \leq t), k = 1, 2$ of working times and $G_k(t) = \mathbf{P}(\beta_k \leq t), k = 1, 2$ of renewal times. The stoppage moment τ is defined when both of renewal processes are in renewal states. The heuristic principle means that the remaining working time α^* of every renewal process in the stationary regime is defined by the stationary distribution (1.23).

Now the stoppage probabilities for every renewal process are defined by the relations:

$$q_1 = \mathbf{P}(\alpha_1^* < \beta_2) = \int_0^\infty \overline{G}_1(t)\overline{F}_2(t)\mathrm{d}t/a_1$$
$$q_2 = \mathbf{P}(\alpha_2^* < \beta_1) = \int_0^\infty \overline{G}_2(t)\overline{F}_1(t)\mathrm{d}t/a_2.$$

or, in the equivalent form

$$q_1 = \mathbf{E}(\alpha_2 \wedge \beta_1)/a_2, \quad q_2 = \mathbf{E}(\alpha_1 \wedge \beta_2)/a_1.$$

According to the heuristic principle of stoppage (see [K89]) the intensity of stoppage time for double renewal system is defined by the following relation:

$$\Lambda = q_1\lambda_1 + q_2\lambda_2, \quad \lambda_k = 1/a_k, \quad a_k = \mathbf{E}\alpha_k,$$

or, in the equivalent form

$$\Lambda = [\mathbf{E}(\alpha_2 \wedge \beta_1) + \mathbf{E}(\alpha_1 \wedge \beta_2)]/\mathbf{E}\alpha_1\mathbf{E}\alpha_2.$$

1.8 Stationary Phase Merging Scheme

Let us consider a system which temporal evolution is described by a semi-Markov process $x(t), t \geq 0$, defined on the following split phase space, given by the semi-Markov kernel $Q(x, B, t)$:

$$E = \bigcup_{k=1}^N E_k, \quad E_k \cap E_{k'} = \emptyset, \quad k \neq k'. \tag{1.24}$$

The phase merging scheme (PMS) is realized on the merged phase space

$$\hat{E} = \{1, 2, ..., N\}. \tag{1.25}$$

In the discrete case we note $Q_{kr}(t)$ instead of $Q(k, \{r\}, t)$, and P_{kr} instead of $P(k, \{r\})$.

The main assumption is that the embedded Markov chain $x_n, n \geq 0$ is ergodic with stationary distribution $\rho(B), B \in \mathcal{E}$, which satisfies the following condition:

C1: $\rho_k := \rho(E_k) > 0,\ 1 \leq k \leq N.$

Definition 1. *The stationary merged MRP, say* $\tilde{x}_n, \tilde{\tau}_n, n \geq 0$, *on the merged phase space* $\hat{E}$ *is determined by the merged semi-Markov kernel*

$$\tilde{Q}_{kr}(t) = \int_{E_k} \rho(\mathrm{d}x) Q(x, E_r, t)/\rho_k = \int_{E_k} \rho_k(\mathrm{d}x) Q(x, E_r, t), \quad k, r \in \hat{E}. \tag{1.26}$$

The transition probabilities of the merged embedded Markov chain $\tilde{x}_n, n \geq 0$, are defined by

$$\tilde{P}_{kr} = \int_{E_k} \rho(\mathrm{d}x) P(x, E_r)/\rho_k, \quad k, r \in \hat{E}, \tag{1.27}$$

and the distribution function of the sojourn time $\tilde{\theta}_k, k \in \hat{E}$, are defined by

$$\tilde{F}_k(t) = \int_{E_k} \rho(\mathrm{d}x) F_x(t)/\rho_k, \quad k \in \hat{E}. \tag{1.28}$$

Remark 8. The stationary merged embedded Markov chain $\tilde{x}_n, n \geq 0$, has the stationary distribution $\rho_k, k \in \hat{E}$.

Remark 9. The stationary merged MRP $\tilde{x}_n, \tilde{\theta}_n, n \geq 0$, has virtual transitions with probabilities $\tilde{P}_{kk}, k \in \hat{E}$. Without virtual transition the MRP $\tilde{x}_n^0, \tilde{\theta}_n^0, n \geq 0$, can be constructed by the following known way:

$$\tilde{P}_{kr}^0 = \tilde{P}_{kr}/(1 - \tilde{P}_{kk}) = \int_{E_k} \rho(\mathrm{d}x) P(x, E_r) / \int_{E_k} \rho(\mathrm{d}x) P(x, E \setminus E_k), \tag{1.29}$$

$$\tilde{F}_k^0(t) = \mathbf{P}(\tilde{\theta}_k^0 \leq t), \quad \tilde{\theta}_k^0 = \sum_{\ell=1}^{\nu_k} \tilde{\theta}_k^{(\ell)}, \tag{1.30}$$

where $\tilde{\theta}_k^{(\ell)}, \ell \geq 1$, are iid random variables with common distribution function $\tilde{F}_k(t)$; the integer valued random variable $\nu_k, k \in \hat{E}$ are geometric distributed with parameter P_{kk}:

$$\mathbf{P}(\nu_k = n) = (1 - \tilde{P}_{kk}) \tilde{P}_{kk}^{n-1}, \quad n \geq 1. \tag{1.31}$$

It is worth noticing that the distribution functions $\tilde{F}_k^0(t)$ are defined by the Laplace transform

$$\tilde{f}_k^0(\lambda) = \int_0^\infty \mathrm{e}^{-\lambda t} \tilde{F}_k^0(\mathrm{d}t) = \tilde{P}_{kk} \tilde{f}_k(\lambda) / [1 - (1 - \tilde{P}_{kk}) \tilde{f}_k(\lambda)], \tag{1.32}$$

where

$$\tilde{f}_k(\lambda) = \int_0^\infty e^{-\lambda t} \tilde{F}_k(dt). \tag{1.33}$$

The stationary merged stochastic system $\tilde{S}$ is defined by the semi-Markov process $\tilde{x}(t), t \geq 0$, on the merged phase space $\hat{E}$ by the merged semi-Markov kernel (1.26).

Relation between the initial semi-Markov process $x(t), t \geq 0$, on the phase space $(E, \mathcal{E})$ and the stationary merged semi-Markov process $\tilde{x}(t), t \geq 0$, on the merged phase space $\hat{E} = \{1, 2, ..., N\}$ is described in Theorem 5 below.

Define the entrance moments into subset $E_k, k \in \hat{E}$:

$$\tau_{m+1}^{(k)} := \inf\{n > \tau_m^{(k)} : x_n \in E_k\}, \quad m \geq 1, \tau_0 = 0, \tag{1.34}$$

and the renewal processes

$$\tilde{x}_m^{(k)} := x_{\tau_m^{(k)}+1}, \quad \tilde{\theta}_m^{(k)} := \theta_{\tau_m^{(k)}+1}, \quad m \geq 0. \tag{1.35}$$

Theorem 5. *([AK80]) Let the embedded Markov chain $x_n, n \geq 0$, defined by the stochastic kernel $P(x, B)$ be ergodic, with the stationary distribution $\rho(B), B \in \mathcal{E}$, that satisfies Condition **C1**. Then the following convergence takes place:*

$$\lim_{m\to\infty} \mathbf{P}(\tilde{x}_m^{(k)} \in E_r, \tilde{\theta}_m^{(k)} \leq t) = \tilde{Q}_{kr}(t), \quad k, r \in \hat{E}. \tag{1.36}$$

Particularly,

$$\lim_{m\to\infty} \mathbf{P}(\tilde{x}_m^{(k)} \in E_r) = \tilde{P}_{kr}, \quad k, r \in \hat{E}, \tag{1.37}$$

and

$$\lim_{m\to\infty} \mathbf{P}(\tilde{\theta}_m^{(k)} \leq t) = \tilde{F}_k(t), \quad k \in \hat{E}. \tag{1.38}$$

That is, the stationary merged semi-Markov kernel (1.26) determines the transition probabilities of the renewal process (1.35) on the merged phase space $\hat{E}$ in the stationary regime.

Let us now consider a stationary phase merging in series scheme. The stationary merged MRP $\tilde{x}_n^\varepsilon, \tilde{\theta}_n^\varepsilon, n \geq 0$, now is considered in the series scheme given by the semi-Markov kernel

$$\tilde{Q}_{kr}^\varepsilon(t) = \begin{cases} \varepsilon \tilde{P}_{kr} \tilde{F}_k(t/\varepsilon), & k, r \in \hat{E}, \quad k \neq r \\ (1 - \varepsilon \tilde{P}_{kk}) \tilde{F}_k(t/\varepsilon), & k, r \in \hat{E}, \quad k = r. \end{cases} \tag{1.39}$$

The stationary merged MRP in series scheme without virtual transitions $\tilde{x}_n^0, \tilde{\theta}_n^0$, $n \geq 0$, is determined by the semi-Markov kernel

$$\begin{aligned} \tilde{Q}_{kr}^{0\varepsilon}(t) &= \tilde{P}_{kr} \tilde{F}_k^{0\varepsilon}(t/\varepsilon), \quad k, r \in \hat{E}, k \neq r, \\ \tilde{P}_{kr}^0 &= P_{kr}/[1 - \tilde{P}_{kk}], \quad k, r \in \hat{E}. \end{aligned} \tag{1.40}$$

The distribution functions $\tilde{F}_k^{0\varepsilon}(t)$ are determined by the Laplace transform

$$\tilde{f}_k^{0\varepsilon}(\lambda) := \int_0^\infty e^{-\lambda t} \tilde{F}_k^{0\varepsilon}(dt) = \varepsilon \tilde{P}_{kk} \tilde{f}_k(\varepsilon\lambda)/[1 - (1 - \varepsilon \tilde{P}_{kk}) \tilde{f}_k(\varepsilon\lambda)]. \tag{1.41}$$

Theorem 6. *Let the supporting EMC $x_n^0, n \geq 0$, is uniformly ergodic with the stationary distributions $\rho_k(B), B \in \mathcal{E}_k, k \in \hat{E}$, and the merged exit probabilities*

$$\hat{q}_k := \int_{E_k} \rho_k(\mathrm{d}x) P_1(x, E \setminus E_k) > 0, \quad k \in \hat{E}, \tag{1.42}$$

and,

$$\max_{1 \leq k \leq N} m_k \leq M < +\infty, \tag{1.43}$$

and moreover the functions $\bar{\bar{F}}_k(t) := 1 - \tilde{F}_k(t), k \in \hat{E}$, are uniformly integrable, then the following weak convergence takes place

$$\tilde{x}_\varepsilon^0(t) \Longrightarrow \hat{x}(t), \quad \varepsilon \to 0.$$

We have

$$\bar{f}_k^{0\varepsilon}(\lambda) \longrightarrow \frac{\Lambda_k}{\lambda + \Lambda_k}, \quad \varepsilon \to 0,$$

where $\Lambda_k = \tilde{P}_{kk}/m_k$, and $m_k = \mathbf{E}\theta_k = \int_0^\infty \bar{F}_k(t)\mathrm{d}t$.

Convergence of sojourn times distributions of finite state space MRP is equivalent to the convergence of compensating operator to the limit Markov process generator, with the same embedded Markov chain kernel [S86, S98]. For more details and proofs on this topic see [KL05a,b, KM03, M04].

References

[AK80] Arjas E., Korolyuk V.S.: Stationary phase merging of Markov renewal processes. DAN of Ukraine, (in Russian), **8**, 3–5 (1980).

[A08] Anisimov A.A.: Switching Processes in Queuing Models. ISTE, Wiley, London (2008).

[F66] Feller W.: An introduction to probability theory and its applications. vol. 2, Wiley, NY (1966).

[G64a] Gnedenko B.V.: On a two-unit redundant system. Izv. Akad. Nauk SSSR, Tehn. Kibernet., **4**, 3–12 (1964).

[G64b] Gnedenko B.V.: Doubling with renewal. Izv. Akad. Nauk SSSR, Tehn. Kibernet., **5**, 111–118 (1964).

[GS74] Gnedenko D.B., Solov'ev A.D.: A general model for standby with renewal. Engrg. Cybernetics 12, no. 6, 82–86.; translated from Izv. Akad. Nauk SSSR Tehn. Kibernet. **6**, 113–118 (1974).

[K65] Korolyuk V.S.: Sojourn time of a semi-Markov process in a fixed set of states. Ukrainian Math. J. **17**(3), 123–128 (1965).

[K69] Korolyuk V.S.: The asymptotic behavior of the sojourn time of a semi-Markov process in a subset of the states. Ukrainian Math. J. **21**, 6, 842–845 (1969).

[K89] Korolyuk V.S.: Stochastic models of systems. Naukova Dumka, (in Russian), Kiev (1989).

[KK99] Korolyuk V.S., Korolyuk V.V.: Stochastic Models of Systems. Kluwer, Dordrecht (1999).

[KKL] Koroliuk V.S., Kutoviy O., Limnios N. Occupation measure functionals in merging phase space. Stochastics, **82**(2), 149–163 (2010).

[KL04] Koroliuk V.S., Limnios N.: Average and diffusion approximation for evolutionary systems in an asymptotic split phase space. Ann. Appl. Probab. **14**(1), 489–516 (2004).

[KL05a] Koroliuk V.S., Limnios N.: Stochastic Systems in Merging Phase Space. World Scientific, Singapore (2005).

[KL05b] Korolyuk, V.S., Limnios, N.: Poisson approximation of increment processes with Markov switching. Theor. Probab. Appl. **49**(4), 629–644 (2005).

[KL00] Korolyuk V.S., Limnios N.: Evolutionary systems in an asymptotic split state space. In: Limnios N. & Nikulin M. (eds) Recent Advances in Reliability Theory: Methodology, Practice and Inference. 145–161, Birkhauser, Boston (2000).

[KS95] Korolyuk V.S., Swishchuk A.: Semi-Markov random evolution. Kluwer, Dordrecht (1995).

[KT82] Korolyuk V.S., Turbin A.F.: Markov Renewal Processes in Reliability Problems of Systems. Naukova Dumka (in Russian) (1982).

[KT93] Korolyuk V.S., Turbin A.F.: Mathematical foundations of the state lumping of large systems. Kluwer, Dordtrecht (1993).

[KM03] Korolyuk V.S., Momonova A.: DAN of Ukraine (2003).

[K90] Kushner H.J.: Weak Convergence Methods and Singular Perturbed Stochastic Control and Filtering Problems. Birkhäuser, Boston (1990).

[LO01] Limnios N., Oprişan G.: Semi-Markov Processes and Reliability. Birkhäuser, Boston (2001).

[LS] Liptser R. Sh., Shiryayev A. N.: Theory of Martingales. Kluwer, Dordrecht (1989).

[M04] Mamonova A.: Preprint, Institute of Mathematics. Kiev (2004).

[SV79] Stroock D.W., Varadhan S.R.S.: Multidimensional Diffusion Processes. Springer, Berlin (1979).

[SK89] Skorokhod A.V.: Asymptotic Methods in the Theory of Stochastic Differential Equations, Vol. 78. AMS, Providence (1989).

[SHS02] Skorokhod A.V., Hoppensteadt F.C., Salehi H.: Random Perturbation Method with Application in Science and Engineering. Springer, New York (2002).

[S04] Silvestrov D. S.: Limit Theorems for Randomly Stopped Stochastic Processes. Series: Probability and its Applications. Springer, New York (2004).

[SL64] Solov'ev A.D.: Asymptotic distribution of the life time of a doubled element (in Russian). Izv. Akad. Nauk SSSR Tehn. Kibernet. **5**, 119–121 (1964).

[SL71] Solov'ev A.D.: Asymptotic behavior of the time of first occurrence of a rare event in a regenerating process. Engrg. Cybernetics, **9**, 6, 1038–1048 (1972); translated from Izv. Akad. Nauk SSSR Tehn. Kibernet. **6**, 79–89 (in Russian) (1971).

[S86] Sviridenko M.N.: Martingale approach to limit theorems for semi-Markov processes. Theor. Probab. Appl. 540–545 (1986).

[S98] Sviridenko M.N.: Martingale characterization of limit distributions in the space of functions without discontinuities of second kind. Math. Notes. **43**, 5, 398–402 (1998).

2

Nonlinearly Perturbed Stochastic Processes and Systems

Dmitrii S. Silvestrov

Stockholm University, Stockholm, Sweden

Abstract: This paper is a survey of results presented in the recent book Gyllenberg and Silvestrov [GS08].[1] This book is devoted to studies of quasi-stationary phenomena for nonlinearly perturbed stochastic processes and systems. New methods of asymptotic analysis for nonlinearly perturbed stochastic processes based on asymptotic expansions for perturbed renewal equation and recurrence algorithms for construction of asymptotic expansions for Markov type processes with absorption are presented. Asymptotic expansions are given in mixed ergodic (for processes) and large deviation theorems (for absorption times) for nonlinearly perturbed regenerative processes, semi-Markov processes, and Markov chains. Applications to analysis of quasi-stationary phenomena in nonlinearly perturbed queueing systems, population dynamics and epidemic models, and for risk processes are presented. The book also contains an extended bibliography of works in the area.

Keywords and phrases: Nonlinear perturbation, Quasi-stationary phenomenon, Pseudo-stationary phenomenon, Stochastic process, Stochastic system, Renewal equation, Asymptotic expansion, Ergodic theorem, Limit theorem, Large deviation, Regenerative process, Regenerative stopping time, Semi-Markov process, Markov chain, Absorption time, Queueing system, Population dynamics, Epidemic model, Lifetime, Risk process, Ruin probability, Cramér-Lundberg approximation, Diffusion approximation

2.1 Introduction

The book mentioned above presents new methods of asymptotic analysis of nonlinearly perturbed stochastic processes and systems with random lifetimes.

[1] Gyllenberg, M., Silvestrov, D.S.: Quasi-Stationary Phenomena in Nonlinearly Perturbed Stochastic Systems. De Gruyter Expositions in Mathematics, **44**, Walter de Gruyter, Berlin, XII + 579 pp. (2008)

V.V. Rykov et al. (eds.), *Mathematical and Statistical Models and Methods in Reliability: Applications to Medicine, Finance, and Quality Control*, Statistics for Industry and Technology, DOI 10.1007/978-0-8176-4971-5_2,

Usually the behaviour of a stochastic system can be described in terms of some Markov type stochastic process $\eta^{(\varepsilon)}(t)$ and its lifetime defined to be the time $\mu^{(\varepsilon)}$ at which the process $\eta^{(\varepsilon)}(t)$ hits a special absorption subset of the phase space of this process for the first time. A typical situation is when the process $\eta^{(\varepsilon)}(t)$ and the absorption time $\mu^{(\varepsilon)}$ depend on a small parameter $\varepsilon \geq 0$ in the sense that some of their local "transition" characteristics depend on the parameter ε. The parameter ε is involved in the model in such a way that the corresponding local characteristics are continuous at the point $\varepsilon = 0$, if regarded as functions of ε. These continuity conditions permit to consider the process $\eta^{(\varepsilon)}(t)$, for $\varepsilon > 0$, as a perturbed version of the process $\eta^{(0)}(t)$.

The object of interest is the joint distribution of the process $\eta^{(\varepsilon)}(t)$ subject to a condition of non-absorption of the process up to a moment t, i.e., the probabilities $\mathsf{P}\{\eta^{(\varepsilon)}(t) \in A,\ \mu^{(\varepsilon)} > t\}$. In models with perturbations, it is natural to study the asymptotic behaviour of these probabilities when the time $t \to \infty$ and the perturbation parameter $\varepsilon \to 0$ simultaneously. The corresponding asymptotic results describe so-called quasi-stationary and pseudo-stationary phenomena for processes $\eta^{(\varepsilon)}(t)$. These phenomena differ by the asymptotic behaviour of the absorption times $\mu^{(\varepsilon)}$. These random variables are stochastically bounded or unbounded as $\varepsilon \to 0$, respectively, in the quasi-stationary and pseudo-stationary cases.

The principal novelty of results presented in book [GS08] is that the models with nonlinear perturbations are studied. Local transition characteristics that were mentioned above are usually some scalar or vector moment functionals $p^{(\varepsilon)}$ of local transition probabilities for the corresponding processes. By a nonlinear perturbation we mean that these characteristics are nonlinear functions of the perturbation parameter ε and that the assumptions made imply that the characteristics can be expanded in an asymptotic power series with respect to ε up to and including some order k, i.e., $p^{(\varepsilon)} = p^{(0)} + p[1]\varepsilon + \cdots + p[k]\varepsilon^k + o(\varepsilon^k)$. The case $k = 1$ corresponds to models with usual linear perturbations while the cases $k > 1$ correspond to models with nonlinear perturbations.

The classes of processes for which this program is realised include nonlinearly perturbed regenerative processes, semi-Markov processes, and continuous time Markov chains with absorption. The approach is based on advanced techniques, developed in the book, of nonlinearly perturbed renewal equations. Applications to the analysis of quasi-stationary phenomena in models of nonlinearly perturbed stochastic systems considered in the book pertain to models of highly reliable queueing systems, M/G queueing systems with quick service, stochastic systems of birth–death type, including epidemic and population dynamics models, metapopulation dynamic models, and perturbed risk processes.

The book [GS08] contains an extended introduction, where the main problems, methods, and algorithms that constitute the content of the book are presented in informal form. In Chaps. 1 and 2, results which deal with a generalisation of the classical renewal theorem to a model of the perturbed renewal equation are presented. These results are interesting by their own and, as we think, can find various applications beyond the areas mentioned in the book. In Chaps. 3–5 quasi- and pseudo-stationary asymptotics is studied for nonlinearly perturbed regenerative processes, semi-Markov processes, and continuous time Markov chains with absorption. Chapters 6 and 7 are devoted to applications of the theoretical results to studies of quasi-stationary phenomena for various nonlinearly perturbed models of stochastic systems. In Chap. 6, quasi-stationary phenomena are studied for highly reliable queueing systems, M/G queueing systems with quick service, stochastic systems of birth–death type, including epidemic and population dynamics models, and metapopulation dynamic models; Chap. 7 deals with perturbed risk processes. Finally, Chap. 8 contains three supplements. The first one gives some basic operation formulas for scalar and matrix asymptotic expansions. In the second supplement, some new prospective directions for future research in the area are discussed. In the last supplement, bibliographical remarks to the bibliography that includes more than 1000 references are given.

2.2 Nonlinearly Perturbed Renewal Equation

Let us consider the family of renewal equations,

$$x^{(\varepsilon)}(t) = q^{(\varepsilon)}(t) + \int_0^t x^{(\varepsilon)}(t-s)F^{(\varepsilon)}(\mathrm{d}s),\ t \geq 0, \tag{2.1}$$

where, for every $\varepsilon \geq 0$, we have the following: (a) $q^{(\varepsilon)}(t)$ is a real-valued function on $[0, \infty)$ that is Borel measurable and locally bounded, i.e., bounded on every finite interval, and (b) $F^{(\varepsilon)}(s)$ is a distribution function on $[0, \infty)$ which is not concentrated at 0 but can be improper, i.e., $F^{(\varepsilon)}(\infty) \leq 1$.

As well known, there exists the unique Borel measurable and bounded on every finite interval solution $x^{(\varepsilon)}(t)$ of (2.1).

In the model of perturbed renewal, the forcing function $q^{(\varepsilon)}(t)$ and distribution $F^{(\varepsilon)}(s)$ depend on some perturbation parameter $\varepsilon \geq 0$ and converge in some sense to $q^{(0)}(t)$ and $F^{(0)}(s)$ as $\varepsilon \to 0$.

The fundamental fact of the renewal theory connected with this equation is the renewal theorem given in its final form by Feller [Fel66]. This theorem describes the asymptotic behavior of solution in the form of asymptotic relation $x^{(0)}(t) \to x^{(0)}(\infty)$ as $t \to \infty$ for non-perturbed renewal equation.

The renewal theorem is a very powerful tool for proving ergodic theorems for regenerative stochastic processes. This class of processes is very broad. It includes Markov processes with discrete phase space. Moreover, Markov processes with a general phase space can be included, under some minor conditions, in a model of regenerative processes with the help of the procedure of artificial regeneration.

Applying the renewal theorem to ergodic theorems for regenerative type processes is based on the well known fact that the distribution of a regenerative process at a moment t satisfies a renewal equation. This makes it possible to apply the renewal theorem and to describe the asymptotic behaviour of the distribution of the regenerative process as $t \to \infty$.

Theorems that generalise the classical renewal theorem to a model of the perturbed renewal equation was proved in papers Silvestrov [Sil76, Sil78, Sil79]. These results are presented in Chap. 1 of De Gruyter Expositions in Mathematics [GS08].

As usual the symbol $F^{(\varepsilon)}(\cdot) \Rightarrow F^{(0)}(\cdot)$ as $\varepsilon \to 0$ means weak convergence of the distribution functions that is, the pointwise convergence in each point of continuity of the limiting distribution function.

Further, the following notations are used,

$$f^{(\varepsilon)} = 1 - F^{(\varepsilon)}(\infty), \quad m_n^{(\varepsilon)} = \int_0^\infty s^n F^{(\varepsilon)}(\mathrm{d}s), \quad n \geq 1.$$

We assume that the functions $q^{(\varepsilon)}(t)$ and the distributions $F^{(\varepsilon)}(s)$ satisfy the following continuity conditions at the point $\varepsilon = 0$, if regarded as functions of ε:

$\mathbf{D_1}$: $F^{(\varepsilon)}(\cdot) \Rightarrow F^{(0)}(\cdot)$ as $\varepsilon \to 0$, where $F^{(0)}(s)$ is a proper non-arithmetic distribution function;

$\mathbf{M_1}$: $m_1^{(\varepsilon)} \to m_1^{(0)} < \infty$ as $\varepsilon \to 0$;

and

$\mathbf{F_1}$: **(a)** $\lim_{u\to 0} \overline{\lim}_{0\leq\varepsilon\to 0} \sup_{|v|\leq u} |q^{(\varepsilon)}(t+v) - q^{(0)}(t)| = 0$ almost everywhere with respect to the Lebesgue measure on $[0, \infty)$;
(b) $\overline{\lim}_{0\leq\varepsilon\to 0} \sup_{0\leq t\leq T} |q^{(\varepsilon)}(t)| < \infty$ for every $T \geq 0$;
(c) $\lim_{T\to\infty} \overline{\lim}_{0\leq\varepsilon\to 0} h \sum_{r\geq T/h} \sup_{rh\leq t\leq (r+1)h} |q^{(\varepsilon)}(t)| = 0$ for some $h > 0$.

It is easy to show that, under conditions $\mathbf{D_1}$, $f^{(\varepsilon)} \to f^{(0)} = 0$ as $\varepsilon \to 0$.

Let also assume the following condition that balances the rate at which time $t^{(\varepsilon)}$ approaches infinity, and the convergence rate of the defect $f^{(\varepsilon)}$ to zero as $\varepsilon \to 0$:

$\mathbf{B}$: $0 \leq t^{(\varepsilon)} \to \infty$ and $f^{(\varepsilon)} \to 0$ as $\varepsilon \to 0$ in such a way that $f^{(\varepsilon)} t^{(\varepsilon)} \to \lambda$, where $0 \leq \lambda \leq \infty$.

The starting point for the research studies presented in book [GS08] is the following theorem (Silvestrov [Sil76, Sil78, Sil79]).

Theorem 1. *Let conditions* $\mathbf{D_1}$, $\mathbf{M_1}$, $\mathbf{F_1}$, *and* $\mathbf{B}$ *hold. Then,*

$$x_\varepsilon(t_\varepsilon) \to \mathrm{e}^{-\lambda/m_1^{(0)}} \frac{\int_0^\infty q^{(0)}(s)\mathrm{d}s}{m_1^{(0)}} \quad \text{as} \quad \varepsilon \to 0. \tag{2.2}$$

Remark 1. It is worth to note that this theorem reduces to the classical renewal theorem in the case of non-perturbed renewal equation, i.e., where the forcing functions $q^{(\varepsilon)}(t) \equiv q^{(0)}(t)$ and distribution functions $F^{(\varepsilon)}(s) \equiv F^{(0)}(s)$ do not depend on ε. In particular, condition $\mathbf{D_1}$ reduces to the assumption that $F^{(0)}(s)$ is a proper non-arithmetic distribution function; $\mathbf{M_1}$ to the assumption that the expectation $m_1^{(0)}$ is finite; and $\mathbf{F_1}$ to the assumption that the function $q^{(0)}(t)$ is directly Riemann integrable on $[0, \infty)$.

In this case, the defect $f^{(\varepsilon)} \equiv 0$ and the balancing condition $\mathbf{B}$ holds for any $t^{(\varepsilon)} \to \infty$ as $\varepsilon \to 0$ with the parameter $\lambda = 0$.

Note that condition $\mathbf{D_1}$ does not require and does not provide that the pre-limit $(\varepsilon > 0)$ distribution functions $F^{(\varepsilon)}(s)$ are non-arithmetic.

Also, condition $\mathbf{F_1}$ does not provide that the pre-limit $(\varepsilon > 0)$ forcing functions $q^{(\varepsilon)}(t)$ are directly Riemann integrable on $[0, \infty)$. However, this condition does imply that the limit forcing functions $q^{(0)}(t)$ has this property.

In the general case, the balancing condition $\mathbf{B}$ restrict the rate of growth for time $t^{(\varepsilon)}$. This restriction becomes unnecessary if an additional Cramér type condition is imposed on the distributions $F_\varepsilon(s)$.

In this case, one can also weaken condition $\mathbf{D_1}$ and accept also the possibility for the limit distribution be improper:

$\mathbf{D_2}$: **(a)** $F^{(\varepsilon)}(\cdot) \Rightarrow F^{(0)}(\cdot)$ as $\varepsilon \to 0$, where $F^{(0)}(t)$ is a non-arithmetic distribution function (possibly improper);
(b) $f^{(\varepsilon)} \to f^{(0)} \in [0, 1)$ as $\varepsilon \to 0$.

The Cramér type condition mentioned above takes the following form:

$\mathbf{C_1}$: There exists $\delta > 0$ such that:
(a) $\overline{\lim}_{0 \leq \varepsilon \to 0} \int_0^\infty e^{\delta s} F^{(\varepsilon)}(ds) < \infty$;
(b) $\int_0^\infty e^{\delta s} F^{(0)}(ds) > 1$.

Let us introduce the moment generation function,

$$\phi^{(\varepsilon)}(\rho) = \int_0^\infty e^{\rho s} F^{(\varepsilon)}(ds), \ \rho \geq 0.$$

Consider the following characteristic equation,

$$\phi^{(\varepsilon)}(\rho) = 1. \tag{2.3}$$

Under conditions $\mathbf{D_2}$ and $\mathbf{C_1}$, there exists $\varepsilon_1 > 0$ such that $\phi^{(\varepsilon)}(\delta) \in (1, \infty)$, and, therefore, (2.3) has a unique non-negative root $\rho^{(\varepsilon)}$ and $\rho^{(\varepsilon)} \leq \delta$, for every $\varepsilon \leq \varepsilon_1$. Also, $\rho^{(\varepsilon)} \to \rho^{(0)}$ as $\varepsilon \to 0$.

Note also that **(a)** $\rho^{(0)} = 0$ if and only if $f^{(0)} = 0$ and **(b)** $\rho^{(0)} > 0$ if and only if $f^{(0)} > 0$.

In this case, condition $\mathbf{F_1}$ takes the following modified form:

$\mathbf{F_2}$: **(a)** $\lim_{u \to 0} \overline{\lim}_{0 \leq \varepsilon \to 0} \sup_{|v| \leq u} |q^{(\varepsilon)}(t + v) - q^{(0)}(t)| = 0$ almost everywhere with respect to the Lebesgue measure on $[0, \infty)$;
(b) $\overline{\lim}_{0 \leq \varepsilon \to 0} \sup_{0 \leq t \leq T} |q^{(\varepsilon)}(t)| < \infty$ for every $T \geq 0$;
(c) $\lim_{T \to \infty} \overline{\lim}_{0 \leq \varepsilon \to 0} h \sum_{r \geq T/h} \sup_{rh \leq t \leq (r+1)h} e^{\gamma t} |q^{(\varepsilon)}(t)| = 0$ for some $h > 0$ and $\gamma > \rho^{(0)}$.

Let us denote,

$$\tilde{x}^{(\varepsilon)}(\infty) = \frac{\int_0^\infty e^{\rho^{(\varepsilon)} s} q^{(\varepsilon)}(s) m(ds)}{\int_0^\infty s e^{\rho^{(\varepsilon)} s} F^{(\varepsilon)}(ds)},$$

where $m(ds)$ is the Lebesgue measure on a real line.

Conditions $\mathbf{D_2}$, $\mathbf{C_1}$, and $\mathbf{F_2}$ imply, due to relation $\rho^{(\varepsilon)} \to \rho^{(0)}$ as $\varepsilon \to 0$, that there exists $0 < \varepsilon_2 \leq \varepsilon_1$ such that $\rho^{(\varepsilon)} < \gamma$ and $\int_0^\infty e^{\rho^{(\varepsilon)} s} |q^{(\varepsilon)}(s)| m(ds) < \infty$ for $\varepsilon \leq \varepsilon_2$. Thus, the functional $\tilde{x}^{(\varepsilon)}(\infty)$ is well defined for $\varepsilon \leq \varepsilon_2$.

The following theorem was also proved in Silvestrov [Sil76, Sil78, Sil79].

Theorem 2. *Let conditions* $\mathbf{D_2}$, $\mathbf{C_1}$, *and* $\mathbf{F_2}$ *hold. Then,*

$$\frac{x^{(\varepsilon)}(t^{(\varepsilon)})}{e^{-\rho^{(\varepsilon)} t^{(\varepsilon)}}} \to \tilde{x}^{(0)}(\infty) \text{ as } \varepsilon \to 0. \tag{2.4}$$

The asymptotic relation (2.4) given in Theorem 2 should be compared with the asymptotic relation (2.2) given in Theorem 1, in the case where $\rho^{(0)} = 0$.

Indeed, relation (2.2) can be re-written in the form given in (2.4), with coefficients $\rho^{(\varepsilon)} = f^{(\varepsilon)}/m_1^{(\varepsilon)}$. The Cramér type condition $\mathbf{C_1}$ makes it possible to use in (2.4) an alternative coefficients $\rho^{(\varepsilon)}$ defined as the solution of the characteristic equation (2.3). The latter coefficients provide better fitting of the corresponding exponential approximation for solution of renewal equation. That is why the asymptotic relation (2.4) does not restrict the rate of growth for time $t^{(\varepsilon)}$ while the asymptotic relation (2.2) does impose such restriction.

Remark 2. It is worth to note that this theorem reduces to the variant of renewal theorem for improper renewal equation in the case of non-perturbed renewal equation, also given in Feller [Fel66]. Condition $\mathbf{D_2}$ reduces to the assumption that $F^{(0)}(s)$ is a non-arithmetic distribution function with defect $f^{(0)} \in [0, 1)$; $\mathbf{C_1}$ to the assumption that the exponential moment $\phi^{(0)}(\delta) \in (1, \infty)$; and $\mathbf{F_2}$ to the assumption that the function $\mathrm{e}^{\gamma t} q^{(0)}(t)$ is directly Riemann integrable on $[0, \infty)$ for some $\gamma > \rho^{(0)}$.

The results formulated in Theorems 1 and 2 created the base for further research studies in the area. For example, Shurenkov [Shu80a, Shu80b] generalised the results of these theorems to the case of perturbed matrix renewal equation using possibility of imbedding the matrix model to the scalar model considered in Theorems 1 and 2.

A new improvement was achieved in the paper Silvestrov [Sil95] and then in the papers Gyllenberg and Silvestrov [GS99a, GS00a]. Under natural additional perturbation conditions, which assume that the defect $f^{(\varepsilon)}$ and the corresponding moments of the distribution $F^{(\varepsilon)}(s)$ can be expanded in power series with respect to ε up to and including an order k, explicit expansions for the corresponding characteristic roots were given, and the corresponding exponential expansions were obtained for solutions of nonlinearly perturbed renewal equations. In [Sil95], the case with asymptotically proper distributions $F^{(\varepsilon)}(s)$ was considered, while, in Gyllenberg and Silvestrov [GS99a, GS00a], the case with asymptotically improper distributions $F^{(\varepsilon)}(s)$ was investigated. These results are presented in Chap. 2 of De Gruyter Expositions in Mathematics [GS08].

Let us introduce the mixed power-exponential moment generating functions,

$$\phi^{(\varepsilon)}(\rho, n) = \int_0^\infty s^n \mathrm{e}^{\rho s} F^{(\varepsilon)}(\mathrm{d}s), \ \rho \geq 0, \ n = 0, 1, \ldots .$$

Note that by the definition $\phi^{(\varepsilon)}(\rho, 0) = \phi^{(\varepsilon)}(\rho)$. Under conditions $\mathbf{D_2}$ and $\mathbf{C_1}$, for any $0 < \delta' < \delta$, there exists $0 < \varepsilon_3 < \varepsilon_2$ such that $\phi^{(\varepsilon)}(\delta', n) < \infty$ for $n = 0, 1, \ldots$ and $\varepsilon \leq \varepsilon_3$. Also, $\phi^{(\varepsilon)}(\rho, n) \to \phi^{(0)}(\rho, n)$ as $\varepsilon \to 0$ for $n = 0, 1, \ldots$ and $\rho \leq \delta'$. Let δ' is chosen such that $\phi^{(0)}(\delta') \in (1, \infty)$. In this case, the characteristic root $\rho^{(0)} < \delta'$ and also there exists $0 < \varepsilon_4 < \varepsilon_3$, such that the characteristic roots $\rho^{(\varepsilon)} < \delta'$ for $\varepsilon \leq \varepsilon_4$.

The basic role plays the following nonlinear perturbation condition:

$\mathbf{P_1^{(k)}}$: $\phi^{(\varepsilon)}(\rho^{(0)}, n) = \phi^{(0)}(\rho^{(0)}, n) + b_{1,n}\varepsilon + \cdots + b_{k-n,n}\varepsilon^{k-n} + o(\varepsilon^{k-n})$ for $n = 0, \ldots, k$, where $|b_{i,n}| < \infty, i = 1, \ldots, k-n, \ n = 0, \ldots, k$.

It is convenient to define $b_{0,n} = \phi^{(0)}(\rho^{(0)}, n), n = 0, 1, \ldots$. From the definition of $\rho^{(0)}$ it is clear that $b_{0,0} = \phi^{(0)}(\rho^{(0)}, 0) = 1$.

It should be noted that, in the case $f^{(0)} = 0$, where characteristic root $\rho^{(0)} = 0$, the perturbation condition $\mathbf{P_1^{(k)}}$ involves usual power moments of distributions $F^{(\varepsilon)}(s)$. While in the case $f^{(0)} > 0$, where characteristic root $\rho^{(0)} > 0$, the perturbation condition involves mixed power-exponential moments of distributions $F^{(\varepsilon)}(s)$.

Let us also formulate the following condition that balances the rate at which time $t^{(\varepsilon)}$ approaches infinity and the convergence rate of perturbation in different asymptotic zones, for $1 \leq r \leq k$:

$\mathbf{B^{(r)}}$: $0 \leq t^{(\varepsilon)} \to \infty$ in such a way that $\varepsilon^r t^{(\varepsilon)} \to \lambda_r$, where $0 \leq \lambda_r < \infty$.

The following theorem is given in Silvestrov [Sil95] and Gyllenberg and Silvestrov [GS99a, GS00a, GS08].

Theorem 3. *Let conditions* $\mathbf{D_2}$, $\mathbf{C_1}$, *and* $\mathbf{P_1^{(k)}}$ *hold. Then,*

(i) *The root* $\rho^{(\varepsilon)}$ *of* (2.3) *has the asymptotic expansion*

$$\rho^{(\varepsilon)} = \rho^{(0)} + a_1\varepsilon + \cdots + a_k\varepsilon^k + o(\varepsilon^k), \tag{2.5}$$

where the coefficients a_n *are given by the recurrence formulas* $a_1 = -b_{1,0}/b_{0,1}$ *and, in general, for* $n = 1, \ldots, k$,

$$\begin{aligned} a_n = & - b_{0,1}^{-1}(b_{n,0} + \sum_{q=1}^{n-1} b_{n-q,1}a_q \\ & + \sum_{2\leq m\leq n} \sum_{q=m}^{n} b_{n-q,m} \cdot \sum_{n_1,\ldots,n_{q-1}\in D_{m,q}} \prod_{p=1}^{q-1} a_p^{n_p}/n_p!), \end{aligned} \tag{2.6}$$

where $D_{m,q}$, *for every* $2 \leq m \leq q < \infty$, *is the set of all nonnegative, integer solutions of the system*

$$n_1 + \cdots + n_{q-1} = m, \quad n_1 + \cdots + (q-1)n_{q-1} = q. \tag{2.7}$$

(ii) *If* $b_{i,0} = 0, i = 1, \ldots, n$, *for some* $1 \leq n \leq k$, *then* $a_1, \ldots, a_n = 0$. *If* $b_{i,0} = 0$, $i = 1, \ldots, n-1$ *but* $b_{n,0} < 0$, *for some* $1 \leq n \leq k$, *then* $a_1, \ldots, a_{n-1} = 0$ *but* $a_n > 0$.

(iii) *If, additionally, conditions* $\mathbf{B^{(r)}}$, *for some* $1 \leq r \leq k$, *and* $\mathbf{F_2}$ *hold, then the following asymptotic relation holds:*

$$\frac{x^{(\varepsilon)}(t^{(\varepsilon)})}{\exp\{-(\rho^{(0)} + a_1\varepsilon + \cdots + a_{r-1}\varepsilon^{r-1})t^{(\varepsilon)}\}} \to \mathrm{e}^{-\lambda_r a_r}\tilde{x}^{(0)}(\infty) \ \text{ as } \ \varepsilon \to 0. \tag{2.8}$$

The asymptotic relation (2.8) given in Theorem 3 should be compared with the asymptotic relation (2.4) given in Theorem 2.

The asymptotic relation (2.4) looks nicely but has actually a serious drawback. Indeed, the exponential normalisation with the coefficient $\rho^{(\varepsilon)}$ is not so effective because of this coefficient is given us only as the root of the nonlinear equation (2.3), for every $\varepsilon \geq 0$.

Relation (2.8) essentially improves the asymptotic relation (2.4) replacing this simple convergence relation by the corresponding asymptotic expansion. The exponential normalisation with the coefficient $\rho^{(0)} + a_1\varepsilon + \cdots + a_{r-1}\ \mathrm{e}^{r-1}$ involves the root $\rho^{(0)}$. To find it one should solve only one nonlinear equation (2.3), for the case $\varepsilon = 0$. As far as the coefficients $a_1, \ldots, a_r$ are concerned, they are given in the explicit algebraic recurrence form.

Moreover, the root $\rho^{(0)} = 0$ in the most interesting case, where $f^{(0)} = 0$, i.e., the limit renewal equation is proper. Here, the non-linear step connected with finding of the root of (2.3) can be omitted.

If there exist $0 < \varepsilon' \leq e_5$ such that the conditions listed in Remark 2 holds for the distribution function $F^{(\varepsilon)}(s)$ and the forcing function $q^{(\varepsilon)}(t)$ for every $\varepsilon \leq \varepsilon'$, then according to Theorem 2, the following asymptotic relation holds for every $\varepsilon \leq \varepsilon'$,

$$\frac{x^{(\varepsilon)}(t)}{\mathrm{e}^{-\rho^{(\varepsilon)}t}} \to \tilde{x}^{(\varepsilon)}(\infty) \quad \text{as } t \to \infty. \tag{2.9}$$

Let us now define mixed power-exponential moment functionals for the forcing functions,

$$\omega^{(\varepsilon)}(\rho, n) = \int_0^\infty s^n \mathrm{e}^{\rho s} q^{(\varepsilon)}(s) m(\mathrm{d}s), \ \rho \geq 0, \ n = 0, 1, \ldots.$$

Under conditions $\mathbf{C_1}$ and $\mathbf{F_2}$, for any $0 < \gamma' < \gamma$, there exists $0 < \varepsilon_6 < \varepsilon_5$ such that $\bar{\omega}^{(\varepsilon)}(\gamma', n) = \int_0^\infty s^n \mathrm{e}^{\gamma' s}|q^{(\varepsilon)}(s)| m(\mathrm{d}s) < \infty$ for $n = 0, 1, \ldots$ and $\varepsilon \leq \varepsilon_6$. Also, $\omega^{(\varepsilon)}(\rho, n) \to \omega^{(0)}(\rho, n)$ as $\varepsilon \to 0$ for $n = 0, 1, \ldots$ and $\rho \leq \gamma'$. Let γ' is chosen such that $\rho^{(0)} < \gamma'$. In this case, there exists $0 < \varepsilon_7 < \varepsilon_6$ such that the characteristic roots $\rho^{(\varepsilon)} < \gamma'$ for $\varepsilon \leq \varepsilon_7$.

Note that the renewal limit $\tilde{x}^{(\varepsilon)}(\infty)$ is well defined for $\varepsilon \leq \varepsilon_7$ even without the non-arithmetic assumption made above in order to provide asymptotic relation (2.9) and, moreover,

$$\tilde{x}^{(\varepsilon)}(\infty) = \frac{\omega^{(\varepsilon)}(\rho^{(\varepsilon)}, 0)}{\phi^{(\varepsilon)}(\rho^{(\varepsilon)}, 1)}. \tag{2.10}$$

Let us now formulate a perturbation condition for mixed power-exponential moment functionals for the forcing functions:

$\mathbf{P_2^{(k)}}$: $\omega^{(\varepsilon)}(\rho^{(0)}, n) = \omega^{(0)}(\rho^{(0)}, n) + c_{1,n}\varepsilon + \cdots + c_{k-n,n}\varepsilon^{k-n} + o(\varepsilon^{k-n})$ for $n = 0, \ldots, k$, where $|c_{i,n}| < \infty, i = 1, \ldots, k-n, n = 0, \ldots, k$.

It is convenient to set $c_{0,n} = \omega^{(0)}(\rho^{(0)}, n), n = 0, 1, \ldots$.

The following theorem supplements Theorem 3.

Theorem 4. *Let conditions* $\mathbf{D_2}$, $\mathbf{C_1}$, $\mathbf{F_2}$, $\mathbf{P_1^{(k+1)}}$, *and* $\mathbf{P_2^{(k)}}$ *hold. Then the functional* $\tilde{x}^{(\varepsilon)}(\infty)$ *has the following asymptotic expansions*:

$$\begin{aligned} \tilde{x}^{(\varepsilon)}(\infty) &= \frac{\omega^{(0)}(\rho^{(0)}, 0) + f'_1\varepsilon + \cdots + f'_k\varepsilon^k + o(\varepsilon^k)}{\phi^{(0)}(\rho^{(0)}, 1) + f''_1\varepsilon + \cdots + f''_k\varepsilon^k + o(\varepsilon^k)} \\ &= \tilde{x}^{(0)}(\infty) + f_1\varepsilon + \cdots + f_k\varepsilon^k + o(\varepsilon^k), \end{aligned} \tag{2.11}$$

where the coefficients f'_n, f''_n *are given by the formulas* $f'_0 = \omega^{(0)}(\rho^{(0)}, 0) = c_{0,0}$, $f'_1 = c_{1,0} + c_{0,1}a_1$, $f''_0 = \phi^{(0)}(\rho^{(0)}, 1) = b_{0,1}, f''_1 = b_{1,1} + b_{0,2}a_1$, *and in general for* $n = 0, \ldots, k$,

$$\begin{aligned} f'_n = {} & c_{n,0} + \sum_{q=1}^n c_{n-q,1}a_q \\ & + \sum_{2 \leq m \leq n} \sum_{q=m}^n c_{n-q,m} \cdot \sum_{n_1, \ldots, n_{q-1} \in D_{m,q}} \prod_{p=1}^{q-1} a_p^{n_p}/n_p!, \end{aligned} \tag{2.12}$$

and

$$f''_n = b_{n,1} + \sum_{q=1}^{n} b_{n-q,2} a_q$$

$$+ \sum_{2 \le m \le n} \sum_{q=m}^{n} b_{n-q,m+1} \cdot \sum_{n_1,\ldots,n_{q-1} \in D_{m,q}} \prod_{p=1}^{q-1} a_p^{n_p}/n_p!, \tag{2.13}$$

and the coefficients f_n *are given by the recurrence formulas* $f_0 = \tilde{x}^{(0)}(\infty) = f'_0/f''_0$ *and in general for* $n = 0, \ldots, k$,

$$f_n = (f'_n - \sum_{q=0}^{n-1} f''_{n-q} f_q)/f''_0. \tag{2.14}$$

It should be noted that the perturbation condition $\mathbf{P}_1^{(\mathbf{k}+1)}$ stronger than $\mathbf{P}_1^{(\mathbf{k})}$ is required in Theorem 4. This is because of the former condition is needed to get the corresponding expansion for $\phi^{(\varepsilon)}(\rho^{(\varepsilon)}, 1)$ in an asymptotic power series with respect to ε up to and including the order k.

Chapter 2 of De Gruyter Expositions in Mathematics [GS08] also contains asymptotic results based on more general perturbation conditions.

It worth to mention that discrete time analogues of some of the results presented above are given in papers by Englund and Silvestrov [ES97], Englund [Eng00, Eng01], and Silvestrov [Sil00b]. Also, exponential asymptotic expansions for renewal equation with non-polynomial perturbations are studied in papers Englund and Silvestrov [ES97, Eng01], and Ni, Silvestrov and Malyarenko [NSM08].

2.3 Nonlinearly Perturbed Regenerative Processes

Method of asymptotic analysis of nonlinearly perturbed renewal equation can be directly used in studies of quasi- and pseudo-stationary asymptotics for nonlinearly perturbed regenerative processes. The corresponding results are presented in Chap. 3 of De Gruyter Expositions in Mathematics [GS08]. This chapter is partly based on the results of the papers Gyllenberg and Silvestrov [GS99a, GS00b].

Let $\xi^{(\varepsilon)}(t), t \ge 0$ be a regenerative process with a measurable phase space X and regeneration times $\tau_n^{(\varepsilon)}, n = 1, 2, \ldots,$ and $\mu^{(\varepsilon)}$ be a regenerative stopping time that regenerates jointly with the process $\xi^{(\varepsilon)}(t)$, at times $\tau_n^{(\varepsilon)}$.

Both the regenerative process $\xi^{(\varepsilon)}(t)$ and the regenerative stopping time $\mu^{(\varepsilon)}$ are assumed to depend on a small perturbation parameter $\varepsilon \ge 0$. The processes $\xi^{(\varepsilon)}(t)$, for $\varepsilon > 0$ are considered as a perturbation of the process $\xi^{(0)}(t)$, and therefore we assume some weak continuity conditions for certain characteristic quantities of these processes regarded as functions of ε at point $\varepsilon = 0$.

As far as the regenerative stopping times are concerned, we consider two cases. The first one is a pseudo-stationary case, where the random variables $\mu^{(\varepsilon)}$ are stochastically unbounded, i.e., $\mu^{(\varepsilon)}$ tend to ∞ in probability as $\varepsilon \to 0$. The second one is the quasi-stationary case, where the random variables $\mu^{(\varepsilon)}$ are stochastically bounded as $\varepsilon \to 0$.

The object of studies is the probabilities $P^{(\varepsilon)}(t, A) = \mathsf{P}\{\xi^{(\varepsilon)}(t) \in A,\ \mu^{(\varepsilon)} > t\}$. These probabilities satisfy the following renewal equation,

$$P^{(\varepsilon)}(t, A) = q^{(\varepsilon)}(t, A) + \int_0^\infty P^{(\varepsilon)}(t - s, A)F^{(\varepsilon)}(\mathrm{d}s),\ t \geq 0, \tag{2.15}$$

where the forcing function $q^{(\varepsilon)}(t, A) = \mathsf{P}\{\xi^{(\varepsilon)}(t) \in A, \tau_1^{(\varepsilon)} \wedge \mu^{(\varepsilon)} > t\}$ and distribution function $F^{(\varepsilon)}(s) = \mathsf{P}\{\tau_1^{(\varepsilon)} \leq s, \mu^{(\varepsilon)} \geq \tau_1^{(\varepsilon)}\}$.

Note that the distribution $F^{(\varepsilon)}(s)$ has the defect $f^{(\varepsilon)} = \mathsf{P}\{\mu^{(\varepsilon)} < \tau_1^{(\varepsilon)}\}$.

In this case, the mixed power-exponential moment generating function $\phi^{(\varepsilon)}(\rho, n) = \mathsf{E}(\tau_1^{(\varepsilon)})^n e^{\rho \tau_1^{(\varepsilon)}} \chi(\mu^{(\varepsilon)} \geq \tau_1^{(\varepsilon)})$ and the characteristic equation (2.3) takes the form $\phi^{(\varepsilon)}(\rho, 0) = 1$.

The corresponding perturbation condition assumes that function $\phi^{(\varepsilon)}(\rho, n)$ (taken in point $\rho^{(0)}$ which is the root of the limit characteristic equation) can be expanded in a power series with respect to ε up to and including the order $k - n$ for every $n = 0, \ldots, k$.

The relationship between the rate with which ε tends to zero and the time t tends to infinity has a delicate influence upon the results. The balance between the rate of perturbation and the rate of growth of time is characterized by the following asymptotic relation $\varepsilon^r t^{(\varepsilon)} \to \lambda_r < \infty$ as $\varepsilon \to 0$ that is assumed to hold for some $1 \leq r \leq k$.

The direct application of Theorem 3 to the renewal equation (2.15) yields, under the corresponding conditions, the following exponential asymptotic expansion,

$$\frac{\mathsf{P}\{\xi^{(\varepsilon)}(t^{(\varepsilon)}) \in A, \mu^{(\varepsilon)} > t^{(\varepsilon)}\}}{\exp\{-(\rho_0 + a_1\varepsilon + \cdots + a_{r-1}\varepsilon^{r-1})t^{(\varepsilon)}\}} \to \tilde{\pi}^{(0)}(A)\mathrm{e}^{-\lambda_r a_r} \text{ as } \varepsilon \to 0, \tag{2.16}$$

where

$$\tilde{\pi}^{(\varepsilon)}(A) = \frac{\int_0^\infty \mathrm{e}^{\rho^{(\varepsilon)} s} q^{(\varepsilon)}(s, A) m(\mathrm{d}s)}{\int_0^\infty s \mathrm{e}^{\rho^{(\varepsilon)} s} F^{(\varepsilon)}(\mathrm{d}s)}.$$

Also, Theorem 4, applied to the renewal equation (2.15), yields, under the corresponding conditions, the asymptotic expansions for the renewal limits $\tilde{\pi}^{(\varepsilon)}(A)$ and then the following asymptotic expansion for the quasi-stationary distributions $\pi^{(\varepsilon)}(A) = \tilde{\pi}^{(\varepsilon)}(A)/\tilde{\pi}^{(\varepsilon)}(X)$,

$$\pi^{(\varepsilon)}(A) = \pi^{(0)}(A) + g_1(A)\varepsilon + \cdots + g_k(A)\varepsilon^k + o(\varepsilon^k). \tag{2.17}$$

Both asymptotic expansions (2.16) and (2.17) are provided by the explicit algorithms for calculating the coefficients in these expansions as rational functions of the coefficients in the expansions involved in the initial perturbation conditions.

The case $\rho_0 = 0$ corresponds to a model with stochastically unbounded random variables $\mu^{(\varepsilon)}$, while the case $\rho_0 > 0$ corresponds to a model with stochastically bounded random variables $\mu^{(\varepsilon)}$. The asymptotic relation (2.16) describes in these cases, respectively, *pseudo-stationary* and *quasi-stationary* phenomena for perturbed regenerative processes.

To clarify the meaning of the asymptotic relation (2.16) let us consider the pseudo-stationary case, where $\rho_0 = 0$. Note that in this case $\tilde{\pi}^{(0)}(X) = 1$ and, therefore, $\tilde{\pi}^{(0)}(A) = \pi^{(0)}(A)$.

If $k = 1$, then the only case $r = 1$ is possible for the above balancing condition for the rate of perturbation and the rate of growth of time. In this case, the asymptotic relation (2.16) is equivalent to the asymptotic relation $\mathsf{P}\{\xi^{(\varepsilon)}(t^{(\varepsilon)}) \in A, \mu^{(\varepsilon)} > t^{(\varepsilon)}\} \to \pi^{(0)}(A)\mathrm{e}^{-\lambda_1 a_1}$ as $\varepsilon \to 0$. It shows that the position of the regenerative process $\xi(t^{(\varepsilon)})$ and the normalised regenerative stopping time $\varepsilon\mu^{(\varepsilon)}$ are asymptotically independent and have, in the limit, a stationary distribution and an exponential distribution, respectively. This can be interpreted as a mixed ergodic theorem (for the regenerative processes) and a limit theorem (for regenerative stopping times).

If $k = 2$, then two cases, $r = 1$ and $r = 2$, are possible for the balancing condition.

The case $r = 1$ was already commented and interpreted above. In this case, relation (2.16) can be given in the equivalent alternative form, $\frac{\mathsf{P}\{\xi^{(\varepsilon)}(t^{(\varepsilon)}) \in A, \mu^{(\varepsilon)} > t^{(\varepsilon)}\}}{\pi^{(0)}(A)\exp\{-a_1\varepsilon t^{(\varepsilon)}\}} \to 1$ as $\varepsilon \to 0$, for non-zero sets such that $\pi^{(0)}(A) \neq 0$. It shows that probability $\mathsf{P}\{\xi^{(\varepsilon)}(t^{(\varepsilon)}) \in A, \mu^{(\varepsilon)} > t^{(\varepsilon)}\}$ can be approximated by the exponential type mixed tail probability $\pi^{(0)}(A)\exp\{-a_1\varepsilon t^{(\varepsilon)}\}$, with the zero asymptotic relative error, in every asymptotic time zone which is determined by the relation $\varepsilon t^{(\varepsilon)} \to \lambda_1$ as $\varepsilon \to 0$, where $0 \leq \lambda_1 < \infty$.

In the case $r = 2$, the asymptotic relation (2.16) reduces to the asymptotic relation $\frac{\mathsf{P}\{\xi^{(\varepsilon)}(t^{(\varepsilon)}) \in A, \mu^{(\varepsilon)} > t^{(\varepsilon)}\}}{\pi^{(0)}(A)\exp\{-a_1\varepsilon t^{(\varepsilon)}\}} \to \mathrm{e}^{-a_2\lambda_2}$ as $\varepsilon \to 0$. It shows that probability $\mathsf{P}\{\xi^{(\varepsilon)}(t^{(\varepsilon)}) \in A, \mu^{(\varepsilon)} > t^{(\varepsilon)}\}$ can be approximated by the the exponential type mixed tail probability $\pi^{(0)}(A)\exp\{-a_1\varepsilon t^{(\varepsilon)}\}$ as $\varepsilon \to 0$, with the asymptotic relative error $1 - \mathrm{e}^{-a_2\lambda_2}$, in every asymptotic time zone which is determined by the relation $\varepsilon^2 t^{(\varepsilon)} \to \lambda_2$ as $\varepsilon \to 0$, where $0 \leq \lambda_2 < \infty$.

If $\lambda_2 = 0$, then $\varepsilon t^{(\varepsilon)} = o(\varepsilon^{-1})$ and the asymptotic relative error is 0. Note that this case also covers the situation where $\varepsilon t^{(\varepsilon)}$ is bounded, which corresponds to the asymptotic relation (2.16) with $k = 1$. This is already an extension of this asymptotic result since it is possible that $\varepsilon t^{(\varepsilon)} \to \infty$.

If $\lambda_2 > 0$, then $\varepsilon t^{(\varepsilon)} = O(\varepsilon^{-1})$, and the asymptotic relative error is $1 - \mathrm{e}^{-\lambda_2 a_2}$. It differs from 0. Therefore, $o(\varepsilon^{-1})$ is an asymptotic bound for the large deviation zone with the asymptotic relative error 0.

To get the approximation with zero asymptotic relative error in the asymptotic time zone which are determined by the relation $\varepsilon^2 t^{(\varepsilon)} \to \lambda_2$ one should approximate the mixed tail probabilities $\mathsf{P}\{\xi^{(\varepsilon)}(t^{(\varepsilon)}) \in A, \mu^{(\varepsilon)} > t^{(\varepsilon)}\}$ by the exponential type mixed tail probabilities $\pi^{(0)}(A)\mathrm{e}^{-(a_1\varepsilon + a_2\varepsilon^2)t^{(\varepsilon)}} \sim \pi^{(0)}(A)\mathrm{e}^{-a_1\varepsilon t^{(\varepsilon)} - a_2\lambda_2}$, i.e., to introduce the corresponding corrections for the parameters in the exponents.

The comments above let one interpret relation (2.16) in the case $r = 2$ as a new type of mixed ergodic and large deviation theorem for the nonlinearly perturbed process $\eta^{(\varepsilon)}(t)$ and the lifetime $\mu_0^{(\varepsilon)}$.

A similar interpretation can be made for the asymptotic relation (2.16) if $k > 2$ and in the quasi-stationary case, where $\rho^{(0)} > 0$.

Finally, the above asymptotic results is expanded to the model of nonlinearly perturbed regenerative processes with transition period.

2.4 Nonlinearly Perturbed Semi-Markov Processes

The asymptotic results obtained in Chaps. 1–3 play the key role in further studies. In Chaps. 4 and 5 of De Gruyter Expositions in Mathematics [GS08], they are applied to analysis of pseudo- and quasi-stationary phenomena for perturbed semi-Markov processes with a finite set of states. The results presented in this chapter are partly based on papers Gyllenberg and Silvestrov [GS99a, GS00a] and Silvestrov [Sil07a, Sil07b].

A semi-Markov process $\eta^{(\varepsilon)}(t), t \geq 0$, with a phase space $X = \{0, \dots, N\}$ and transition probabilities $Q_{ij}^{(\varepsilon)}(u)$ is considered. The first hitting time $\mu_0^{(\varepsilon)}$ to the state 0 plays the role of an absorption time. Asymptotic behaviour of probabilities $P_{ij}^{(\varepsilon)}(t) = \mathsf{P}_i\{\eta^{(\varepsilon)}(t) = j, \mu_0^{(\varepsilon)} > t\}$ is an object of studies.

This can be done by using the facts that a semi-Markov process can be considered as a regenerative process with regeneration times which are subsequent return moments to any fixed state $j \neq 0$ and the first hitting time to the absorption state 0 is a regenerative stopping time. The asymptotic results mentioned above are obtained by applying the corresponding results for regenerative processes given in Chap. 3.

Not only the generic case, where the limiting semi-Markov process has one communication class of recurrent-without absorption states, is considered in details, but also the case, where the limiting semi-Markov process has one communication class of recurrent-without absorption states and, additionally, the class of non-recurrent-without absorption states. The latter model covers a significant part of applications.

In this case, the distribution function ${}_0G_{jj}^{(\varepsilon)}(t)$ of the return-without absorption time in a state $j \neq 0$ generates the renewal equation. The corresponding characteristic equation takes the form $\int_0^\infty \mathrm{e}^{\rho s}\, {}_0G_{jj}^{(\varepsilon)}(\mathrm{d}s) = 1$. It is shown that the characteristic root $\rho^{(\varepsilon)}$ of this equation does not depend on the choice of a recurrent-without absorption state $j \neq 0$.

It is natural to formulate the perturbation conditions in terms of transition probabilities $Q_{ij}^{(\varepsilon)}(u)$. In particular, nonlinear perturbation conditions are imposed on these transition probabilities, which assume that mixed power-exponential moment generation functions $p_{ij}^{(\varepsilon)}[\rho, n] = \int_0^\infty s^n \mathrm{e}^{\rho s} Q_{ij}^{(\varepsilon)}(\mathrm{d}s), i \neq 0, j \in X$ (taken in point $\rho^{(0)}$ which is the root of the corresponding characteristic equation for the limit case $\varepsilon = 0$) can be expanded in a power series with respect to ε up to and including the order $k - n$ for every $n = 1, \dots, k$.

Conditions and expansions formulated for regenerative processes are specified in terms of expansions for the moments of regeneration times. As was pointed above the return times play the role of regeneration moments for semi-Markov processes. Therefore, the corresponding asymptotic expansions for absorption probabilities and the moments of return and hitting times for perturbed semi-Markov processes must be derived from the nonlinear perturbation conditions imposed on transition probabilities $Q_{ij}^{(\varepsilon)}(u)$. Then, the corresponding asymptotic results for regenerative processes can be applied.

Thus, as the first step, asymptotic expansions for hitting probabilities, power and mixed power-exponential moments of hitting times are constructed using a procedure that is based on recursive systems of linear equations for hitting probabilities and moments of hitting times. These moments satisfy recurrence systems of linear equations with the same perturbed coefficient matrix and the free terms connected by special recurrence systems of relations. In these relations, the free terms for the moments of a given order are given as polynomial functions of moments of lower orders. This permits to build an effective recurrence algorithm for constructing the corresponding asymptotic expansions. Each sub-step in this recurrence algorithm is of a matrix but linear type, where the solution of the system of linear equations with nonlinearly perturbed coefficients and free terms should be expanded in asymptotic series. These expansions are also provided with a detailed analysis of their pivotal properties. These results have their own values and possible applications beyond the problems studied in the book.

As soon as asymptotic expansions for moments of return-without absorption times are constructed, the second nonlinear but scalar step of construction asymptotic expansions expansions for the characteristic root $\rho^{(\varepsilon)}$ can be realised.

The separation of two steps described above, the first one matrix and recurrence but linear and the second one nonlinear but scalar, significantly simplify the whole algorithm.

The asymptotic expansions for the quasi-stationary distributions require two more steps, which are needed for constructing asymptotic expansions at the point $\rho^{(\varepsilon)}$ for the corresponding moment generation functions, giving expressions for quasi-stationary probabilities in the quotient form, and for transforming the corresponding asymptotic quotient expressions to the form of power asymptotic expansions.

As a result, one get an effective algorithm for a construction of the asymptotic expansions given in relations (2.16) and (2.17). It seems, that the method used for obtaining the expansions mentioned above has its own value and great potential for future studies.

As a particular but important example, the model of nonlinearly perturbed continuous time Markov chains with absorption is also considered. In this case, it is more natural to formulate the perturbation conditions in terms of generators of the perturbed Markov chains. Here, an additional step in the algorithms is needed, since the initial perturbation conditions for generators must be expressed in terms of the moments for the corresponding semi-Markov transition probabilities. Then, the basic algorithms obtained for nonlinearly perturbed semi-Markov processes can be applied.

Chapters 1–5 present a theory that can be applied in studies of pseudo- and quasi-stationary phenomena in nonlinearly perturbed stochastic systems.

2.5 Nonlinearly Perturbed Stochastic Systems

Chapter 6 of book [GS08] deals with applications of the results obtained in Chaps. 1–5 to an analysis of pseudo- and quasi-stationary phenomena in nonlinearly perturbed stochastic systems. This chapter is partly based on the results of the papers Gyllenberg and Silvestrov [GS94, GS99a, GS00a].

Examples of stochastic systems under consideration are queueing systems, epidemic, and population dynamics models with finite lifetimes. In queueing systems, the lifetime is usually the time at which some kind of a fatal failure occurs in the system. In epidemic models, the time of extinction of the epidemic in the population plays the role of the lifetime, while in population dynamics models, the lifetime is usually the extinction time for the corresponding population.

Several classical models being the subject of long term research studies were selected. These models serve nowadays mainly as platforms for demonstration of new methods and innovation results. Our goal also is to show what kind of new types results related to quasi-stationary asymptotics can be obtained for such models with nonlinearly perturbed parameters.

As the first example, a M/M queueing system with highly reliable main servers is considered. This queueing system is our first choice because of its function can be described by some nonlinearly perturbed continuous time Markov chains with absorption. Here, all conditions take a very explicit and clear form.

Also a M/G queueing system with quick service and a bounded queue buffer is considered. In this case, the perturbed stochastic processes, which describe the dynamics of the queue in the system, belong to the class of so-called stochastic processes with semi-Markov modulation. These processes admit a construction of imbedded semi-Markov processes and are more general than semi-Markov processes. This example was chosen because it shows in which way the main results obtained in the book can be applied to stochastic processes more general than semi-Markov processes, in particular, to stochastic processes with semi-Markov modulation.

The next example is based on classical semi-Markov and Markov birth-and-death type processes. Some classical models of queueing systems, epidemic or population dynamic models can be described with the use of such processes. We show in which way nonlinear perturbation conditions should be used and what form will take advanced quasi- and pseudo-stationary asymptotics developed in Chaps. 1–5.

Finally, an example of nonlinearly perturbed metapopulation model is considered. This example is interesting since it brings, for the first time, the discussion on advanced quasi- and pseudo-stationary asymptotics in this actual area of research in mathematical biology.

2.6 Nonlinearly Perturbed Risk Processes

The classical risk processes are still the object of intensive research studies as show, for example, references given in the bibliography of De Gruyter Expositions in Mathematics [GS08]. Of course, the purpose of these studies is not any more to derive formulas relevant for field applications. These studies intend to illustrate new methods and types of results that can later be expanded to more complex models. The same approach was used by us when choosing this model. The aim was to show that the innovative methods of analysis for nonlinearly perturbed processes developed in the book can yield new results for this classical models.

Chapter 7 of De Gruyter Expositions in Mathematics [GS08] contains results that extend the classical Cramér–Lundberg and diffusion approximations for the ruin probabilities to a model of nonlinearly perturbed risk processes. Both approximations are presented in a unified way using the techniques of perturbed renewal equations developed in Chaps. 1 and 2. This chapter is partly based on the results of the papers Gyllenberg and Silvestrov [GS99a, GS99b, GS00b] and Silvestrov [Sil00a, Sil07b].

The main new element in the results presented in Chap. 7 is a high order exponential asymptotic expansion in these approximations for nonlinearly perturbed risk processes. Correction terms are obtained for the Cramér-Lundberg and diffusion type approximations, which provide the right asymptotic behaviour of relative errors in the perturbed model. We study the dependence of these correction terms on the relations between the rate of perturbation and the rate of growth of the initial capital.

Also, various variants of the diffusion type approximation, including the asymptotics for increments and derivatives of the ruin probabilities are given.

Finally, we give asymptotic expansions in the Cramér–Lundberg and diffusion type approximations for distribution of the capital surplus prior and at ruin for nonlinearly perturbed risk processes.

It seems to us that results presented in Chaps. 6 and 7 illustrate well a potential of asymptotic methods developed in the book.

The works of Englund [Eng99a,Eng99b,Eng01] and Ni, Silvesxtrov, and Malyarenko [NSM08] may also be mentioned. They also deal with applications of methods based on perturbed renewal to asymptotic analysis of nonlinearly perturbed queuing systems and nonlinearly perturbed risk processes, but for models with non-polynomial nonlinear perturbations.

2.7 Conclusion

The last Chap. 8 of De Gruyter Expositions in Mathematics [GS08] contains three supplements.

The first supplement presents some basic arithmetic operation formulas for scalar and matrix asymptotic expansions.

In the second supplement, some new directions in the research concerned pseudo- and quasi-stationary phenomena for perturbed stochastic processes and systems that relate to the theory developed in this book are discussed and commented on. There is a hope that this discussion will be especially useful for young researchers and stimulate their interest to research studies in these areas. The corresponding extended comments can also be found in Silvestrov [Sil08].

The third supplement in Chap. 8 contains the brief bibliographical remarks.

The extended and carefully gathered bibliography has more than 1000 references to works in related areas, dealing with ergodic and quasi-ergodic theorems, stability theorems, limit and large deviation theorems for lifetime-type functionals and asymptotic aggregation theorems for regenerative, Markov, and semi-Markov type processes, as well as applications of such theorems to queueing systems, models of population dynamics, epidemic models, and other stochastic systems.

Here, we would like to point some originating and survey papers and books related to perturbation problems for stochastic processes. These are Vere-Jones [Ver62], Hanen [Han63], Kingman [Kin63], Kato [Kat66], Korolyuk and Turbin [KT76], [KT78], Wentzell and Freidlin [WF79], Silvestrov [Sil80], Seneta [Sen81], Solov'ev [Sol83], Asmussen [Asm87], [Asm00], Kalashnikov and Rachev [KR88], Stewart and Sun Ji

Guang [SS90], Ho and Cao [HC91], Meyn and Tweedie [MT93], Kalashnikov [Kal94], [Kal97], Kovalenko [Kov94], Kartashov [Kar96], Embrechts, Klüppelberg, and Mikosch [EKM97], Kijima [Kij97], Kovalenko, Kuznetsov, and Pegg [KKP97], Borovkov [Bor98], Stewart [Ste98], [Ste01], Yin and Zhang [YZ98], Korolyuk V.S. and Korolyuk, V.V. [KK99], Latouche and Ramaswami [LR99], Bening and Korolev [BK02], Whitt [Whi02], Silvestrov [Sil04], Koroliuk and Limnios [KL05], Anisimov [Ani08], and Gyllenberg and Silvestrov [GS08].

Quasi-stationary phenomena and related problems are a subject of intensive studies during several decades. However, the development of theory of quasi-stationary phenomena is still far from its completion. The part of the theory related to conditions of existence of quasi-stationary distributions is comparatively well developed while computational aspects of the theory are underdeveloped. The content of the book [18] is concentrated in this area. The book presents new effective methods for asymptotic analysis of pseudo- and quasi-stationary phenomena for nonlinearly perturbed stochastic processes and systems. Moreover, the results presented in the book unite, for the first time, research studies of pseudo- and quasi-stationary phenomena in the frame of one theory. Methods of asymptotic analysis for nonlinearly perturbed stochastic processes and systems developed in the book have their own values and possible applications beyond the problems studied in the book.

The results presented in the book will be interesting to specialists, who work in such areas of the theory of stochastic processes as ergodic, limit, and large deviation theorems, analytical and computational methods for Markov chains, regenerative, Markov, semi-Markov, risk and other classes of stochastic processes, renewal theory, and their queueing, reliability, population dynamics, and other applications. There is a hope that the book will also attract attention of those researchers, who are interested in new analytical methods of analysis for nonlinearly perturbed stochastic processes and systems, especially those who like serious analytical work.

References

[Ani08] Anisimov, V.V.: Switching Processes in Queuing Models. Applied Stochstic Models Series, ISTE and Wiley, London (2008)

[Asm87] Asmussen, S.: Applied Probability and Queues. Wiley Series in Probability and Mathematical Statistics, Wiley, New York and Stochastic Modelling and Applied Probability, **51**, Springer, New York (1987, 2003)

[Asm00] Asmussen, S.: Ruin Probabilities. Advanced Series on Statistical Science & Applied Probability, **2**, World Scientific, Singapore (2000)

[BK02] Bening, V.E., Korolev, V.Yu.: Generalized Poisson Models and their Applications in Insurance and Finance. Modern Probability and Statistics, VSP, Utrecht (2002)

[Bor98] Borovkov, A.A.: Ergodicity and Stability of Stochastic Processes. Wiley Series in Probability and Statistics, Wiley, New York (1998)

[EKM97] Embrechts, P., Klüppelberg, C., Mikosch, T.: Modelling Extremal Events for Insurance and Finance. Applications of Mathematics, **33**, Springer, Berlin (1997)

[Eng99a] Englund, E.: Perturbed renewal equations with application to M/M queueing systems. 1, Teor. Ĭmovirn. Math. Stat., **60**, 31–37 (1999a) (Also in Theory Probab. Math. Stat., **60**, 35–42)

[Eng99b] Englund, E.: Perturbed renewal equations with application to M/M queueing systems. 2. Teor. Ĭmovirn. Math. Stat., **61**, 21–32 (1999b) (Also in Theory Probab. Math. Stat., **61**, 21–32)

[Eng00] Englund, E.: Nonlinearly perturbed renewal equations with applications to a random walk. Theory Stoch. Process., **6(22)**, no. 3–4, 33–60 (2000)

[Eng01] Englund, E.: Nonlinearly perturbed renewal equations with applications. Doctoral Dissertation, Umeå University (2001)

[ES97] Englund, E., Silvestrov, D.S.: Mixed large deviation and ergodic theorems for regenerative processes with discrete time. Theory Stoch. Process., **3(19)**, no. 1–2, 164–176 (1997)

[Fel66] Feller, W.: An Introduction to Probability Theory and Its Applications. Vol. II. Wiley, New York (1966)

[GS94] Gyllenberg, M., Silvestrov, D.S.: Quasi-stationary distributions of a stochastic metapopulation model. J. Math. Biol., **33**, 35–70 (1994)

[GS99a] Gyllenberg, M., Silvestrov, D.S.: Quasi-stationary phenomena for semi-Markov processes. In: Janssen, J., Limnios, N. (eds) Semi-Markov Models and Applications. Kluwer, Dordrecht, 33–60 (1999a)

[GS99b] Gyllenberg, M., Silvestrov, D.S.: Cramér-Lundberg and diffusion approximations for nonlinearly perturbed risk processes including numerical computation of ruin probabilities. Theory Stoch. Process., **5(21)**, no. 1–2, 6–21 (1999b)

[GS00a] Gyllenberg, M., Silvestrov, D.S.: Nonlinearly perturbed regenerative processes and pseudo-stationary phenomena for stochastic systems. Stoch. Process. Appl., **86**, 1–27 (2000a)

[GS00b] Gyllenberg, M., Silvestrov, D.S.: Cramér–Lundberg approximation for nonlinearly perturbed risk processes. Insur. Math. Econom., **26**, no. 1, 75–90 (2000b)

[GS08] Gyllenberg, M., Silvestrov, D.S.: Quasi-Stationary Phenomena in Nonlinearly Perturbed Stochastic Systems. De Gruyter Expositions in Mathematics, **44**, Walter de Gruyter, Berlin (2008)

[Han63] Hanen, A.: Théorèmes limites pour une suite de chaînes de Markov. Ann. Inst. H. Poincaré, **18**, 197–301 (1963)

[HC91] Ho, Y.C., Cao, X.R.: Perturbation Analysis of Discrete Event Dynamic Systems. Kluwer International Series in Engineering and Computer Science, Kluwer, Boston (1991)

[Kal94] Kalashnikov, V.V.: Mathematical Methods in Queuing Theory. Mathematics and its Applications, **271**, Kluwer, Dordrecht (1994)

[Kal97] Kalashnikov, V.V.: Geometric Sums: Bounds for Rare Events with Applications. Mathematics and its Applications, **413**, Kluwer, Dordrecht (1997)

[KR88] Kalashnikov, V.V., Rachev, S.T.: Mathematical Methods for Construction of Queueing Models. Nauka, Moscow (1988) (English edition: The Wadsworth & Brooks/Cole Operations Research Series, Wadsworth & Brooks/Cole, Pacific Crove, CA (1990)).

[Kar96] Kartashov, M.V.: Strong Stable Markov Chains. VSP, Utrecht and TBiMC, Kiev (1996)

[Kat66] Kato, T.: Perturbation Theory for Linear Operators. Springer, Berlin (1966, 1976, 1995)

[Kij97] Kijima, M.: Markov Processes for Stochastic Modelling. Stochastic Modeling Series, Chapman & Hall, London (1997)

[Kin63] Kingman, J.F.: The exponential decay of Markovian transition probabilities. Proc. Lond. Math. Soc., **13**, no. 3, 337–358 (1963)

[KK99] Korolyuk, V.S., Korolyuk, V.V.: Stochastic Models of Systems. Mathematics and its Applications, **469**, Kluwer, Dordrecht (1999)

[KL05] Koroliuk, V.S., Limnios, N.: Stochastic Systems in Merging Phase Space. World Scientific, Singapore (2005)

[KT76] Korolyuk, V.S., Turbin, A.F.: Semi-Markov Processes and Its Applications. Naukova Dumka, Kiev (1976)

[KT78] Korolyuk, V.S., Turbin, A.F.: Mathematical Foundations of the State Lumping of Large Systems. Naukova Dumka, Kiev (1978) (English edition: Mathematics and its Applications, **264**, Kluwer, Dordrecht (1993))

[Kov94] Kovalenko, I.N.: Rare events in queuing theory – a survey. Queuing Syst. Theory Appl., **16**, no. 1–2, 1–49 (1994)

[KKP97] Kovalenko, I.N., Kuznetsov, N.Yu., Pegg, P.A.: Mathematical Theory of Reliability of Time Dependent Systems with Practical Applications. Wiley Series in Probability and Statistics, Wiley, New York (1997)

[LR99] Latouche, G., Ramaswami, V.: Introduction to Matrix Analytic Methods in Stochastic Modeling. ASA-SIAM Series on Statistics and Applied Probability, Society for Industrial and Applied Mathematics (SIAM), Philadelphia, PA and American Statistical Association, Alexandria, VA (1999)

[MT93] Meyn, S.P., Tweedie, R.L.: Markov Chains and Stochastic Stability. Communications and Control Engineering Series, Springer, London (1993)

[NSM08] Ni, Y., Silvetrov, D., Malyarenko, A.: Exponential asymptotics for nonlinearly perturbed renewal equation with non-polynomial perturbations. J. Numer. Appl. Math. **1(96)**, 173–197 (2008)

[Sen81] Seneta, E.: Non-negative Matrices and Markov Chains. Springer Series in Statistics, Springer, New-York (1981, 2006)

[Shu80a] Shurenkov, V.M.: Transition phenomena of the renewal theory in asymptotical problems of theory of random processes. I. Mat. Sbornik, **112(154)**, no. 1 (5), 115–132 (1980a) (English translation in Math. USSR–Sbornik, **40**, no. 1, 107–123).

[Shu80b] Shurenkov, V.M.: Transition phenomena of the renewal theory in asymptotical problems of theory of random processes. II. Mat. Sbornik, **112(154)**, no. 2(6), 226–241 (1980b) (English translation in Math. USSR–Sbornik, **40**, no. 2, 211–225).

[Sil76] Silvestrov, D.S.: A generalization of the renewal theorem. Dokl. Akad. Nauk. Ukr. SSR, Ser. A, no. 11, 978–982 (1976)

[Sil78] Silvestrov, D.S.: The renewal theorem in a series scheme 1. Teor. Veroyatn. Mat. Stat., **18**, 144–161 (1978) (English translation in Theory Probab. Math. Statist. **18**, 155–172)

[Sil79] Silvestrov, D.S.: The renewal theorem in a series scheme 2. Teor. Veroyatn. Mat. Stat., **20**, 97–116 (1979) (English translation in Theory Probab. Math. Statist. **20**, 113–130).

[Sil80] Silvestrov, D.S.: Semi-Markov Processes with a Discrete State Space. Library for an Engineer in Reliability, Sovetskoe Radio, Moscow (1980)

[Sil95] Silvestrov, D.S.: Exponential asymptotic for perturbed renewal equations. Teor. Ĭmovirn. Mat. Stat., **52**, 143–153 (1995) (English translation in Theory Probab. Math. Statist., **52**, 153–162).

[Sil00a] Silvestrov, D.S.: Perturbed renewal equation and diffusion type approximation for risk processes. Teor. Ĭmovirn. Mat. Stat., **62**, 134–144 (2000a) (English translation in Theory Probab. Math. Statist., **62**, 145–156).

[Sil00b] Silvestrov, D.S.: Nonlinearly perturbed Markov chains and large deviations for lifetime functionals. In: Limnios, N., Nikulin, M. (eds) Recent Advances in Reliability Theory: Methodology, Practice and Inference. Birkhäuser, Boston, 135–144 (2000b)

[Sil04] Silvestrov, D.S.: Limit Theorems for Randomly Stopped Stochastic Processes. Probability and Its Applications, Springer, London (2004)

[Sil07a] Silvestrov, D.S.: Asymptotic expansions for quasi-stationary distributions of nonlinearly perturbed semi-Markov processes. Theory Stoch. Process., **13(29)**, no. 1–2, 267–271 (2007a)

[Sil07b] Silvestrov, D.S.: Asymptotic expansions for distributions of the surplus prior and at the time of ruin. Theory Stoch. Process., **13(29)**, no. 4, 183–188 (2007b)

[Sil08] Silvestrov, D.S.: Nonlinearly perturbed stochstic processes. Theory Stoch. Process., **14(30)**, no. 3–4, 129–164 (2008)

[Sol83] Solov'ev, A.D.: Analytical methods for computing and estimating reliability. In: Gnedenko, B.V. (ed) Problems of Mathematical Theory of Reliability. Radio i Svyaz', Moscow, 9–112 (1983)

[Ste98] Stewart, G.W.: Matrix Algorithms. Vol. I. Basic Decompositions. Society for Industrial and Applied Mathematics, Philadelphia, PA (1998)

[Ste01] Stewart, G.W.: Matrix Algorithms. Vol. II. Eigensystems. Society for Industrial and Applied Mathematics, Philadelphia, PA (2001)

[SS90] Stewart, G.W., Sun, Ji Guang: Matrix Perturbation Theory. Computer Science and Scientific Computing, Academic Press, Boston (1990)

[Ver62] Vere-Jones, D.: Geometric ergodicity in denumerable Markov chains. Quart. J. Math., **13**, 7–28 (1962)

[WF79] Wentzell, A.D., Freidlin, M.I.: Fluctuations in Dynamical Systems Subject to Small Random Perturbations. Probability Theory and Mathematical Statistics, Nauka, Moscow (1979) (English edition: Random Perturbations of Dynamical Systems. Fundamental Principles of Mathematical Sciences, **260**, Springer, New York (1998)).

[Whi02] Whitt, W.: Stochastic-Process Limits. An Introduction to Stochastic-Process Limits and their Application to Queues. Springer Series in Operations Research, Springer, New York (2002)

[YZ98] Yin, G.G., Zhang, Q.: Continuous-time Markov Chains and Applications. A Singular Perturbation Approach. Applications of Mathematics, **37**, Springer, New York (1998)

3

On a Copula for Failure Times of System Elements

Alexander Andronov

Riga Technical University, Riga, Latvia Aleksandrs.Andronovs@rtu.lv

Abstract: A considered reliability system consists from n identical elements. Each element can be available or failed. A failure rate of element equals λ/i, where i is a number of available elements. A time till a failure of an element has an exponential distribution. Independence property has place too. Therefore, a total failure rate is the constant λ, until at least one element is available. Initially, all elements are available. The main aim of the paper is to determine a joint cumulative distribution function of the available time for all elements, as well as a corresponding copula. It allows us to generalize the received results on a nonexponential case.

Keywords and phrases: Multicomponent system, Reliability function, Copula

3.1 Introduction

Usually, reliability systems described in the literature are considered under supposition that system's elements fail independently [Ger00, LN00]. However, often a case takes place when some element failure increases a load on worked elements, so element available times are dependent random variables. In connection with that a problem of choosing a corresponding multidimensional distribution to fit the given statistical data is urgent.

In econometrics, such problem's solution often uses the so-called *copulas* [Nel06, Emb]. Lately, the copulas have been used widely in the reliability theory [AN08,BMN07, Spi07, Spi09]. Joint distribution function $C(u_1, u_2, ..., u_n) = P\{U_1 \leq u_1, U_2 \leq u_2, ..., U_n \leq u_n\}$ is called a *copula* if the marginal distributions of all components $U_1, U_2, ..., U_n$ are uniform on [0, 1]. The following fact is basic [Skl59]: any multivariate continuous distribution function $G(x_1, x_2, ..., x_n) = P\{X_1 \leq x_1, X_2 \leq x_2, ..., X_n \leq x_n\}$ can be presented uniquely via cumulative distribution function of its component $F_i(x_i) = P\{X_i \leq x_i\}$ by corresponding copula C: $G(x_1, x_2, ..., x_n) = C(F_1(x_1), F_2(x_2), ..., F_n(x_n))$.

An aim of the current paper is using a copula-based approach to a description of a reliability function of the system that elements have dependent available times.

V.V. Rykov et al. (eds.), *Mathematical and Statistical Models and Methods in Reliability: Applications to Medicine, Finance, and Quality Control*, Statistics for Industry and Technology, DOI 10.1007/978-0-8176-4971-5_3,

We consider a concrete system of such a sort and derive corresponding copula $C(u_1, u_2, ..., u_n)$. Then we can use this copula with various marginal distributions of components getting a distribution family for fitting aim.

Let us describe a considered system. The one consists of n elements. Each element can be in two states: available or failed. A failure rate for each element is λ/i, where i is a number of available elements, with that corresponding time till a failure has an exponential distribution. Independence property has place too. Therefore a total failure rate is constant λ, until at least one element is available.

We would like to answer the following questions:

1. What is a distribution of time moment T when the last element fails?
2. What is a marginal distribution of available time X for a fixed element?
3. What is an joint distribution of available times for two and more fixed elements?
4. What is a corresponding copula?

To get answers to the first two questions, we should have in mind the following. (1) A failure flow is a Poisson process with intensity λ, till such a time moment when the last element fails. (2) The time moment of the i-th failure has an Erlang distribution with parameters λ and i [Ros96]. (3) Each such moment is a failure time moment of some element. The fixed element fails as i-th in succession to probability $1/n$.

Time moment T distribution when the last element fails. Here, we have the Erlang distribution [Ros96] with parameters λ and n:

$$R\{T \leq t\} = 1 - \sum_{i=0}^{n-1} \frac{(\lambda t)^i}{i!} e^{-\lambda t}, \quad t > 0. \tag{3.1}$$

The marginal probability density function for available time X of a fixed element. Here, we have such an expression:

$$p_\lambda(x) = \frac{1}{n} \sum_{i=1}^{n} \lambda \frac{(\lambda x)^{i-1}}{(i-1)!} e^{-\lambda x}, \quad x \geq 0. \tag{3.2}$$

A corresponding cumulative distribution function is

$$\begin{aligned} P_\lambda(x) = P\{X \leq x\} &= \int_0^x p_\lambda(z) dz = \frac{1}{n} \sum_{i=1}^{n} \left(1 - \sum_{j=0}^{i-1} \frac{(\lambda x)^j}{j!} e^{-\lambda x}\right) \\ &= 1 - \frac{1}{n} \sum_{j=0}^{n-1} \sum_{i=j+1}^{n} \frac{(\lambda x)^j}{j!} e^{-\lambda x} \\ &= 1 - \frac{1}{n} \sum_{j=0}^{n-1} (n-j) \frac{(\lambda x)^j}{j!} e^{-\lambda x} \\ &= 1 - (\lambda x)^{n-1} \frac{1}{(n-1)!} e^{-\lambda x} \\ &\quad - \left(1 - \frac{\lambda x}{n}\right) \sum_{i=0}^{n-2} \frac{(\lambda x)^i}{i!} e^{-\lambda x}, \quad x \geq 0. \end{aligned} \tag{3.3}$$

Further, we consider multidimensional distribution, marginal bivariate distribution and corresponding copula. Numerical example and final remarks conclude the paper.

3.2 Multidimensional Distribution

A joint probability density function for $X_1, X_2, ..., X_n$ obviously is

$$g(x_1, x_2, ..., x_n) = \frac{1}{n} x_{i^*}^{-(n-1)} \frac{1}{(n-1)!} \lambda (\lambda x_{i^*})^{n-1} \mathrm{e}^{-\lambda x_{i^*}} = \frac{1}{n!} \lambda^n \mathrm{e}^{-\lambda x_{i^*}}, \ \forall\ 0 \le x_i \le x_{i^*}, \tag{3.4}$$

where $x_{i^*} = \max\{x_1, x_2, ..., x_n\}$.

Let us verify the normalization condition. For $i^* = 1$ we have

$$\int_0^\infty \int_0^z \cdots \int_0^z g(z, x_2, ..., x_n) \mathrm{d}x_2 ... \mathrm{d}x_n \mathrm{d}z = \int_0^\infty z^{n-1} \frac{1}{n!} \lambda^n \mathrm{e}^{-\lambda z} \mathrm{d}z = \frac{1}{n}.$$

This result can be multiplying by n because full integral contains possibilities that maximal component can take any place from n ones ($i^* = 1, 2, ..., n$). So the normalization condition is fulfilled.

One can get that an joint probability density function for $X_1, X_2, ..., X_k$, $1 < k \le n$, has the following form:

$$g(x_1, x_2, ..., x_k) = \frac{1}{n} x_{i^*}^{-(k-1)} \sum_{i=k}^{n} \frac{(i-1)(i-2)...(i-k+1)}{(n-1)(n-2)...(n-k+1)!} \frac{1}{(i-1)!} \lambda (\lambda x_{i^*})^{i-1} \mathrm{e}^{-\lambda x_{i^*}}, \tag{3.5}$$

where $x_{i^*} = \max\{x_1, x_2, ..., x_k\}$.

The last expression is valid for $k = 1$ too, if we set $(i-1)(i-2)...i = 1$, and $(n-1)(n-2)...n = 1$.

Now we intend to obtain an expression for joint cumulative distribution function $G(x_1, x_2, ..., x_n) = P\{X_1 \le x_1, X_2 \le x_2, ..., X_n \le x_n\}$. At first we consider a case $x_1 < x_2 < \cdots < x_n$. The event $\{X_1 \le x_1, X_2 \le x_2, ..., X_n \le x_n\}$ consists of n disjoint events, when the random variable X_i, $i = 1, 2, ..., n$, takes the maximal value. If we denote this maximal value by z then

$$\begin{aligned} G(x_1, x_2, ..., x_n) \\ = \int_0^{x_1} \int_0^z \cdots \int_0^z g(z, \nu_2, ..., \nu_n) \mathrm{d}\nu_n ... \mathrm{d}\nu_2 \mathrm{d}z \\ + \sum_{i=2}^{n} \Bigg\{ \int_0^{x_1} \int_0^z \cdots \int_0^z g(\nu_1, \nu_2, ..., \nu_{i-1}, z, \nu_{i+1}, ..., \nu_n) \mathrm{d}\nu_n \mathrm{d}\nu_{n-1} ... \mathrm{d}\nu_{i+1} \mathrm{d}\nu_{i-1} ... \mathrm{d}\nu_1 \mathrm{d}z \\ + \int_{x_1}^{x_2} \int_0^{x_1} \int_0^z \cdots \int_0^z g(\nu_1, \nu_2, ..., \nu_{i-1}, z, \nu_{i+1}, ..., \nu_n) \mathrm{d}\nu_n \mathrm{d}\nu_{n-1} ... \mathrm{d}\nu_{i+1} \mathrm{d}\nu_{i-1} ... \mathrm{d}\nu_1 \mathrm{d}z + \cdots \end{aligned}$$

$$+\int_{x_2}^{x_3}\int_0^{x_1}\int_0^{x_2}\int_0^{z}\dots\int_0^{z} g(\nu_1,\nu_2,...,\nu_{i-1},z,\nu_{i+1},...,\nu_n)\mathrm{d}\nu_n\mathrm{d}\nu_{n-1}...\mathrm{d}\nu_{i+1}\mathrm{d}\nu_{i-1}...\mathrm{d}\nu_1\mathrm{d}z+\cdots$$

$$\cdots+\int_{x_{i-1}}^{x_i}\int_0^{x_1}\dots\int_0^{x_{i-1}}\int_0^{z}\dots\int_0^{z} g(\nu_1,\nu_2,...,\nu_{i-1},z,\nu_{i+1},...,\nu_n)\mathrm{d}\nu_n\mathrm{d}\nu_{n-1}$$
$$...\mathrm{d}\nu_{i+1}\mathrm{d}\nu_{i-1}...\mathrm{d}\nu_1\mathrm{d}z\Bigg\}.$$

A substitution (3.4) in the integral gives the following expression:

$$G(x_1,x_2,...,x_n)=\frac{1}{n}\left(1-\sum_{j=0}^{n-1}\frac{1}{j!}(\lambda x_1)^j\mathrm{e}^{-\lambda x_1}\right)+\sum_{i=2}^{n}\left\{\frac{1}{n}\left(1-\sum_{j=0}^{n-1}\frac{1}{j!}(\lambda x_1)^j\mathrm{e}^{-\lambda x_1}\right)\right.$$
$$+\int_{x_1}^{x_2}\int_0^{x_1} z^{n-2}\frac{\lambda^n}{n!}\mathrm{e}^{-\lambda z}\mathrm{d}\nu_1\mathrm{d}z+\int_{x_2}^{x_3}\int_0^{x_1}\int_0^{x_2} z^{n-3}\frac{\lambda^n}{n!}\mathrm{e}^{-\lambda z}\mathrm{d}\nu_2\mathrm{d}\nu_1\mathrm{d}z+\cdots$$
$$\left.\cdots+\int_{x_{i-1}}^{x_i}\int_0^{x_1}\dots\int_0^{x_{i-1}} z^{n-i}\frac{\lambda^n}{n!}\mathrm{e}^{-\lambda z}\mathrm{d}\nu_{i-1}...\mathrm{d}\nu_1\mathrm{d}z\right\}$$
$$=1-\sum_{j=0}^{n-1}\frac{1}{j!}(\lambda x_1)^j\mathrm{e}^{-\lambda x_1}+\sum_{i=2}^{n}\left\{x_1\frac{\lambda}{n(n-1)}\left[\sum_{j=0}^{n-2}(x_1^j\mathrm{e}^{-\lambda x_1}-x_2^j\mathrm{e}^{-\lambda x_2})\frac{\lambda^j}{j!}\right]\right.$$
$$+x_1x_2\frac{\lambda^2}{n(n-1)(n-2)}\left[\sum_{j=0}^{n-3}(x_2^j\mathrm{e}^{-\lambda x_2}-x_3^j\mathrm{e}^{-\lambda x_3})\frac{\lambda^j}{j!}\right]+\cdots$$
$$\left.\cdots+x_1x_2...x_{i-1}\frac{\lambda^{i-1}}{n(n-1)...(n-i+1)}\left[\sum_{j=0}^{n-i}(x_{i-1}^j\mathrm{e}^{-\lambda x_{i-1}}-x_i^j\mathrm{e}^{-\lambda x_i})\frac{\lambda^j}{j!}\right]\right\}.$$

A final formula has the following view:

$$G(x_1,x_2,...,x_n)=1-\sum_{j=0}^{n-1}\frac{1}{j!}(\lambda x_1)^j\mathrm{e}^{-\lambda x_1}$$
$$+\sum_{i=2}^{n}x_1x_2...x_{i-1}\frac{\lambda^{i-1}}{n(n-1)...(n-i+2)}$$
$$\times\sum_{j=0}^{n-i}\left(x_{i-1}^j\mathrm{e}^{-\lambda x_{i-1}}-x_i^j\mathrm{e}^{-\lambda x_i}\right)\frac{\lambda^j}{j!}. \tag{3.6}$$

Now we consider a case when the sequence $x_1,x_2,...,x_n$ is not ordered. We should have in mind that gotten distribution is exchangeable [Nel06, p. 38]. Let $\pi=(\pi(1),\pi(2),...,\pi(n))$ be a permutation of $\{1,2,...,n\}$. We consider $\pi(i)$ as a number of the element $x_{\pi(i)}$ that take i-th place in the ordered sequence $x^{(1)} < x^{(2)} < \cdots < x^{(n)} : x_{\pi(i)} = x^{(i)}$. Then the previous formula takes place if we change the index i by $\pi(i)$. So we have

$$G(x_1, x_2, ..., x_n) = 1 - \sum_{j=0}^{n-1} \frac{1}{j!}(\lambda x_{\pi(1)})^j e^{-\lambda x_{\pi(1)}}$$
$$+ \sum_{i=2}^{n} x_{\pi(1)} x_{\pi(2)} ... x_{\pi(i-1)} \frac{\lambda^{i-1}}{n(n-1)...(n-i+2)}$$
$$\times \sum_{j=0}^{n-i} \left(x^j_{\pi(i-1)} e^{-\lambda x_{\pi(i-1)}} - x^j_{\pi(i)} e^{-\lambda x_{\pi(i)}} \right) \frac{\lambda^j}{j!}.$$

In fact, to calculate the distribution function $G(x_1, x_2, ..., x_n)$ for arbitrary order of $x_1, x_2, ..., x_n$, we must range this sequence, get ordered one $x^{(1)} < x^{(2)} < \cdots < x^{(n)}$, and use formula (3.6) for $x_i = x^{(i)}$.

3.3 Different Properties

In this section, we discuss different properties of the distribution (3.6), namely the marginal bivariate distribution, the covariance *Cov* of available times X and X' for two fixed different components and maximum and minimum distributions are considered.

3.3.1 Marginal Bivariate Distribution

Bivariate cumulative distribution function $F(x_1, x_2) = P\{x \le x_1, X' \le x_2\}$ of the available times X and X' for two fixed different components is calculated from (3.6). Setting $x_3 = \cdots = x_n = \infty$, $x^{(1)} = \min\{x_1, x_2\}$, $x^{(2)} = \max\{x_1, x_2\}$, we have for $0 \le x^{(1)} \le x^{(2)}$:

$$F(x_1, x_2) = 1 - \sum_{i=0}^{n-1} \frac{(\lambda x^{(1)})^i}{i!} e^{-\lambda x^{(1)}}$$
$$+ \frac{1}{n} \lambda x^{(1)} \sum_{j=0}^{n-2} \left\{ \frac{[\lambda x^{(1)}]^j}{j!} e^{-\lambda x^{(1)}} - \frac{[\lambda x^{(2)}]^j}{j!} e^{-\lambda x^{(2)}} \right\}$$
$$+ \frac{1}{n(n-1)} \lambda^2 x^{(1)} x^{(2)} \sum_{j=0}^{n-3} \left\{ \frac{[\lambda x^{(2)}]^j}{j!} e^{-\lambda x^{(2)}} \right\}$$
$$= 1 - \sum_{i=0}^{n-1} \frac{(\lambda x^{(1)})^i}{i!} e^{-\lambda x^{(1)}} \left[1 - \frac{i}{n} \right]$$
$$- \frac{1}{n} \lambda x^{(1)} \sum_{j=0}^{n-2} \frac{[\lambda x^{(2)}]^j}{j!} e^{-\lambda x^{(2)}} \left[1 - \frac{j}{n-1} \right]. \tag{3.7}$$

Now it is possible to calculate corresponding probability density function for $0 \leq x^{(1)} \leq x^{(2)}$:

$$\frac{\partial^2}{\partial x^{(1)} \partial x^{(2)}} F(x^{(1)} x^{(2)})$$
$$= -\frac{1}{n}\lambda^2 \sum_{j=0}^{n-2} \left[\frac{(\lambda x^{(2)})^{j-1}}{(j-1)!} - \frac{(\lambda x^{(2)})^j}{j!} \right] \mathrm{e}^{-\lambda x^{(2)}} \left[1 - \frac{j}{n-1} \right] + \frac{1}{n}\lambda^2 \mathrm{e}^{-\lambda x^{(2)}}$$
$$= -\frac{\lambda^2}{n} \left(1 - \frac{(\lambda x^{(2)})^{n-2}}{(n-2)!} \right) \mathrm{e}^{-\lambda x^{(2)}} + \frac{\lambda^2}{n(n-1)} \sum_{j=1}^{n-2} \left[\frac{(\lambda x^{(2)})^{j-1}}{(j-1)!} j - \frac{(\lambda x^{(2)})^j}{(j-1)!} \right] \mathrm{e}^{-\lambda x^{(2)}}$$
$$+ \frac{1}{n}\lambda^2 \mathrm{e}^{-\lambda x^{(2)}}$$
$$= \frac{1}{n}\lambda^2 \frac{(\lambda x^{(2)})^{n-2}}{(n-2)!} \mathrm{e}^{-\lambda x^{(2)}} + \frac{1}{n(n-1)}\lambda^2 \sum_{j=1}^{n-2} \left[j - \lambda x^{(2)} \right] \frac{(\lambda x^{(2)})^{j-1}}{(j-1)!} \mathrm{e}^{-\lambda x^{(2)}}. \tag{3.8}$$

A view of the cumulative distribution function (3.7) says that one can contain a singular component [MO67, Nel06, KG10] concentrated on the line $x^{(1)} = x^{(2)}$. To verify such a possibility we must calculate a full probabilistic mass of the absolutely continuous component (3.8). With that end in view let us calculate a two-dimensional integral of (3.8) on the triangle with vertices $(0, 0)$, $(0, \infty)$, (∞, ∞) where $x^{(1)} \leq x^{(2)}$. This integral gets a half of the full mass of (3.8). If it equals 1/2 then the distribution (3.7), (3.8) is an absolutely continuous one. The Appendix contains a proof of this fact.

3.3.2 Covariance

Now we wish to calculate the covariance *Cov* of available times X and X' for two fixed different components. As a corresponding bivariate distribution is absolutely continuous, the standard formulas can be applied. Unfortunately, such a way is too long. We prefer a probabilistic reasoning.

Let $X^{(1)} < X^{(2)} < \cdots < X^{(n)}$ is an ordered sequence of $X_1, X_2 < \cdots < X_n$. Then for the unordered pair (X, X') for $k < m$:

$$P\{X = X^{(k)}, X' = X^{(m)}\} = \frac{2}{n(n-1)}.$$

In this case, $X = Z_1 + \cdots + Z_k$, $X' = X + Z_{k+1} + \cdots + Z_m$ where $\{Z_i\}$ are independent identically and exponentially distributed with parameter λ random variables. As here we have the Erlang distributions then (see, e.g., [Sle06, p. 254])

$$E\left(X|X = X^{(k)}\right) = \frac{k}{\lambda}, \quad E\left(X^2|X=X^{(k)}\right) = k(k+1)\frac{1}{\lambda^2}, \quad E(Z_{k+1} + \cdots + Z_m) = \frac{m-k}{\lambda}.$$

Therefore,

$$E(XX'|X = X^k, X = X^{(m)})$$
$$= E(X(X + Z_{k+1} + \cdots + Z_m)|X = X^{(k)}, X = X^{(m)})$$
$$= E(X^2|X = X^{(k)}) + E(X|X = X^{(k)})E(Z_{k+1} + \cdots + Z_m)$$
$$= \frac{1}{\lambda^2}k(k+1) + \frac{1}{\lambda^2}k(m-k) = \frac{1}{\lambda^2}k(m+1).$$

Further, the following equations will be used:

$$\sum_{i=1}^{n} i^2 = \frac{1}{6}n(n+1)(2n+1), \quad \sum_{i=1}^{n} i^3 = \frac{1}{4}(n(n+1))^2.$$

Now we have

$$\begin{aligned}
E(XX') &= \sum_{m=2}^{n}\sum_{k=1}^{m-1} P\{X = X^{(k)}, X' = X^{(m)}\} E(XX' | X = X^{(k)}, X' = X^{(m)}) \\
&= \frac{2}{n(n-1)} \sum_{m=2}^{n}\sum_{k=1}^{m-1} \frac{1}{\lambda^2} k(m+1) \\
&= \frac{2}{n(n-1)\lambda^2} \sum_{m=2}^{n} (m+1)\frac{1}{2}(m-1)m \\
&= \frac{1}{n(n-1)\lambda^2} \left[\sum_{m=2}^{n} m^3 - \sum_{m=2}^{n} m\right] \\
&= \frac{1}{n(n-1)\lambda^2} \left[\frac{1}{4}(n(n+1))^2 - 1 - \frac{1}{2}(n-1)(n+2)\right] \\
&= \frac{1}{4\lambda^2}(n+1)(n+2).
\end{aligned}$$

Mean, second moment and variance for the distribution (3.2) are:

$$\begin{aligned}
E(X) &= \frac{1}{n}\sum_{i=1}^{n} \frac{1}{\lambda} i = \frac{1}{2\lambda n} n(n+1) = \frac{1}{2\lambda}(n+1), \\
E(X^2) &= \frac{1}{n}\sum_{i=1}^{n} \frac{1}{\lambda^2} i(i+1) = \frac{1}{n}\sum_{i=1}^{n} \frac{1}{\lambda^2} i^2 + \frac{1}{2\lambda^2}(n+1) \\
&= \frac{1}{2\lambda^2}(n+1) + \frac{1}{n\lambda^2}\frac{1}{6} n(n+1)(2n+1) = \frac{1}{3\lambda^2}(n+1)(n+2), \\
\mathrm{Var}(X) &= E(X^2) - E(X)^2 = \frac{1}{12\lambda^2}(n+1)(n+5).
\end{aligned}$$

Finally, the following equations for the covariance and correlation coefficient take place:

$$\mathrm{Cov} = E(XX') - E(X)^2 = \frac{1}{4\lambda^2}(n+1)(n+2) - \frac{1}{4\lambda^2}(n+1)^2 = \frac{1}{4\lambda^2}(n+1), \quad (3.9)$$

$$\rho(X) = \frac{\mathrm{Cov}}{\mathrm{Var}(X)} = \frac{3}{n+5}.$$

3.3.3 Maximum and Minimum Distributions

Let us find distributions of the maximum $X^* = \max\{X, X'\}$ and minimum $X_* = \min\{X, X'\}$ of available times X and X' for two fixed different components with joint distribution (3.7). For X^* we have:

$$
\begin{aligned}
&P\{X^* \le x\} \\
&= F(x,x) = 1 - \sum_{i=0}^{n-1} \frac{(\lambda x)^i}{i!} \mathrm{e}^{-\lambda x} \left[1 - \frac{i}{n}\right] - \frac{1}{n} \lambda x \sum_{j=0}^{n-2} \frac{[\lambda x]^j}{j!} \mathrm{e}^{-\lambda x} \left[1 - \frac{j}{n-1}\right] \\
&= 1 - \sum_{i=0}^{n-1} \frac{(\lambda x)^i}{i!} \mathrm{e}^{-\lambda x} + \frac{1}{n} \sum_{i=1}^{n-1} \frac{(\lambda x)^i}{(i-1)!} \mathrm{e}^{-\lambda x} - \frac{1}{n} \sum_{j=0}^{n-2} \frac{[\lambda x]^{j+1}}{j!} \mathrm{e}^{-\lambda x} \\
&+ \frac{1}{n(n-1)} \sum_{j=1}^{n-2} \frac{[\lambda x]^{j+1}}{(j-1)!} \mathrm{e}^{-\lambda x} \\
&= 1 - \frac{(\lambda x)^{n-1}}{(n-1)!} \mathrm{e}^{-\lambda x} - \frac{(\lambda x)^{n-2}}{(n-2)!} \mathrm{e}^{-\lambda x} - \left(1 - \frac{1}{n(n-1)} (\lambda x)^2\right) \sum_{j=1}^{n-2} \frac{(\lambda x)^{j-1}}{(j-1)!} \mathrm{e}^{-\lambda x}.
\end{aligned}
$$

For X_* the following result takes place:

$$
\begin{aligned}
&P\{X_* \le x\} \\
&= P\{X \le x\} + P\{X' \le x\} - P\{X^* \le x\} = 2P\{X \le x\} - P\{X^* \le x\} \\
&= 2\left(1 - \sum_{j=0}^{n-1} \frac{(\lambda x)^j}{j!} \mathrm{e}^{-\lambda x} + \sum_{j=1}^{n-1} \frac{1}{n} \frac{(\lambda x)^j}{(j-1)!} \mathrm{e}^{-\lambda x}\right) - P\{X^* \le x\} \\
&= 1 - \sum_{j=0}^{n-1} \frac{(\lambda x)^j}{j!} \mathrm{e}^{-\lambda x} + \frac{1}{n}\left(2 - \frac{\lambda x}{n-1}\right) \sum_{j=1}^{n-2} \frac{(\lambda x)^j}{(j-1)!} \mathrm{e}^{-\lambda x} + 2 \frac{(\lambda x)^{n-1}}{n(n-2)!} \mathrm{e}^{-\lambda x}.
\end{aligned}
$$

3.4 Copula

Our last aim is to find an expression for a copula corresponding to the joint distribution (3.8). Let $q(p) > 0$ be the root of the equation

$$
p = 1 - \frac{1}{n} \sum_{j=0}^{n-1} (n-j) \frac{1}{j!} x^j \mathrm{e}^{-x}, \quad 0 < p < 1. \tag{3.10}
$$

Obviously, it is the p-quantile of the distribution (3.3) for $\lambda = 1$:

$$
P_1(q(p)) = 1 - \frac{1}{n} \sum_{j=0}^{n-1} (n-j) \frac{1}{j!} q(p)^j \mathrm{e}^{-q(p)} = p, \quad 0 < p < 1.
$$

Substituting $p = P_\lambda(x)$ gives

$$
P_1(q(P_\lambda(x))) = 1 - \frac{1}{n} \sum_{j=0}^{n-1} (n-j) \frac{1}{j!} q(P_\lambda(x))^j \mathrm{e}^{-q(P_\lambda(x))} = P_\lambda(x) \quad x > 0.
$$

By a comparison with (3.3) we get

$$\lambda x = q(P_\lambda(x)).$$

It allows us to represent (3.6) for $x_1 < x_2 < \cdots < x_n$ by such a way:

$$G(x_1, x_2, ..., x_n) = 1 - \sum_{j=0}^{n-1} \frac{1}{j!} q(P_\lambda(x_1))^j \mathrm{e}^{-q(P_\lambda(x_1))} + \sum_{i=2}^{n} q(P_\lambda(x_1))...q(P_\lambda(x_{i-1}))$$

$$\times \frac{1}{n(n-1)...(n-i+2)} \sum_{j=0}^{n-i} \left(q(P_\lambda(x_{i-1}))^j \mathrm{e}^{-q(P_\lambda(x_{i-1}))} - q(P_\lambda(x_i))^j \mathrm{e}^{-q(P_\lambda(x_i))} \right) \frac{1}{j!}. \tag{3.11}$$

Therefore, a copula of interest for $0 < u_1 < u_2 < \cdots < u_n < 1$ has the following form:

$$\begin{aligned} C(u_1, u_2, ..., u_n) = 1 &- \sum_{j=0}^{n-1} \frac{1}{j!} q(u_1)^j \mathrm{e}^{-q(u_1))} \\ &+ \sum_{i=2}^{n} q(u_1)...q(u_{i-1}) \frac{1}{n(n-1)...(n-i+2)} \\ &\times \sum_{j=0}^{n-i} (q(u_{i-1})^j \mathrm{e}^{-q(u_{i-1})} - q(u_i)^j \mathrm{e}^{-q(u_i)}) \frac{1}{j!}. \end{aligned} \tag{3.12}$$

Using this formula, one can represent the joint distribution (3.6) for $x_1 < x_2 < \cdots < x_n$ in the form

$$G(x_1, x_2, ..., x_n) = C(P_\lambda(x_1), P_\lambda(x_2), ..., P_\lambda(x_n)),$$

where $P_\lambda(x_i)$ is calculated by (3.3).

Now we are able to generalize our results and consider a family of reliability functions of the system for $x_1 < x_2 < \cdots < x_n$ as

$$R(x_1, x_2, ..., x_n) = C(F(x_1), F(x_2), ..., F(x_n)),$$

where $F(x)$ is an arbitrary reliability function of elements.

Note a case of unordered sequence $x_1, x_2, ..., x_n$ is considered as earlier.

3.5 Example

We consider our basic model for the following data: $n = 4$, $\lambda = 2$. Therefore, for initial failure rate $\lambda/n = 0.5$, a fixed element has the following values of the mean μ and the standard deviation σ of a time till a failure: $\mu = \sigma = n/\lambda = 2$. The table contains values of the joint distribution function (3.6) for different values of argument $x = (x_1\ x_2\ x_3\ x_4)^T$.

Now we show how the same copula (3.12) can be used for another marginal distributions $F(x)$. As the last uniform distribution $U(x)$ and lognormal distribution $L(x)$ are chosen:

$$U(x) = \begin{cases} 0 & \text{if} \quad x < 0, \\ \frac{x}{2\mu} & \text{if} \quad 0 < x < 2\mu, \\ 1 & \text{if} \quad x > 2\mu, \end{cases} \qquad L(x) = \begin{cases} 0 & \text{if} \quad x < 0, \\ \Phi\left(\frac{\ln(x)-a}{s}\right) & \text{if} \quad x > 0, \end{cases}$$

where $\Phi(z)$ is the cumulative distribution function of the standard normal distribution, a and $s > 0$ are parameters of lognormal distribution.

The parameters of both distribution $U(x)$ and $L(x)$ were chosen in such a way that expectation μ coincide for all three distributions. For the lognormal distribution standard deviation σ coincides too. Let us remember [see, e.g., Sle06 p. 305] that μ and σ of the lognormal distribution are calculated with respect to formulas

$$\mu = \exp\left(a + \frac{1}{2}s^2\right), \quad \sigma^2 = (\exp(s^2) - 1)\exp(2a + s^2).$$

For our numerical data $a = 0.3464$ and $s = 0.8326$.

The corresponding values of the joint cumulative distribution functions $GU(x)$ and $GL(x)$ are represented in the table too (Table 3.1). A comparison of all three distributions shows that the considered copula (3.12) generates different distributions. It allows us to suggest using the considered model for a description of complex systems reliability.

Table 3.1. Values of the joint distribution functions

$x = (x_1, x_2, x_3, x_4)^T$	$(0.5, 1, 1.5, 2)^T$	$(0.5, 1.5, 2, 3)^T$	$(1, 1.5, 3, 3.5)^T$	$(1, 2, 3, 3.5)^T$	$(2, 2, 2, 2)^T$	$(1.5, 2.5, 3.5, 4)^T$	$(2, 3, 3.5, 4)^T$	$(3, 3, 3, 3)^T$	$(3, 3.5, 4, 4)^T$	$(6, 6.5, 6.5, 6.5)^T$
$G(x)$	0.122	0.185	0.363	0.415	0.567	0.621	0.767	0.849	0.912	0.999
$GU(x)$	0.021	0.047	0.107	0.136	0.159	0.264	0.400	0.473	0.690	1.000
$GL(x)$	0.032	0.060	0.198	0.236	0.341	0.410	0.534	0589	0677	0.904

3.6 Conclusion

In the paper, a new family of multivariate distribution functions for nonnegative random variables has been suggested. All distributions have the same copula and differ from one another by marginal distributions. The last allows choosing a multivariate distribution that in best way fits the given data of system reliability.

3.7 Appendix

For the probability density function (3.8) we have:

$$
\begin{aligned}
&\int_0^\infty \int_0^{x^{(2)}} \frac{\partial^2}{\partial x^{(1)} \partial x^{(2)}} F(x^{(1)}, x^{(2)}) \mathrm{d}x^{(1)} \mathrm{d}x^{(2)} \\
&= \int_0^\infty x^{(2)} \left[\frac{1}{n} \lambda^2 \frac{(\lambda x^{(2)})^{n-2}}{(n-2)!} \mathrm{e}^{-\lambda x^{(2)}} \right. \\
&\quad \left. + \frac{1}{n(n-1)} \lambda^2 \sum_{j=1}^{n-2} [j - \lambda x^{(2)}] \frac{(\lambda x^{(2)})^{j-1}}{(j-1)!} \mathrm{e}^{-\lambda x^{(2)}} \right] \mathrm{d}x^{(2)} \\
&= \int_0^\infty (n-1) \left[\frac{1}{n} \lambda \frac{(\lambda x^{(2)})^{n-1}}{(n-1)!} \mathrm{e}^{-\lambda x^{(2)}} \right] \mathrm{d}x^{(2)} \\
&\quad + \int_0^\infty \left[\frac{1}{n(n-1)} \lambda \sum_{j=1}^{n-2} j^2 \frac{(\lambda x^{(2)})^j}{j!} \mathrm{e}^{-\lambda x^{(2)}} \right] \mathrm{d}x^{(2)} \\
&\quad - \int_0^\infty \left[\frac{1}{n(n-1)} \lambda \sum_{j=2}^{n-1} j(j-1) \frac{(\lambda x^{(2)})^j}{j!} \mathrm{e}^{-\lambda x^{(2)}} \right] \mathrm{d}x^{(2)} \\
&= \int_0^\infty (n-1) \left[\frac{1}{n} \lambda \frac{(\lambda x^{(2)})^{n-1}}{(n-1)!} \mathrm{e}^{-\lambda x^{(2)}} \right] \mathrm{d}x^{(2)} \\
&\quad + \int_0^\infty \left[\frac{1}{n(n-1)} \lambda^2 x^{(2)} \mathrm{e}^{-\lambda x^{(2)}} \right] \mathrm{d}x^{(2)} \\
&\quad - \int_0^\infty \left[\frac{1}{n(n-1)} \lambda (n-1)^2 \frac{(\lambda x^{(2)})^{n-1}}{(n-1)!} \mathrm{e}^{-\lambda x^{(2)}} \right] \mathrm{d}x^{(2)} \\
&\quad + \int_0^\infty \left[\frac{1}{n(n-1)} \lambda \sum_{j=2}^{n-1} j \frac{(\lambda x^{(2)})^j}{j!} \mathrm{e}^{-\lambda x^{(2)}} \right] \mathrm{d}x^{(2)} \\
&= \int_0^\infty \left[\frac{1}{n(n-1)} \lambda^2 x^{(2)} \mathrm{e}^{-\lambda x^{(2)}} \right] \mathrm{d}x^{(2)} + \frac{1}{n(n-1)} \sum_{j=2}^{n-1} j \\
&= \frac{1}{n(n-1)} \left[1 + \frac{1}{2}(n-2)(n+1) \right] = \frac{2 + n^2 + n - 2n - 2}{2n(n-1)} = \frac{1}{2}.
\end{aligned}
$$

References

[AN08] Abegaz, F. and Naik-Nimbalkar, U.V. Modeling statistical dependence of Markov chains via copula models, Journal of statistical planning and inference, **138**, 1131–1146 (2008).

[BMN07] Bagdonavicius, V., Malov, S. and Nikulin, M. Testing of the homogeneity of marginal distribution in copula model, Comptes Rendus Mathematique, **346**, 445–450 (2008).

[Ger00] Gertsbakh, I. Reliability theory: with applications to preventive maintenance, Springer, Berlin (2000).

[Emb] Embrechts P. Some Selected Papers. `http://www.math.ethz.ch/$\sim$baltes/ftp/papers.html`.

[KG10] Kundu, D. and Gupta, R.D. Modify Sarhan-Balakrishnan singular bivariate distribution, Journal of statistical planning and inference, **140**, 526–538 (2010).

[LN00] Limnios, N. and Nikulin, M. Recent Advances in Reliability Theory: Methodology, Practice, and Inference, Birkhauser, Boston (2000).

[MO67] Marshall, A.W. and Olkin, I. A multivariate exponential distribution, Journal of American Statistical Association, **62**, 30–44 (1967).

[Nel06] Nelsen, R.B. An introduction to copulas, Second edition, Springer, New York (2006).

[Ros96] Ross, S.M. Stochastic Processes, Wiley, New York (1996).

[Skl59] Sklar, A. Fonctions de repartition a n dimensions et leurs marges. Publications de l'Institut de Statistique de L'Universit de Paris, **8**, 229–231 (1959).

[Sle06] Sleeper, A. Six Sigma Distribution Modeling, McGrawHill, New York (2006).

[Spi07] Spizzichino, F.L. Ageing and positive dependence. Encyclopedia of Statistics for Quality and Reliability. Ruggeri, F., Kennett, R. and Faltin, F.W. (Eds). Wiley, Chichester, UK, 82–95 (2007).

[Spi09] Spizzichino, F.L. A concept of duality for multivariate exchangeable survival models. Fuzzy Sets and Systems, **160**, 325–333 (2009).

4

On One Method of Reliability Coefficients Calculation for Objects in Non-Homogeneous Event Flows

Alexander Antonov*, Kristina Belova, and Valeriy Chepurko

Obninsk State Technical University for Nuclear Power Engineering (INPE), Obninsk, Russia, antonov@iate.obninsk.ru

Abstract: The paper considers a new mathematical model for calculating reliability coefficients of the systems (or elements) which probabilistic characteristics can vary in time. The systems with the operable state and the down state are considered. The new mathematical model can take into account possible "distortions" of an event flows by means of a normalizing flow function Ψ. The normalizing flow function model is presented. The equations for failure flow parameter, distribution density of operating time between failures, resource reliability characteristics, and availability function are deduced.

Keywords and phrases: Failure and repair flows, Failure flow parameter, Joint failure–repair flow, Leading flow function, Non-homogeneity process, Normalizing flow function model

4.1 Introduction

Functioning process of a technical equipment can be divided into three operating periods. There are a burn-in period, a normal life or useful life period and an ageing period. Reliability coefficients of equipment and a methods for their calculating depend on the operating period. The failure flow parameter is relatively constant on the useful life period. But we should take into consideration the burn-in period with a decreasing failure flow parameter and the ageing period that exhibits an increasing failure flow parameter. In general, more complex time dependences can take place. Let's consider a repair time (a repair of system elements or a system repair). We can assume that the repair flow consists of the "non-homogeneous" (concerning a distribution) repair time. For instance, a mean time to repair can gradually rise since an equipment ages and a fault location time and a repair complication rise.

The aim of this paper is to develop the new mathematical model that can take into account possible "distortions" of an event flows and allow to calculate reliability coefficients of the systems, where probabilistic characteristics can vary in time.

V.V. Rykov et al. (eds.), *Mathematical and Statistical Models and Methods in Reliability: Applications to Medicine, Finance, and Quality Control*, Statistics for Industry and Technology, DOI 10.1007/978-0-8176-4971-5_4,

4.2 General Concepts

Let μ_i, $i = 1, 2, \ldots, n$ be the random instant of event occurrence. Let's consider the stochastic process that consists of a sequence $\{\mu_i\}$, $i = 1, 2, \ldots, n$.

Let $\xi_i = \mu_i - \mu_{i-1}$, $i = 1, 2, \ldots$, $\mu_0 = 0$ be the interval between μ_i and μ_{i-1}. Now we introduce the following notions. The stochastic process is called the homogeneous event flow in time if the following conditions hold:

- ξ_i, $i = 1, 2, \ldots, n$ – independent identically distributed random variables, i.e.

$$\mathsf{P}\left(\xi_1 < x_1; \xi_2 < x_2; ...; \xi_n < x_n\right) = \prod_{i=1}^{n} \mathsf{P}\left(\xi_i < x_i\right), \, n \geq 2$$

- The distribution function for ξ_i, $i = 1, 2, \ldots, n$ is defined as

$$F_i(x) = \mathsf{P}\left(\xi_i < x\right) = F(x),$$

We prefer the notion "homogeneous event flow" though the conditions given above define a recurrent flow. Statistical characteristics of the process did't change in time, if the process is homogeneous. And characteristics of the process depend on an observation period location on the time base, if the process is non-homogeneous. In this cases the notions "stationary event flow" and "non-stationary event flow" are mostly used. Nevertheless, we assume the notions "homogeneous event flow" and "non-homogeneous event flow" (in time) more convenient. The latter notion is explained by the following example. Points of failure instants on the burn-in period are depleted and at the system ageing stage are thickened. Considering the useful life period, "exhaustion" or "thickening" ranges will absence in a failure flow. Hence, the failure flow is supposed homogeneous. In the general case, an event flow non-homogeneity nature can be arbitrary.

4.3 Review of Models Taking into Account Non-Homogeneity

One of the methods for taking into account a non-homogeneity of failure and repair flows was proposed in the paper [Sae94]. Let's consider the event flow (see the Fig. 4.1) that consists of failure and repair events. By ξ^0 denote the operating time between failures and by ξ^1 denote the repair time.

Suppose that the random quantities ξ_i^0, $i = 1, 2, \ldots$ are independent and the random variables ξ_i^1, $i = 1, 2, \ldots$ are also independent.

The basic assumption of this model is

$$\xi_{n+1}^0 \stackrel{d}{=} \gamma^n \xi_1^0, \; \xi_{n+1}^1 \stackrel{d}{=} \beta^n \xi_1^1,$$

where by $\stackrel{d}{=}$ denote identically distributed random quantities (equality by distribution), $\gamma \in (0;1)$ is an ageing factor (each sequential operating time between failures ξ_{n+1}^0

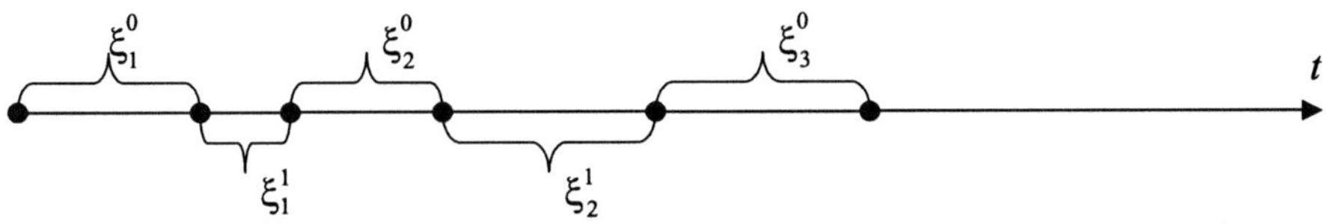

Figure 4.1. Failure and repair flow

becomes stochastically smaller than preceding ξ_n^0 by means of the ageing factor), $\beta \in (1; \infty)$ – a repair incompleteness factor (each sequential repair time ξ_{n+1}^1 becomes stochastically greater than preceding ξ_n^1). In the paper, the new type integral equations with “advanced argument” are presented. This equations establish the relations between an availability function of the system $P_1(t)$ and an inverse Laplace transform L^{-1} of two variables function.

$$P_1(t) = 1 - F(t) + \int_0^t L^{-1}\left(\hat{P}_2(s; s)\right) \mathrm{d}t,$$

where $F(t)$ is the distribution function ξ_1^0, $P_2(x; y)$ is the solution of equation with “advanced argument”:

$$P_2(x; y) = \varphi(x; y) + \int_0^x \int_0^y \frac{1}{\gamma\beta} P_2\left(\frac{x - u_1}{\gamma}; \frac{y - u_2}{\beta}\right) f(u_1) g(u_2) \mathrm{d}u_1 \mathrm{d}u_2,$$

$$\varphi(x; y) = \left(f(x) - \frac{1}{\gamma} \int_0^x f\left(\frac{x - u}{\gamma}\right) f(u) \mathrm{d}u \right) g(y),$$

where $f(x)$ and $g(y)$ – the distribution density of the operating time between failures ξ_1^0 and the repair time ξ_1^1, respectively.

If the operating times between failures and the repair times have exponential distribution, the recurrent algorithm for solving this equations was proposed. The author of the paper was noted, that Laplace transform finding for more complicated distribution law is incorrect problem.

Mention the other models presented in the papers [BF88,Ven91,Max01]. The rates of failure and repair flows are arbitrary functions in these models. Considering a thinning event flow the particular case of such models can be obtained. The thinning event flow consist of events that can occur with some probability. Flow events occur more rarely if the probability decreases. Thus, the system behaviour at the ageing period can be simulated.

4.4 Normalizing Flow Function Model

The basic point of the normalizing flow function model is that the continuous strictly monotone increasing mapping Ψ of an abstract recurrent (or homogeneous) event flow into a "real" (or non-homogeneous) event flow is constructed (see the papers [VC02, NC04, AKC06, SOC07, Che07, IC09, SC09, Ski09]). Note that the abstract event flow has the dimension of function $\Psi^{-1}(t)$, where t is time.

Let us suppose a system repair occurs immediately (i.e. the repair time may be considered the negligible quantity against the operating time between failures). By μ_k denote the instant of kth event occurrence of the abstract event flow. Than $\mu_k = \sum_{i=1}^{k} \xi_i$, where ξ_i is the interval between consistent events of the failure flow (the operating time between failures). Considering one (recurrent) failure flow all ξ_i, $i = 1, 2, \ldots$ are independent, identically distributed random variables.

"Real" event flow instants are defined by the formula

$$\tilde{\mu}_n = \Psi\left(\sum_{i=1}^{n} \xi_i\right) = \Psi\left(\mu_n\right); \; n = 1, 2, \ldots; \; \mu_0 = 0,$$

where $\Psi(\cdot)$ is the continuously differentiable strictly monotone increasing on the interval $[0; \infty)$ function and $\Psi(0) = 0$.

Then the ith operating time between failures is

$$\zeta_i = \tilde{\mu}_i - \tilde{\mu}_{i-1} = \Psi\left(\mu_i\right) - \Psi\left(\mu_{i-1}\right), \tag{4.1}$$

the variables $\zeta_1, \zeta_2, \ldots$ are dependent if $\Psi(x) \neq \text{const} \cdot x$.

Denote by ζ_i, $i = 1, 2, \ldots, n$ the system operation cycle time or the operability cycle, i.e. the distance between consistent system failures.

The conversion homogeneous event flow into "arbitrary" flow by means of function $\Psi(t)$ is represented on the Fig. 4.2. The instants of homogeneous event flow μ_i are depicted on the abscissa. The instants of non-homogeneous event flow $\tilde{\mu}_i$ are depicted on the ordinate axis.

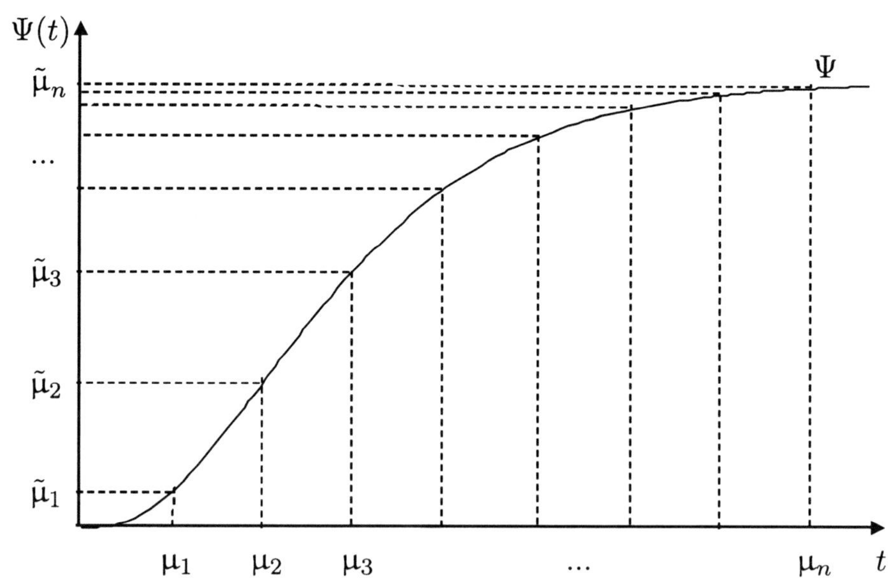

Figure 4.2. Conversion homogeneous event flow into "arbitrary" event flow

4.5 Leading Flow Function and Failure Flow Parameter

Let $N(t)$ be the number of flow events that have occurred up to time t (starting from time 0). Then the leading flow function, denoted by $\Omega(t)$, is the expected value of $N(t)$ or $\mathsf{E}N(t)$. Consider the following theorem.

Theorem 1.

$$\Omega(t) = \nu\left(\Psi^{-1}(t)\right), \tag{4.2}$$

where $\nu(t)$ *is the solution of renewal equation*

$$\nu(t) = F_{\xi}(t) + \int_0^t \nu(t-\tau) f_{\xi}(\tau)\mathrm{d}\tau,$$

where $F_{\xi}(t)$ *and* $f_{\xi}(t)$ *are the distribution function and the distribution density of random quantity* ξ, *respectively.*

Let $\lim\limits_{t\to\infty} \Psi^{-1}(t) = \infty$ *then*

$$\lim_{t\to\infty} \frac{\Omega(t)}{\Psi^{-1}(t)} = \frac{1}{\mathsf{E}\xi}, \tag{4.3}$$

i.e. $\Omega(t) \sim \frac{\Psi^{-1}(t)}{\mathsf{E}\xi}$.

The proof of Theorem 1 will be given in Appendix. Note the asymptotic behaviour of an average failure number is non-linearly and is defined by the normalizing flow function. This means that we have an ample opportunities for real process simulating.

Derivative of leading flow function (4.2) or the failure flow parameter $\omega(t)$ is defined as

$$\omega(t) = \Omega'(t) = \left[\Psi^{-1}(t)\right]' \nu'\left(\Psi^{-1}(t)\right) = \left[\Psi^{-1}(t)\right]' \upsilon\left(\Psi^{-1}(t)\right),$$

where $\upsilon(t)$ is the solution of renewal equation

$$\upsilon(t) = f_{\xi}(t) + \int_0^t \upsilon(t-\tau) f_{\xi}(\tau)\mathrm{d}\tau.$$

We can obtain the equation similar to (4.3) for the failure flow parameter

$$\lim_{t\to\infty} \frac{\omega(t)}{\left[\Psi^{-1}(t)\right]'} = \frac{1}{\mathsf{E}\xi}.$$

Example 1. Let the abstract homogeneous failure flow be the simplest flow with the rate λ. In this case $\nu(t) = \frac{t}{\mathsf{E}\xi} = \lambda t$ and $\Omega(t) = \lambda\Psi^{-1}(t)$.

If $\Psi(t) = \alpha t^{\gamma}$, then $\Omega(t) = \lambda\left(\frac{t}{\alpha}\right)^{\frac{1}{\gamma}}$ and $\omega(t) = \frac{\lambda}{\alpha\gamma}\left(\frac{t}{\alpha}\right)^{\frac{1-\gamma}{\gamma}}$.

It may be concluded, analysing the last equations, that the failure flow parameter increases by power law if $\gamma \in (0;1)$. Hence, it can expected that an ageing type system is observed.

Failure flow parameter decreases by power law if $\gamma > 1$. This parameter behaviour is common to the systems on burn-in period.

4.6 Getting of Distribution Law for the Operability Cycle

Suppose the distribution density $f_\xi(t)$ (therefore the distribution function $F_\xi(t)$) of random quantity ξ and the normalizing flow function $\Psi(t)$ are known. Let's find the distribution law for the ith operability cycle ζ_i. Then (4.1) can be written as

$$\zeta_i = \Psi\left(\mu_{i-1} + \xi_i\right) - \Psi\left(\mu_{i-1}\right).$$

The equation for the distribution law of random quantity ζ_1 is defined as

$$F_{\zeta_1}(t) = \mathsf{P}\left(\Psi\left(\xi_1\right) < t\right) = F_\xi\left(\Psi^{-1}(t)\right).$$

By definition for independent random variables the distribution law of random quantity ζ_i is defined as

$$F_{\zeta_i}(t) = \mathsf{P}\left(\Psi\left(\mu_{i-1} + \xi_i\right) - \Psi\left(\mu_{i-1}\right) < t\right),$$

The last formula can be rewritten as

$$F_{\zeta_i}(t) = \iint\limits_{\Psi(u+v)+\Psi(v)<t} f_{\mu_{i-1}}(u) f_\xi(v) \mathrm{d}u \mathrm{d}v.$$

Integrating with respect to v we get

$$F_{\zeta_i}(t) = \int\limits_0^\infty f_{\mu_{i-1}}(u) F_\xi\left(\Psi^{-1}\left(t + \Psi(u)\right) - u\right) \mathrm{d}u, \tag{4.4}$$

where the distribution density $f_{\mu_{i-1}}(t)$ is defined as

$$f_{\mu_{i-1}}(t) = \int\limits_0^t f_{\mu_{i-2}}(t-u) f_\xi(u) \mathrm{d}u,$$

since $\mu_{i-1} = \mu_{i-2} + \xi_{i-1}$.

Example 2. Let $f_\xi(t)$ be the uniform on $[0; T]$ law density

$$f_\xi(t) = \begin{cases} \frac{1}{T}, & 0 \le t \le T \\ 0, & t < 0 \text{ or } t > T \end{cases}$$

Suppose that $\Psi(t) = \alpha t^\gamma$. The system type, which we simulate, will depends on quantity γ. Consider two special cases for normalizing function power index: $\gamma = 0.5$ (ageing system) and $\gamma = 2$ ("rejuvenescent" system). Let's find the distribution density for random quantities ζ_1 and ζ_2 analytically. Omitting intermediate computations finally we obtain the following equations.

If $\gamma = 0.5$, then

$$f_{\zeta_1}(t) = \begin{cases} \frac{2t}{\alpha^2 T}, & 0 < t \le \alpha\sqrt{T}; \\ 0, & t < 0 \text{ or } t > \alpha\sqrt{T}. \end{cases}$$

$$f_{\zeta_2}(t) = \begin{cases} \frac{2t}{\alpha^2 T} + \frac{4}{3\alpha^2\sqrt{T}}, & 0 < t \le \alpha\sqrt{T}\left(\sqrt{2}-1\right); \\ \frac{\left(\alpha^2 T - t^2\right)^2}{\alpha^4 T^2} \cdot \left(\frac{1}{2t} + \frac{\alpha^2 T - t^2}{6t^3}\right), & \alpha\sqrt{T}(\sqrt{2}-1) < t \le \alpha\sqrt{T}; \\ 0, & t < 0 \text{ or } t > \alpha\sqrt{T}. \end{cases}$$

If $\gamma = 2$, then

$$f_{\zeta_1}(t) = \begin{cases} \frac{1}{2\sqrt{\alpha t T}}, & 0 < t \le \alpha T^2; \\ 0, & t < 0 \quad t > \alpha T^2. \end{cases}$$

$$f_{\zeta_2}(t) = \begin{cases} \frac{\ln\left(\sqrt{\alpha}T + \sqrt{\alpha T^2 + t}\right) - 0.5\ln(t)}{2\alpha T^2}, & 0 < t \le \alpha T^2; \\ \frac{\ln\left(\sqrt{\alpha}T + \sqrt{\alpha T^2 + t}\right) - \ln\left(\frac{t}{\sqrt{\alpha T}}\right)}{2\alpha T^2}, & \alpha T^2 < t \le 3\alpha T^2; \\ 0, & t < 0 \text{ or } t > 3\alpha T^2. \end{cases}$$

The Fig. 4.3 illustrates the distribution density curves for the first and the second operability cycles when $\gamma = 0.5$ (right figure) and $\gamma = 2$ (left figure). Note that for "rejuvenescent" system the density function support expands and the second operating time between failures ζ_2 is stochastically greater than ζ_1. The tendency for an ageing system is opposite, i.e. the density function support compresses and the second operating time between failures is stochastically less than the first one.

Example 3. Let $f_\xi(t)$ be the exponential law density with rate λ. Then the quantities μ_i have the gamma distribution with the shape parameter i and the scale parameter λ as is well known.

Let $\Psi(t) = \alpha \ln(1 + \beta t)$, assuming $\alpha, \beta > 0$. In this case, the ageing system is simulated. Then inverse normalizing flow function $\Psi^{-1}(t)$ is defined as

$$\Psi^{-1}(t) = \frac{1}{\beta}\left(e^{\frac{t}{\alpha}} - 1\right), \quad \Psi^{-1}(t + \Psi(u)) - u = \kappa(t)\left(u + \frac{1}{\beta}\right),$$

where $\kappa(t) = \left(e^{\frac{t}{\alpha}} - 1\right)$.

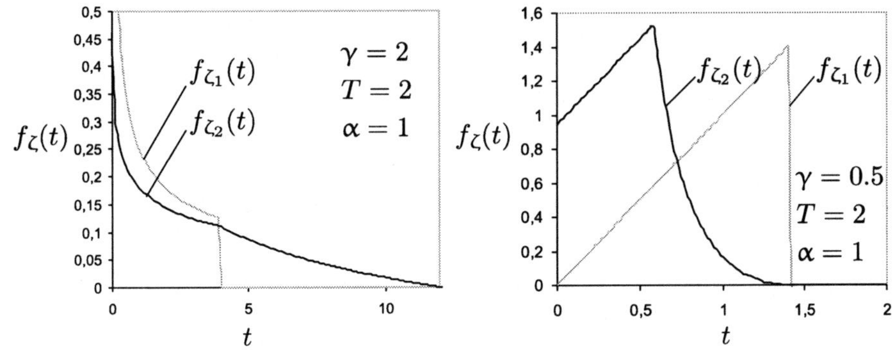

Figure 4.3. Distribution density change for operability cycles

Substituting in (4.4), we get

$$F_{\zeta_i}(t) = 1 - \frac{1}{(\kappa(t)+1)^{i-1}} e^{-\frac{\lambda\kappa(t)}{\beta}} = 1 - \exp\left(-\frac{\lambda}{\beta}\left(e^{\frac{t}{\alpha}} - 1\right) - \frac{t(i-1)}{\alpha}\right).$$

Consider the ratio of survival functions on two successive operability cycles

$$\frac{\bar{F}_{\zeta_{i+1}}(t)}{\bar{F}_{\zeta_i}(t)} = \frac{1 - F_{\zeta_{i+1}}(t)}{1 - F_{\zeta_i}(t)} = \frac{1}{\kappa+1} = e^{-\frac{t}{\alpha}} < 1,\ i = 1, 2, \ldots$$

if $t \neq 0$. Therefore, simulated system belong to the ageing system class.

4.7 Getting of Resource Reliability Characteristics

The back residual time and the straight residual time presented in [BF88] are employed as resource reliability characteristics. The back residual time R_t is the system operation time from system operation commencement or system operation resumption after the last repair to time t. The straight residual time V_t is the system operation time from time t, in which the system is operable, to the next failure.

The subscript t denotes the time dependence for convenience.

As before suppose that the system repair occurs immediately. And let $N(t)$ be the the number of repairs (failures) that have occurred up to time t. By $\tilde{\mu}_k$ denote the instant of kth event occurrence of non-homogeneous repair (failure) flow.

Then the back residual time is defined as

$$R_t = t - \tilde{\mu}_{N(t)};$$

And the straight residual time is defined as

$$V_t = \tilde{\mu}_{N(t)+1} - t;$$

Figure 4.4 illustrates the aforesaid.

Let us introduce the index Z_t that characterizes the system operation cycle time on time t. Is is defined as

$$Z_t = R_t + V_t.$$

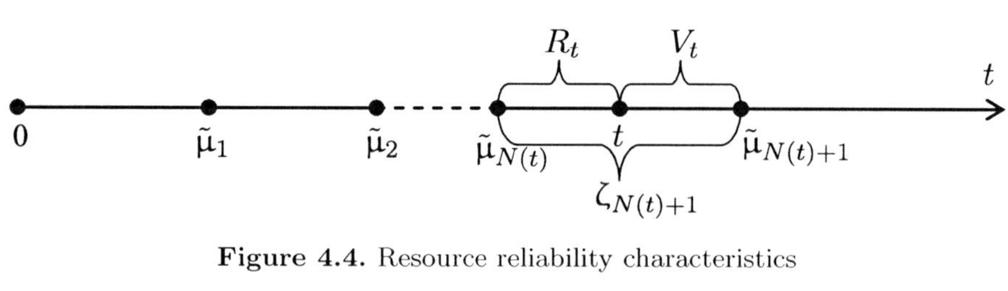

Figure 4.4. Resource reliability characteristics

Theorem 2. *Distribution function for the back residual time is defined as*

$$F_{R_t}(x) = \begin{cases} \int\limits_{\Psi^{-1}(t-x)}^{\Psi^{-1}(t)} \upsilon(u)\bar{F}_\xi\left(\Psi^{-1}(t) - u\right) \mathrm{d}u, & x < t \\ 1, \quad x \geq t. \end{cases}$$

Distribution function for the straight residual time is defined as

$$F_{V_t}(x) = F_\xi\left(\Psi^{-1}(t+x)\right) - \int\limits_0^{\Psi^{-1}(t)} \upsilon(u)\bar{F}_\xi\left(\Psi^{-1}(t+x) - u\right) \mathrm{d}u.$$

The proof of Theorem 2 will be given in Appendix.

Let us define the average back and straight residual times in the following theorem.

Theorem 3. *The average back residual time is defined as*

$$\mathsf{E}R_t = t\bar{F}_\xi\left(\Psi^{-1}(t)\right) + \int\limits_0^\infty g_R(x;t) f_\xi(x)\mathrm{d}x,$$

where $g_R(x;t) = \int\limits_{(\Psi^{-1}(t)-x)\vee 0}^{\Psi^{-1}(t)} (t - \Psi(\tau))\,\upsilon(\tau)\mathrm{d}\tau.$

The average straight residual time is defined as

$$\mathsf{E}V_t = \int\limits_{\Psi^{-1}(t)}^\infty (\Psi(x) - t) f_\xi(x)\mathrm{d}x + \int\limits_0^\infty g_V(x;t) f_\xi(x)\mathrm{d}x,$$

where $g_V(x;t) = \int\limits_{(\Psi^{-1}(t)-x)\vee 0}^{\Psi^{-1}(t)} (\Psi(\tau + x) - t)\,\upsilon(\tau)\mathrm{d}\tau.$

Example 4. Let $f_\xi(t)$ be the exponential law density with the rate λ. Let $\Psi(x) = \alpha\sqrt{\lambda x}$. In this case, $\upsilon(t) = \lambda$ and

$$F_{R_t}(x) = \begin{cases} 1 - \exp\left(\frac{x^2 - 2tx}{\alpha^2}\right), & x < t; \\ 1, \quad x \geq t. \end{cases}$$

$$\mathsf{E}R_t = t + \alpha e^{-\left(\frac{t}{\alpha}\right)^2} \gamma\left(\frac{3}{2}; -\left(\frac{t}{\alpha}\right)^2\right),$$

where $\gamma(a;x)$ is incomplete gamma-function.

$$F_{V_t}(x) = 1 - \exp\left(\frac{-2tx - x^2}{\alpha^2}\right).$$

$$\mathsf{E}V_t = \frac{\sqrt{\pi}}{2}\alpha \exp\left(\frac{t}{\alpha}\right)^2 \operatorname{erfc}\left(\frac{t}{\alpha}\right).$$

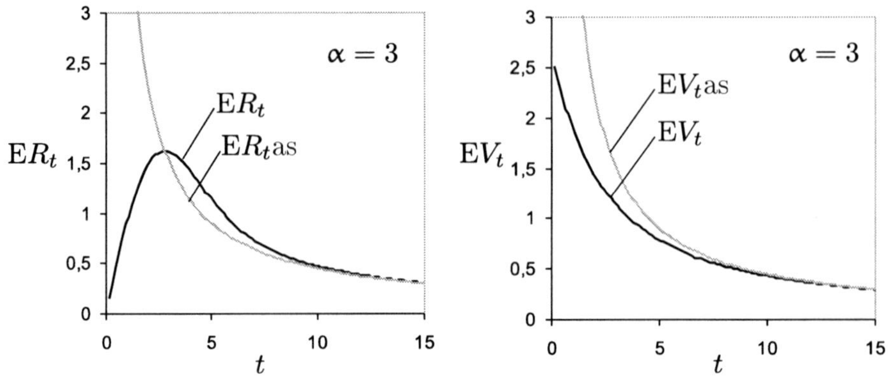

Figure 4.5. Average straight and back residual times

It can be shown that asymptotically as $t \to \infty$

$$\mathsf{E}V_t \sim \frac{\alpha^2}{2t}, \quad \mathsf{E}R_t \sim \frac{\alpha^2}{2t}.$$

The curves for average residual times and their asymptotic value are shown in Fig. 4.5 (left figure for the average back residual times and right figure for the average straight residual times). If the failure flow is homogeneous, the average back and straight residual times tend to the constant asymptotically. In our case, both characteristics tend to 0 as $t \to \infty$. It is typical for ageing system.

4.8 Availability Function Calculation

In this section, we develop a mathematical model for calculating availability function provided that the autonomous failure and repair flows are the mappings of two abstract independent recurrent flows.

Consider the process with alternate intervals of operability and repair. Suppose the system is operational during the random time ζ_i^0. Then the system fail and the system repair last the random time ζ_i^1 and so on.

Let $\tilde{\mu}_i^0 = \Psi_0\left(\mu_i^0\right), i = 0, 1, \ldots$ be the autonomous failure flow. And denote by $\tilde{\mu}_i^1 = \Psi_1\left(\mu_i^1\right), i = 0, 1, \ldots$ the autonomous repair flow. By $\left\{\mu_i^0\right\}$ and $\left\{\mu_i^1\right\}, i = 0, 1, \ldots$ denote the abstract independent recurrent (homogeneous) flows of failures and repairs, respectively. The functions $\Psi_0(x)$ and $\Psi_1(x)$ satisfy the conditions:

- $\Psi_i(0) = 0, i = 0, 1;$
- $\Psi_i(x) \in C^1_{(0;\infty)}$, moreover $\Psi_i'(x) > 0, i = 0, 1.$

The availability function is defined as

$$P_1(t) = \mathsf{P}\left(Q^+ \ni t\right) = \sum_{i=0}^{\infty} \mathsf{P}\left(\tau_i^1 \le t < \tau_{i+1}^0\right),$$

where Q^+ is the union of operability intervals, $\tau_i^0 = \tilde{\mu}_i^0 + \tilde{\mu}_{i-1}^1$ is the ith failure instant, $\tau_i^1 = \tilde{\mu}_i^0 + \tilde{\mu}_i^1$ is the ith repair instant, $\tilde{\mu}_i^1 = \Psi\left(\sum_{j=1}^{i} \xi_j^1\right)$, $\tilde{\mu}_i^0 = \Psi\left(\sum_{j=1}^{i} \xi_j^0\right)$, $i = 1, 2, \dots,$ $\left(\tilde{\mu}_0^1 = \tilde{\mu}_0^0 = 0\right)$ – the instants of event occurrence of autonomous flows, i.e. either only repair flow $\left\{\tilde{\mu}_i^1\right\}$ or only failure flow $\left\{\tilde{\mu}_i^0\right\}$.

Then the equations for availability function can be rewritten as

$$P_1(t) = \sum_{i=0}^{\infty} F_{\tau_i^1}(t) - \sum_{i=0}^{\infty} F_{\tau_{i+1}^0}(t).$$

Consider the new flow characteristics below. The leading function of joint failure–repair flow with equal number of failures and repairs $\Omega_+(t_0; t_1)$ is defined as

$$\Omega_+(t_0; t_1) = \sum_{i=0}^{\infty} \mathsf{P}\left(\tilde{\mu}_i^0 \le t_0; \tilde{\mu}_i^1 \le t_1\right) = \sum_{i=0}^{\infty} F_{\tilde{\mu}_i^0; \tilde{\mu}_i^1}(t_0; t_1). \tag{4.5}$$

And the leading function of joint failure–repair flow with failure number greater than repair number by 1 is defined as

$$\Omega_-(t_0; t_1) = \sum_{i=0}^{\infty} \mathsf{P}\left(\tilde{\mu}_{i+1}^0 \le t_0; \tilde{\mu}_i^1 \le t_1\right) = \sum_{i=0}^{\infty} F_{\tilde{\mu}_{i+1}^0; \tilde{\mu}_i^1}(t_0; t_1)$$

As a matter of fact (4.5) is the average number of pairs $\left(\tilde{\mu}_i^0; \tilde{\mu}_i^1\right) = \left(\sum_{j=1}^{i} \xi_j^0; \sum_{j=1}^{i} \xi_j^1\right)$, that satisfy the inequality set $\tilde{\mu}_i^0 \le t_0$; $\tilde{\mu}_i^1 \le t_1$ (see the Fig. 4.6). In the figure the number of pairs that satisfy the inequalities equals 3. These pairs is depicted by solid line. The fourth pair depicted dash line is ignored since $\tilde{\mu}_3^1 > t_1$.

If the necessary derivatives exist, then the failure–repair flow parameter with equal number of failures and repairs is defined as

$$\omega_+(t_0; t_1) = \frac{\mathrm{d}^2\Omega_+(t_0; t_1)}{\mathrm{d}t_0 \mathrm{d}t_1} = \sum_{i=0}^{\infty} f_{\tilde{\mu}_i^0; \tilde{\mu}_i^1}(t_0; t_1) \tag{4.6}$$

and the failure–repair flow parameter with failure number greater than repair number by 1 is defined as

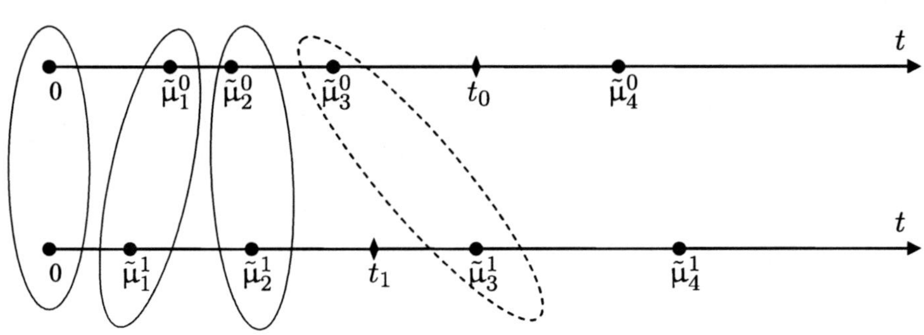

Figure 4.6. Joint failure–repair flow

$$\omega_{-}(t_0;t_1)=\frac{\mathrm{d}^2\Omega_{-}(t_0;t_1)}{\mathrm{d}t_0\mathrm{d}t_1}=\sum_{i=0}^{\infty}f_{\tilde{\mu}_{i+1}^0;\tilde{\mu}_i^1}(t_0;t_1) \tag{4.7}$$

Consider the properties of characteristics for joint event flow.

1. $\Omega_{+}(t_0;t_1)$, $\Omega_{-}(t_0;t_1)$ – nondecreasing functions with respect to each arguments (identically cumulative distribution function property);

2. Limit properties:

a. $\lim\limits_{t_0\to\infty}\Omega_{+}(t_0;t_1)=\lim\limits_{t_0\to\infty}\Omega_{-}(t_0;t_1)=\Omega_{\tilde{\mu}^1}(t_1)=\sum\limits_{i=0}^{\infty}F_{\tilde{\mu}_i^1}(t_1)$, where $F_{\tilde{\mu}_i^1}(t_1)$ is the leading function of autonomous repair flow;

b. $\lim\limits_{t_1\to\infty}\Omega_{+}(t_0;t_1)=\Omega_{\mu^0}(t_0)=\sum\limits_{i=0}^{\infty}F_{\tilde{\mu}_i^0}(t_1)$, where $F_{\tilde{\mu}_i^0}(t_1)$ is the leading function of autonomous failure flow, including zero instant;

c. $\lim\limits_{t_1\to\infty}\Omega_{-}(t_0;t_1)=\Omega_{\tilde{\mu}_{i+1}^0}(t_0)=\sum\limits_{i=0}^{\infty}F_{\tilde{\mu}_{i+1}^0}(t_1)$, where $F_{\tilde{\mu}_{i+1}^0}(t_1)$ is the leading function of autonomous failure flow, excepting zero instant;

3. Boundary conditions:

a. $\Omega_{+}(0;t_1)=\Omega_{+}(t_0;0)=1$;

b. $\Omega_{-}(0;t_1)=0$;

c. $\Omega_{-}(t_0;0)=F_{\tilde{\mu}_1^0}(t_0)$.

Theorem 4. *The availability function is defined as*

$$P_1(t)=\int\limits_{\Psi_0(x_1)+\Psi_1(x_2)\le t}\left[\omega_{+}(x_1;x_2)-\omega_{-}(x_1;x_2)\right]\mathrm{d}x_1\mathrm{d}x_2. \tag{4.8}$$

If the failure and repair flows are homogeneous, i.e. $\Psi_0(x)=\Psi_1(x)=x$, *then the last equation may be rewritten as*

$$P_1(t)=\int\limits_0^t\int\limits_0^{t-x_1}\left[\omega_{+}(x_1;x_2)-\omega_{-}(x_1;x_2)\right]\mathrm{d}x_2\mathrm{d}x_1. \tag{4.9}$$

4. Let us obtain the renewal equation analogue for introduced characteristic, if event flows are homogeneous. Then $\Psi_j(x)=x$, i.e. $\tilde{\mu}_i^j=\mu_i^j$, $i=0,1,\ldots,$ $j=0,1$.

Let us execute double integral Laplace transform for (4.6) with respect to variables t_0 and t_1:

$$\overline{\overline{\omega}}_{+}(p_0;p_1)=\int\limits_0^\infty\int\limits_0^\infty\exp\left(-p_0t_0-p_1t_1\right)\omega_{+}(t_0;t_1)\,\mathrm{d}t_0\mathrm{d}t_1=\sum_{i=0}^{\infty}\overline{\overline{f}}_{\mu_i^0;\mu_i^1}(p_0;p_1)\,.$$

After simple transformations we get the renewal equation analogue in images

$$\overline{\overline{\omega}}_{+}(p_0;p_1)=\frac{1}{1-\overline{\overline{f}}_{\xi^0;\xi^1}(p_0;p_1)}.$$

Passing to originals, we obtain the two-dimensional integral Volterra equations of the second kind:

$$\omega_+(t_0;t_1) = \delta(t_0)\,\delta(t_1) + \int_0^{t_0}\int_0^{t_1} f_{\xi^0;\xi^1}(t_0-x_0;t_1-x_1)\,\omega_+(x_0;x_1)\,\mathrm{d}x_0\mathrm{d}x_1, \quad (4.10)$$

$$\Omega_+(t_0;t_1) = \mathrm{H}(t_0)\,\mathrm{H}(t_1) + \int_0^{t_0}\int_0^{t_1} f_{\xi^0;\xi^1}(t_0-x_0;t_1-x_1)\,\Omega_+(x_0;x_1)\,\mathrm{d}x_0\mathrm{d}x_1. \quad (4.11)$$

Executing analogous calculations for (4.7), we get

$$\omega_-(t_0;t_1) = f_{\xi^0}(t_0)\,\delta(t_1) + \int_0^{t_0}\int_0^{t_1} f_{\xi^0;\xi^1}(t_0-x_0;t_1-x_1)\,\omega_-(x_0;x_1)\,\mathrm{d}x_0\mathrm{d}x_1, \quad (4.12)$$

$$\Omega_-(t_0;t_1) = F_{\xi^0}(t_0)\,\mathrm{H}(t_1) + \int_0^{t_0}\int_0^{t_1} f_{\xi^0;\xi^1}(t_0-x_0;t_1-x_1)\,\Omega_-(x_0;x_1)\,\mathrm{d}x_0\mathrm{d}x_1. \quad (4.13)$$

Remarks: In (4.10)–(4.13), by $\delta(t)$ denote the generalised Dirac delta function and by $\mathrm{H}(t)$ denote the Heaviside function.

Example 5. Let ξ^0 and ξ^1 be the exponential distributed random variables with the rates λ_0 and λ_1, respectively. Suppose the event flows are homogeneous.

Then

$$\overline{\overline{\omega}}_+(p_0;p_1) = \frac{(p_0+\lambda_0)(p_1+\lambda_1)}{p_0p_1+\lambda_0p_1+\lambda_1p_0} = 1 + \frac{\lambda_0\lambda_1}{p_1+\lambda_1}\frac{1}{p_0+\frac{\lambda_0p_1}{p_1+\lambda_1}}$$

and

$$\omega_+(t_0;t_1) = \delta(t_0)\,\delta(t_1) + \lambda_0\lambda_1\exp[-\lambda_0t_0-\lambda_1t_1]\,I_0\left(2\sqrt{\lambda_0\lambda_1t_0t_1}\right),$$

where $I_0(x)$ is the modified zero-order Bessel function.

We can similarly get that

$$\omega_-(t_0;t_1) = \lambda_0\exp(-\lambda_0t_0-\lambda_1t_1)\left[\delta(t_1) + \sqrt{\frac{\lambda_0\lambda_1t_0}{t_1}}I_1\left(2\sqrt{\lambda_0\lambda_1t_0t_1}\right)\right],$$

where $I_1(x)$ is modified Bessel functions of the first kind.

Then

$$\Omega_+(t_0;t_1) = \mathrm{H}(t_0)\,\mathrm{H}(t_1) + \int_0^{\lambda_0t_0}\int_0^{\lambda_1t_1} e^{-x_0-x_1}I_0\left(2\sqrt{x_0x_1}\right)\mathrm{d}x_0\mathrm{d}x_1$$

$$= \mathrm{H}(t_0)\,\mathrm{H}(t_1) + \lambda_1t_1\left(1-e^{-\lambda_0t_0}\right) + \int_0^{\lambda_1t_1} e^{-\lambda_0t_0-x}(\lambda_1t_1-x)\sqrt{\frac{\lambda_0t_0}{x}}I_1\left(2\sqrt{\lambda_0x_0x}\right)\mathrm{d}x.$$

$$\Omega_{-}(t_0;t_1) = \left(1 - e^{-\lambda_0 t_0}\right) \mathrm{H}(t_1) + \int_0^{\lambda_0 t_0} \int_0^{\lambda_1 t_1} e^{-x_0 - x_1} I_0\left(2\sqrt{x_0 x_1}\right) \mathrm{d}x_0 \mathrm{d}x_1$$

$$= \left(1 - e^{-\lambda_0 t_0}\right) \left(\mathrm{H}(t_1) + \lambda_1 t_1\right) + \int_0^{\lambda_1 t_1} e^{-\lambda_0 t_0 - x} \left(\lambda_1 t_1 - x\right) \sqrt{\frac{\lambda_0 t_0}{x}} I_1\left(2\sqrt{\lambda_0 x_0 x}\right) \mathrm{d}x.$$

The last integrals are evaluated by means of numerical methods. The calculation results of availability function by (4.9) agree with results obtained by well-known equation

$$P_1(t) = \frac{\lambda_1}{\lambda_0 + \lambda_1} + \frac{\lambda_0}{\lambda_0 + \lambda_1} e^{-(\lambda_0 + \lambda_1)t}$$

for exponential distributed operating and repair times (see [BF88, Ven91, Max01]).

The availability function may be calculated by (4.8) for a wide class of distributions. However, the main advantage of the joint flow model is that the availability function can be assessed in case of non-homogeneous failure and repair flows.

Example 6. Suppose that the failure and repair flows are regular, i.e. the random variables ξ^0 and ξ^1 are degenerate. Then executing the simple transformation we get

$$P_1(t) = \sum_{k=0}^{\infty} \left[\mathrm{H}\left(t - \Psi_0\left(k\xi^0\right) - \Psi_1\left(k\xi^1\right)\right) - \mathrm{H}\left(t - \Psi_0\left((k+1)\xi^0\right) + \Psi_1\left(k\xi^1\right)\right)\right].$$

4.9 Numerical Computation of Availability Function

Consider the following example. Let us calculate the availability function if non-homogeneity is given by power function $\Psi_i(t) = t^{\gamma_i}$, $\gamma_i > 0$, $i = 0, 1$. Denote by ξ_j^0 and ξ_j^1 the intervals between two successive events of recurrent (homogeneous) failure flow $\{\mu_i^0\}$ and repair flow $\{\mu_i^1\}$, respectively. The random variables ξ_j^0 and ξ_j^1 are distributed by Weibull law with the shape parameters $m^0 = m^1 = 2$, the scale parameters $\theta^0 = 5$, $\theta^1 = 2$. In this case, $\mathsf{E}\xi^0 = 4.43$ and $\mathsf{E}\xi^1 = 1.77$.

Consider the following variants:

1. Event flows are not "distorted": $\gamma_0 = 1$, $\gamma_1 = 1$.
2. Failure flow is depleted (failures occur more rarely), repair flow is homogeneous: $\gamma_0 = 1.2$, $\gamma_1 = 1$.
3. Failure flow is thickened (failures become more frequent), repair flow is homogeneous:
   a. $\gamma_0 = 0.8$, $\gamma_1 = 1$;
   b. $\gamma_0 = 0.5$, $\gamma_1 = 1$.
4. Failure flow is thickened (failures become more frequent), repair flow is depleted (repairs occur more rarely): $\gamma_0 = 0.8$, $\gamma_1 = 1.2$.

The curves of availability function for variants 1–4 are presented in the Fig. 4.7.

The curve 1 display the situation when the event flow are not "distortion". The curve 2 of availability function lies above the curve 1, since in case 2 failure flow is

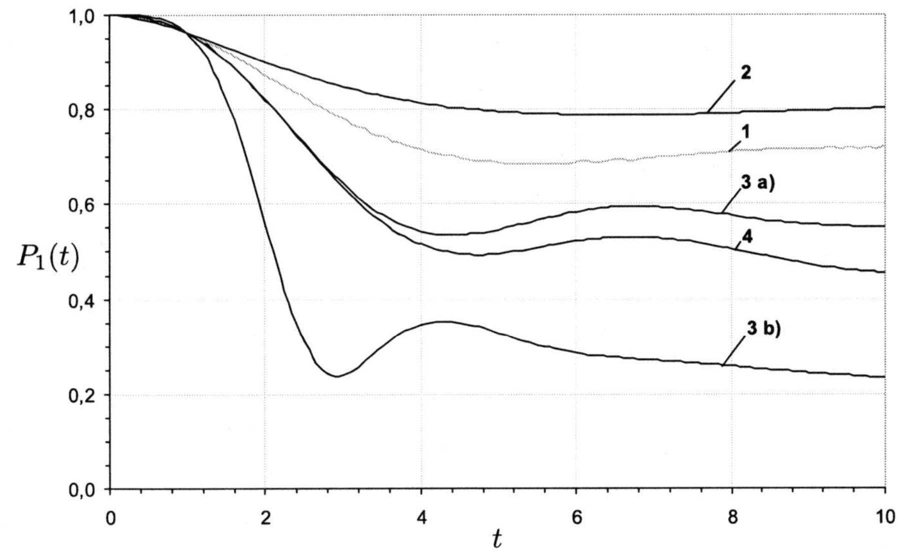

Figure 4.7. The availability function for variants 1–4

depleted. Consider the curves 3a and 3b. The parameter γ_0 in case 3b is smaller then the γ_0 in case 3a, i.e. failure flow in case 3b becomes more frequent. Hence, the curve 3b of availability function lies below the curve 3a. The local extremums of the curve 3b are due to specific Weibull distribution shape at given parameters. So the local minimum point (approximately equal 3) correspond to the most probable instant of the first failure and the local maximum point (approximately equal 4.5) correspond to the most probable instant of the first repair.

References

[Sae94] Saenko N.B. Reliability calculation for the systems taking into account repair incompleteness of elements. Proceedings of institute of higher education. Instrumentation, **11–12**, 76–79 (1994)

[BF88] Baihelt F., Franken P. Reliability and maintenance. Radio i svyaz, Moscow (1988)

[Ven91] Ventzel .S., Ovcharov L.. Theory of random processes and its engineering applications. Nauka, Moscow (1991)

[Max01] Probabilistic mathematics branches. Textbook for bachelors of technology path /Under the editorship of Maximov U.D. Ivan Fedorov, Sankt-Peterburg (2001)

[VC02] Volnikov I.S., Chepurko V.A. Non-homogeneous failure and repair flow. Collected papers of Automated Control Systems department. Obninsk, INPE, **14**, 36–44 (2002)

[NC04] Nikitina A.G., Chepurko V.A. Taking into account of the failure flow non-homogeneity. Collected papers of Automated Control Systems department. Obninsk, INPE, **15**, 31–36 (2004)

[AKC06] Antonov A.V., Karaulov I.N., Chepurko V.A. Optimization of preventive measures implementation taking into account equipment degradation and ageing. Collected papers of Automated Control Systems department. Obninsk, INPE, **16**, 31–38 (2006)

[SOC07] Saakyan S.P., Ostreikovsky V.A., Chepurko V.A. Processing technique of statistical equipment reliability data in operational process of nuclear power station. Proceedings of institute of higher education. Nuclear power engineering, **3.1**, 30–37 (2007)

[Che07] Chepurko V.A. Systems reliability characteristics taking into account failure flow non-homogeneity. Collected papers of Automated Control Systems department. Obninsk, INPE, **17**, 29–40 (2007)

[IC09] Ivanova K.A., Chepurko V.A. Mathematical model of joint event flow. Collected papers of Automated Control Systems department. Obninsk, INPE, **18**, 64–79 (2009)

[SC09] Skiba M.A., Chepurko V.A. On one reliability coefficient of the system in time-varying operating conditions. Collected papers of Automated Control Systems department. Obninsk, INPE, **18**, 29–41 (2009)

[Ski09] Skiba M.A. On the mean life and residual times non-homogeneous failure flow. Collected papers of Automated Control Systems department. Obninsk, INPE, **18**, 41–56 (2009)

Appendix

Proof of Theorem 1.

The leading flow function $\Omega(t) = \mathsf{E}N(t) = \sum_{i=1}^{\infty} \mathsf{P}(\tilde{\mu}_i \le t) = \sum_{i=1}^{\infty} F_{\tilde{\mu}_i}(t)$. Then $\Omega(\Psi(t)) = \nu(t) = \sum_{i=1}^{\infty} F_{\mu_i}(t)$.

The integral renewal equation holds for $\nu(t)$. Furthermore, elementary renewal theorem is satisfied, i.e., $\lim_{t\to\infty} \frac{\nu(t)}{t} = \frac{1}{\mathsf{E}\xi}$.

Proof of Theorem 2.

Using the total probability formula, let us find the distribution function of back residual time

$$F_{R_t}(x) = \mathsf{P}\left(\tilde{\mu}_{N(t)} > t - x\right) = \sum_{k=0}^{\infty} \mathsf{P}\left(\tilde{\mu}_k > t - x;\, N(t) = k\right)$$

$$= \sum_{k=0}^{\infty} \mathsf{P}\left(\Psi^{-1}(t-x) < \mu_k \le \Psi^{-1}(t) < \mu_k + \xi_{k+1}\right).$$

The summand corresponding $k = 0$ equals $\mathrm{H}(x-t)\overline{F}\left(\Psi^{-1}(t)\right)$, where H(x) is Heaviside function.

Finding the probabilities expressed in terms of integrals, we get

$$F_{R_t}(x) = \mathrm{H}(x-t)\overline{F}\left(\Psi^{-1}(t)\right) + \sum_{k=1}^{\infty} \int\limits_{0\vee(\Psi^{-1}(t-x))}^{\Psi^{-1}(t)} f_{\mu_k}(u)\overline{F}\left(\Psi^{-1}(t)-u\right)\mathrm{d}u$$

$$= \mathrm{H}(x-t)\overline{F}\left(\Psi^{-1}(t)\right) + \int\limits_{0\vee(\Psi^{-1}(t-x))}^{\Psi^{-1}(t)} \upsilon(u)\overline{F}\left(\Psi^{-1}(t)-u\right)\mathrm{d}u,$$

where $a \vee b = \max(a;b)$.

If $x \geq t$ then $F_{R_t}(x) = \overline{F}_{\xi}(z) + \int\limits_0^z \upsilon(u)\overline{F}(z-u)\,\mathrm{d}u$, where $z = \Psi^{-1}(t)$.

Using Laplace transform with respect to z on the right side , we obtain

$$\hat{F}_{R_t}(p) = \frac{1-\hat{f}(p)}{p} + \frac{\hat{f}(p)}{1-\hat{f}(p)} \cdot \frac{1-\hat{f}(p)}{p} = \frac{1}{p}.$$

The original of obtained equation equals 1.

The straight residual time distribution function can be fined by a similar way.

Proof of Theorem 4.

Let us derive the equation for availability function calculation.

$$P_1(t) = \sum_{i=0}^{\infty} F_{\tau_i^1}(t) - \sum_{i=0}^{\infty} F_{\tau_{i+1}^0}(t)$$

$$= \sum_{i=0}^{\infty} \left[\mathsf{P}\left(\Psi_0\left(\mu_i^0\right) + \Psi_1\left(\mu_i^1\right) \leq t\right) - \mathsf{P}\left(\Psi_0\left(\mu_{i+1}^0\right) + \Psi_1\left(\mu_i^1\right) \leq t\right)\right]$$

$$= \sum_{i=0}^{\infty} \left[\int\limits_{\Psi_0(x_1)+\Psi_1(x_2)\leq t} f_{\mu_i^0;\mu_i^1}(x_1;x_2) - \int\limits_{\Psi_0(x_1)+\Psi_1(x_2)\leq t} f_{\mu_{i+1}^0;\mu_i^1}(x_1;x_2)\right]\mathrm{d}x_1\mathrm{d}x_2.$$

Swapping summation and integration (suppose that this mathematical operation is possible) and using independent property, we get which required to be proved.

The equations for availability function in case of homogeneous flow is obtained by the substitution of $\Psi_i(x)$ for x.

5

A New Approach to Maintenance Optimization by Modeling Intensity Control

Reza Ahmadi* and Martin Newby

City University, London, UK, ahmadi_stat@yahoo.com

Abstract: In this paper, we present an optimal-control-based framework for deteriorating systems that are subject to repair and inspection. The model is based on two common assumptions that include the maintenance process is adapted to partial information including history of inspection events, and inspections do not impact on the failure characteristics of components. The latter means, at inspection times the system is repaired minimally that brings the system back to the operating condition just previous to inspection (as-bad-as-old) otherwise the operating system with slight overhaul is left to continue to operate. To model the unobservable damage (state) process which measures the effect of operating environment on the system, non-homogeneous Markov process with state space $S = \{1, 2\}$ is applied. By projection on the observed history, the partial information control problem is converted into a complete information problem. To put the maintenance model in an intensity control framework, the transition rate of Markov process driven by control process u is addressed as a measure to control not only intensity of inspection, but also the flow of revenue associated with the damage process. To solve optimal inspection problem, and to optimally adjust the state of the process over inter-arrival time of inspections (repair), an evolution of optimal control process which is solution of Hamilton–Jacobi equations is derived. To tackle the maintenance optimization problem numerically, an example is given. Provided optimal control process, a sequence of optimal inspection times and corresponding inspection intensity and conditional survival function of the system are obtained. Also, to predict the system failure, conditional mean times to failure (CMTTF) of the system indexed by inspection events are derived. Finally, a failure intensity based optimal stopping rule to replace the system is proposed.

Keywords and phrases: Control process, Cox process, Degradation model, Filtering theorem, Hamilton–Jacobi equation, Inspection intensity, Intensity control model, Maintenance, Mean residual, Minor failure, Non-homogeneous Markov process, Optimal inspection time, Optimization, Partial information, Proportional intensity model

V.V. Rykov et al. (eds.), *Mathematical and Statistical Models and Methods in Reliability: Applications to Medicine, Finance, and Quality Control*, Statistics for Industry and Technology, DOI 10.1007/978-0-8176-4971-5_5,

5.1 Introduction

For systems (e.g., manufacturing systems) which resulting output (revenue) depend on the system performance, an optimal inspection policy and also an ideal adjustment of health state are of the essence. On one hand, insufficient inspection leads to some system malfunction and the complete breakdown of the system. On the other hand frequent inspections of system to rectify defects results in more inspection costs. So, to maximize the revenue from the system which is in continuous operation, in addition to optimal control of the system state, an inspection strategy to give a correct balance between frequency and time of system inspections, and the resulting output (revenue) is required.

The literature on the optimal inspection problem is vast. Barlow et al. [BHP63] under some assumptions presents an optimal inspection policy for systems which are subject to non-self announced failure. Based on Barlow's maintenance model, inspections don't affect on the failure characteristics, inspection ceases upon detection of failure and at inspection times no repair takes place. Given above assumptions it is shown that optimal inspection times are the solution of system of equations remarked by BHP algorithm. The extreme sensitivity of the algorithm to initial value t_1 (first inspection time) is termed as the major problem of the Barlow et al. model. Keller [K74], and Kaio and Osaki [KO84] model the optimum inspection problem with respect to the inspection density that measures the number of checks per unit of time. Both Keller's model and the method of Kaio and Osaki to evaluate the inspection time sequence use the assumption that the time between the failure and its detection is half of the inspection interval. Lack of required accuracy resulting from the assumption is termed as both models problem. Munford and Shahani [MS74] define an inspection sequence characterized by the conditional failure probability of the system. The sequence of inspection times is optimized by minimizing the expected total cost. Chelbi and Ait-Kad [CA99] introduce an improved inspection model that the conditional failure probability should be an increasing function of the inspection number. To consider the appropriateness of optimum inspection models above, Jiang and Jardine [JJ05] present two optimization models. They show that optimal inspection time sequence derived from the proposed models is relatively accurate and computationally simple.

Following by taking advantages of optimal intensity control [B81], a new maintenance optimization model subject to revenue is proposed. The model presented here in preference to typical inspection models has potential not only to tackle optimal inspection problem but also to optimally drive the flow of revenue over inter-arrival times of inspection. The latter results from adjusting the state of the system. The model is based on a set of assumptions that include inspections do not impact on the systems failure characteristics. That means, at inspection time the detected defect of component is rectified by minimal repair that brings the component back to the operating condition just previous to failure (as-bad-as-old) otherwise the operating component with slight overhaul is left to continue to operate. Also, it is assumed that the maintenance model is given partial information $\mathcal{F}_t^N (t \geq 0)$ including just the history of inspection times. To model the evolution of the system failure influenced by environmental factors, the proportional intensity model (PIM) which is a generalized case of the proportional hazard model (PHM) [JM92] is applied. Also, the effect of environmental factors referred by damage process is described by the non-homogeneous Markov process $X_t(t \geq 0)$. Using the same approach as Jensen [JG93], the intensity of inspections or equivalently the inspection frequency of system is modelled by a stochastic measure characterized by the Markov process X_t. To evaluate the sequence of inspection times we apply the Cox process [DJ03] driven by the stochastic process X_t. By bringing the filtering theorem [B81] to bear given partial information an explicit solution to the estimation of the underlying process X_t is derived. It is shown above solution which is an extended version of the Jensen model's [JG93] transforms the inspection problem into the model adapted to the observable history $\mathcal{F}_t^N$. To represent the maintenance model in an intensity control-based framework, the transition rate of Markov process driven by control process u is addressed as a measure to control not only intensity of inspection, but also the flow of revenue associated with the damage process. To adjust the state of the system and give a correct balance between frequency of inspections and the resulting output (revenue), an optimal control strategy by the Deterministic Hamilton–Jacobi equations is obtained.

The paper is organized as follows. Section 2 deals with modelling damage and inspection process. By using filtering theorem an estimation of the underlying process X_t given observable history $\mathcal{F}_t^N$ is derived. The section ends with modelling failure of the system through proportional intensity process. The next section is devoted to modelling intensity control adapted to partial information. In Sect. 4 to numerically tackle the optimal intensity control problem an example is given. At the end some approaches to extend the maintenance optimization are proposed.

5.2 Model

5.2.1 Damage and Inspection Process Modelling

To model system's deteriorating process, let the effect of operating environment on the system be described by a (damage) process (X_t), $(t \in \mathbb{R}_+)$ with the state space $S = \{1, 2\}$ (X can be the level of metal particle in engine oil, vibration level, or temperature). It is assumed that the transition between states is controlled by the non-homogeneous Markov process with time dependent transition probability and intensity matrix $P(t) = (p_{ij}(t))$ and $Q(t) = (q_{ij}(t))$, $(i, j \in S)$, respectively, that for $s < t$

$$q_{ij}(t) = \lim_{t \to s} \tfrac{p_{ij}(s,t)}{t-s}, \quad i, j \in S$$

and $q_i(t) = -q_{ii}(t) = \sum_{i \neq j} q_{ij}(t)$.

Clearly, in terms of $q_{12}(t)$ the waiting time distribution in state one can be written as

$$S(s,t) = \exp\left\{-\int_s^t q_{12}(v)\mathrm{d}v\right\}, \quad s < t \tag{5.1}$$

To model inspection process of the system which is subject to the random environment, let the counting process (N_t), $(t \in \mathbb{R}_+)$, $N_t = \sum_{n \geq 1} I_{\{T_n \leq t\}}$ defined on a measurable space $(\Omega, \mathcal{F})$ where $T_1 < T_2 < ...$, $\lim_{n \to \infty} T_n = \infty$ (nonexplosive) imply the sequence of inspection times. To consider the effect of environmental factors on inspections frequency of the system, using the same approach as Jensen [JG93], let the intensity of inspections be linked by the $\mathcal{F}$-adapted state process (X_t). More precisely, it is assumed that (N_t) admits the following semimartingale representation:

$$N_t = \int_0^t \gamma_{X_s} \mathrm{d}s + M_t, \quad (t \in \mathbb{R}_+), M \in \mathcal{M}_0 \tag{5.2}$$

Where $0 < \gamma_1 < \gamma_2 < \infty$, $\gamma_{X_t}, (t \in \mathbb{R}_+)$ is $\mathcal{F}$-progressive and $\mathcal{M}_0$ refers to the class of $\mathcal{F}$-martingales with paths which are right-continuous and left-limited (with $M_0 = 0$). Note that the condition $\gamma_1 < \gamma_2$ assures that the inspections frequency of the system with raising damage level (state of the process) increases.

In the rest of section, we focus on modelling maintenance process through alternative definition of doubly stochastic poisson process (Cox process) [B81]. To model the sequence of inspection times, let $\bar{F}_n(v)$, $(n \in \mathbb{N}_0)$ be the regular conditional distribution of the inter-arrival times $V_{n+1} = T_{n+1} - T_n$, $(T_0 = 0)$. By using the Cox process which is subject to $\mathcal{F}_t^N$-adapted measure $\hat{\gamma}(n,t) = \sum_{i \in S} \gamma_i \hat{\varphi}(n,t;i)$, $\bar{F}_n(v)$ is given by

$$\begin{aligned} \bar{F}_n(v) &= p[V_{n+1} \geq v | \hat{\gamma}_t(n)] \\ &= \exp\left(- \int_{T_n}^{T_n+v} \sum_{i \in S} \gamma_i \hat{\varphi}(n,t;i) \mathrm{d}t \right) \end{aligned} \tag{5.3}$$

Where $T_n \leq t < T_{n+1}$, $\mathcal{F}_t^N$ refers to the partial information including observed history of inspection events that is $\mathcal{F}_t^N = \sigma\{N_s : 0 \leq s \leq t\}$, $\hat{\gamma}(n,t) = \hat{\gamma}_t(n)$ and $\hat{\varphi}(n,t)$ denote the intensity of inspection and the probability measure of the state of the process X (see Sect. 1.2.2) at time t following n^{th} inspection respectively.

In the special case when $n = 0$, then (5.3) reduces to the first inspection time law. That is,

$$\begin{aligned} \bar{F}_0(v) &= p[V_1 \geq v | \hat{\gamma}_t(0)] \\ &= \exp\left(- \int_0^v \sum_{i \in S} \gamma_i \hat{\varphi}(n,t;i) \mathrm{d}t \right) \end{aligned} \tag{5.4}$$

Where (5.4) can be considered as the probability measure of the random time V_1 to the occurrence of the first non-homogeneous Poisson event. In the sequel, let η_{n+1} denote the expected value of $(n+1)$th inter-arrival inspection time. Then by using (5.3) we have

$$\begin{aligned} \eta_{n+1} &= \int_0^\infty \bar{F}_n(v) \mathrm{d}v \\ &= \int_0^\infty \exp\left(- \int_{T_n}^{T_n+v} \sum_{i \in S} \gamma_i \hat{\varphi}(n,t;i) \mathrm{d}t \right) \mathrm{d}v \end{aligned} \tag{5.5}$$

Clearly, in light of the fact that above integral term depends on the inspection time $T_n (n \geq 0)$, η_{n+1} is a random measure. To settle this problem, an estimated version of η_{n+1} is applied. In other words,

$$\begin{aligned} \hat{\eta}_{n+1} &= \int_0^\infty \hat{\bar{F}}_n(v)\mathrm{d}v \\ &= \int_0^\infty \exp\left(-\int_{\mu_n}^{\mu_n+v} \sum_{i\in S} \gamma_i \hat{\varphi}(n,t;i)\mathrm{d}t\right)\mathrm{d}v \end{aligned} \tag{5.6}$$

For $n \geq 1$, and

$$\begin{aligned} \hat{\mu}_1 &= \int_0^\infty \hat{\bar{F}}_0(v)\mathrm{d}v \\ &= \int_0^\infty \exp\left(-\int_0^v \sum_{i\in S} \gamma_i \hat{\varphi}(n,t;i)\mathrm{d}t\right)\mathrm{d}v \end{aligned} \tag{5.7}$$

where μ_{n+1} is the $(n+1)^{th}$ expected inspection times, i.e., $\mu_{n+1} = E(T_{n+1}) = \sum_{k=1}^{n+1} \hat{\eta}_{n+1}$.

Following section is devoted to the estimation of the underlying process X_t through the filtering theorem [B81].

5.2.2 Partial Information Based Estimation of the Underlying Process X_t

To get an evolution of environmental factors represented by the stochastic process X_t, let $\hat{\varphi}_t(i)$ $(i \in S)$ denote the stochastic indicator function of X_t, i.e., $\varphi_t(i) = I(X_t = i)$. From filtering theorem and given the partial information $\mathcal{F}_t^N$ it can be shown that

$$\begin{aligned} \hat{\varphi}_t(j) &= \hat{\varphi}_0(j) \\ &\quad + \int_0^t \left(\sum_{i\in S} \hat{\varphi}_s(i)\left\{q_{ij}(s) + \hat{\varphi}_s(j)(\gamma_i - \gamma_j)\right\}\right)\mathrm{d}s \\ &\quad + \sum_{n\geq 1}\left(-\hat{\varphi}_{T_n^-}(j) + \frac{\gamma_j \hat{\varphi}_{T_n^-}(j)}{\sum_{i=1}^m \gamma_i \hat{\varphi}_{T_n^-}(i)} I_{\{T_n \geq t\}}\right), \end{aligned} \tag{5.8}$$

Or, equivalently, over inter-arrival inspection times: $T_n \leq t < T_{n+1}$,

$$\begin{aligned} \hat{\varphi}(n,t;j) &= \hat{\varphi}(n,T_n;j) \\ &\quad + \int_{T_n}^t \left(\sum_{i\in S} \hat{\varphi}(n,s;i)\left\{q_{ij}(s) + \hat{\varphi}(n,s;j)(\gamma_i - \gamma_j)\right\}\right)\mathrm{d}s \end{aligned} \tag{5.9}$$

and at inspection times,

$$\hat{\varphi}_{T_n}(j) = \frac{\gamma_j \hat{\varphi}_{T_n^-}(j)}{\sum_{i\in S} \gamma_i \hat{\varphi}_{T_n^-}(i)}, \quad j \in S, \tag{5.10}$$

where $\hat{\varphi}_{t^-}(j)$ refers to the left limit.

In the sequel, to get an explicit solution of $\hat{\varphi}_t(i)$ let $Q(t) = \begin{pmatrix} -q_1(t) & q_{12}(t) \\ 0 & 0 \end{pmatrix}$ where $q_{12}(t)$ is an increasing function of t, i.e., over time the sojourn time rate in the state one decreases. From (5.9) and using the differential equation an explicit solution of $\hat{\varphi}(n, t; 2)$ is given by

$$\hat{\varphi}(n,t;2) = 1 - \left\{ \frac{\exp\left(-\int (\bar{\gamma} + q_1(t))\,\mathrm{d}t\right)\big|_{t=T_n}}{1 - \hat{\varphi}_{T_n}(2)} - \int_{T_n}^{t} \bar{\gamma} \exp\left(-\int_0^u (\bar{\gamma} + q_1(v))\,\mathrm{d}v\right)\mathrm{d}u \right\}^{-1} \times \exp\left(-\int (\bar{\gamma} + q_1(t))\mathrm{d}t\right) \tag{5.11}$$

where $\bar{\gamma} = \gamma_1 - \gamma_2$.

Now in the sense that the transition between states is controlled by homogeneous Markov process, i.e., $q_{12}(t) = q_{12}, \forall t \geq 0$, then (5.11) reduces to the probability measure introduced in Jensen model [JG93]. To get a immediate solution for above measure, assume that $q_{12}(t) = t$ for $t \geq 0$. So, from (5.1), the waiting time distribution in the state one is given by

$$S(0,t) = S(t) = \exp\left(-\frac{t^2}{2}\right), \tag{5.12}$$

On the other hand, the other measure which can lend itself beautifully to describe the effect of the system's age on the deterioration of the system is mean residual time in the state one. So, if $m(t), t \geq 0$ is treated as the mean residual time in state one, then it follows

$$m(t) = \sqrt{2\pi} \exp\left(\frac{t^2}{2}\right)[1 - \phi(t)], \tag{5.13}$$

and $\phi(t), t \in \mathbb{R}_+$ denotes the standardized normal density. Moreover, from (5.11) it is easy to show that the probability measure $\hat{\varphi}(n, t; 2)$ is represented as

$$\hat{\varphi}(n,t;2) = 1 - \left\{ \frac{\exp\left(-(\bar{\gamma}T_n + \frac{T_n^2}{2})\right)}{1 - \hat{\varphi}_{T_n}(2)} - \sqrt{2\pi}\bar{\gamma}\exp(\frac{\bar{\gamma}^2}{2})[\phi(t+\bar{\gamma}) - \phi(T_n + \bar{\gamma})] \right\}^{-1} \times \exp\left(-(\bar{\gamma} + \frac{t^2}{2})\right) \tag{5.14}$$

The next section is devoted to evaluating both survival and conditional survival function of the system subject to inspection. To tackle this problem, the proportional intensity process is used as a tool to measure the failure rate of the system.

5.2.3 Failure Modelling

To model the effect of the damage process X_t $(t \geq 0)$ on failure rate, we address $\mathcal{F}_t$-adapted proportional intensity model (PIM) which is widely used in survival analysis. It is assumed that the failure rate is the product of a baseline failure rate dependent only on the age of the system and a positive function ψ dependent on the value of the state of the $\mathcal{F}$-adapted process X. In other words,

$$\lambda(t, X_t) = \lambda_0(t)\psi(X_t), \quad t \in \mathbb{R}_+ \tag{5.15}$$

where both the baseline intensity $\lambda_0(t)$ and $\psi(x)$ are bounded and increasing so that the system deteriorates with age and increases in the state X_t. More precisely, it is supposed that $I(\xi \leq t)$, the indicator process of the system lifetime, admits the semimartingale representation:

$$I(\xi \leq t) = \int_0^t I(\xi > s)\lambda_s \mathrm{d}s + M_t^\lambda, \quad t \in \mathbb{R}_+ \tag{5.16}$$

where $(\lambda_t), t \in \mathbb{R}_+$ is a progressively measurable with respect to the filtration $\mathcal{F}_t$ with $E\left[\int_0^t |\lambda_s| \mathrm{d}s\right] < \infty$ for all $t \in \mathbb{R}_+$ and $M^\lambda = (M_t^\lambda) \in \mathcal{M}_0$.

Now to predict the failure of the system adapted to the partial information $\mathcal{F}_t^0 = \sigma\{T_1 \wedge t\} = \sigma\{I(T_1 > s), 0 \leq s \leq t\}$, let $R(0,t)$ $(0 \leq t < T_1)$ denotes the survival function of the system up to the first inspection time. Using the projection theorem [JG93] from (5.16) we obtain

$$E[Z_t|\mathcal{F}_t^0] = 1 - R(0,t) = \int_0^t R(0,s)\hat{\lambda}_s(0)\mathrm{d}s + \bar{M}_t^\lambda, \tag{5.17}$$

where $\lambda_t(0) = \lambda(0,t)$ and $S(0,t)$ denote the failure rate and the survival function of the system at time t $(0 \leq t \leq T_1)$ adapted to the filtration $\mathcal{F}_t^0$,

$$S(0,t) = E[I(\xi > t)|\mathcal{F}_t^0] = P(\xi > t|\mathcal{F}_t^0),$$

and $\bar{M}_t^\lambda$ is an $\mathcal{F}_t^0$ martingale.

By using the fact that $S(0,t)$ over interval $[0, T_1)$ has continuous paths of bounded variation, then the martingale term $\bar{M}_t^\lambda$ is identically 0 and solution of the resulting integral equation (5.17) is

$$S(0,t) = \exp\left(-\int_0^t \sum_{i \in S} \lambda(s,i)\hat{\varphi}(0,s;i)\mathrm{d}s\right) \quad 0 \leq t < T_1. \tag{5.18}$$

where $\hat{\varphi}(0,t;i) = E[\varphi(0,t;i)|\mathcal{F}_t^0]$ $(i \in S)$ represents the probability measure of the state of the process given the filtration $\mathcal{F}_t^0$.

In the sequel, let $S(n,t)$ denote the survival function of the system following nth $(n > 0)$ inspection event at time t $(0 \leq t < T_{n+1} - T_n)$. It can be easily shown that,

$$S(n,t) = R(n,t)\prod_{i=1}^{n} R(i-1, T_i - T_{i-1}) \prod_{0<s\leq t+T_n} \exp\left(-\Delta\hat{\lambda}_s\right), \tag{5.19}$$

where $\Delta L_s = L_s - L_{s^-}$ denotes the jump height at time s caused by jumps of the martingale part of the state of the process X adapted to observable history $\mathcal{F}_t^N$ (see (5.17)) and $R(n,t)$ refers to the conditional survival function of the system at time t $(0 \leq t < T_{n+1} - T_n)$ following nth $(n \geq 0)$ inspection. In other words,

$$
\begin{aligned}
R(n,t) = P(\xi > t + T_n | \xi > T_n, \mathcal{F}_{T_n}^N) &= \frac{P(\xi > t + T_n | \mathcal{F}_{T_n}^N)}{P(\xi > T_n | \mathcal{F}_{T_n}^N)} = \frac{S(n,t)}{S(n,0)} \\
&= \exp\left(-\int_{T_n}^{T_n+t} \hat{\lambda}(n,s)\mathrm{d}s\right) \\
&= \exp\left(-\int_{T_n}^{T_n+t} \lambda_0(s) \sum_{i\in S} \psi(i)\hat{\varphi}(n,s,;i)\mathrm{d}s\right)
\end{aligned}
\tag{5.20}
$$

So, $R(n,t)$ condition on subfiltration $\mathcal{F}_{T_n}^N$ gives the probability of surviving the system beyond $(t + T_n)$ $(n \geq 0)$ provided that the system at nth inspection time has been in operating state.

Next section describes the modelling intensity control of inspection times provided the partial information which is the history of the inspection events. Also, the mean residual waiting time $m^u(t)$ and the probability measure $\hat{\varphi}^u(0,t;1)$ are addressed as measures to show how the flow of the $\mathcal{F}^N$-adapted damage process X is influenced by control process.

5.3 Modelling Intensity Control Given Partial Information

Before introducing the modelling intensity control formally, consider an example on evolution of both mean residual waiting time and waiting time distribution influenced by the control measure u_t $(t \geq 0)$. The following example shows how the process u_t through the $\mathcal{F}_t^N$-adapted measure $\hat{\varphi}_t$ plays a main role to control the state of the process and inspection frequency.

Example 1. Suppose that $q_1(t) = u.t$ and $S^u(t)$, $m^u(t)$ denote the waiting time distribution and mean residual waiting time in the state one associated with the control process u_t. Clearly,

$$
\begin{aligned}
m^u(t) &= \frac{\int_t^\infty S^u(v)\mathrm{d}v}{S^u(t)} \\
&= \sqrt{\frac{2\pi}{u}} \exp\left(\frac{ut^2}{2}\right)[1 - \phi(\sqrt{u}t)],
\end{aligned}
\tag{5.21}
$$

By using (5.21) an evolution of $m^u(t)$ is illustrated (see Fig. 5.1). As shown at fixed time t with increasing the control value u the mean residual waiting time in the state one

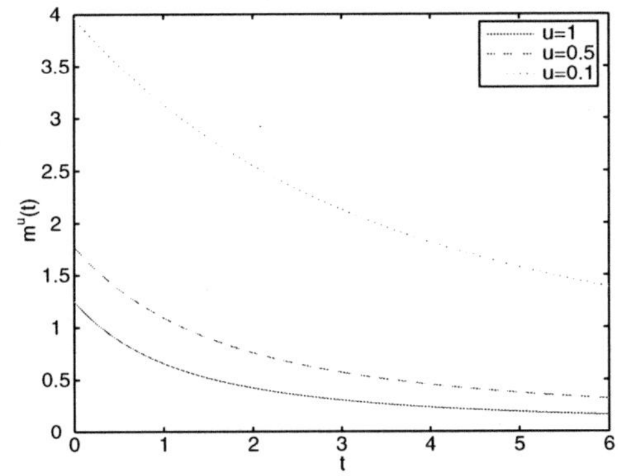

Figure 5.1. An evolution of the mean residual waiting times in the state one given $q_1^u(t) = u.t$ and control values $u = 0.1, 0.5, 1$

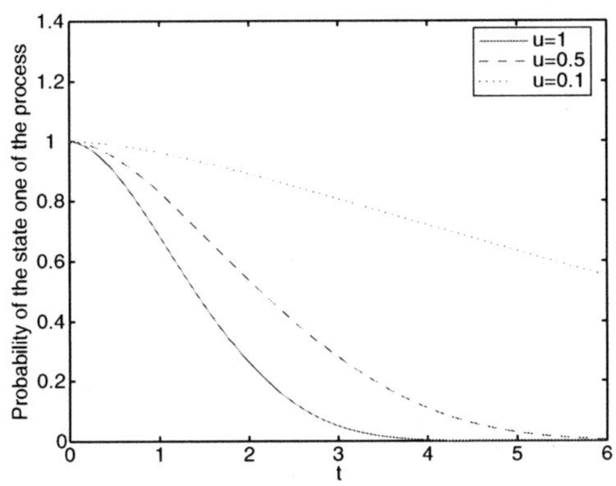

Figure 5.2. An evolution of the probability of the state one of the process given $q_1^u(t) = u.t$, $\lambda_{21} = 1$ and the control values u=0.1,0.5,1

decreases that means larger value of the control process u_t, the more deterioration of the system. Also, if $\hat{\varphi}^u(0, t; 1)$ refers to the probability of the system state one indexed by the control process u_t by applying (5.14) it is easy to show that

$$\hat{\varphi}^u(0, t; 1) = \left\{1 - \bar{\gamma}\sqrt{\frac{2\pi}{u}}\exp(\frac{\bar{\gamma}^2}{2u})\left[\phi\left(\sqrt{u}(t + \frac{\bar{\gamma}}{u})\right) - \phi\left(\frac{\bar{\gamma}}{\sqrt{u}}\right)\right]\right\}^{-1} \times \exp\left(-(\bar{\gamma}t + \frac{ut^2}{2})\right), \quad 0 \leq t < T_1 \tag{5.22}$$

where T_1 is the time to the first inspection. As shown (see Fig. 5.2) with increasing the intensity of leaving the state one controlled by u the deterioration process of the system intensifies.

Now to represent the inspection model in a intensity control set up and optimize the maintenance process, assume that $\mathcal{U}$ is the set of $\mathbb{R}_+$-valued measurable control processes of the form $u_t = u(t, N_t(\omega))$ where for each $n \in N_+$, the mapping $t \to u(t, n)$ is $\mathcal{F}_t^N$-predictable and $u_t \in \mathcal{U}, t \geq 0, \omega \in \Omega$. To each control $u \in \mathcal{U}$ we associate a probability measure (*control dynamics*) $P_u, u \in \mathcal{U}$ on $(\Omega, \mathcal{F})$. To control the occurrence of repairs over time, it is assumed that N_t through transition rate $q_1(t)$ admits a $(P_u, \mathcal{F}_t^N)$-intensity $\gamma_t(u)$ of the form

$$\gamma_t(\omega, u) = \gamma(t, N_t(\omega), u_t(\omega)), \tag{5.23}$$

so that $q_1^u(t) = q_1(t, u)$.

Thus, $q_1^u(t)$ can be regarded as a key tool to turn the maintenance template into the intensity control model. In other words, the complete-information maintenance model by means of the probability measure $\hat{\varphi}_t^u$ regulated by the control process u_t reduces to the intensity control pattern given partial information. In addition, to each $u \in \mathcal{U}$ we correspond a nonnegative measure $J(u)$:

$$J(u) = E_u \left[\int_0^T (\mu_{X_t}^u - k_\varepsilon(t, u)) - \sum_{n>0} k_{T_n}(u) - \phi_{X_T}^u \right] < \infty \tag{5.24}$$

That T is a positive time, $\mu_{X_t}^u$ is a nonnegative $\mathcal{F}_t^N$-progressive process and $k_{T_n}(u)$, $\phi_{X_T}^u$ are nonnegative $\mathcal{F}_t^N$ predictable, and $\mathcal{F}_T$-measurable random variable respectively. The measure $J(u)$ associated to u is the value function, $\mu_{X_t}^u$, indexed by the system state, denotes the reward per unit of time such that $(\mu_2 < \mu_1)$. That means with raising the wear level of the system the revenue obtained over inter-arrival time decreases. The term $k_\varepsilon(t, u) = \varepsilon(1 - u)t$ with scale parameter ε $(\varepsilon > 0)$ denotes the cost per unit of time to adjust the deterioration level of the system through the control process $u \in \mathcal{U}$. As shown given the control value u, with process of time the repair cost increases. On the other hand, at any fixed time $0 < t < T$, with decreasing control process u, $k_\varepsilon(t, u)$ increases. In special case, if the control process u takes the value 1, that is, the system is left to operate without repair and maintenance action, then $k_\varepsilon(t, u) = 0$. Also, $k_{T_n}(u)$ represents the inspection cost at n^{th} inspection action time and $\phi_{X_T}^u$ is the final cost for replacement, and inspection at terminal time T. On the other hand, since N_t admits $\mathcal{F}_t^N$-intensity γ_t and $k_t^u = k(t, N_t, u)$ is $\mathcal{F}_t^N$-predictable process, then the traditional definition of the stochastic intensity [AJ98] follows

$$J(u) = E_u \left[\int_0^T (\mu_{X_t}^u - k_\varepsilon(t, u) - k(t, N_t, u)\gamma_t(u))\, \mathrm{d}t - \phi_{X_T}^u \right] \tag{5.25}$$

Finally, to meet the control problem assumptions, let the above adjusted value function be converted into the $\mathcal{F}_t^N$-adapted measure. By projection on the observed history of the process $\mathcal{F}_t^N$-adapted version of $\hat{J}(u)$ is

$$\begin{aligned}
\hat{J}(u) &= E_u \left[\int_0^T \left(\sum_{i \in I} \mu_i \hat{\varphi}_t^u(i) - k_\varepsilon(t, u) - k(t, N_t, u)\gamma_t(u) \right) \mathrm{d}t - \sum_{i \in S} \phi_i^u \hat{\varphi}_T(i) \right] \\
&= E_u \left[\int_0^T \left(\sum_{i \in S} \mu_i \hat{\varphi}_t^u(i) - k_\varepsilon(t, u) - k(t, N_t, u) \sum_{i \in S} \gamma_i \hat{\varphi}_t^u(i) \mathrm{d}t - \sum_{i \in S} \phi_i^u \hat{\varphi}_T(i) \right) \right]
\end{aligned} \tag{5.26}$$

As seen, the intensity control problem is subject to *Markovian Controls* [B81]. More precisely, $J(u)(\forall u \in \mathfrak{U})$ is characterized just with respect to $\mathcal{F}_t^N$-adapted measure $\hat{\varphi}_t$, or equivalently N_t. That means the history of the process ($\omega \in \Omega$) doesn't play a pure role in the equation.

In the following, we restrict ourselves to solving the optimal control problem over a finite horizon in a *Markovian Control* manner. To achieve this aim which by selecting an optimum strategy $\{u_t^* \equiv u_t^*(t,n) : u_t^* \in \mathfrak{U}\}$:

$$\hat{J}(u^*) = \sup_{u \in \mathfrak{U}} \hat{J}(u)$$

the Deterministic Hamilton–Jacobi equations [B81] is employed.

Corollary 1. *Let for the measures* $\hat{\gamma}(t,n,u)$, $\hat{\varphi}_t^u(i) = \hat{\varphi}(t,n,u)(i)$, *and* $k(t,n,u)$ *where are independent of* ω *there exists for each* $n \in N_+$ *a function* $V(t,n)$ *such that*

$$\begin{aligned} \frac{\partial V(t,n)}{\partial t} &+ Sup_{u \in U_t} \{\hat{\gamma}(t,n,u)\,[V(t,n) - V(t,n-1) \\ &\qquad -k(t,n,u)] + \sum_{i \in S} \mu_i \hat{\varphi}(t,n,u)(i) - k_\varepsilon(t,u) \Big\} = 0, \\ V(T,n) &= \inf_{u \in \mathfrak{U}} \phi(T,n,u) \end{aligned} \tag{5.27}$$

where

$$\phi(T,n,u) = \textstyle\sum_{i \in S} \phi_i \hat{\varphi}_T^u(i).$$

Suppose also that there exists for each $n \in N_+$ *a measurable* R_+*−valued function* $u^*(t,n)$ *such that*

$$u^*(t,n) \in \mathfrak{U}, \quad t \in [0,T],$$

and

$$\begin{aligned} u^*(t,n) = argmax_{u \in U_t} &\{\hat{\gamma}(t,n,u)\,[V(t,n) - V(t,n-1) \\ &-k(t,n,u)] + \sum_{i \in S} \mu_i \hat{\varphi}(t,n,u)(i) - k_\varepsilon(t,u) \Big\}, \end{aligned} \tag{5.28}$$

Then u_t^* *defined by*

$$u_t^*(\omega) = u^*(t, N_t(\omega)) \tag{5.29}$$

for $\omega \in F_t^N$ *is an optimal solution.*

In the next section, a numerical example to solve above optimal intensity control is given. With respect to the cost structure (5.24), it is shown how the solution of Hamilton–Jacobi equation (5.27) subject to control process $u \in \mathfrak{U}$ provides an optimal policy not only to control frequency of inspections, but also to optimally derive the flow of revenue over consecutive inspection times. Also, given the optimal control process u^*, an optimal stopping rule to replace the system is obtained.

5.4 Numerical Example

To obtain an optimal control solution for the deteriorating model presented above, it is assumed $q_1(t) = t$ and the inspection cost is linked by the probability measure of the system state as $k(t,n,u) = K - C\hat{\varphi}(t,n,u)(1)$ $(0 < C < K)$. It means with raising the impairment level of the system, the inspection cost decreases.

Thus, on the basis of theses measures the Hamilton–Jacobi equation [B81] reduces to

$$\begin{aligned}&\frac{\partial V}{\partial t}(t,n) + \sup_{u \in U_t} \{C\bar{\gamma}(\hat{\varphi}(t,n,u)(1))^2 \\ &+ \hat{\varphi}(t,n,u)(1)\left[(\mu_1 - \mu_2) + C\gamma_2 + (V(t,n) - V(t,n-1) - K)\,\bar{\gamma}\right] \\ &\quad +\lambda_2\,(V(t,n) - V(t,n-1) - K) - k_\varepsilon(t,u) + \mu_2\} = 0,\end{aligned} \tag{5.30}$$

and the optimal control process $u^*(t,n)$ is given by

$$\begin{aligned}&u^*(t,n) = argmax_{u \in \mathcal{U}_t} \\ &\Big\{C\bar{\gamma}\,(\hat{\varphi}(t,n,u)(1))^2 + \hat{\varphi}(t,n,u)(1)\left[((\mu_1 - \mu_2) + C\gamma_2)\,(V(t,n) - V(t,n-1)\right. \\ &\quad \left.-K)\,\bar{\gamma}\right] \quad +\lambda_2\,(V(t,n) - V(t,n-1) - K) - k_\varepsilon(t,u) + \mu_2\}\end{aligned} \tag{5.31}$$

To solve the ordinary differential equation (5.30) subject to optimal expected revenue $V(t,n)$ and corresponding optimal control process $u^*_{(t,n)}$, $t \in [T_n, 15]$ and $(n \geq 0)$, let $K = 2$, $C = 1$, $\mu_1 = 2$, $\mu_2 = 1$, $\gamma_2 = 2$, $\gamma_1 = 1$, $\alpha = 2$, $\beta = \sqrt{2}$, $\phi_1 = 1$, $\phi_2 = 2$ and $\varepsilon = 0.15$. By taking advantages of (5.6) and the Euler method with step size $h = 0.1$ an evolution of the optimal expected revenue $V(t,n)$ for $0 \leq n \leq 12$ is derived. As illustrated in Fig. 5.3, the optimal expected revenue $V(t,n)$ for $t \in [0, 8.6.3]$ is non-decreasing in the number of inspections, at 11th inspection the revenue reaches to the maximum value, then for $t \in (8.63, 15]$ it follows a decreasing trend.

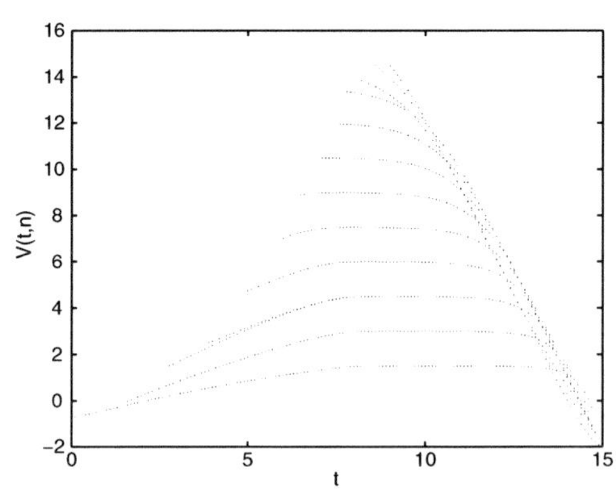

Figure 5.3. An evolution of the optimal expected revenue $V(t,n), t \in [0,15], (0 \leq n \leq 12)$

Also, corresponding to nth inspection event ($n = 0, 1, 2, ...12$), the following table illustrates a sequence of the expected revenue V_n^*, expected inspection times $\hat{\mu}_n$, mean time between inspections (MTBI) $\Delta\hat{\mu}_n$, and the optimal control process $u^*_{(t,n)}$ where

$$V_n^* = \max_{\hat{\mu}_n \leq t \leq 15} V(t, n), \quad 0 \leq n \leq 12$$

As seen, the optimal control sequence $u^*_{(t,n)}$ for $n = 0, 1, ..., 4$ take the boundary values $\{0.1, 1\}$ of the constraint set $[0.1, 1]$, and from fifth inspection on, i.e., $(5 \leq n \leq 12)$, $u^*_{(t,n)}$ chooses just the upper endpoint 1. The second row of the table describes an evolution of the expected revenue V_n^* which is concave in the number of inspections n. As shown, V_n^* achieves to its maximum value at 11th inspection, that is, $V^* = \max_n V_n^* = 14.8487$ $\forall n \geq 0$, then for $(n \geq 12)$ follows an decreasing trend. So, the sequence of V_n^* not only gives us a solution to the optimal control process u_t^* and optimal inspection problem which is the optimum number of inspections, but also provides a solution to the optimal replacement problem. More precisely, the optimum maintenance policy includes an optimal stopping time to replace the system at 11th inspection event which is $T^* = \hat{\mu}_{11} = 8.63$ and a sequence of the optimal inspection times $\hat{\mu}_n$ for $n = 1, 2, ..., 11$ driven by the optimal control process u_t^*, $t \in [0, 8, 63]$ which is bang-bang in the sense that it takes the boundary values $\{0.1, 1\}$ that is

$$u_t^* = 0.1I(0 \leq t \leq 7.9) + I(7.9 < t \leq 8.63).$$

Also, a sequence of the mean time between inspections $\Delta\hat{\mu}_n = \hat{\mu}_n - \hat{\mu}_{n-1}$ $(0 \leq n \leq 12)$ $(\hat{\mu}_0 = 0)$ is shown (see Fig. 5.5). As expected, mean time between inspections decreases as the number of inspections increases. This results from control process dependent inspection intensity $\hat{\gamma}(n, t, u^*)$ being increasing in the number of inspections n and time t (see Fig. 5.4).

n	V_n^*	$\hat{\mu}_n$	$\Delta\hat{\mu}_n$	$u^*_{(t,n)}$
0	1.5	0	–	$0.1I(0 \leq t < 8.8) + I(8.8 \leq t < 15)$
1	3	1.36	1.36	$0.1I(1.36 \leq t < 8.4) + I(8.4 \leq t < 15)$
2	4.4998	2.63	1.269	$0.1I(2.63 \leq t < 8.4) + I(8.4 \leq t < 15)$
3	5.9992	3.79	1.165	$0.1I(3.79 \leq t < 8.3) + I(8.3 \leq t < 15)$
4	7.4974	4.87	1.081	$0.1I(4.87 \leq t < 7.9) + I(7.9 \leq t < 15)$
5	8.9937	5.88	1.017	$I(5.88 \leq t < 15)$
6	10.4878	6.36	0.48	$I(6.36 \leq t < 15)$
7	10.9757	6.83	0.452	$I(6.83 \leq t < 15)$
8	11.9662	7.28	0.451	$I(7.28 \leq t < 15)$
9	13.3437	7.73	0.451	$I(7.73 \leq t < 15)$
10	13.8464	8.18	0.451	$I(8.18 \leq t < 15)$
11	**14.4887**	**8.63**	**0.451**	$I(\mathbf{8.63} \leq t < 15)$
12	14.2907	9.087	0.45	$I(9.08 \leq t < 15)$

To have a realization of the prediction of the system failure, let $\psi(x) = x$ and the baseline function be distributed Weibull with intensity function

$$\lambda_0(t) = \frac{\alpha t^{\alpha-1}}{\beta^\alpha} \quad t \geq 0$$

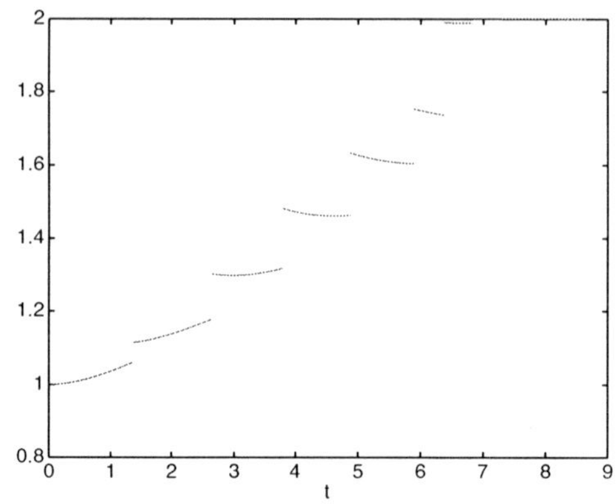

Figure 5.4. An evolution of the inspection intensity $\hat{\gamma}(n, t, u^*)$ given the optimal control process u_t^*

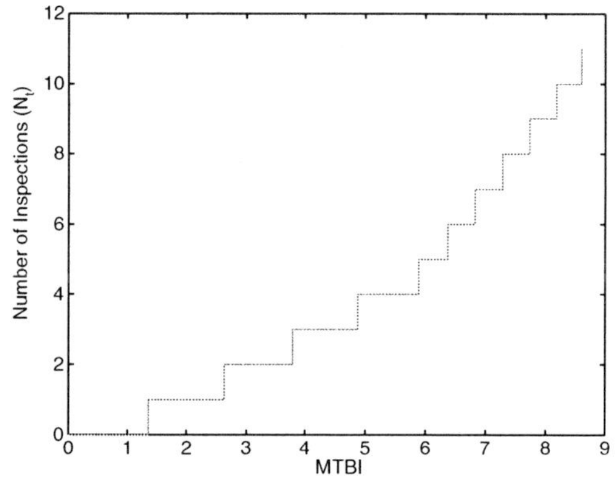

Figure 5.5. An evolution of the mean time between inspections (MTBI) given the optimal control process u_t^*

By applying (5.10) an evolution of the conditional survival function of the system (see Figs. 5.5–5.7) and a decreasing sequence of corresponding conditional mean time to failure (CMTTF) (see the following table) are illustrated. For example, at initial time $t = 0$, it is predicted after $t = 1.22$ unit of time the system fails. In addition, given that the system is in operating state at inspection time $\hat{\mu}_1 = 1.36$, it is expected after 0.491 unit of time the system is in failure state.

n	0	1	2	3	4	5	6	7	8	9	10
CMTTF	1.22	0.491	0.262	0.166	0.118	0.090	0.073	0.067	0.063	0.059	0.055

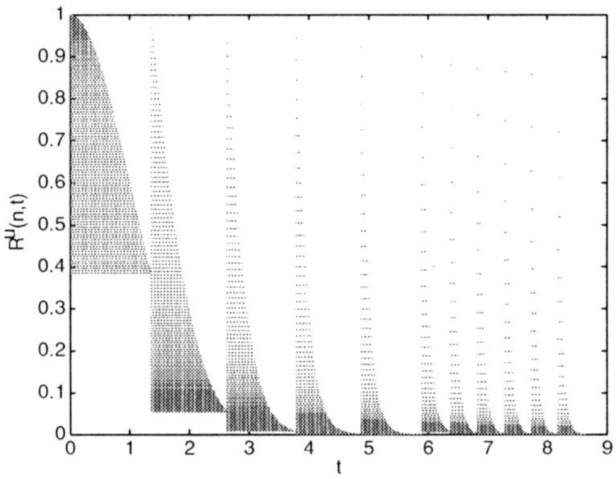

Figure 5.6. An evolution of conditional survival function given the optimal control process u_t^*

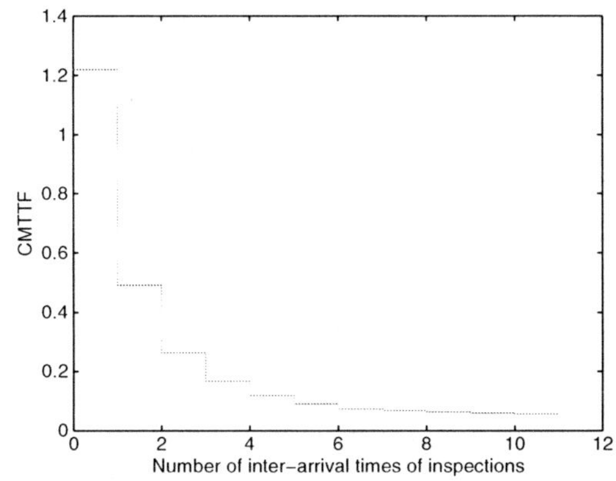

Figure 5.7. An evolution of conditional mean time to failure (CMTTF) given the optimal control process u_t^*

Finally, Fig. 5.8 beautifully give us a rule to optimal replacement of the system based on the failure intensity $\hat{\lambda}(n,t,u^*)$ that is

$$T^* = \inf\left\{t \geq 0 : \hat{\lambda}(n,t,u^*) \geq c, \hat{\mu}_n \leq t < \hat{\mu}_{n+1}\right\}, (n \geq 0), \tag{5.32}$$

where c denotes the optimum threshold deterioration level at which the system is replaced. Subject to the optimal replacement time $T^* = 8.63$, the threshold level is given by

$$c = \hat{\lambda}(10, 8.63, u^*) = 17.26.$$

5.4.1 Perspectives

In this paper, we presented an optimal-control-based framework to the adjustment of the damage process and the determination of optimal inspection time sequence for systems which are subject to repair and inspection. The model rests on two common assumptions which are inspections do not impact on the failure characteristics and the

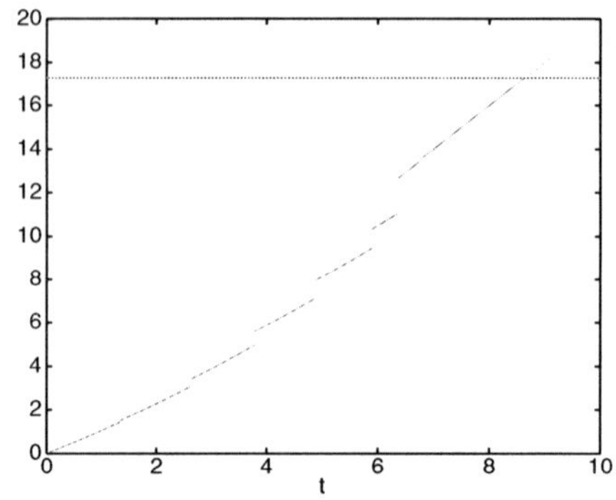

Figure 5.8. An evolution of failure intensity $\hat{\lambda}(n, t, u^*)$ given the optimal control process u_t^*

process is adapted to partial information including evolution of inspection events. To optimize the damage process based maintenance process, the intensity control model was employed as a key tool not only to adjust the $\mathcal{F}^N$-adapted state of the process over inter-arrival times of inspections (repair), but also to optimally control the frequency of inspections.

To numerically solve the maintenance optimization problem an example with realistic results on scheduling optimal inspection times was given. In addition, to get an insight into prediction of system failure an evolution of conditional survival function and conditional mean time to failure of the system were provided. Also, an optimal stopping rule to replace the system was obtained.

The maintenance optimization model presented here has potential to consider the optimal inspection problem of a variety of systems which are subject to repair and inspections. In such a case, the problem of controlling the intensity of point process N_t is generalized to multivariate point process case, i.e., $\big(N_1(t), N_2(t), ..., N_k(t)\big)$ which $N_i(t)$ $(i = 1, 2, ..., k)$ refers to the number of inspections of system type i. Also, it is anticipated the optimal control model discussed here in the same approach as Davis model [D93] can be extended to the case that the repair and maintenance action at intervention times is not restricted to minimal repairs. In the generalized case, the flow of the process which will consist of random jumps resulting from repair and maintenance action, and continuous motion between consecutive jumps is controlled in such away that the maximum expected value including continuous revenue, jump and terminal cost is derived. Besides, to reflect the degree of repairs, virtual age (VA) processes (see [LBJ04,K89]) as age reduction factors can be considered as a key tool. In this way, VA process is beautifully reflected in the $\mathcal{F}^N$-adapted damage process $\hat{\varphi}_t$ through a change of the time origin.

References

[AJ98] Aven, T., Jensen, U.: *Stochastic Models In Reliability*, Springer, Berlin (1998)

[B81] Bremaud, P.: *Point Processes and Queues: Martingale Dynamics*, Springer, Berlin (1981)

[BHP63] Barlow, R.E., Hunter L.C., Proschan, F.: Optimum checking procedures. J Soc Ind Appl Math, **11** (1963) pp 1078–1095

[CA99] Chelbi, A., Ait-Kadi, D.: A nearly optimal inspection policy. Opl Res Q, **23** (1999) pp 373–379

[DJ03] Dassios A., Jang J.: Pricing of catastrophic reinsurance and derivatives using the Cox process with shot intensity noise. Finance Stochast, **7(1)** (2003) pp 73–95

[D93] Davis, M.H.A.: *Markov Models and Optimization*, Chapman and Hall, London (1993)

[JM92] Jardine, A., Makis,V.: Optimal replacement in the proportional hazard models. INFOR, **20(1)** (1992) pp 172–183

[JJ05] Jiang, R., Jardine, A.: Two optimization models of the optimum inspection problem. J Oper Res Soc, **56** (2005) pp 1176–1183

[JG93] Jensen, U., Hsu, J.G.: Optimal stopping by means of point process observations with application in reliability, Manag Oper Res, **18** (1993) pp 645–657

[KO84] Kaio, N., Osaki, S.: Some remarks on optimum inspection policies. IEEE Trans Reliab, **R-33** (1984) pp 273–279

[K74] Keller, J.B.: Optimum checking schedules for systems subject to random failure. Mngt Sci, **21** (1974) pp 256–260

[K89] Kijima, M.: Some results for repairable systems with general repairs. J Appl Probab, **26** (1989) pp 89–102

[LBJ04] Lugtigheid, D. , Banjevic, D., Jardine, A.K.S.: Modelling repairable system reliability with explanatory variables and repair and maintenance actions. IMA J Manag Math, **15** (2004) pp 89–110

[MS74] Munford, A.G., Shahani, A.K.: An optimal inspection strategy for randomly failing equipment. Reliab Eng Syst Saf, **63** (1974) pp 127–131

6

Longitudinal Latent Markov Processes Observable Through an Invariant Rasch Model

Moussedek Bousseboua[1] and Mounir Mesbah[2,*]

[1] University Mentouri Constantine, Algérie
[2] University Pierre et Marie Curie, Paris 6, France, mounir.mesbah@upmc.fr

Abstract: In this chapter, we introduce a class of longitudinal latent processes where, at any time, the latent variable is assumed to be indirectly observable, through a set of categorical binary variables (items). We assume that the measurement model relating the observed items and the latent variable, at each time, is a Rasch model. Besides the description of these models, the objective aimed in this work, is the estimation of the parameters of the model by maximum likelihood method via an EM algorithm. We consider more deeply two classes of distribution for the longitudinal latent process: (1) the General Latent Markov process and a special case, (2) the latent autoregressive process of order one.

Keywords and phrases: Longitudinal data, Rasch model, Markov process, AR process, EM algorithm, Marginal likelihood, Quality of life, Unobserved process, Latent variable models, Measurement model

6.1 Introduction

In this paper, we present the class of latent processes in general and latent Markov processes and latent autoregressive processes in particular. The latter can be regarded like a particular case of the class of latent Markov processes. Latent processes has been considered extensively in the past [WIG73, VER86, BAR06, LAN94, POL86] in various fields of application and for various purposes. For instance, in psychology, measure of the latent ability of individual in educational context is crucial. In Health Related Quality of Life Research, one use questionnaires to assess various latent trait, in retrospective as well in prospective (i.e., longitudinal) studies [MES09]. One of the basic hypotheses of the latent models is local independence. This hypothesis postulates that the observed variables used to measure a specific latent variable of interest, are independent conditionally that latent variable. This condition is generally assumed in all latent variable models. It is also included in the Rasch model [ADM06, FIS95], which is the most popular Item Response Theory (IRT) model. This model possesses some interesting properties and permits an explicit expression of the likelihood function of

V.V. Rykov et al. (eds.), *Mathematical and Statistical Models and Methods in Reliability: Applications to Medicine, Finance, and Quality Control*, Statistics for Industry and Technology, DOI 10.1007/978-0-8176-4971-5_6,

observations. This paper includes a summary description of latent process models in general and of latent Rasch model in particular. Then, one focuses our work to the case of Rasch Latent Markov processes and Rasch Latent autoregressive models under the hypothesis of the local independence. For each model, we derive the likelihood function and solve the problem of parameter estimation by the method of maximum likelihood using an EM algorithm. Feddag and Mesbah (2005) [FED05] have considered similar situation in a more limited context with only few times of observation (no more than three). They used GEE methods to estimate item as well as latent distribution parameters. In Sect. 6.2, the general form of our longitudinal process is presented, including its Rasch measurement part, and the probability law of the observed responses is given. Section 6.3 is devoted to the presentation of an application of our method to two important classes of latent processes: a General First Order Markov Latent process and a First Order Autoregressive Latent process (AR(1) Latent process). Latent parameters Estimation is considered in Sect. 6.4, while its application to our specific Rasch Latent process is detailed in Sect. 6.5.

6.2 Description of the Latent Process

The model consists of a finite trajectory of a multivariate process:

$$(X_{i,k}(t), \Theta_i(t)) : 1 \leq k \leq q, 1 \leq i \leq n, 1 \leq t \leq T$$

where $(X_{i,k}(t))$ is the longitudinal process of observations, assumed with a finite support and, for all i, $(\Theta_i(t))$ is an unobservable (latent) process. The variable $X_{i,k}(t)$ represents the response at the instant t of the person i to the item k.

In most applications, the q variables $X_{i,1}(t), X_{i,2}(t), ..., X_{i,q}(t)$ are, for every occasion t, the responses of the individual i to a dichotomous questionnaire including $q-$items (questions). The same questionnaire is submitted on different opportunities to the same individuals. A characteristic of these models is that, whatever the value of k, every response variable $X_{i,k}(t)\}$ depends only on the corresponding latent individual variable $\Theta_i(t)$.

First, let's determine the law of the latent process $(\Theta_i(t))$ in view of a finite trajectory of the observation process $(X_{i,k}(t))$. In our context, the model can be represented at all times t by

$$\begin{array}{ccc} \underbrace{\{X_{1,1}(t) \quad \cdots \quad X_{1,q}(t)\}} & \cdots & \underbrace{\{X_{n,1}(t) \quad \cdots \quad X_{n,q}(t)\}} \\ \uparrow & \cdots & \uparrow \\ \Theta_1(t) & \cdots & \Theta_n(t) \end{array}$$

The process of observations $\{X_{i,k}(t)\}\}$ have values in $\{0,1\}$, where value 1 corresponds to a correct (or a positive) answer of the i individual to the k item. The word "correct" is used in the context of education, while the word "positive" can be used in more general context like Quality of Life, or any other Psychological, Sociological context. The latent variable $\Theta_i(t)$ depends only on the individual i. It can be interpreted as a measure of his ability, in the context of education, and more generally as a measure of his position in a scalar axis corresponding to the latent unidimensional trait measured by the questionnaire.

We note $\underline{\mathbf{X}}_i(t) = (X_{i,1}(t), ..., X_{i,q}(t))'$ the response vector of the individual i at the instant t and by $\pi(x_{i,k}/\theta_i(t))$ the conditional probability $\mathbf{P}(X_{i,k}(t) = x_{i,k}/\Theta_i(t) = \theta_i(t))$, with $i = \overline{1,n}$, $k = \overline{1,q}$ and $t = \overline{1,T}$. The probability distribution of the response vector of the individual i, in view of the observation $\underline{\mathbf{x}}_i = (\underline{\mathbf{x}}_i(1), ..., \underline{\mathbf{x}}_i(T))$ of $\underline{\mathbf{X}}_i = (\underline{\mathbf{X}}_i(1), ..., \underline{\mathbf{X}}_i(T))$ during the whole period $\{1, ..., T\}$, can be written:

$$p(\underline{\mathbf{x}}_i) = \int \cdots \int p((\underline{\mathbf{x}}_i/(\underline{\theta}_i) \times g_i(\underline{\theta}_i)\mathrm{d}\theta_i(1) \cdots \mathrm{d}\theta_i(T), \tag{6.1}$$

where g_i is the probability density of the latent vector $(\Theta_i(1), ..., \Theta_i(T))'$, $\underline{\theta}_i = (\theta_i(1), ..., \theta_i(T))'$ and $\underline{\mathbf{x}}_i(t) = (x_{i,1}(t), ..., x_{i,q}(t))'$ an observation of vector $\underline{\mathbf{X}}_i(t)$.

In our current work, we assume that the probability distribution of the latent variable belongs to a known parametric family of distributions. One of our main goal is to estimate its unknown parameters using the observations $x_{i,k}(t)$ (item responses).

Assuming local independence of the items, and Rasch measurement model to relate latent parameter to observed responses will allow us a simplification of the conditional probabilities

$$p(\underline{\mathbf{x}}_i(t)/\theta_i(t)) = P(\underline{\mathbf{X}}_i(t) = \underline{\mathbf{x}}_i(t)/\Theta_i(t) = \theta_i(t)).$$

Local independence of the items means that the response variables $X_{i,1}(t), ...,$ $X_{i,q}(t)$ of the person i, are, conditionally to their corresponding latent variable $\Theta_i(t)$, independent. It is one of the underlying Rasch model assumptions. It is also an assumption included in all Item Response Theory family models.

This assumption will help us to simplify the writing of $p(\underline{\mathbf{x}}_i(t)/\theta_i(t))$ by giving us a factorization of the likelihood which will make easier the estimation process of the parameters.

6.2.1 The Rasch Model

The most popular measurement model, i.e., a model defining *the distribution of observed items conditional to the latent unidimensional person variable* is the Rash model (Fisher and Molenaar (1995)). This model is extensively used in various psychometric fields such educational research, and more recently, the analysis of health related Quality of Life [7]. Let the following assumptions:

(h_1) The conditional law of $X_i(t)$ with respect to latent variables $\Theta_i(t) : t = 1..., T$, depends only on the corresponding latent variable:

$$P(\underline{\mathbf{X}}_i(t) = \underline{\mathbf{x}}_i(t)/\underline{\mathbf{\Theta}}_i = \underline{\theta}_i) = P(\underline{\mathbf{X}}_i(t) = \underline{\mathbf{x}}_i(t)/\Theta_i(t) = \theta_i(t)). \tag{6.2}$$

We note $p(\underline{\mathbf{x}}_i(t)/\theta_i(t))$ this conditional probability. This assumption means that the response variable at the instant t depends only on the latent variable $\Theta(t)$.

(h_2) The random vectors $\underline{\mathbf{X}}_i(1), \underline{\mathbf{X}}_i(2), \ldots, \underline{\mathbf{X}}_i(T)$ are conditionally independents to the latent vector $\{\Theta_i(1) = \theta_i(1), ..., \Theta_i(T) = \theta_i(T)\}$:

$$p(\underline{\mathbf{x}}_i(1), ..., \underline{\mathbf{x}}_i(T)/\theta_i(1), ..., \theta_i(T)) = \prod_{t=1}^{T} p(\underline{\mathbf{x}}_i(t)/\theta_i(t)), \tag{6.3}$$

where $p(\underline{\mathbf{x}}_i(1), ..., \underline{\mathbf{x}}_i(T)/\theta_i(1), ..., \theta_i(T))$ is the joint law of $\underline{\mathbf{X}}_i(1), \underline{\mathbf{X}}_i(2), ..., \underline{\mathbf{X}}_i(T)$ conditionally to $\{\Theta_i(1) = \theta_i(1), ..., \Theta_i(T) = \theta_i(T)\}$. This is a longitudinal conditional independence assumption of observed items relatively to the latent process.

(h_3) The response variables $X_1(t), ..., X_q(t)$ are conditionally independent relatively to the latent variable $\Theta_i(t)$:

$$p(\underline{\mathbf{x}}_i(t)/\theta_i(t)) = \prod_{k=1}^{q} \pi(x_{i,k}(t)/\theta_i(t)), \tag{6.4}$$

where

$$\pi(x_{i,k}(t)/\theta_i(t)) = P(X_{i,k}(t) = x_{i,k}(t)/\Theta_i(t) = \theta_i(t)),$$

and

$$\underline{\mathbf{x}}_i(t) = (x_{i,1}(t), ..., x_{i,q}(t))' \in \{0,1\}^q.$$

This hypothesis corresponds to a questionnaire including q items conditionally independent relatively to the corresponding latent variable. This is the local independence assumption.

(h_4) The conditional probabilities $\pi(x_{i,k}(t)/\theta_i(t))$ are given by

$$\pi(x_{i,k}(t)/\theta_i(t)) = \frac{\exp(x_{i,k}(t)(\theta_i(t) - \beta_k))}{1 + \exp((\theta_i(t) - \beta_k))}, \tag{6.5}$$

where $\theta_i(t)$ is the latent feature or the ability of the i individual (to measure) and β_k a real parameter called difficulty parameter tied to the k item. The introduction of the difficulty parameter β_k brings us to note this conditional law by $\pi(x_{i,k}(t)/\theta_i(t), \beta_k)$. Formula (6.5) is known as the Rasch model.

On the other hand, individuals are, as usual, considered as independently sampled, in other terms, the latent processes $(\Theta_i(t) : i = \overline{1,n})$ are independents. Also, it is sufficient in this section to describe the model and the likelihood function for only one individual. To this end, and by convenience, we omit the relative indication to individuals and we note by $\{\underline{\mathbf{X}}_t : 1 \leq t \leq T\}$ a finite trajectory of the observation process and where $\underline{\mathbf{X}}_t = (X_1(t), ..., X_q(t))'$ is the vector of response variables at time t relatively to only one individual and $(\Theta(t) : 1 \leq t \leq T)$ the corresponding latent process.

6.2.2 Probability Law of the Observations

Let $\underline{\mathbf{x}}(1), ..., \underline{\mathbf{x}}(T)$ be a trajectory of observation vectors $\underline{\mathbf{X}}(1), ..., \underline{\mathbf{X}}(T)$ relative to one individual only, during the period $\{1, ..., T\}$. We have the following result:

Proposition 1. *Under the hypotheses h_1, h_2, and h_3 above, the conditional law $p(\underline{\mathbf{x}}(1), ..., \underline{\mathbf{x}}(T))$ of the observations relative to one individual only, during the period $\{1, ..., T\}$, can be written*

$$p(\underline{\mathbf{x}}(1),...,\underline{\mathbf{x}}(T)) = \int ... \int_{\mathbf{R}^T} \prod_{t=1}^{T} (1-\alpha_1(\theta(t)))^q (\frac{\alpha_1(\theta(t))}{1-\alpha_1(\theta(t))})^{r(t)} g(\underline{\theta}) \mathrm{d}(\underline{\theta}), \quad (6.6)$$

where $\alpha_1(\theta(t)) = \pi(1/\theta(t))$, $r(t) = \sum_{k=1}^{q} x_k(t)$ *the score (number of correct responses) of individual at the* t^{-th} *opportunity and* g *the density of latent vector* $(\Theta(1), ..., \Theta(T))'$.

Proof. Looking at the law of $(\underline{\mathbf{X}}(1), ..., \underline{\mathbf{X}}(T))$ as the marginal law of the joint law of $(\underline{\mathbf{X}}(1), ..., \underline{\mathbf{X}}(T))$ and the $(\Theta(1), ..., \Theta(T))$, and taking into account the hypotheses h_1, h_2, we can write

$$p(\underline{\mathbf{x}}(1),...,\underline{\mathbf{x}}(T)) = \int ... \int_{\mathbf{R}^T} \prod_{t=1}^{T} p(\underline{\mathbf{x}}(t)/(\Theta(t) = \theta(t))) g(\underline{\theta}) \mathrm{d}(\underline{\theta}). \quad (6.7)$$

Then according to the hypothesis h_3 of the local independence on the one hand, and as $\pi(x_k(t)/\theta(t)) = (1-\alpha_1(\theta(t))).[\frac{\alpha_1(\theta(t))}{1-\alpha_1(\theta(t))}]^{x_k(t)}$ on the other hand, we deduce the result (6.6).

The probability $p(\underline{\mathbf{x}}(1), ..., \underline{\mathbf{x}}(T))$ depends on the transition probabilities $\alpha(1/\theta(t))$ which constitute as many of parameters to estimate in order to identify the model. Moreover, if we suppose this model satisfies the hypothesis h_4 (Rasch model), then the number of parameters can be reduced substantially and the probability of the observation vector relative to one individual only, becomes relatively simple.

Corollary 1. *If the model* $((\underline{\mathbf{X}}(t), \Theta(t)) : 1 \leq t \leq T)$ *is a latent Rasch model, then the law of probability* $p(\underline{\mathbf{x}}(1), ..., \underline{\mathbf{x}}(T))$ *of observations relative to one individual only during the period* $\{1, ..., T\}$, *can be written*

$$p(\underline{\mathbf{x}}(1),...,\underline{\mathbf{x}}(T)) = \int ... \int \frac{\exp(\sum_{t=1}^{T} r(t)\theta(t) - \sum_{k=1}^{q} \beta_k r_k)}{\prod_{t=1}^{T} \prod_{k=1}^{q} [1+\exp((\theta(t)-\beta_k))]} g(\underline{\theta}) \mathrm{d}(\underline{\theta}), \quad (6.8)$$

where $r(t)$ *is the score at the* t^{-th} *opportunity and* $r_k = \sum_{t=1}^{T} x_k(t)$ *the number of correct responses to the* k *item over the* $T-$*opportunities.*

Proof. It is sufficient to replace in (6.7), $\pi(x_k(t)/\theta(t))$ by

$$\frac{\exp(x_k(t)(\theta(t) - \beta_k))}{1+\exp((\theta(i) - \beta_k))}.$$

6.3 Application to Markov Latent Processes

In this section, we consider, as a direct application of the previous sections, two classes of latent processes: a General First Order Markov Latent process and a First Order Autoregressive Latent process (AR(1) Latent process).

6.3.1 First Order Markov Latent Process

This model consider the latent process $(\Theta(t) : 1 \leq t \leq T)$ as being a Markov chain of one order. We suppose the chain $(\Theta(t) : 1 \leq t \leq T)$ has a real support, and invariant Gaussian centered law with variance σ^2 and has a kernel of transition $\mathcal{N}$ Gaussian of variance σ^2.

Proposition 2. *If the process* $\{\Theta(t) : 1 \leq t \leq T\}$ *has a real support, invariant Gaussian centered law with variance* σ^2*, with a kernel of transition* $\mathcal{N}$ *Gaussian of variance* σ^2*, then the vector* $(\Theta(1), ..., \Theta(T))$ *is Gaussian and has a probability density* g *given by*

$$g(\theta(1), ..., \theta(T)) = \frac{1}{\sigma^T \sqrt{(2\pi)^T}} \exp\{-\frac{1}{2\sigma^2}[\theta(1)^2 + \sum_{t=2}^{T}(\theta(t) - \theta(t-1))^2]\}. \qquad (6.9)$$

Proof. By conditionality, the joint law of variables $\Theta(1), ..., \Theta(T)$ is deduced easily:

$$g(\theta(1), ..., \theta(T)) = \lambda(\theta(1)) \prod_{t=2}^{T} \mathcal{N}(\theta(t-1), \theta(t)), \qquad (6.10)$$

where λ is the invariant law of the process $\{\Theta(t) : 1 \leq t \leq T\}$. And from here we deduce the expression of g.

Remark. If we suppose the chain having a finite number of states $\theta_1, ..., \theta_m$, then the matrix of transition probabilities of $\mathbf{P}$ has coefficients $\pi(\theta_u(t)/\theta_v(t-1)) = P(\Theta(t) = \theta_u(t)/\Theta(t-1) = \theta_v(t-1)), 1 \leq u, v \leq m$. And so besides, we suppose that the chain $(\Theta(t) : 1 \leq t \leq T)$ is homogeneous, irreducible and possess an invariant law, then the matrix of transition probabilities is independent of t, and the probability law of the vector $(\Theta(t) : 1 \leq t \leq T)$ can be written

$$P(\Theta(1) = \theta_{i_1}, ..., \Theta(T) = \theta_{i_T}) = \lambda(\theta_{i_1}) \prod_{t=2}^{T} \pi(\theta_{i_t}/\theta_{i_{t-1}}), \qquad (6.11)$$

where $i_t \in \{1, ..., m\}$.

In what follows, we suppose that the chain $(\Theta(t))$ has a real support, and is homogeneous and irreducible. The following result specifies the law of observations relative to one individual only.

6.3.2 Rasch First Order Markov Latent Process

We have the following result that is deduced directly from the proposition 1, by replacing in this relation the density g by its expression.

Corollary 2. *Under the hypotheses* h_1, h_2 *and* h_3 *above, the law of* $\underline{\mathbf{X}} = (\underline{\mathbf{X}}(1), ..., \underline{\mathbf{X}}(T))$ *can be written in view of a trajectory* $\underline{\mathbf{x}} = (\underline{\mathbf{x}}(1), ..., \underline{\mathbf{x}}(T))$:

$$p(\underline{\mathbf{x}}) = \frac{1}{\sigma^T \sqrt{(2\pi)^T}} \int ... \int_{\mathbf{R}^T} \prod_{t=1}^{T} [1 - \alpha_0(\theta(t))]^q [\frac{\alpha_0(\theta(t))}{1 - \alpha_0(\theta(t))}]^{r(t)} h(\underline{\theta}) \mathrm{d}(\underline{\theta}), \qquad (6.12)$$

where $\alpha_0(\theta(t)) = \pi(1/\theta_t), h((\underline{\theta}) = h(\theta(1), ..., \theta(T)) = \exp\{-\frac{1}{2\sigma^2}[\theta(1)^2 + \sum_{t=2}^{T}(\theta(t) - \theta(t-1))^2]$ *and* $r(t) = \sum_{k=1}^{q} x_k(t)$ *the score obtained at the* t^{-th} *occasion.*

The probability $p(\underline{\mathbf{x}}(1), ..., \underline{\mathbf{x}}(T))$ depends on transition probabilities $\pi(x_k(t)/\theta(t))$, also, if we suppose in addition that this model is a Rasch model, then we deduce from corollary 1 and the corollary 2 above, the following result:

Corollary 3. *Under the hypothesis of latent Rasch model, the law* $p(\underline{\mathbf{x}}_1, ..., \underline{\mathbf{x}}_T)$ *of the observations relative to one individual only during the period* $\{1, ..., T\}$, *becomes*

$$p(\underline{\mathbf{x}}(1), ..., \underline{\mathbf{x}}(T)) = \frac{1}{\sigma^T \sqrt{(2\pi)^T}} \int ... \int \phi(\underline{\mathbf{x}}, \underline{\theta}, \underline{\beta}) h(\underline{\theta}) \mathrm{d}(\underline{\theta}), \tag{6.13}$$

where $\phi(\underline{\mathbf{x}}, \underline{\theta}, \underline{\beta}) = \frac{\exp(\sum_{t=1}^{T} \theta_t . r(t) - \sum_{k=1}^{q} \beta_k . r_k)}{\prod_{t=1}^{T} \prod_{k=1}^{q} [1+\exp(\theta_{i_t} - \beta_k)]}$, $r_k = \sum_{t=1}^{T} x_k(t)$, $\underline{\theta} = (\theta_1 \ldots, \theta_T)$ *and* $\underline{\beta} = (\beta_1 \ldots, \beta_q)$.

Remark. If the chain is of finite support $\{\theta_1, ..., \theta_m\}$, then the law (6.13) becomes

$$p(\underline{\mathbf{x}}) = \sum_{i_1=1}^{m} ... \sum_{i_T=1}^{m} \prod_{t=1}^{T} \lambda(\theta_{i_1}).\pi(\theta_{i_t}/\theta_{i_{t-1}}).[1 - \alpha_0(\theta_{i_t})]^q.[\frac{\alpha_0(\theta_{i_t})}{1 - \alpha_0(\theta_{i_t})}]^{r(t)}, \tag{6.14}$$

where $\alpha_0(\theta_{i_t}) = \pi(1/\theta_{i_t})$ and $r(t) = \sum_{k=1}^{q} x_k(t)$ the score obtained at the tth occasion. And in a Rasch model this probability is written

$$p(\underline{\mathbf{x}}) = \sum_{i_1=1}^{m} ... \sum_{i_T=1}^{m} \prod_{t=1}^{T} \lambda(\theta_{i_1}).\pi(\theta_{i_t}/\theta_{i_{t-1}}) \frac{\exp\{\sum_{t=1}^{T} \theta_t . r(t) - \sum_{k=1}^{q} \beta_k . r_k\}}{\prod_{t=1}^{T} \prod_{k=1}^{q} [1 + \exp(\theta_{i_t} - \beta_k)]}. \tag{6.15}$$

6.3.3 $\mathcal{AR}(1)$ Latent Process

We suppose that the latent process $(\Theta(t) : 1 \le t \le T)$ is an autoregressive process of one order

$$\Theta(t) = \rho\Theta(t-1) + \varepsilon(t), \tag{6.16}$$

where $(\varepsilon(t))$ is a white Gaussian noise of variance σ^2 and ρ a real constant. In order to ensure the stationarity to the second order of the latent process $(\Theta(t))$, it is sufficient to consider the parameter ρ of modulus strictly lower than 1. On the other hand, it is easy to verify that this process is Gaussian and defines a Markov chain of one order. Concerning this last point, we can express it

Proposition 3. *If the latent process* $(\Theta(t))$ *is autoregressive of one order, then this process is a Markov process of the first order. Besides, if* $|\rho| < 1$, *this process is stationary at the second order with initial centered normal law with variance* $\frac{\sigma^2}{1-\rho^2}$ *and the conditional law of* $\Theta(t)$ *knowing* $\Theta(t-1) = \theta_0$ *is Gaussian with mean* $\rho\theta_0$ *and variance* σ^2.

The join law of variables $(\Theta(1), ..., \Theta(T))$ is deduced easily by conditioning

$$g(\theta(1), ..., \theta(T)) = f_0(\theta(1)) \prod_{t=2}^{T} f^{\theta(t-1)}(\theta(t)), \tag{6.17}$$

which gives

$$g(\underline{\theta}) = \frac{\sqrt{1-\rho^2}}{\sqrt{(2\pi)^T}\sigma^T} \exp\{-\frac{1}{2\sigma^2}[(1-\rho^2)\theta^2(1) + \sum_{t=2}^{T}(\theta(t) - \rho\theta(t-1))^2]\}, \tag{6.18}$$

where f_0 is the density of the initial law of the chain $(\Theta(t))$ and $f^{\theta(t-1)}(\theta(t))$ the conditional density of $\Theta(t)$ knowing $\Theta(t-1)$. We notice again that the Markovian character of the latent process ensures that $\Theta(t)$ is conditionally independent of $\Theta(t-2), ..., \Theta(1)$ knowing $\Theta(t-1)$.

6.3.4 Rasch $\mathcal{AR}(1)$ Latent Process

As for the Markovian case, we deduce here once again

Corollary 4. *Under the hypotheses h_1, h_2 and h_3 above, the law of the $\mathcal{AR}(1)$ model is written for one individual only in view of a trajectory $\underline{\mathbf{x}}(1), ..., \underline{\mathbf{x}}(T)$ of the process $(\underline{\mathbf{X}}(t) : 1 \leq t \leq T)$:*

$$p(\underline{\mathbf{x}}) = \frac{\sqrt{1-\rho^2}}{\sqrt{(2\pi)^T.\sigma^T}} \int_{-\infty}^{+\infty} ... \int_{-\infty}^{+\infty} \psi(\underline{\mathbf{x}}, \underline{\theta}, \rho).H(\underline{\mathbf{x}}, \underline{\theta}) \mathrm{d}\theta(T) \cdots \mathrm{d}\theta(2)\mathrm{d}\theta(1), \quad (6.19)$$

where

$$\psi(\underline{\mathbf{x}}, \underline{\theta}, \rho) = (1 - \alpha_0(\theta(1)))^q e^{-\frac{1}{2\sigma^2}(1-\rho^2)\theta(1)^2} (\frac{\alpha_0(\theta(1))}{1 - \alpha_0(\theta(1))})^{r(1)},$$

$$H(\underline{\mathbf{x}}, \underline{\theta}, \rho) = \prod_{t=2}^{T} (1 - \alpha_0(\theta(t)))^q (\frac{\alpha_0(\theta(t))}{1 - \alpha_0(\theta(t))})^{r(t)}.e^{-\frac{1}{2\sigma^2}[(\theta(t) - \rho\theta(t-1))^2]},$$

$$\alpha_0(\theta(t)) = \pi(1/\theta(t)) \quad r(t) = \sum_{k=1}^{q} x_k(t).$$

Proof. It is sufficient to write

$$p(\underline{\mathbf{x}}) = \int ... \int_{R^T} p(\underline{\mathbf{x}}/\underline{\theta}) g(\underline{\theta}) \mathrm{d}\theta(1) \cdots \mathrm{d}\theta(T), \quad (6.20)$$

and taking into account the hypotheses h_1 and h_3, we can write

$$p(\underline{\mathbf{x}}) = \int ... \int_{\mathbf{R}}^{T} \prod_{t=1}^{T} p(\underline{\mathbf{x}}(t)/\theta(t)).g(\underline{\theta}) \mathrm{d}\theta(1) \cdots \mathrm{d}\theta(T), \quad (6.21)$$

which gives

$$p(\underline{\mathbf{x}}) = = \int ... \int_{R^T} \prod_{t=1}^{T} \prod_{k=1}^{q} \pi(x_k(t)/\theta(t)) g(\theta(1), ..., \theta(T)) \mathrm{d}\theta(1) \cdots \mathrm{d}\theta(T), \quad (6.22)$$

where considering the definition of the g density of the latent vector and that $\pi(x_k(t)/\theta(t)) = [\alpha_0(\theta(t))]^{x_k(t)}.[1 - \alpha_0(\theta(t))]^{1-x_k(t)}$, the expression of the likelihood stated.

The probability $p(\underline{\mathbf{x}}_1, ..., \underline{\mathbf{x}}_T)$ depends of the transition probabilities $\pi(x_k(t)/\theta(t))$, also, if we suppose that the model is a Rasch model, then we have

Corollary 5. *If the conditional probabilities $\pi(x_k(t)/\theta t)$ follow the Rasch model, then the law of the model for one individual only, is written*

$$p(\underline{\mathbf{x}}) = \frac{\sqrt{1-\rho^2}}{\sqrt{(2\pi)^T.\sigma^T}} \int_{-\infty}^{+\infty} ... \int_{-\infty}^{+\infty} \frac{F(\underline{\mathbf{x}}, \underline{\theta})}{\prod_{t=1}^{T} \prod_{k=1}^{q} (1 + \exp(\theta(t) - \beta_k))} \mathrm{d}\theta(T) \cdots \mathrm{d}\theta(1), \quad (6.23)$$

where $F(\underline{\mathbf{x}}, \underline{\theta}) = \exp\{-\frac{1}{2\sigma^2}(1-\rho^2)\theta(1)^2 + \sum_{t=2}^{T}((\theta(t) - \rho\theta(t-1))^2) + \sum_{t=1}^{T}(\theta(t).r(t) - s(t))\}$ *and* $s(t) = \sum_{k=1}^{q} x_k(t)\beta_k$.

Proof. It is sufficient to replace in (6.22) the probabilities of transition $\pi(x_k(t)/\theta(t)) = [\alpha_0(\theta(t))]^{x_k(t)}[1-\alpha_0(\theta(t))]^{1-x_k(t)}$ by their equivalents of Rasch and to deduce the result.

6.4 Estimation

The two classes of models considered in this section, already defined in the previous section, satisfy the hypotheses of the Rasch model.

We suppose that the latent process is such that $(\Theta_i(t) : 1 \leq t \leq T), i = \overline{1,n}$, are independents and normally distributed with variance σ_i^2.

Depending on whether the latent process is a first Order Markov process or an $\mathcal{AR}(1)$ process, we want to estimate the parameter vector $\eta = (\beta, \sigma^2)$ (resp. $\eta = (\beta, \rho, \sigma^2)$) where $\beta = (\beta_1, ..., \beta_q)'$ is the vector of item difficulty parameters and $\sigma^2 = (\sigma_1^2, ..., \sigma_n^2)'$ where σ_i^2 is the variance of $\Theta_i(t)$ (resp. $\beta = (\beta_1, ..., \beta_q)'$ is the vector of item difficulty parameters, $\rho = (\rho_1, ..., \rho_n)$ the vector of autoregressive parameters and $\sigma^2 = (\sigma_1^2, ..., \sigma_n^2)'$ where σ_i^2 the variance of the white noise $(\varepsilon_i(t))$.

6.4.1 The Marginal Likelihood

The latent Rasch model can be considered as a particular case of more general models for incomplete data analysis. The vector $\underline{\mathbf{x}}_i$ of observations can be interpreted as incomplete because it misses the latent feature. This incomplete vector can be considered as an observable function of the vector $(\underline{\mathbf{x}}_i, \theta_i)$ of complete data (partially observed). The likelihood function of such incomplete data, called Marginal likelihood, can be written:

$$p(\eta;\underline{\mathbf{x}}) = \prod_{i=1}^{n} \int \cdots \int_{\mathbf{R}^T} p(\underline{\mathbf{x}}_i/\underline{\theta}_i; \eta_i) g_i(\theta_i(1), ..., \theta_i(T)) \mathrm{d}\theta_i(1) \cdots \mathrm{d}\theta_i(T), \tag{6.24}$$

where $\underline{\theta}_i = (\theta_i(1), ..., \theta_i(T))', g_i$ the probability density of the latent vector $\underline{\Theta}_i = (\Theta_i(1), ..., \Theta_i(T))'$, $\underline{\mathbf{x}} = ((\underline{\mathbf{x}}_1(t), ..., \underline{\mathbf{x}}_n(t))' : 1 \leq t \leq T)$ a realization of the observation process $\{(\underline{\mathbf{X}}_1(t), ..., \underline{\mathbf{X}}_n(t))' : 1 \leq t \leq T\}$ and where $\underline{\mathbf{x}}_i = (\underline{\mathbf{x}}_i(1), ..., \underline{\mathbf{x}}_i(T))'$. Using Rasch assumptions, this marginal likelihood becomes:

$$p(\eta;\underline{\mathbf{x}}) = \prod_{i=1}^{n} \int \cdots \int_{\mathbf{R}^T} \psi(\eta_i;\underline{\mathbf{x}_i}) g_i(\theta_i(1), ..., \theta_i(T)) \mathrm{d}\theta_i(1) \cdots \mathrm{d}\theta_i(T), \tag{6.25}$$

where

$$\psi(\eta_i;\underline{\mathbf{x}_i}) = \frac{\exp(\sum_{t=1}^{T} r_i(t)\theta_i(t) - \sum_{k=1}^{q} \beta_k . r_{i,k})}{\prod_{t=1}^{T} \prod_{k=1}^{q} [1 + \exp((\theta_i(t) - \beta_k))]}. \tag{6.26}$$

The method that consists of estimating the parameter η in maximizing this marginal likelihood (called Marginal Maximum Likelihood method) is computationally difficult, also among approaches commonly used to maximize this function, the approach based on the application of the EM algorithm [DEM77] is generally privileged. Below, this approach is described in details.

6.4.2 The EM Algorithm

The EM algorithm consists of maximizing iteratively, neither the log-likelihood of incomplete data, but the mean of the log-likelihood of the complete data with respect to the observations. In fact, it is about maximizing the following expression of the log-likelihood function:

$$\mathbf{E}\{\log[f(\underline{\mathbf{x}}, \underline{\theta}; \eta)]/\underline{\mathbf{x}}, \eta^{(p)}\}, \tag{6.27}$$

where $f(\underline{\mathbf{x}}, \underline{\theta}, \eta) = \prod_{i=1}^{n} p(\underline{\mathbf{x}_i}/\underline{\theta}_i; \eta_i) g_i(\underline{\theta}_i, \eta_i)$, $\underline{\theta} = (\underline{\theta}_1, ..., \underline{\theta}_n)'$ and $\underline{\theta}_i = (\theta_i(1), ..., \theta_i(T))'$, $\underline{\mathbf{x}} = (\underline{\mathbf{x}}_1, ..., \underline{\mathbf{x}}_n)'$ and $\underline{\mathbf{x}}_i = (\underline{\mathbf{x}}_i(1), ..., \underline{\mathbf{x}}_i(T))'$. More precisely, in the stage $p+1$, given the value $\eta^{(p)}$ of the estimate obtained at the p stage, the algorithm is as follows:

Stage E (expectation):
This stage consists of calculating the conditional mean of the complete log-likelihood knowing the current values of the estimators

$$\mathbf{Q}(\eta/\eta^{(p)}) = \mathbf{E}(\log f(\underline{\mathbf{x}}, \underline{\theta}; \eta)/\underline{\mathbf{x}}, \eta^{(p)}). \tag{6.28}$$

We have the following result:

Proposition 4. *The conditional mean* $\mathbf{Q}(\eta/\eta^{(p)})$ *is given by:*

$$\sum_{i=1}^{n} \int \cdots \int [\log\{g_i(\underline{\theta}_i, \eta_i)\} + \log\{p(\underline{\mathbf{x}_i}, \underline{\theta}_i; \eta_i/\underline{\mathbf{x}}, \eta^{(p)})\}]\pi_i(\underline{\theta}_i/\underline{\mathbf{x}_i}, \eta^{(p)}) d\underline{\theta}_i, \tag{6.29}$$

where

$$\pi_i(\underline{\theta}_i/\underline{\mathbf{x}_i}, \eta^{(p)}) \propto p(\underline{\mathbf{x}_i}, \eta^{(p)}/\theta_i(1), ..., \theta_i(T)).g_i(\theta_i(1), ..., \theta_i(T)/\eta^{(p)}),$$

and

$$\log\{p(\underline{\mathbf{x}_i}, \underline{\theta}_i; \eta_i/\underline{\mathbf{x}}, \eta^{(p)})\} = \sum_{t=1}^{T} \sum_{k=1}^{q} \log\{\pi(x_{i,k}(t)/\underline{\mathbf{x}}, \theta_i(t), \eta_i^{(p)})\}.$$

Proof. We have $\mathbf{Q}(\eta/\eta^{(p)}) = \sum_{i=1}^{n} \int \cdots \int_{R^T} [\log(g_i(\theta_i, \eta_i)) + \log[p(\mathbf{x}_i, \theta_i, \eta_i/\mathbf{x}, \eta^p)] \times q_i(\theta_i/\mathbf{x}_i, \eta^p)] d\theta_i$ where $q_i(\theta_i/\mathbf{x}_i, \eta^p)] d\theta_i$ is the conditional density of latent vector $\Theta_i = (\Theta_i(1), ..., \Theta_i)(T)$ knowing the observation vector $\mathbf{X}$. by a standard calculation we find that

$$q_i(\theta_i/\mathbf{x}_i, \eta^p)] d\theta_i = \frac{p(\mathbf{x_i 8 \theta_i}, \eta^{\mathbf{P}})}{p(\mathbf{x}, \eta^p) \times g(\theta)_i, \eta^p}$$

And as

$$p(\mathbf{x_i}, \eta^{(p)}) = \int \int \prod_{t=1}^{T} \prod_{k=1}^{q} \pi(x_{i,k}/\theta_i(t), \eta^p) d\theta_i(1) ... d\theta_i(T)$$

it followed that $q_i(\theta_i/\mathbf{x}_i, \eta^{(p)})$ coincides with a nearly constant $p(x_i/\theta_i(t), \eta^p)(\theta_i, \eta^{(p)})$. More like

$$\log\{p(\underline{\mathbf{x}}, \underline{\theta}, \eta/\underline{\mathbf{x}}, \eta^{(p)})\} = \sum_{i=1}^{n} \sum_{t=1}^{T} \sum_{k=1}^{q} \log\{\pi(x_{i,k}(t)/\theta_i, \eta_i^p) \pi_i \theta_i/\mathbf{x}_i \eta^{(p)}\},$$

the result of the proposltion follows.

Stage M (maximization):

This second stage consists of finding the value $\eta^{(p+1)}$ that achieves the maximum of the quantity $Q(\eta/\eta^{(p)})$:

$$Q(\eta^{(p+1)}) = \arg\max_{\eta} Q(\eta/\eta^{(p)}). \tag{6.30}$$

This stage requires explicit expressions of formulas (6.25) in order to highlight the dependence of $Q(\eta/\eta^{(p)})$ as a function of the parameters model. To calculate the value $\eta^{(p+1)}$, we apply the Newton–Raphson algorithm and the integrals are approximated numerically using the Gauss–Hermite quadrature formulas. The E and M stages are repeated alternately until the difference $Q(\eta^{p+1}) - Q(\eta^{p})$ is less than a prior fixed quantity.

6.5 Estimation of Rasch Latent Markov Processes

We apply the results above to the cases of latent Markov processes and latent $\mathcal{AR}(1)$ processes.

Rasch Latent Markov Processes

In this case, the parameter $\eta = (\beta, \sigma^2)$ with $\beta = (\beta_1, ..., \beta_q)'$ the vector of the difficulty parameters and $\sigma^2 = (\sigma_1^2, ..., \sigma_n^2)'$, with σ_i^2 is the variance of the Markov process Θ_i. The conditional mean of the complete log likelihood knowing the current value of the estimator $\eta^{(p)}$ is written in this case

Corollary 6. *Under the hypotheses of the Rasch model, we have*

$$\mathbf{Q}(\eta/\eta^{(p)}) = -\frac{T}{2}\sum_{i=1}^{n} \log(2\pi\sigma_i^2) + H_1 + H_2, \tag{6.31}$$

where

$$\begin{aligned} H_1 &= -\sum_{i=1}^{n} \frac{1}{2\sigma_i^2} \int \cdots \int [\theta_i^2(1) + \sum_{t=2}^{T} (\theta_i(t) - \theta_i(t-1))^2] \\ &\quad \times \pi_i(\underline{\theta}_i/\underline{\mathbf{x}}, \beta^{(p)}, \sigma^{2(p)}) \mathrm{d}\theta_i(1) \cdots \mathrm{d}\theta_i(T), \\ H_2 &= \sum_{i=1}^{n}\sum_{t=1}^{T}\sum_{k=1}^{q} \int \cdots \int \log\{\frac{\exp(x_{i,k}(t)(\theta_i(t) - \beta_k))}{1 + \exp((\theta_i(t) - \beta_k))}\} \pi_i(\underline{\theta}_i/\underline{\mathbf{x}_i}, \eta^{(p)}) \mathrm{d}\theta_i(1) \cdots \mathrm{d}\theta_i(T), \end{aligned}$$

and

$$\pi_i(\underline{\theta}_i/\underline{\mathbf{x}},, \eta^{(p)}) \propto \frac{\exp\{\sum_{t=1}^{T} \theta_i(t).r_i(t) - \sum_{k=1}^{q} \beta_k^{p}.r_i(k) - \frac{1}{2\sigma^{2(p_i)}}[\theta_i^2(1) + \sum_{t=2}^{T}(\theta_i(t) - \theta_i(t-1))^2]\}}{\prod_{t=1}^{T}\prod_{k=1}^{q}[1 + \exp(\theta_i(t) - \beta_k^{(p)})]}.$$

Proof. This comes directly from the relation (6.25) and the hypothesis of local independence.

The maximization with respect to the σ_i^2 concerns only the first two sums of this last expression and the maximization with respect to the β_k involves only the double sum over on i and t of the last term of this same expression. Which brings us to calculate

$$\sigma_i^{2(p+1)} = \frac{1}{T} \int \dots \int [\theta_i^2(1) + \sum_{t=2}^{T} (\theta_i(t) - \theta_i(t-1))^2] \pi_i(\underline{\theta}_i / \underline{\mathbf{x}}, \beta^(p), \sigma^{2(p)}) \mathrm{d}\theta_i(1) \cdots \mathrm{d}\theta_i(T),$$

for all i and to resolve the q following equations

$$\sum_{i=1}^{n} \sum_{t=1}^{T} x_{i,k}(t)$$
$$= \sum_{i=1}^{n} \sum_{t=1}^{T} \int \dots \int [1 + \exp((\beta_k - \theta_i(t)))]^{-1} \pi_i(\underline{\theta}_i / \underline{\mathbf{x}}, \eta^{(p)}) \mathrm{d}\theta_i(1) \cdots \mathrm{d}\theta_i(T) \quad 1 \le k \le q,$$

and the integrals are again approximated numerically using the Raphson–Newton algorithm and the Gauss–Hermite quadrature approximation formulas.

6.5.1 Rasch Latent $\mathcal{AR}(1)$ Processes

In this case, we recall that the parameter is $\eta = (\beta, \rho, \sigma^2)$ with $\beta = (\beta_1, ..., \beta_q)$, $\rho = (\rho_1, ..., \rho_n)$, and $\sigma^2 = (\sigma_1^2, ..., \sigma_n^2)$.

Corollary 7. *Under the hypotheses of the Rasch model, we have*

$$\mathbf{Q}(\theta/\eta^{(p)}) = -\frac{T}{2} \sum_{i=1}^{n} \log(2\pi\sigma_i^2) + \frac{1}{2} \sum_{i=1}^{n} \log(1 - \rho_i^2) + G_1 + G_2, \tag{6.32}$$

where

$$G_1 = -\sum_{i=1}^{n} \frac{1}{2\sigma_i^2} \int \dots \int \varphi(\underline{\theta}_i, \underline{\rho}_i) \pi_{\mathbf{i}}(\underline{\theta}_i / \underline{x}_i, \eta^{(p)}) \mathrm{d}\theta_i(1) \cdots \mathrm{d}\theta_i(T),$$
$$G_2 = +\sum_{i=1}^{n} \sum_{t=1}^{T} \sum_{k=1}^{q} \int \dots \int \psi(\underline{x}_i, \underline{\theta}_i) \pi_{\mathbf{i}}(\underline{\theta}_i / \underline{x}_i, \eta^{(p)}) \mathrm{d}\theta_i(1) \cdots \mathrm{d}\theta_i(T)$$

with

$$\varphi(\underline{\theta}_i, \underline{\rho}_i) = (1 - \rho_i^2)\theta_1^2 + \sum_{t=2}^{T} (\theta_i(t) - \rho_i \theta_i(t-1))^2,$$
$$\psi(\underline{x}_i, \underline{\theta}_i) = \ln \frac{\exp(x_{i,k}(t)(\theta_i(t) - \beta_k))}{1 + \exp((\theta_i(t) - \beta_k))} \quad \textit{and } \pi_i(\underline{\theta}_i / \underline{\mathbf{x}}, , \eta^{(p)}) \propto C,$$

and where

$$C = \frac{\exp\{\sum_{t=1}^{T} \theta_i(t).r_i(t) - \sum_{k=1}^{q} \beta_k^{(p)}.r_i(k) - \frac{1}{2\sigma_i^{2(p)}} [(1-\rho_i^{2(p)})\theta_i^2(1) + \sum_{t=2}^{T} (\theta_i(t) - \rho_i^{(p)} \theta_i(t-1))^2]\}}{\prod_{t=1}^{T} \prod_{k=1}^{q} [1 + \exp(\theta_i(t) - \beta_k^{(p)})]}.$$

Proof. The conditional mean of the complete log-likelihood take the form (6.33) with

$$\pi_{\mathbf{i}}(\underline{\theta}_i/\underline{\mathbf{x}}, \eta^{(p)}) = \pi_i(\underline{(\theta_i)}/\underline{\mathbf{x}}, \eta^{(p)}),$$

and we see that

$$\pi_{\mathbf{i}}(\underline{\theta}_i/\underline{\mathbf{x}}, \eta^{(p)}) \propto C.$$

Here, the maximization with respect to σ_i^2 concerns only the first and the third term of this last expression and the maximization with respect to ρ_i takes on the second and third term of this same expression. It brings us to solve the following system of equations:

$$\sigma_i^{2(p+1)} = \frac{1}{T}\sum_{i=1}^{n}\int \dots \int [(1-\rho_i^2)\theta_i^2(1) + \sum_{t=2}^{T}(\theta_i(t) - \rho_i\theta_i(t-1)^2]$$
$$\times \pi_i(\underline{\theta}_i/\underline{\mathbf{x}}, \eta^{(p)})\mathrm{d}\theta_i(1)\cdots\mathrm{d}\theta_i(T),\ i = 1, ..., n,$$

and also

$$\frac{\rho_i}{1-\rho_i^2} = \frac{1}{\sigma^2}\int \dots \int [\rho_i\theta_i^2(1) + \sum_{t=2}^{T}\theta_i t - 1(\theta_i(t) - \rho_i\theta_i(t-1))]$$
$$\times \pi_i(\underline{\theta}_i/\underline{\mathbf{x}}, \beta^{(p)}, \rho^{(p)}, \sigma^{2(p)})\mathrm{d}\theta_i(1)\cdots\mathrm{d}\theta_i(T),\ i = 1, ..., n,$$

and the maximization with respect to β_k involves only the double sum over i and t of the last term of this expression. That leads to solve the q following equations

$$\sum_{i=1}^{n}\sum_{t=1}^{T} x_{i,k} = \sum_{i=1}^{n}\sum_{t=1}^{T}\int \dots \int [1 + \exp((\beta_k - \theta_i(t)))]^{-1}$$
$$\times \pi_{\mathbf{i}}(\underline{(\theta_i)/\underline{\mathbf{x}}, \eta^{(p)}})\mathrm{d}\theta_i(1)\cdots\mathrm{d}\theta_i(T),\ k = 1, ..., q.$$

In the two cases, the integrals are again approximated numerically using the Gauss–Hermite quadrature formulas.

References

[ADM06] Admane, O., Mesbah M. Estimation des paramètres du modèle Rasch dichotomique. Annales de L'ISUP. Vol 1-Fas 1-2 pp. 3–30 (2006)

[BAR06] Bartolucci, F. Likelihood inference for a class of latent Markov models under linear hypotheses on the transition probabilities. Journal of the royal Statistical Society, Series B, 68, 155–178 (2006)

[DEM77] Dempstair, A.P., Laird N.M., Rubin, D.B. Maximum likelihood from incomplete data via the EM algorithme. Journal of royal statistical Society, Ser. B., 39, 1–38. (1977)

[FED05] Feddag, M.L., Mesbah, M. Generalized estimating equations for longitudinal mixed Rasch model. Journal of Statistical Planning and Inference, 129, 159–179 (2005)

[FIS95] Fisher, G.H. and Molenaar, I.W. Rasch models, foundations, recent developments and applications. Springer-Verlag, New York (1995)

[LAN94] Langeheine, R. and Van de Pol, F. Discrete-time mixed Markov latent class models. In: Dale, A., Davies, R.B. (eds) Analysing social and political change. A casebook of methods. Sage, London, pp. 170–197 (1994)

[MES09] Mesbah, M. Statistical quality of life. In: Balakrishnan, N., Campbell, B.R., Vidakovic, B., Kotz, S., Johnson, N.L. (eds) Methods and applications of statistics in the life and health sciences. Chapter 73. Wiley, New York (2009)

[POL86] Van de Pol, F., de Leeuw, J. A latent Markov model to correct for measurement error. Sociological Methods and Research, 15, 118–141 (1986)

[VER86] Vermunt, J.R., Langeheine, R., Böckenholt, U. Discrete-time discrete-state latent Matkov models withtime-constant and time-varying covariates. Journal of Educational and Behavioral Statistics, 24, 179–207 (1999)

[WIG73] Wiggins, L.M. Panel analysis. Elsevier, Amsterdam (1973)

7

Dynamics of Dependence Properties for Lifetimes Influenced by Unobservable Environmental Factors

Rachele Foschi and Fabio Spizzichino*

Department of Mathematics, Università degli Studi di Roma "La Sapienza", P.le A. Moro, 2, 00185, Roma, Italy,
foschi@mat.uniroma1.it; Fabio.Spizzichino@uniroma1.it

Abstract: We consider non-negative conditionally independent and identically distributed random variables and analyze conditions for monotonicity of survival copulas of residual lifetimes. Concentrating attention on the bivariate copula, we compare its behavior at the instant of default with its evolution between two defaults. The assumptions for our results will be expressed in terms of conditional hazard rates.

Keywords and phrases: Survival Copulas of residual lifetimes, Default contagion, Multivariate stochastic orders, Longitudinal observations, Posterior distributions

7.1 Introduction

Tail dependence and *Default contagion* emerge as two subjects of interest in multivariate survival models and are typically related with the analysis of non-negative random variables, that have the meaning of waiting times until top events for different *units*.

An interesting fact is that such phenomena are equally relevant in the two different fields of Reliability and Finance where, respectively, components in a system or firms in a same market can be represented as units that are prone to default.

Tail dependence refers to those aspects of stochastic dependence (among the different units) that are related with a high frequency of simultaneous occurrence of extreme events for the units themselves (see e.g. [CJ06, MFE05]).

Roughly speaking, the term *default contagion* describes the cases when the default of a unit causes a decrease in the joint survival probability of residual lifetimes of the remaining units (see in particular [MFE05] and references cited therein).

Under a different language, specific notions related with such a general idea of default contagion had already been introduced and studied, in the past, within the frame of reliability theory (see e.g. [AN84, SSh07, SSp98], and, for the specific case of exchangeable lifetimes, [Spi01]). One of such notions is Weakened by Failures. Such notions were formulated in terms of different multivariate stochastic orderings and are related to corresponding notions of positive dependence. In order to describe these notions of dependence, it was used the term *dependence of dynamic type*.

V.V. Rykov et al. (eds.), *Mathematical and Statistical Models and Methods in Reliability: Applications to Medicine, Finance, and Quality Control*, Statistics for Industry and Technology, DOI 10.1007/978-0-8176-4971-5_7, © Springer Science+Business Media, LLC 2010

One can then say that both tail dependence and default contagion are related to stochastic dependence and to evolution of it, at certain instants. For this reason we can expect that the two phenomena are linked together. Such links are not immediately clear however and have not been completely investigated in the literature.

Several notions of stochastic dependence are conveniently described in terms of copulas and, more precisely, in terms of survival copulas in the cases of non-negative variables (see e.g. [Joe97, Nel06]). It is then interesting to analyze the time-evolution of survival copulas under the *longitudinal* observation of lifetimes.

Actually, in the literature, evolution of dependence has been mostly considered in absence of defaults (see e.g. [CJ06] and [FS08]); and tail dependence can just be seen as a limiting concept in such a frame.

Time evolution of survival copulas can then be considered a more general topic than tail dependence, comprehending both the behavior of copulas at default times and, as a limiting case of their behavior between two subsequent default times, the tail dependence.

In this paper, we present a study of the evolution of survival copulas in both the cases characterized by the presence or absence of defaults. We aim at comparing the two different types of analysis and at pointing out the link between them.

It will also turn out that the behavior of copulas at default times triggers properties of default contagion.

A prime source of stochastic dependence among random variables (and, in particular, among lifetimes) is just created by the influence of common factors that are not directly observable. In particular, a very basic case of stochastic dependence is the one of conditional independence.

The literature about the role of conditional independence, in several different fields of application, is huge and well-known and there is no need to refer to it, here. Let us only mention that, in the field of reliability, the unobservable factors often have the meaning of *environmental conditions*. In the field of financial risk, this type of situation is related with the so-called *information-induced dependence* (see e. g. [MFE05]).

What is important to notice here is that different problems related with the analysis of dependence take a very special and expressive form, under the assumption that dependence is just produced by conditional independence. This circumstance especially holds for what concerns the problem of evolution of dependence, under defaults and survivals, that is addressed to in this paper.

For this reason, we concentrate attention on the case when the observable lifetimes are conditionally independent and identically distributed (= conditionally i.i.d.) given a vector of non-negative random variables $\boldsymbol{\Theta} \equiv (\theta_1, \ldots, \theta_d)$.

In such a situation, the use of a Bayesian approach turns out as a completely natural one and the evolution of the conditional density of $\boldsymbol{\Theta}$ given observed histories becomes a central object of interest.

A preliminary version of this analysis, with Θ a scalar random variable, had already been presented in [Fos08]. We expand here the topic of the evolution of dependence and, in particular, we point out some specific aspects of the extension to the multivariate case.

In order to proceed with our analysis, we introduce now the following notation.

We consider non-negative random variables $T_1, \ldots, T_n$, to be seen as the waiting times to n (stochastically dependent) top events, such as the failures of components operating in a same system or the defaults of firms in a same market.

We assume $T_1, \ldots, T_n$ to be exchangeable and denote by

$$\overline{F}(x_1, \ldots, x_n) = P(T_1 > x_1, \ldots, T_n > x_n),$$
$$\overline{G}(x) = \overline{F}(x, 0, \ldots, 0) = P(X > x),$$
$$\hat{C}(u_1, \ldots, u_n) = \overline{F}\{\overline{G}^{-1}(u_1), \ldots, \overline{G}^{-1}(u_n)\}$$

the corresponding joint survival function, marginal univariate survival function and survival copula, respectively. We shall refer to $\overline{F}$ as an *exchangeable survival model.*

It is also natural analyzing the above model from a dynamic point of view. With the vector $(T_1, \ldots, T_n)$, we associate the counting process

$$N_t = \sum_{i=1}^{\infty} \mathbf{1}_{\{T_i \le t\}},$$

whose jump times are just the order statistics $T_{(1)}, \ldots, T_{(n)}$ of $(T_1, \ldots, T_n)$, letting $T_{(0)} = T_0 = 0$ a.s..

We assume that the available observation up to a time $t > 0$ is an event of the form

$$E_t \equiv \{T_{(1)} = t_1, \ldots, T_{(k)} = t_k, T_{(k+1)} > t\}. \tag{7.1}$$

Of course such events belong to the σ-algebra $\mathcal{F}_t$, where $\{\mathcal{F}_t\}_{t \ge 0}$ is the filtration representing the *internal history* of the process $\{N_t\}_{t \ge 0}$. Notice that the event E_t implies $\{N_t = k\}$.

Given the event E_t, it is useful for our purposes to consider the ordered residual lifetimes at time t of the surviving units, i.e. the random variables

$$(T_{(k+1)} - t, \ldots, T_{(n)} - t)|\mathcal{F}_t.$$

In order to deal, at any t, with exchangeable random variables, we define the following vector $\mathbf{X}_t \equiv (X_t^1, \ldots, X_t^{n-k})$ of *exchangeable residual lifetimes.*

Definition 1. *The exchangeable residual lifetimes of* $(T_1, \ldots, T_n)$ *at time* t *are the exchangeable random variables* $X_t^1, \ldots, X_t^{n-k}$ *admitting* $(T_{(k+1)} - t, \ldots, T_{(n)} - t)|\mathcal{F}_t$ *as order statistics.*

Concerning the distribution of $(X_t^1, \ldots, X_t^{n-k})$, we put

$$\overline{F}_t(x_1, \ldots, x_{n-k}) = P\left(X_t^1 > x_1, \ldots, X_t^{n-k} > x_{n-k}|\mathcal{F}_t\right) \tag{7.2}$$
$$\overline{G}_t(x) = \overline{F}_t(x, 0, \ldots, 0) = P\left(X_t^1 > x|\mathcal{F}_t\right)$$
$$\hat{C}_t(u_1, \ldots, u_{n-k}) = \overline{F}_t\{\overline{G}_t^{-1}(u_1), \ldots, \overline{G}_t^{-1}(u_{n-k})\}. \tag{7.3}$$

$\hat{C}_t(u_1, \ldots, u_{n-k})$ is the *survival copula* of the random vector $\left(X_t^1, \ldots, X_t^{n-k}\right)$.

As said before, a problem of interest is the evolution of stochastic dependence properties of $\overline{F}_t$, as t increases. In particular, we restrict ourselves to the case when $T_1, \ldots, T_n$ are conditional independent given a random vector $\boldsymbol{\Theta} = (\Theta_1, \ldots, \Theta_d)$.

Furthermore, we consider the dependence between just two variables, and thus we study the evolution of the family $\{\hat{C}_t^{(2)}\}$, where $\hat{C}_t^{(2)}$ is the bivariate margin of the survival copula in (7.3).

Our results will be based on some notions of multivariate stochastic ordering and will be exposed in Sect. 7.3. Therein we study some aspects of the monotonicity behavior of the family $\{\hat{C}_t^{(2)}\}$, both at default times and between two subsequent default times. More precisely, we find out conditions for stochastic dependence being decreasing at default times and progressively increasing between two subsequent default times. To this purpose we will present a preliminary result (see Proposition 1) that relates monotonicity properties of the posterior densities of $\boldsymbol{\Theta}$ to properties of hazard rates of $T_1, \ldots, T_n$ given $\boldsymbol{\Theta}$.

Section 7.2 contains some preliminary computations: namely, for the conditionally i.i.d. case, we shall write down for $\overline{F}_t$, $\overline{G}_t$, $\hat{C}_t$ some explicit expressions that will be used in Sect. 7.3. The paper will end with a final discussion, presenting a brief summary and a few concluding remarks, and an Appendix where we collect some needed notions about positive dependence and multivariate stochastic orderings.

7.2 Some Basic Facts

Let $\boldsymbol{\Theta}$ be a random vector with joint density π_0. We denote by $\boldsymbol{\theta} = (\theta_1, \ldots, \theta_d)$ its realization. Let $T_1, \ldots, T_n$ be conditionally i.i.d. given $\boldsymbol{\Theta}$, with conditional survival function $\overline{G}(x|\boldsymbol{\theta})$ and conditional density $g(x|\boldsymbol{\theta})$. For our purposes, our first step is the computation of the survival copula $\hat{C}_t$. To this aim, we adapt the previous formula (7.2) for $\overline{F}_t$ to the present case of conditional independence and identical distribution:

$$\overline{F}(x_1, \ldots, x_n) = \int_{\mathbb{R}_+^d} \overline{G}(x_1|\boldsymbol{\theta}) \cdots \overline{G}(x_n|\boldsymbol{\theta}) \pi_0(\boldsymbol{\theta}) d\boldsymbol{\theta} =$$

$$\int_{\mathbb{R}_+^d} \int_{x_1}^{\infty} \cdots \int_{x_n}^{\infty} g(\xi_1|\boldsymbol{\theta}) \cdots g(\xi_n|\boldsymbol{\theta}) d\xi_1 \cdots d\xi_n \pi_0(\boldsymbol{\theta}) d\boldsymbol{\theta}.$$

In particular, the one-dimensional *predictive* survival function and probability density, respectively, become

$$\overline{G}(x) = \int_{\mathbb{R}_+^d} \overline{G}(x|\boldsymbol{\theta}) \pi_0(\boldsymbol{\theta}) d\boldsymbol{\theta},$$

$$g(x) = \int_{\mathbb{R}_+^d} g(x|\theta) \pi_0(\theta) d\theta.$$

$r(x)$ and $r(x|\boldsymbol{\theta})$, respectively, denote the *predictive* hazard rate of each T_i and the *conditional* hazard rate, i.e.

$$r(x) = \frac{g(x)}{\overline{G}(x)}, \quad r(x|\boldsymbol{\theta}) = \frac{g(x|\boldsymbol{\theta})}{\overline{G}(x|\boldsymbol{\theta})}.$$

In terms of the conditional univariate survival functions $\overline{G}(x|\boldsymbol{\theta})$, we want to write down, for any x, $2 \le j \le n-k$, the joint survival function $\overline{F}_t$.

In this respect, it is important to notice that, if $T_1, \ldots, T_n$ are conditionally i.i.d. given $\boldsymbol{\Theta}$, it can be shown that $X_t^1, \ldots, X_t^{n-k}$ are conditionally i.i.d. given $\boldsymbol{\Theta}$ as well. In particular, each X_t^i has conditional univariate survival function $\overline{G}_t(x|\boldsymbol{\theta})$ (details are given in [Fos10]).

We can now proceed with the computation of $\overline{F}_t$. For any t, conditionally on the history $\mathfrak{F}_t$, $\boldsymbol{\Theta}$ admits density

$$\pi_t(\boldsymbol{\theta}) \equiv \pi(\boldsymbol{\theta}|\mathfrak{F}_t).$$

If the observation up to t is E_t, defined in (7.1), we can write, by applying the Bayes' formula,

$$\pi_t(\boldsymbol{\theta}) \propto [\overline{G}(t|\boldsymbol{\theta})]^{n-k} g(t_1|\boldsymbol{\theta}) \cdots g(t_k|\boldsymbol{\theta}) \pi_0(\boldsymbol{\theta}). \tag{7.4}$$

Furthermore

$$\overline{F}_t(x_1, \ldots, x_{n-k}) = \int_{\mathbb{R}_+^d} \frac{\overline{G}(x_1 + t|\boldsymbol{\theta}) \cdots \overline{G}(x_{n-k} + t|\boldsymbol{\theta})}{[\overline{G}(t|\boldsymbol{\theta})]^{n-k}} \pi_t(\boldsymbol{\theta}) \mathrm{d}\boldsymbol{\theta}. \tag{7.5}$$

Thus, the univariate margin becomes

$$\overline{G}_t(x) \equiv P(X > t + x|\mathfrak{F}_t) =$$

$$\overline{F}_t(x, 0, \ldots, 0) = \int_{\mathbb{R}_+^d} \frac{\overline{G}(x + t|\boldsymbol{\theta})}{\overline{G}(t|\boldsymbol{\theta})} \pi_t(\boldsymbol{\theta}) \mathrm{d}\boldsymbol{\theta}. \tag{7.6}$$

We can now write, for any t, the survival copula $\hat{C}_t$. By combining (7.5) and (7.6), we obtain

$$\hat{C}_t(u_1, \ldots, u_{n-k}) = \int_{\mathbb{R}_+^d} \frac{\overline{G}(\overline{G}_t^{-1}(u_1) + t|\boldsymbol{\theta})}{\overline{G}(t|\boldsymbol{\theta})} \cdots \frac{\overline{G}(\overline{G}_t^{-1}(u_{n-k}) + t|\boldsymbol{\theta})}{\overline{G}(t|\boldsymbol{\theta})} \pi_t(\boldsymbol{\theta}) \mathrm{d}\boldsymbol{\theta}. \tag{7.7}$$

Notice that π_t in (7.5)–(7.7) is given by (7.4).

In particular, as we have said in the Introduction, we will study the evolution of the bivariate copulas, obtained as the bivariate margin of the copula in the previous formula and therefore given by

$$\hat{C}_t(u_1, u_2) = \int_{\mathbb{R}_+^d} \frac{\overline{G}(\overline{G}_t^{-1}(u_1) + t|\boldsymbol{\theta})}{\overline{G}(t|\boldsymbol{\theta})} \frac{\overline{G}(\overline{G}_t^{-1}(u_2) + t|\boldsymbol{\theta})}{\overline{G}(t|\boldsymbol{\theta})} \pi_t(\boldsymbol{\theta}) \mathrm{d}\boldsymbol{\theta}. \tag{7.8}$$

Remark 1. As to the univariate survival function, it is interesting to notice the difference between $\overline{G}_t(x)$ in (7.6), i.e. the univariate survival function conditional on the history up to t of all the variables in the model, and

$$\overline{H}_t(x) \equiv P(X > t + x|X > t) = \frac{\overline{G}(x + t)}{\overline{G}(t)},$$

i.e. the univariate survival function of one variable, conditional on the survival at t of only that variable. On the other hand, since conditionally on $\boldsymbol{\Theta}$ the variables are independent, we can notice that, given $\boldsymbol{\Theta}$, conditioning on $\{X > t\}$ is equivalent to conditioning on $\mathfrak{F}_t$. Therefore, the two univariate conditional survival functions $\overline{G}_t(x|\boldsymbol{\theta})$ and $\overline{H}_t(x|\boldsymbol{\theta})$ do coincide:

$$\overline{G}_t(x|\boldsymbol{\theta}) \equiv P(X > t + x|\mathfrak{F}_t, \boldsymbol{\theta}) = P(X > t + x|X > t, \boldsymbol{\theta}) = \overline{H}_t(x|\boldsymbol{\theta}).$$

7.3 Monotonicity Properties of Survival Copulas

In order to analyze the evolution of dependence when t elapses, we consider, in particular, the family of survival copulas $\hat{C}_t^{(2)}(u_1, u_2)$, defined in (7.8).

As declared in the Introduction, we aim at obtaining conditions for monotonicity properties of $\left\{\hat{C}_t^{(2)}\right\}$.

On the set of bivariate copulas the following partial order, called *more PQD* order or *concordance* order, is defined (see e.g. [Joe97, LX06]):

Definition 2. *$\hat{C}_{t''}$ is more PQD than $\hat{C}_{t'}$, written as $\hat{C}_{t'} \preceq_{PQD} \hat{C}_{t''}$, if*

$$\hat{C}_{t'}(u_1, u_2) \leq \hat{C}_{t''}(u_1, u_2)$$

for any $u_1, u_2 \in [0, 1]$.

With respect to this order, we can define a notion of monotonicity:

Definition 3. *The map $t \rightarrow \hat{C}_t$ is increasing (in the PQD order) if $\hat{C}_{t'} \preceq_{PQD} \hat{C}_{t''}$ for any $t' < t''$.*

Remark 2. We point out that $t \rightarrow \hat{C}_t$ increasing means that residual lifetimes become more and more dependent at increase of age

It is natural to split the analysis of $\left\{\hat{C}_t^{(2)}\right\}$ into two different stages, namely:

(a) at default times $T_{(k)}$'s

or

(b) between two of them, i.e. within the intervals $(T_{(k)}, T_{(k+1)})$, for $k = 0, \ldots, n-2$.

Under the hypothesis that $T_1, \ldots, T_n$ are conditionally i.i.d., we obtain that the survival copulas can be given rather explicit expression. Therefore, in the following, we can study the monotonicity of the process $\left\{\hat{C}_t^{(2)}\right\}$ by means of direct comparisons.

In both the two cases (a) and (b), monotonicity properties of $t \rightarrow \hat{C}_t^{(2)}$ will be easily achieved by imposing suitable monotonicity assumptions on the conditional hazard rate $r(t|\boldsymbol{\theta})$.

More precisely, monotonicity properties of $t \rightarrow \hat{C}_t$ can be obtained from monotonicity properties of $t \rightarrow \pi_t$, as stated in Propositions 2 and 3 below. On their turn, monotonicity properties of $t \rightarrow \pi_t$ can be traced back to monotonicity properties of $\boldsymbol{\theta} \rightarrow r(t|\boldsymbol{\theta})$, as illustrated in the following Proposition 1. We also remark that this proposition has some connections with the notion of default contagion (see Remark 4 below).

For our purposes, we need some notions about dependence and multivariate stochastic orders, such as the likelihood-ratio order, the weak likelihood-ratio order, MTP_2 dependence (see e.g. [SSh07]); these notions will also be recalled in the Appendix.

Proposition 1. *Let us assume the condition*

(a) $r(t|\boldsymbol{\theta}) \uparrow \boldsymbol{\theta}$.

Then we have

(b) $\pi_{T_{(k)}} \geq_{wlr} \pi_{T^-_{(k)}}$ *a.s.;*

(c) $\pi_{t'} \geq_{wlr} \pi_{t''}$ *a.s., for any* $t' < t''$, $t', t'' \in [T_{(k)}, T_{(k+1)})$, $k = 0, \ldots, n-2$.

Proof. The implication (a) $\Rightarrow$ (b) is obvious by taking into account (7.15) and the identity $\dfrac{\pi_{T_{(k)}}(\boldsymbol{\theta})}{\pi_{T^-_{(k)}}(\boldsymbol{\theta})} = r(T_{(k)}|\boldsymbol{\theta})$.

In order to prove that (a) implies (c), we notice that

$$r(t|\boldsymbol{\theta}) \uparrow \boldsymbol{\theta} \Leftrightarrow \int_{t'}^{t''} r(t|\boldsymbol{\theta})\mathrm{d}t \uparrow \boldsymbol{\theta} \quad \forall\, t' < t''.$$

Since $\overline{G}(x|\boldsymbol{\theta}) = \exp\left\{-\int_0^x r(t|\boldsymbol{\theta})dt\right\}$,

$$\frac{\overline{G}(t''|\boldsymbol{\theta})}{\overline{G}(t'|\boldsymbol{\theta})} = \exp\left\{-\int_{t'}^{t''} r(t|\boldsymbol{\theta})\mathrm{d}t\right\} \downarrow \boldsymbol{\theta},$$

and the same holds for $\dfrac{\pi_{t''}(\boldsymbol{\theta})}{\pi_{t'}(\boldsymbol{\theta})} = \left(\dfrac{\overline{G}(t''|\boldsymbol{\theta})}{\overline{G}(t'|\boldsymbol{\theta})}\right)^{n-k}$.

Therefore, by definition of wlr order, $\pi_{t''} \leq_{wlr} \pi_{t'}$.

Remark 3. It can also be proven, by means of technical arguments, that (b) and (c) are equivalent (see [Fos10]).

Remark 4. It can be easily shown that the assumption (a) in the former Proposition 1 implies a condition of default contagion, in the following sense:

$$\overline{F}^{(n-k)}_{T^-_{(k)}}(x_1, \ldots, x_{n-k}) \geq \overline{F}^{(n-k)}_{T_{(k)}}(x_1, \ldots, x_{n-k}), \quad \forall\, x_1, \ldots, x_{n-k} \geq 0. \tag{7.9}$$

It is then interesting to notice that the assumption guaranteeing default contagion also implies, under our conditions, a jump downward of the copulas of the surviving units, at default times (see Proposition 3 below).

Notice that, in the proofs of Propositions 2 and 3 below, we need the assumption π_t MTP_2 for any t. As a matter of fact, this hypothesis is rather strong. Sufficient conditions for it can however be formulated in terms of $r(t|\boldsymbol{\theta})$ and $\overline{G}(t|\boldsymbol{\theta})$ by taking into account the expression (7.4) (see also [SSp98]).

However, $t \longrightarrow \pi_t$ monotonic in the weak likelihood ratio order (as in Proposition 1) and π_t MTP_2 for any t guarantees $t \longrightarrow \pi_t$ being monotonic in the likelihood ratio order (see Lemma 1 in the Appendix). The latter condition implies the monotonicity of $t \longrightarrow \pi_t$ with respect to the usual stochastic order.

We are now in a position to state and prove the following results about the monotonicity of the family $\left\{\hat{C}^{(2)}_t\right\}$ of the survival copulas.

Proposition 2. *Let $r(t|\boldsymbol{\theta})$ be monotonic w.r.t. $\boldsymbol{\theta}$, π_t MTP$_2$ for any t and*

$$\frac{\overline{G}\left(\overline{G}_t^{-1}(u)+t|\boldsymbol{\theta}\right)}{\overline{G}\left(t|\boldsymbol{\theta}\right)} \quad \textit{increasing w.r.t. } t.$$

Then $t \mapsto \hat{C}_t^{(2)}$ is increasing in the intervals between two jumps.

Proof. Assume, for instance, $r(t|\boldsymbol{\theta})$ increasing in $\boldsymbol{\theta}$. The proof when $r(t|\boldsymbol{\theta})$ is decreasing in $\boldsymbol{\theta}$ is analogous. Let us denote

$$\rho(\boldsymbol{\theta},t)=\frac{\overline{G}(\overline{G}_t^{-1}(u_1)+t|\boldsymbol{\theta})}{\overline{G}(t|\boldsymbol{\theta})}\frac{\overline{G}(\overline{G}_t^{-1}(u_2)+t|\boldsymbol{\theta})}{\overline{G}(t|\boldsymbol{\theta})}. \tag{7.10}$$

In view of (7.8), the condition $t \mapsto \hat{C}_t^{(2)}$ being increasing reads as

$$\int_{\mathbb{R}_+^d}\rho(\boldsymbol{\theta},t')\pi_{t'}(\boldsymbol{\theta})d\boldsymbol{\theta} \le \int_{\mathbb{R}_+^d}\rho(\boldsymbol{\theta},t'')\pi_{t''}(\boldsymbol{\theta})\mathrm{d}\boldsymbol{\theta} \tag{7.11}$$

for any $t' \le t''$. Thus, we want to prove (7.11). By Proposition 1, $\boldsymbol{\theta} \to r(t|\boldsymbol{\theta})$ increasing implies $\pi_{t'} \ge_{wlr} \pi_{t''}$. Since π_t is MTP$_2$ for any t, by Lemma 1, $\pi_{t'} \ge_{lr} \pi_{t''}$ and, therefore, in particular, it will be $\pi_{t'} \ge_{st} \pi_{t''}$. By Proposition 1, $\boldsymbol{\theta} \to r(t|\boldsymbol{\theta})$ increasing implies $\boldsymbol{\theta} \to \rho(\boldsymbol{\theta},t)$ decreasing. Therefore, by (7.14) in the Appendix, (7.11), $\pi_{t'} \ge_{st} \pi_{t''}$ implies

$$\int_{\mathbb{R}_+^d}\rho(\boldsymbol{\theta},t')\pi_{t'}(\boldsymbol{\theta})\mathrm{d}\boldsymbol{\theta} \le \int_{\mathbb{R}_+^d}\rho(\boldsymbol{\theta},t')\pi_{t''}(\boldsymbol{\theta})\mathrm{d}\boldsymbol{\theta}.$$

Since, by hypothesis, $t \to \rho(\boldsymbol{\theta},t)$ is increasing,

$$\int_{\mathbb{R}_+^d}\rho(\boldsymbol{\theta},t')\pi_{t''}(\boldsymbol{\theta})\mathrm{d}\boldsymbol{\theta} \le \int_{\mathbb{R}_+^d}\rho(\boldsymbol{\theta},t'')\pi_{t''}(\boldsymbol{\theta})\mathrm{d}\boldsymbol{\theta},$$

and therefore

$$\int_{\mathbb{R}_+^d}\rho(\boldsymbol{\theta},t')\pi_{t'}(\boldsymbol{\theta})\mathrm{d}\boldsymbol{\theta} \le \int_{\mathbb{R}_+^d}\rho(\boldsymbol{\theta},t'')\pi_{t''}(\boldsymbol{\theta})\mathrm{d}\boldsymbol{\theta},$$

that is the thesis.

Proposition 3. *Let $r(t|\boldsymbol{\theta})$ be a monotonic function of $\boldsymbol{\theta}$ and π_t MTP$_2$ for any t. Then, with probability 1,*

$$\hat{C}_{T_{(k)}}^{(2)} \prec_{PQD} \hat{C}_{T_{(k)}^-}^{(2)},$$

i.e., for any $u_1, u_2 \in [0,1]$,

$$\hat{C}_{T_{(k)}}^{(2)}(u_1,u_2) \le \hat{C}_{T_{(k)}^-}^{(2)}(u_1,u_2). \tag{7.12}$$

Proof. Let us define, as in (7.10),

$$\rho(\boldsymbol{\theta},T_{(k)})=\frac{\overline{G}(\overline{G}_{T_{(k)}}^{-1}(u_1)+T_{(k)}|\boldsymbol{\theta})\,\overline{G}(\overline{G}_{T_{(k)}}^{-1}(u_2)+T_{(k)}|\boldsymbol{\theta})}{\overline{G}(T_{(k)}|\boldsymbol{\theta})\qquad\qquad\overline{G}(T_{(k)}|\boldsymbol{\theta})}$$

and

$$\rho(\boldsymbol{\theta}, T_{(k)}^-) = \frac{\overline{G}(\overline{G}_{T_{(k)}^-}^{-1}(u_1) + T_{(k)}|\boldsymbol{\theta})}{\overline{G}(T_{(k)}|\boldsymbol{\theta})} \frac{\overline{G}(\overline{G}_{T_{(k)}^-}^{-1}(u_2) + T_{(k)}|\boldsymbol{\theta})}{\overline{G}(T_{(k)}|\boldsymbol{\theta})}.$$

We want to prove that

$$\int_{\mathbb{R}_+^d} \rho(\boldsymbol{\theta}, T_{(k)}) \pi_{T_{(k)}}(\boldsymbol{\theta}) \mathrm{d}\boldsymbol{\theta} \leq \int_{\mathbb{R}_+^d} \rho(\boldsymbol{\theta}, T_{(k)}^-) \pi_{T_{(k)}^-}(\boldsymbol{\theta}) \mathrm{d}\boldsymbol{\theta}.$$

Assume $\boldsymbol{\theta} \to r(t|\boldsymbol{\theta})$ to be increasing. The proof when $\boldsymbol{\theta} \to r(t|\boldsymbol{\theta})$ is decreasing is analogous.
By Proposition 1, $\pi_{T_{(k)}} \geq_{wlr} \pi_{T_{(k)}^-}$ and then, by Lemma 1, $\pi_{T_{(k)}} \geq_{lr} \pi_{T_{(k)}^-}$ holds.
On the other hand, $r(t|\boldsymbol{\theta}) \uparrow \boldsymbol{\theta}$ implies $\dfrac{\overline{G}(x+t|\boldsymbol{\theta})}{\overline{G}(t|\boldsymbol{\theta})} \downarrow \boldsymbol{\theta}$ and therefore $\boldsymbol{\theta} \to \rho(\boldsymbol{\theta}, t)$ decreasing. We recall that

$$\overline{G}_{T_{(k)}}(u) = \int_{\mathbb{R}_+^d} \frac{\overline{G}(x + T_{(k)}|\boldsymbol{\theta})}{\overline{G}(T_{(k)}|\boldsymbol{\theta})} \pi_{T_{(k)}}(\boldsymbol{\theta}) \mathrm{d}\boldsymbol{\theta}, \ \overline{G}_{T_{(k)}^-}(u) = \int_{\mathbb{R}_+^d} \frac{\overline{G}(x + T_{(k)}|\boldsymbol{\theta})}{\overline{G}(T_{(k)}|\boldsymbol{\theta})} \pi_{T_{(k)}^-}(\boldsymbol{\theta}) \mathrm{d}\boldsymbol{\theta}.$$

Since $\pi_{T_{(k)}} \geq_{lr} \pi_{T_{(k)}^-}$ implies $\pi_{T_{(k)}} \geq_{st} \pi_{T_{(k)}^-}$, by condition (7.14), it follows that $\overline{G}_{T_{(k)}}(u) \leq \overline{G}_{T_{(k)}^-}(u)$. This implies

$$\frac{\overline{G}(\overline{G}_{T_{(k)}}^{-1}(u) + T_{(k)}|\boldsymbol{\theta})}{\overline{G}(T_{(k)}|\boldsymbol{\theta})} \leq \frac{\overline{G}(\overline{G}_{T_{(k)}^-}^{-1}(u) + T_{(k)}|\boldsymbol{\theta})}{\overline{G}(T_{(k)}|\boldsymbol{\theta})}, \tag{7.13}$$

that is $\rho(\boldsymbol{\theta}, T_{(k)}) \leq \rho(\boldsymbol{\theta}, T_{(k)}^-)$. Hence

$$\int_{\mathbb{R}_+^d} \rho(\boldsymbol{\theta}, T_{(k)}) \pi_{T_{(k)}}(\boldsymbol{\theta}) d\boldsymbol{\theta} \leq \int_{\mathbb{R}_+^d} \rho(\boldsymbol{\theta}, T_{(k)}) \pi_{T_{(k)}^-}(\boldsymbol{\theta}) \mathrm{d}\boldsymbol{\theta} \leq \int_{\mathbb{R}_+^d} \rho(\boldsymbol{\theta}, T_{(k)}^-) \pi_{T_{(k)}^-}(\boldsymbol{\theta}) \mathrm{d}\boldsymbol{\theta}.$$

Thus, we can conclude that, for any u_1, u_2 in $[0,1]$, $\hat{C}_{T_{(k)}}(u_1, u_2) \leq \hat{C}_{T_{(k)}^-}^{(2)}(u_1, u_2)$.

The analysis of a simple and very well known model allows us to show now an example of application of both Proposition 2 and 3 in the case $\theta \in \mathbb{R}_+$.

Example 1. Consider $\overline{G}(t|\theta) = e^{-\theta t}$, so that $r(t|\theta) = \theta$ (i.e. $r(t|\theta)$ is constant w.r.t. t and increasing w.r.t. θ) and the assumptions of Proposition 2 and Proposition 3 are satisfied. In particular, if the initial distribution of Θ is Gamma(α_0, β_0), then, at any time $t > 0$, the conditional distribution of Θ is again Gamma and $\hat{C}_t^{(2)}$ is a Clayton copula. More precisely,

$$\pi_t(\theta) = \frac{\beta_t^{\alpha_t}}{\Gamma(\alpha_t)} \theta^{\alpha_t - 1} e^{-\beta_t \theta},$$

with $\alpha_t = \alpha_0 + N_t$, $\beta_t = \beta_0 + \sum_{k=1}^n \min(T_{(k)}, t)$, and

$$\hat{C}_t^{(2)}(u_1, u_2) = \left(u_1^{-\frac{1}{\alpha_t}} + u_2^{-\frac{1}{\alpha_t}} - 1 \right)^{-\alpha_t}.$$

We obtain that $t \to \hat{C}_t^{(2)}$ remains constant between two subsequent default times and makes a jump downward at instants of default.

7.4 Concluding Remarks

In this section, we present some remarks and comments about the results we have obtained. As we have already said, our main results are contained in the Propositions 2 and 3.

Proposition 2 states that, if the conditional hazard rate $r(t|\boldsymbol{\theta})$ is monotonic in $\boldsymbol{\theta}$ and if the components of $\boldsymbol{\Theta}$ satisfy a suitable positive dependence property, namely MTP_2, dependence among residual lifetimes continuously increases at the increase of survival time.

In particular, since residual lifetimes become more and more dependent, Proposition 2 gives conditions for the phenomenon of tail dependence. The latter circumstance means that extremal events are more dependent than non-extremal ones. In other words, we can intuitively expect that, eventually, failures will occur each close to the other ones.

Proposition 3 states instead that, under the same conditions on $r(t|\boldsymbol{\theta})$ and on $\boldsymbol{\Theta}$, inequality (7.12) holds, i.e. the dependence among residual lifetimes discontinuously decreases when a failure occurs.

Notice that (7.12), i.e. $\hat{C}_{T_{(k)}} \preceq_{PQD} \hat{C}_{T^-_{(k)}}$, and default contagion are two phenomena referring to the behaviour of residual lifetimes at the instants of defaults. Actually, they appear to be two different phenomena; they are however related in some way. More precisely, it can be shown that $r(t|\boldsymbol{\theta})$ monotonic in $\boldsymbol{\theta}$ also implies default contagion. Still remaining in the present case of conditionally i.i.d. observations, we can however say more: it can be shown that $\hat{C}_{T_{(k)}} \preceq_{PQD} \hat{C}_{T^-_{(k)}}$ implies default contagion in the sense of (7.9) (see [Fos10] for details).

Let us now come to comment on some technical assumptions in Propositions 2 and 3.

Notice that, even if the likelihood ratio order is a very well known and most used notion, in Proposition 1, it is sufficient requiring the weak likelihood ratio order. In fact, as it happens in the univariate case, we actually use the condition $\dfrac{\pi_t(\boldsymbol{\theta})}{\pi_s(\boldsymbol{\theta})}$ being monotonic in $\boldsymbol{\theta}$. When $\boldsymbol{\theta}$ is univariate, such a condition corresponds just to a characterization of the likelihood ratio order; in the multivariate case, it gives the definition of weak likelihood ratio order. The point is that, in order to prove Propositions 2 and 3, we need monotonicity of the family $\{\pi_t\}$ with respect to the usual stochastic order. In the multivariate case, the usual stochastic order is implied by the likelihood ratio order, but not by the weak likelihood ratio order. In order to retrieve the monotonicity of the family $\{\pi_t\}$ with respect to the usual stochastic order, it is necessary imposing, further, the condition that $\boldsymbol{\Theta}$ is MTP_2.

As a matter of fact, in the multivariate case, we distinguish between weak and strong notions of a same stochastic ordering, whereas, in the univariate case, the two notions do coincide. Adding the condition $\boldsymbol{\Theta}$ MTP_2 allows us to obtain, from the weaker notion of stochastic order, the stronger one.

More in general, one could argue that MTP_2, representing a strong notion of dependence, allows us to treat the random vector $\boldsymbol{\Theta}$ like a scalar random variable. On its turn, this fact makes it possible to automatically extend many results, valid for the case of a univariate non-observable factor, to the multivariate case.

Acknowledgements. This research work has been developed, and partially supported, in the frame of Progetto di Ricerca 2009: "Modelli e Algoritmi Stocastici - Università "La Sapienza" " and Progetto di Ricerca GNAMPA 2009: "Successioni Scambiabili, Copule e Relative Generalizzazioni".

References

[AN84] Arjas, E. and Norros, I.: Life lengths and association: a dynamic approach. Mathematics of Operations Research, **9**, 151–158 (1984).

[BMRS09] Belzunce, F., Mercader, J.-A., Ruiz, J.-M., Spizzichino, F.: Stochastic comparisons of multivariate mixture models. Journal of Multivariate Analysis, **100**(8), 1657–1669 (2009).

[CJ06] Charpentier, A. and Juri, A.: Limiting dependence structures for tail events, with applications to credit derivatives. Journal of Applied Probability, **43**, 563–586 (2006).

[Fos08] Foschi, R.: Evolution of survival copulas under longitudinal observation of dependent default times. The case of conditional independence. Proceedings of the International Workshop on Applied Probability. Université de Technologie de Compiègne (2008).

[Fos10] Foschi, R.: Evolution of conditional dependence of residual lifetimes. PhD Thesis, University La Sapienza, Rome (2010).

[FS08] Foschi, R. and Spizzichino, F.: Semigroups of semicopulas and evolution of dependence at increase of age. Mathware & Soft Computing, **XV**, 95–111 (2008).

[Joe97] Joe, H.: Multivariate Models and Dependence Concepts. Chapman & Hall, London (1997).

[Koc99] Kochar, S.: On stochastic orderings between distributions and their sample spacings. Statistics & Probability Letters, **42**, 345–352 (1999).

[LX06] Lai, C.-D. and Xie, M.: Stochastic Ageing and Dependence for Reliability. Springer, New York (2006).

[MFE05] McNeil, A. J., Frey, R., Embrechts, P.: Quantitative Risk Management: Concepts, Techniques and Tools. Princeton University Press, New York (2005).

[Nel06] Nelsen, R. B.: An Introduction to Copulas, Springer, Heidelberg (2006).

[SSh07] Shaked, M. and Shanthikumar, J. G.: Stochastic Orders, Springer Series in Statistics, Springer, New York (2007).
Stochastic Orders", Springer Series in Statistics, Springer, New York (2007).

[SSp98] Shaked, M. and Spizzichino, F.: Positive dependence properties of conditionally independent random lifetimes. Mathematics of Operations Research, **23**, 944–959 (1998).

[Spi01] Spizzichino, F.: Subjective Probability Models for Lifetimes. Chapman & Hall/CRC, Boca Raton, FL, New York (2001).

Appendix

We assume that the reader is familiar with the most common notions about univariate stochastic orderings and report here those essential notions about multivariate stochastic orderings, that are needed for out treatment (see [SSh07] for a complete review about all these topics. See also [BMRS09] for specific aspects of multivariate stochastic orderings of vectors of conditionally independent variables).

Let U, V be two random vectors with joint survival functions $\overline{F}_U$, $\overline{F}_V$ and probability densities f_U, f_V respectively.

On $\mathbb{R}^d$ we consider the usual componentwise partial order, defined as follows: let $\mathbf{u} = (u_1, \ldots, u_d)$, $\mathbf{v} = (v_1, \ldots, v_d)$ be two vectors in $\mathbb{R}^d$; then we write $\mathbf{u} \leq \mathbf{v}$ if $v_i \leq v_i$ for $i = 1, \ldots, d$.
A set $B \subseteq \mathbb{R}^d$ is called an *upper set* if $\mathbf{v} \in B$ whenever $\mathbf{u} \leq \mathbf{v}$ and $\mathbf{u} \in B$.

We write $f_U \leq_{st} f_V$ if $P(U \in B) \leq P(V \in B)$, for any upper set $B \subseteq \mathbb{R}^d$; in this case we also say that *U is smaller than V in the usual stochastic order.*

$f_U \leq_{st} f_V$ implies $\overline{F}_U(\mathbf{t}) \leq \overline{F}_V(\mathbf{t})$ for any $\mathbf{t} \in \mathbb{R}^d$, as it is immediately seen by letting the upper set B to be a d-rectangle. The condition $\overline{F}_U \leq \overline{F}_V$ is called *upper orthant order* and it is denoted by $f_U \leq_{uo} f_V$. A characterization of the usual stochastic order, analogous to the one in the one-dimensional case, is given by the condition

$$\int_{\mathbb{R}^d} \rho(\mathbf{t}) f_U(\mathbf{t}) d\mathbf{t} \geq \int_{\mathbb{R}^d} \rho(\mathbf{t}) f_V(\mathbf{t}) d\mathbf{t} \tag{7.14}$$

for any decreasing function $\rho : \mathbb{R}^d \to \mathbb{R}$.

We also use the symbols $\vee$ and $\wedge$: $\mathbf{u} \vee \mathbf{v} = (u_1 \vee v_1, \ldots, u_d \vee v_d)$ where $u_i \vee v_i \equiv \max\{u_i, v_i\}$; analogously $\mathbf{u} \wedge \mathbf{v} = (u_1 \wedge v_1, \ldots, u_d \wedge v_d)$ where $u_i \wedge v_i \equiv \min\{u_i, v_i\}$. The multivariate likelihood ratio order $f_U \leq_{lr} f_V$ is defined by means of the following condition:

$$f_U(\mathbf{u}) f_V(\mathbf{v}) \leq f_U(\mathbf{u} \wedge \mathbf{v}) f_V(\mathbf{u} \vee \mathbf{v})$$

for every $\mathbf{u}, \mathbf{v} \in \mathbb{R}^d$.

The likelihood ratio (= lr) order $f_U \leq_{lr} f_V$ implies the so-called *weak likelihood ratio* (= wlr) order. The latter notion is defined by imposing a condition that is analogous to the characterization of the likelihood ratio order in the univariate case:

Definition 4. *We say that U is smaller than V in the wlr order, and we denote it by $f_U \leq_{wlr} f_V$, if*

$$\frac{f_U(\mathbf{t})}{f_V(\mathbf{t})} \downarrow \mathbf{t}. \tag{7.15}$$

We recall that $f_U \leq_{lr} f_V$ implies $f_U \leq_{st} f_V$ and $f_U \leq_{wlr} f_V$ implies $f_U \leq_{uo} f_V$, but $f_U \leq_{wlr} f_V$ does not imply $f_U \leq_{st} f_V$.

We say that a function $f : \mathbb{R}^d \to \mathbb{R}$ is MTP$_2$ if

$$f(\mathbf{u}) f(\mathbf{v}) \leq f(\mathbf{u} \wedge \mathbf{v}) f(\mathbf{u} \vee \mathbf{v})$$

for any $\mathbf{u}, \mathbf{v} \in \mathbb{R}^d$.

Lemma 1 ([Koc99]). *If $\dfrac{f_V(\mathbf{t})}{f_U(\mathbf{t})} \uparrow \mathbf{t}$ and f_U is MTP$_2$, then $f_U \leq_{lr} f_V$.*

8

On Alternative of Choice for a Prophylaxis Problem

B.P. Harlamov

Institute of Problems of Mechanical Engineering, Russian Academy of Science, St. Petersburg, Russia, b.p.harlamov@gmail.com

Abstract: A stationary server system with observable degradation is considered. A simple Cox model for the hazard rate dependency on degradation is used. An optimization problem for choice of a Markov time to begin prophylactic repair of the system is being investigated. For non-monotone processes of degradation there exists a sequence of Markov times which some local condition of optimality is fulfilled at. In this case the following alternative arises: either the first time of this sequence is the best, or the global optimal time does not exist. A regenerative degradation process is shown to illustrate this alternative.

Keywords and phrases: Degradation, Cox model, Random process, Markov time, Repair, Hazard rate

8.1 Renewal System

This work is a supplement to the paper [RH09]. We give here additional explanations of some aspects of the problem, e.g. a localizing Markov time, a trajectory of choice, an alternative of choice, and show an example of the visually proven alternative for a twice regenerative server system.

We consider a repairable server system with one channel. Its action consists of work and repair periods. The repair periods are of two types: a repair after a failure, and a prophylactic repair. After repair the system is as good as new. We describe this server system as a regenerative process, generated by a sequence of i.i.d. cycles of activity $(X_n, \zeta_n, \tau_n, M_n, R_n)_{n=1}^{\infty}$. Elements of every cycle are

- $X_n = (X_n(t))$ $(t \geq 0)$, an one-dimensional random process which is actually considered on a random interval $[0, \zeta_n \wedge \tau_n)$;
- ζ_n, a random time to failure, connected with the process X_n with the help of a hazard rate $h(X_n)$, where $h(x)$ $(x \in \mathbb{R})$ is a non-decreasing positive function;

V.V. Rykov et al. (eds.), *Mathematical and Statistical Models and Methods in Reliability: Applications to Medicine, Finance, and Quality Control*, Statistics for Industry and Technology, DOI 10.1007/978-0-8176-4971-5_8,

- τ_n, a time of beginning of the prophylactic repair, which is a Markov time with respect to the process X_n and the natural filtration $(\mathcal{F}_t)$ on the set of trajectories;
- M_n, a random duration of a repair period.
- R_n, a random value of loss due to stopping.

We interpret $X_n(t)$ as observable degradation of the system in the n-th cycle of regeneration. The condition survival function of ζ is

$$P(\zeta_1 > t|\, X_1) = \exp\left(- \int_0^t h(X_1(s))\,\mathrm{d}s \right).$$

Thus, the survival function is assumed to be described with Cox model, which is recently rather popular among specialists on reliability theory; see, for example, [BN00], and the newest one [BMN09] and [HNR09]. The prophylactic repair in the nth cycle begins at the instant, τ_n, defined by an operator of the system, which have to optimize it according to some loss function. Let $T_n = \zeta_n \wedge \tau_n + M_n$ be a duration of the n-th cycle of regeneration. We assume the system has a loss during one regeneration cycle as follows

$$f_n = - \int\limits_0^{\zeta_n \wedge \tau_n} A(X_n(t))\,\mathrm{d}t + R_n,$$

where $A(\cdot)$ is a non-decreasing positive function; we interpret $A(X_n(t))\,dt$ as a useful output during an infinitesimal interval dt. In what follows we use conditional expectations:

$$m_1 \equiv E(M_1|\, \zeta_1 \le \tau_1), \quad m_2 \equiv E(M_1|\, \zeta_1 > \tau_1),$$

$$r_1 \equiv E(R_1|\, \zeta_1 \le \tau_1), \quad r_2 \equiv E(R_1|\, \zeta_1 > \tau_1),$$

and suppose that $m_1 > m_2$ and $r_1 > r_2$. We will use a loss function of the optimization problem, $L \equiv L_\tau$, coinciding with average loss of the system during a long period of action, i.e.

$$L \equiv \lim_{t\to\infty} \frac{1}{t} \sum_{k=1}^{N_t} f_k \overset{\text{a.s.}}{=} \frac{Ef_1}{ET_1}, \tag{8.1}$$

where N_t is the corresponding counting process. We have

$$Ef_1 \equiv Y_\tau = -E \int\limits_0^{\zeta_1 \wedge \tau_1} A(X_1(t))\,dt + r_1 P(\zeta_1 \le \tau_1) + r_2 P(\zeta_1 > \tau_1),$$

$$ET_1 \equiv Z_\tau = E(\zeta_1 \wedge \tau_1) + m_1 P(\zeta_1 \le \tau_1) + m_2 P(\zeta_1 > \tau_1).$$

Further, we will drop index 1 in denotations of random processes and values of the first regeneration cycle if it does not lead to ambiguity, and use denotation X_t instead of $X(t)$, when it seems convenient.

Using positivity of ζ, we can obtain

$$Ef_1 = -W_\tau + r_1 - (r_1 - r_2)U_\tau, \quad ET_1 = V_\tau + m_1 - (m_1 - m_2)U_\tau,$$

where

$$U_\tau = E\exp\left(-\int_0^\tau h(X_s)\,\mathrm{d}s\right),$$

$$V_\tau = E\int_0^\tau \exp\left(-\int_0^t h(X_s)\,\mathrm{d}s\right)\,\mathrm{d}t,$$

$$W_\tau = E\int_0^\tau A(X_t)\exp\left(-\int_0^t h(X_s)\,\mathrm{d}s\right)\,\mathrm{d}t.$$

8.2 Optimization Problem

8.2.1 Local Optimality Equation

We will investigate a classical optimization problem to find τ_* such that

$$L_{\tau_*} = \min_{\tau\in\mathcal{T}} L_\tau,$$

where $\mathcal{T}$ is the set of Markov times with respect to filtration $(\mathcal{F}_t)$. In [RH09], we show that a necessary condition for a Markov time τ to be a local minimum of L_τ is the following a.s. equation

$$\frac{-A(X_\tau) + (r_1 - r_2)h(X_\tau)}{1 + (m_1 - m_2)h(X_\tau)} = \frac{Y_\tau}{Z_\tau} \quad (\tau < \infty). \tag{8.2}$$

Moreover, this local minimum exists any time when the left hand side of (8.2) reaches the level, represented by the right hand side of this equation, with up-crossing this level.

8.2.2 Localizing Markov Times

Since the left hand side of (8.2) is valid under any appropriate functions A and h, and the right hand side is not a random value, we conclude that almost all the trajectories of the process have identical meanings at the point τ. Thus, we consider a subclass $\mathcal{T}' \subset \mathcal{T}$ with a property: $(\forall\tau \in \mathcal{T}')$ $(\exists a \in \mathbb{R})$ $X_\tau \overset{\text{a.s.}}{=} a$ on the set $\{\tau < \infty\}$. We call $\mathcal{T}'$ a class of localizing Markov times. We have $\mathcal{T}' = \bigcup_{a\in\mathbb{R}} \mathcal{T}_a$, where $\mathcal{T}_a$ is a subclass of all the localizing Markov times, corresponding to the point a. Hence (8.2) can be rewritten as

$$\frac{-A(a) + (r_1 - r_2)h(a)}{1 + (m_1 - m_2)h(a)} = \frac{Y_{\tau_a}}{Z_{\tau_a}} \quad (\tau_a \in \mathcal{T}_a,\ \tau_a < \infty). \tag{8.3}$$

A trivial example of localizing Markov time for a process with a.s. continuous trajectories and the initial point $x_0 < a$ is the first exit time from the interval

$(-\infty, a)$ (we denote this as $\mu_a \equiv \sigma_{(-\infty,\, a)}$, where $\sigma_\Delta(\xi) = \inf\{t \ge 0 : \xi(t) \notin \Delta\}$). In this case, a trivial solution (if any) can be obtained as a unique solution of the equation

$$\frac{-A(a) + (r_1 - r_2)h(a)}{1 + (m_1 - m_2)h(a)} = \frac{Y_{\mu_a}}{Z_{\mu_a}} \quad (\mu_a < \infty).$$

Both parts of this equation are functions of a. A crossing point of the corresponding graphs represents a solution of this equation. For a continuous not-decreasing process X_t the solution of this equation coincides with the unique solution of (8.3). In case of non-monotone process X_t a solution of this equation can be not unique. Namely, one can construct a composition of Markov times which contains a shifted μ_a. Since for $\tau_1 \equiv \tau \dot{+} \mu_a$ the identity $X_{\tau_1} = X_{\mu_a} \circ \theta_\tau = a$ on the set $\{\tau_1 < \infty\}$ holds the set of possible candidates on a role of the best solutions of (8.3) is very rich (here and in what follows we denote $\tau_1 \dot{+} \tau_2 \equiv \tau_1 + \tau_2 \circ \theta_{\tau_1}$, where θ_t is the shift operator on the set of cadlag functions [Har07]). In order to find the best localizing time in [RH09] we used so called trajectories of choice which were constructed for a sequence of levels (a_n) with the help of corresponding localizing Markov times μ_a and ν_a ($\nu_a \equiv \sigma_{(a,\, \infty)}$). Our aim was to prove so called alternative of choice.

Definition 1 *The following two assertions are called to be alternative of choice for a given problem:*

1. *either the trivial solution is the best,*
2. *or the optimal solution does not exist.*

It was motivated by the situation where the right hand side of (8.2) is a constant. In [RH09a], we prove that in this case for a Markov process (X_t) this alterative is fair. In order to prove this alternative for more general situation in [RH09], we used representation of (8.3) in form of equality of two functions of argument a with certain direction of varying. For example, the left hand side is decreasing function, and the right hand side is increasing function of a, which implies existence of their unique crossing point. Usually one part of this equality does not depend on our choice, and the other part is rather arbitrary. This part can be chosen by the operator of the system if he (she) hopes to decrease the loss function. This part determines so called a trajectory of choice. There exist some properties of trajectories of choice which restrict a set of possible solutions. In [RH09], we used a special function, G, on a set of parameters for to prove some sufficient conditions of choice. For a process X_t of diffusion type we had shown that if the sign of G is negative everywhere in the set of parameters the first assertion of the alternative is true. If the sign of G is positive everywhere in the set of parameters the second assertion of the alternative is true. Uncertainty remains when this function changes its sign inside the set of parameters. Thus, we did not receive the full proof of the alternative of choice for a diffusion process (X_t). In this case, the problem requires further investigation.

8.3 Alternative of Choice for Twice Regenerative System

In this paper, we present an example where the alternative of choice is proved. In what follows we will consider an a.s. piece-wise continuous non-decreasing process X_t with negative jumps. Let X_t be a regenerative process. Thus, the system we investigate is twice regenerative. The first order of regeneration is the system as a whole with the infinite sequence of work–repair cycles. And the second order of regeneration is the action period inside one cycle. This internal regenerative process is to be terminated due to failure of the system or stopping for prophylaxis repair. An interpretation of this second order regenerative process of degradation is discussed, for example, in [BN00].

Let $(\sigma_n)_1^\infty$ be points of jumps of a homogeneous renewal process and $T_n = \sigma_n - \sigma_{n-1}$. It means that (T_n) is a sequence of i.i.d. positive random variables. Let there exist a value $c > 0$ such that on every interval (σ_{n-1}, σ_n) the process (X_t) increases continually from zero to c; at points of discontinuity, σ_n, the process is assumed to be right continuous. This process need not be Markov, but it is assumed to have points of regeneration (σ_n). Let us denote $B(X_\tau)$ the left hand part of (8.2):

$$B(a) \equiv \frac{-A(a) + \Delta r\, h(a)}{1 + \Delta m\, h(a)},$$

where $\Delta r = r_1 - r_2$, $\Delta m = m_1 - m_2$. Further we will assume $A(0) = h(0) = 0$. In [RH09], the following equivalent form of this equation is proved.

$$r_2 - B(X_\tau)m_2 = H_\tau(X_\tau) \quad (\tau < \infty), \tag{8.4}$$

where

$$H_\tau(a) = E\int_0^\tau \exp\left(-\int_0^t h(X_s)\,ds\right)(1 + \Delta m\, h(X_t))(B(a) - B(X_t))\,dt$$

$(a \geq 0)$. Hence, (8.3) can be represented as

$$r_2 - B(a)m_2 = H_{\tau_a}(a). \tag{8.5}$$

We can see, that this equation does not require restriction $\tau_a < \infty$ [RH09]. Moreover, the left hand side of this equation is a decreasing function of a, and the right hand side of this equation is an increasing function of a if $\tau_a \equiv \mu_a$ and

$$\frac{h'(x)}{1 + \Delta m\, h(x)} \geq \frac{A'(x)}{\Delta r + \Delta m\, A(x)} \quad (x > 0);$$

(see [RH09]). Further, we will assume that this condition is fulfilled in a strengthen form:

$$\frac{h'(x)}{1 + \Delta m\, h(x)} \geq \frac{A'(x)}{\Delta r + \Delta m\, A(x)} + \frac{r_2(1 + \Delta m\, h(x)}{c\, m_2\,(\Delta r + \Delta m\, A(x))} \quad (0 < x < c). \tag{8.6}$$

Since $B(0) = 0$, this condition follows from the restriction

$$(r_2 - B(x)m_2)' \leq -r_2/c,$$

which supplies existence a point of intersection, $a_1 \in (0, c)$, of curves determined by the left and right hand sides of (8.5), if $\tau_a \equiv \mu_a$.

Let us consider trajectories of choice corresponding to the family of Markov times $\tau_{k,a} = \sigma_{k-1}\dot{+}\mu_a$ ($0 < a < c$, $k \ge 1$, $\sigma_0 = 0$). It means we test any such a time on a role of a solution of (8.5). There exists the first solution, μ_{a_1}, corresponding to the equation

$$r_2 - B(a) = H_{\mu_a}(a).$$

We have $H_{\tau_{1,0}}(0) \equiv H_{\mu_0}(0) = 0$, and for all $k \ge 2$

$$H_{\tau_{k,0}}(0) = H_{\sigma_{k-1}}(0) = H_{\sigma_{k-2}}(0) -$$

$$-E\int_{\sigma k-2}^{\sigma k-1} \exp\left(-\int_0^t h(X_s)\,\mathrm{d}s\right)(1 + \Delta m\, h(X_t))B(X_t)\,\mathrm{d}t \le H_{\tau_{k-1,0}}(0).$$

On the other hand

$$H_{\tau_{k,c-0}}(c) \equiv H_{\sigma_{k+1}}(c) = H_{\sigma_k}(c) +$$

$$+E\int_{\sigma_k}^{\sigma k+1} \exp\left(-\int_0^t h(X_s)\,\mathrm{d}s\right)(1 + \Delta m\, h(X_t))(B(c) - B(X_t))\,\mathrm{d}t \ge H_{\tau_{k-1,c-0}}(c).$$

Since every function $H_{\tau_{k,a}}(a)$ is continuous for all $k \ge 2$ there exists a point of intersection, $a_k \in (0, c)$, of curves determined by the left and right hand sides of (8.5), if $\tau_a \equiv \tau_{k,a}$.

Since the condition of a local minimum (8.2) can be rewritten as $B(a) = L_{\tau_a}$ and $B(a)$ is an increasing function we conclude that $L_{\tau_{k,a_k}}$ increases if the sequence (a_k) increases, and it decreases if (a_k) decreases. In order to prove the alternative of choice we can do it for the sequence (a_k).

We denote $H_\tau(a) \equiv B(a)S_\tau - T_\tau$, where

$$S_\tau = E\int_0^\tau \exp\left(-\int_0^t h(X_s)\,\mathrm{d}s\right)(1 + \Delta m\, h(X_t))\,\mathrm{d}t,$$

$$T_\tau = E\int_0^\tau \exp\left(-\int_0^t h(X_s)\,\mathrm{d}s\right)(1 + \Delta m\, h(X_t))B(X_t)\,\mathrm{d}t.$$

Using the regeneration property we have

$$S_{\sigma_k} = S_{\sigma_{k-1}} + E\int_{\sigma_{k-1}}^{\sigma_k} \exp\left(-\int_0^t h(X_s)\,\mathrm{d}s\right)(1 + \Delta m\, h(X_t))\,\mathrm{d}t = S_{\sigma_{k-1}} +$$

$$+E\exp\left(-\int_0^{\sigma_{k-1}} h(X_s)\,\mathrm{d}s\right)\int_{\sigma_{k-1}}^{\sigma_k} \exp\left(-\int_{\sigma_{k-1}}^t h(X_s)\,\mathrm{d}s\right)(1 + \Delta m\, h(X_t))\,\mathrm{d}t =$$

$$= S_{\sigma_{k-1}} + U_{\sigma_{k-1}} E\int_0^{\sigma_1} \exp\left(-\int_0^t h(X_s)\,\mathrm{d}s\right)(1 + \Delta m\, h(X_t))\,\mathrm{d}t =$$

$$= S_{\sigma_{k-1}} + U_{\sigma_{k-1}} S_{\sigma_1}.$$

Since $U_{\sigma_k} = f^k$, where

$$f \equiv E\exp\left(-\int_0^{\sigma_1} h(X_s)\,\mathrm{d}s\right),$$

we obtain

$$S_{\sigma_k} = S_{\sigma_1}(1 + f + f^2 + \cdots + f^{k-1}) = S_{\sigma_1}\frac{1 - f^k}{1 - f}.$$

Similarly, we obtain

$$T_{\sigma_k} = T_{\sigma_1}\frac{1 - f^k}{1 - f}.$$

Hence,

$$H_{\sigma_k}(a) = (B(a)S_{\sigma_1} - T_{\sigma_1})\frac{1 - f^k}{1 - f}.$$

Further, for $k \geq 2$ we have

$$H_{\tau_{k,a}}(a) = H_{\sigma_{k-1}}(a) + f^{k-1}H_{\mu_a}(a) = (B(a)S_{\sigma_1} - T_{\sigma_1})\frac{1 - f^{k-1}}{1 - f} + f^{k-1}H_{\mu_a}(a).$$

Hence, we can rewrite (8.5) in terms of $H_{\mu_a}(a)$

$$r_2 - m_2 B(a) = (B(a)S_{\sigma_1} - T_{\sigma_1})\frac{1 - f^{k-1}}{1 - f} + f^{k-1}H_{\mu_a}(a),$$

which is equivalent to

$$\left(r_2 + T_{\sigma_1}\frac{1 - f^{k-1}}{1 - f}\right)f^{1-k} - B(a)\left(m_2 + S_{\sigma_1}\frac{1 - f^{k-1}}{1 - f}\right)f^{1-k} = H_{\mu_a}(a).$$

The left hand side of this equation is a linear function with respect to $B(a) : y = \alpha_k - B(a)\beta_k$, where

$$\alpha_k = \left(r_2 + T_{\sigma_1}\frac{1 - f^{k-1}}{1 - f}\right)f^{1-k},$$

$$\beta_k = \left(m_2 + S_{\sigma_1}\frac{1 - f^{k-1}}{1 - f}\right)f^{1-k}.$$

In order to know how the solution of this equation, a_k, is situated with respect to a_1, we find a crossing point of this strait line with the line $y = r_2 - B(a)m_2$. We have

$$\left(r_2 + T_{\sigma_1}\frac{1 - f^{k-1}}{1 - f}\right)f^{1-k} - x\left(m_2 + S_{\sigma_1}\frac{1 - f^{k-1}}{1 - f}\right)f^{1-k} = r_2 - x\,m_2,$$

hence

$$x = \frac{r_2 + T_{\sigma_1}/(1 - f)}{m_2 + S_{\sigma_1}/(1 - f)}.$$

The essence of this relation is that its right hand side does not depend on k. It is easy to show that $\alpha_k < \alpha_{k+1}$ $(k \geq 1)$. Hence, the sequence $B(a_k)$ of crossing points of lines $y = \alpha_k - \beta_k B(a)$ with the graph of the function $H_{\mu_a}(a)$ either increases (when $x > a_1$), or decreases (when $x < a_1$) (see Fig. 8.1). In the first case, μ_{a_1} is the best Markov time (the point of the global minimum of the loss function). In the second case, the optimal time does not exist, since every next local minimum is less then preceding one. It proves the alternative of choice. Thus, we have proved the theorem.

Theorem 1 *Let the server system be a twice regenerative system, where the first order regenerative process corresponds to a sequence of i.i.d. work-repair cycles described above. Let X_t be a regenerative process of the second order with points of regeneration (σ_n), such that X_t increases continuously from zero up to $c > 0$ on every interval $[\sigma_{n-1}, \sigma_n)$, and let the process be right continuous at each point σ_n. Let the loss function of the system be*

$$L_\tau = \frac{-W_\tau + r_1 - (r_1 - r_2)U_\tau}{V_\tau + m_1 - (m_1 - m_2)U_\tau},$$

where

$$U_\tau = E\exp\left(-\int_0^\tau h(X_s)\,\mathrm{d}s\right),$$

$$V_\tau = E\int_0^\tau \exp\left(-\int_0^t h(X_s)\,\mathrm{d}s\right)\,\mathrm{d}t,$$

$$W_\tau = E\int_0^\tau A(X_t)\exp\left(-\int_0^t h(X_s)\,\mathrm{d}s\right)\,\mathrm{d}t.$$

Let the inequality

$$\frac{h'(x)}{1 + \Delta m\, h(x)} \geq \frac{A'(x)}{\Delta r + \Delta m\, A(x)} + \frac{r_2(1 + \Delta m\, h(x)}{c\, m_2\,(\Delta r + \Delta m\, A(x))} \qquad (0 < x < c).$$

be fulfilled. Then the alternative of choice (Definition 1) is fair.

Figure 8.1 illustrates two variants of the solution, A and B, implying two different decisions for choice a time of switching the system to the prophylaxis repair. The lines going through the point A one can interpret as four positions of a lever of the first kind relative to four crossing points with the curve $H_{\mu_a}(a)$. The lines going through the point B one can interpret as four positions of a lever of the second kind relative to four crossing points with this curve. It is clear that x-coordinates of these crossing points go in opposite directions from the point a_1 when the ends of these levers go up along the y-axis.

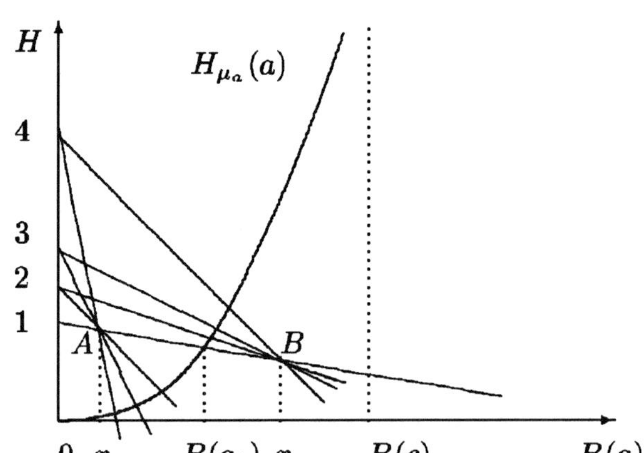

Figure 8.1. Proof of the alternative

8.3.1 Decision Under the Alternative

In order to compare x and a_1 (i.e. to choose a variant of the lever), one has to use a more complete information on the process X_t. Let (σ_n) be a sequence of jump times of a homogeneous Poisson process with the intensity λ, and let X_t be piece-wise linear function: i.e. $X_t = c(t - \sigma_{n-1})/(\sigma_n - \sigma_{n-1})$ if $\sigma_{n-1} \le t < \sigma_n$. We assume $A \equiv 0$, and $h(x) = x$ $(0 \le x \le c)$. In this case, we can use more simple form of the optimality equation, and the corresponding trajectory of choice. One can derive from (8.3) the equation

$$\frac{r_2 - a(r_1 m_2 - r_2 m_1)}{\Delta r} = \widetilde{H}_{\tau_a}(a) \equiv aV_{\tau_a} + U_{\tau_a} - 1. \tag{8.7}$$

In this case, the function $\widetilde{H}$ can be evaluated analytically. We obtain

$$\widetilde{H}_{\mu_a}(a) = \frac{a}{\sqrt{2\lambda c}} \arctan \frac{a}{\sqrt{2\lambda c}}.$$

The time of the first local minimum is the first crossing time of the level a_1, which is found from the equation

$$\frac{a}{\sqrt{2\lambda\, c}} \arctan \frac{a}{\sqrt{2\lambda\, c}} = \alpha + \beta a, \tag{8.8}$$

where

$$\alpha = \frac{r_2}{\Delta r}, \quad \beta = \frac{-r_1 m_2 + r_2 m_1}{\Delta r}.$$

In order to decide if this is a time of the global minimum one has to check fulfillment at least one from the following three inequalities

(1) $a_1 < a_2$,

(2) $a_1 < x$, where x is the x-coordinate of the crossing point of left sides of all the corresponding equations,

(3) $\alpha + \beta a_1 < \frac{x}{\sqrt{2\lambda\, c}} \arctan \frac{x}{\sqrt{2\lambda\, c}}$,

which are equivalent. For arbitrary parameters α and β an analytical solution of (8.8) does not exist. Putting $c = 1$ and $\beta = 0$ and using computer we obtain partition of the space of parameters (λ, α) on three regions:

(1) inadmissible combination of parameters,

(2) region $A = \{x < a_1\}$,

(3) region $B = \{x > a_1\}$, see Fig. 8.2

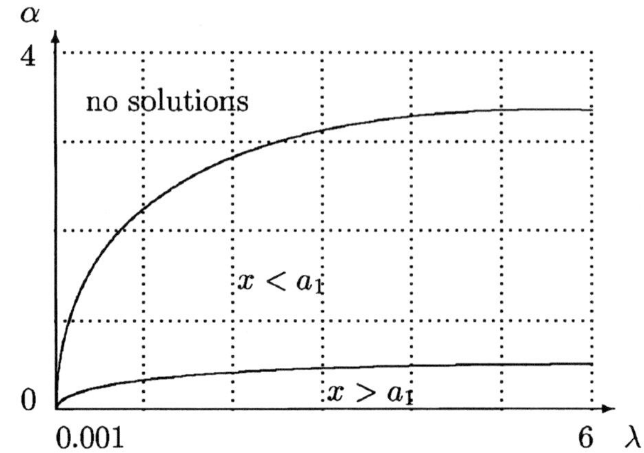

Figure 8.2. Regions of different prophylaxis policies

References

[RH09a] Rasova, S.S., Harlamov, B.P. Optimal stopping time under observation on non-monotone degradation. In MMR 2009 - Mathematical Methods in reliability. Theory. Methods. Applications. VI International Conference. Extended Abstracts. Moscow, 22–29 June, 2009, 71–75.

[RH09] Rasova, S.S., Harlamov, B.P. Optimal prophylaxis policy under non-monotone degradation. In: Nikulin M. et al. (eds.) Advances in Degradation Modeling Applications to Reliability, Survival Analysis, and Finance, pp. 181–194. Birkhauser, Boston (2010).

[Har07] Harlamov, B.P. Continuous Semi-Markov Processes. ISTE & Wiley, London (2007).

[BN00] Bagdonavicius, V., Nikulin, M. Statistical analysis of degradation data in dynamic environment. In: Dipartimento di matematica "Guido Castelnuovo", 2000, 1–20.

[BMN09] Bagdonavicius, V., Masiulaityte, I., Nikulin, M. Reliability estimation from faliure-degradation data with covariates. In: Nikulin M. et al. (eds.) Advances in Degradation Modeling Applications to Reliability, Survival Analysis, and Finance, pp. 275–291. Birkhauser, Boston (2010).

[HNR09] Firoozeh Hagigi, Nazanin Nooraee, and Narges Nazeri Rad. On the general degradation path model: review and simulation. In: Nikulin M. et al. (eds.) Advances in Degradation Modeling Applications to Reliability, Survival Analysis, and Finance, pp. 147–155. Birkhauser, Boston (2010).

9

Optimal Incomplete Maintenance for Systems with Discrete Time-to-Failure Distribution

Waltraud Kahle

Institute of Mathematical Stochastics, Otto-von-Guericke-University, D-39016 Magdeburg, Germany, waltraud.kahle@ovgu.de

Abstract: In this research, we are concerned with the statistical modeling of repairable systems. We consider an incomplete maintenance model, that is, the impact of a maintenance action is not minimal (as bad as old) and not perfect (as good as new) but lies between these boundary cases. Further, we assume that after a failure the system is repaired minimally. Cost optimal maintenance policies for various cost functions are considered.

Keywords and phrases: Incomplete repair, Optimal maintenance, Simulation of failure-repair processes

9.1 Introduction

In many applications, the clock time is not the best scale in which to describe lifetimes. As example, if a unit has a sequence of tasks to perform, then its lifetime is measured by the number of tasks performed before its failure. In this paper, we study a failure-repair process when the time to failure is discrete. First, we describe some discrete distributions and introduce their failure rate and their cumulative hazard. Further, we introduce the concept of failure rate optimal maintenance for discrete lifetime distributions, that is, a maintenance policy where the rate of occurrence of failures (ROCOF) under preventive maintenance is constant. It is described how to simulate such failure-repair processes and some examples for simulated data are given. In the last section, the model with minimal repairs after a failure and block incomplete maintenance is described.

9.2 Discrete Lifetime Distributions and Their Failure Rate

There are a number of possible for consideration discrete lifetime distributions. In this paper, the Poisson distribution and the discrete Weibull distribution are considered.

V.V. Rykov et al. (eds.), *Mathematical and Statistical Models and Methods in Reliability: Applications to Medicine, Finance, and Quality Control*, Statistics for Industry and Technology, DOI 10.1007/978-0-8176-4971-5_9, © Springer Science+Business Media, LLC 2010

9.2.1 The Poisson Distribution

Let the lifetime T be shifted Poisson distributed, that is $T = X + 1$ where

$$\mathrm{P}(X = t) = \frac{\lambda^t}{t!} e^{-\lambda} , \quad t = 0, 1, \ldots . \tag{9.1}$$

The failure rate for discrete distributions (and especially for the Poisson distribution) is given by

$$h(t) = \frac{\mathrm{P}(T = t)}{\mathrm{P}(T \geq t)} = \frac{\lambda^t / t!}{\sum_{k=t}^{\infty} \lambda^k / k!}, \quad t = 0, 1, \ldots .$$

Since

$$\sum_{k=t+1}^{\infty} \frac{\lambda^k}{k!} = \sum_{k=t}^{\infty} \frac{\lambda^{k+1}}{(k+1)!} \leq \frac{\lambda}{t+1} \sum_{k=t}^{\infty} \frac{\lambda^k}{k!},$$

we get

$$\frac{\lambda^t / t!}{\sum_{k=t}^{\infty} \lambda^k / k!} \leq \frac{\lambda^{t+1} / (t+1)!}{\sum_{k=t+1}^{\infty} \lambda^k / k!},$$

that is, the Poisson distribution has an increasing failure rate. We use the shifted version of this distribution, that is, the item is absolutely reliable at time 0 and the first failure may appear at time $t = 1$.

9.2.2 The Discrete Weibull Distribution

There are different *discrete Weibull distributions*, as example introduced by Padgett, Spurrier [PS85] or the Nakagawa–Osaki model [NO75]. In this paper, we use the following model. Let X be Weibull distributed with density function

$$f(x) = \frac{\beta}{\alpha} (\frac{x}{\alpha})^{\beta - 1} \exp(-(\frac{t}{\alpha})^\beta), \quad x \geq 0 .$$

Now, for the discrete version we put the probability mass of the interval $(t-1, t]$ into the point t, that is

$$\mathrm{P}(T = t) = \int_{s=t-1}^{t} f(s) ds = \exp(-(\frac{t-1}{\alpha})^\beta) - \exp(-(\frac{t}{\alpha})^\beta) \, , t = 1, 2, \ldots . \tag{9.2}$$

If $\beta = 1$, then we get the geometric distribution

$$\mathrm{P}(T = t) = (\exp(-\frac{1}{\alpha}))^{t-1} - (\exp(-\frac{1}{\alpha}))^{t-1} \cdot \exp(-\frac{1}{\alpha}) = p^{t-1}(1-p) .$$

Note that for discrete failure time distributions the failure rate is a probability unlike for continuous distributions. Further, the cumulative hazard is

$$H(t) = -\ln R(t) = -\ln(1 - F(t)) \neq \sum_{i=1}^{t} h(t)$$

where $F(t)$ and $R(t)$ are distribution function and survival function, respectively. It is convex if $h(t)$ is increasing. For an other definition of failure rates for discrete distributions we refer to Xie, Gaudoin, Bracquemond [XGB02] and the references given in their paper.

9.3 Kijima Type Repairs

Consider the impact of repairs. A system (machine) starts working with an initial prescribed failure rate $\lambda_1(t) = \lambda(t)$. Let t_1 denote the random time of the first sojourn. At this time t_1 the item will be repaired with the degree ξ_1. When the system is minimally repaired then the degree is equal to one, and if the repair makes the system as good as new then this degree is zero. The virtual age of the system at the time t_1, following the repair, is $v_1 = \xi_1 t_1$, implying the age of the system is reduced by maintenance actions. The distribution of the time until the next sojourn then has failure intensity $\lambda_2(t) = \lambda(t - t_1 + v_1)$. Assume now that t_k is the time of the kth $(k \geq 1)$ sojourn and that ξ_k is the degree of repair at that time. We assume that $0 \leq \xi_k \leq 1$, for $k \geq 1$ (see [Kij89] and [KMS88]).
After repair the failure intensity during the $(k+1)^{th}$ sojourn is determined by

$$\lambda_{k+1}(t) = \lambda(t - t_k + v_k) \quad , \quad t_k \leq t < t_{k+1}, k \geq 0,$$

where the virtual age v_k is for Kijima's Type II imperfect repair model

$$v_k = \xi_k(v_{k-1} + (t_k - t_{k-1})),$$

that is, the repair resets the intensity of failure proportional to the virtual age.

Kijima's Type I imperfect repair model suggests that upon failure, the repair undertaken could serve to reset the intensity only as far back as the virtual age at the start of working after the last failure. That is:

$$v_k = t_{k-1} + \xi_k(t_k - t_{k-1}).$$

The process defined by $v(t, \xi_k, k = 1, 2, \dots) = t - t_k + v_k$, $t_k \leq t < t_{k+1}$, $k \geq 0$ is called the *virtual age process* (Last, Szekli [LS98]).

In Gasmi, Love, Kahle [GLK03], a generalized Kijima type model was considered, where a major repair gives an additional impact. It was shown that the likelihood function can be developed from the general likelihood function for observation of point processes (Liptser and Shiryayev [LS78]). Further, the likelihood ratio statistic can be used to find confidence estimates for the unknown parameters.

The numerical results for this data file are surprising: Under different assumptions about the repair actions (renewals, Kijima type I or II, mixture of Kijima type repairs and renewals in dependence on the time required for repair) a value for β was estimated approximately to be 1, see [GLK03] and [KL03]. That is, the failure intensity is more or less constant. But in this case the failure behavior does not depend on maintenance actions.

The results suggest, that in practice the engineers make a good maintenance policy, that is, they make repairs in connection with the state of the system. The idea is that such a policy makes the apparent failure behavior of a system to be that of an exponential distribution. This is consistent with our data. In Fig. 9.1, we see the cumulative distribution function of the operating time between failures together with the fitted CDF of an exponential distribution and the Q-Q plot (observed quantiles against the quantiles of the exponential model). These plots suggest reasonable agreement with the exponential model if we consider only the failure process and ignore all maintenance events.

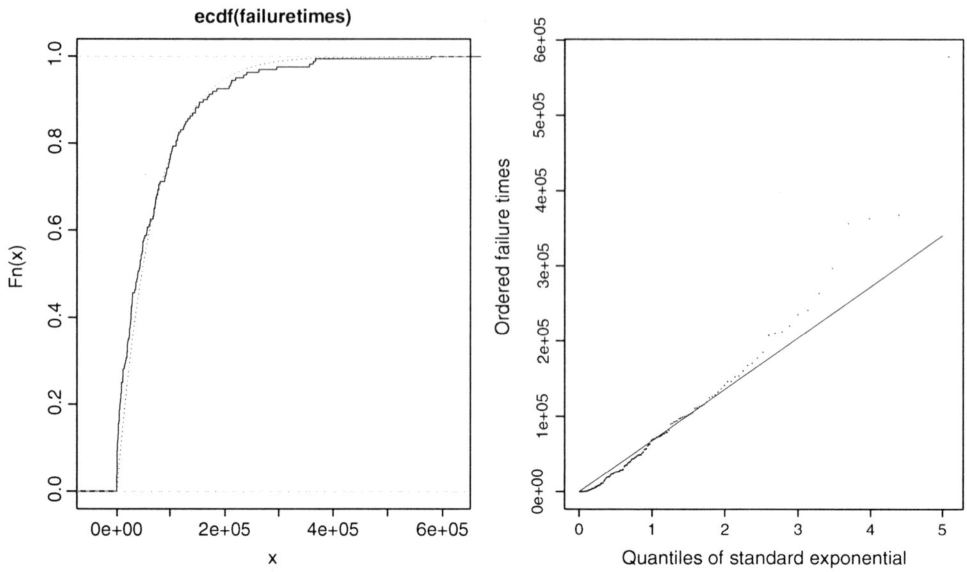

Figure 9.1. Operating time between failures: CDF and exponential Q–Q plot

Definition 1. *A maintenance policy is called* **failure rate optimal**, *if the state dependent preventive maintenance actions lead to a constant ROCOF of the failure process.*

In the following, we assume that preventive maintenance actions are undertaken at fixed times τ, 2τ, ... Failures between maintenances are removed by minimal repairs. The maintenances, however, are not a renewals. After a maintenance, the virtual age of the system is smaller than the real time, but not 0.

9.4 Optimal Maintenance as Time Scale Transformation

First, we consider failure rate optimal maintenances. Following an idea in Finkelstein [Fin00] we assume that by repair actions, the time scale is transformed by a function $W(t)$. Let $H_0(t)$ be the baseline cumulative hazard function and let $H_1(t) = H_0(W(t))$ be the resulting hazard after a transformation of the time scale. For the Weibull hazard

$$H_0(t) = (t/\alpha)^\beta,$$

and $W(t) = t^{1/\beta}$ we get

$$H_1(t) = H_0(t^{1/\beta}) = \frac{t}{\alpha^\beta} ,$$

that is, the hazard function of an exponential distribution with parameter $\lambda_1 = 1/\alpha^\beta$.

In practice, we have repair actions at discrete time points, which lead to the question of the degrees of repair at these points. Let us consider two examples. In both examples we assume that after a failure the system is repaired minimally. Additionally, maintenance decisions were regularly carried out. We assume that maintenance actions served to adjust the virtual age of the system in a Kijima type manner.

Example 1. Assume that the distances between maintenance actions are constant and all repair actions follow the Kijima type I repair process. Let $\tau, 2\tau, \ldots$ be the time points of maintenance actions. Then it is possible to find a discrete time transformation which consists of different degrees of repair. Let the sequence of degrees be

$$\xi_k = \frac{k^{1/\beta} - (k-1)^{1/\beta}}{\tau^{1-1/\beta}} .$$

Then the virtual age v_n of the system at time $t_n = n \cdot \tau$ can be found to be

$$v_n = \tau \sum_{k=1}^{n} \xi_k = \tau \sum_{k=1}^{n} \frac{k^{1/\beta} - (k-1)^{1/\beta}}{\tau^{1-1/\beta}} = (n \cdot \tau)^{1/\beta} .$$

Example 2. Again we assume that the distances between maintenance actions are constant, but now we consider the Kijima type II repair process. In this case, the appropriate sequence of degrees of repair is

$$\xi_k = \frac{k^{1/\beta}}{(k-1)^{1/\beta} + \tau^{1-1/\beta}} .$$

In both cases, the sequence is decreasing, that is, with increasing time the repairs must become better.

It should be noted that in case of time scale transformation it is not necessary to make a difference between Kijima type I and II. In both examples, the virtual age at maintenance points was reseted to those of the continuous time transformation as it is shown in Fig. 9.2.

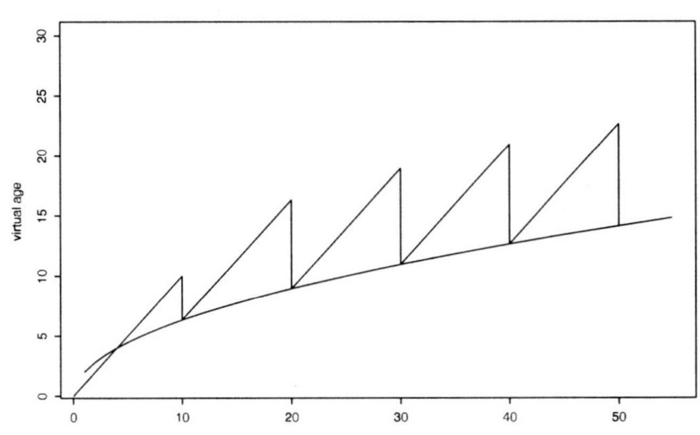

Figure 9.2. A discrete time transformation

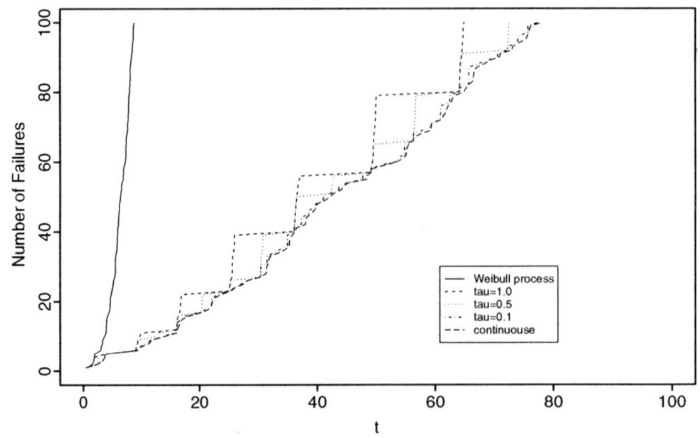

Figure 9.3. Weibull process without and with preventive maintenance actions

In Fig. 9.3, are shown the cumulative hazard functions for an Weibull process without maintenance (solid line) and for maintenance actions every $\tau = 1, .5, .1$ time units. For this, a Weibull process with parameters $\alpha = 1$ and $\beta = 2$ and 100 failures was simulated. If the difference τ between maintenance actions is relatively small, then the empirical cumulative hazard function of the process with preventive maintenance is closed to that of a Poisson process. The dotted line shows the theoretical cumulative hazard function of an homogeneous Poisson process.

There are many other possibilities for finding failure rate optimal maintenance policies. One other very simple policy is to consider constant degrees of repair. It is easy to see that in this case the repair actions must take place more often with increasing time.

9.5 Simulation of Failure–Repair Processes

9.5.1 Simulation of Weibull Processes

In this section, we describe the simulation of a failure process when failures are removed by minimal repairs and the transformation of such failure processes by preventive maintenance actions. First, we want simulate a so called Weibull process, that is, a non homogeneous Poisson process with a power law mean function

$$H(t) = (\frac{t}{\alpha})^{\beta} \; . \tag{9.3}$$

For this, let be given a sequence of iid rectangular distributed random numbers $z_1, z_2, \ldots, z_n$. We make use of the connection between distribution function and cumulative hazard function

$$F(t) = \exp(-H(t)) \; .$$

Now, the first failure time t_1 can be found to be

$$z_1 := \mathrm{P}(T > t_1) = \exp(-(\frac{t_1}{\alpha})^\beta), \quad \text{and therefore}$$
$$t_1 = \alpha(-\log(z_1))^{1/\beta} .$$

The time t_2 of the next failure has the conditional distribution

$$\mathrm{P}(T > t_2 | T > t_1) = \frac{\mathrm{P}(T > t_2)}{\mathrm{P}(T > t_1)}$$

Therefore, we can simulate t_2 by

$$z_2 := \mathrm{P}(T > t_2 | T > t_1) = \frac{\exp(-(\frac{t_2}{\alpha})^\beta)}{\exp(-(\frac{t_1}{\alpha})^\beta)} \quad \text{and it follows that}$$
$$t_2 = \alpha(-\log(z_2) + (\frac{t_1}{\alpha})^\beta)^{1/\beta} .$$

Following this procedure we get for the kth failure time

$$t_k = \alpha(-\log(z_k) + (\frac{t_{k-1}}{\alpha})^\beta)^{1/\beta} . \tag{9.4}$$

Remark: It is interesting, that we can get a Weibull process also by a simple time transformation. Again, let be given a sequence of iid rectangular distributed random numbers $z_1, z_2, \ldots, z_n$. Then

$$y_k = -\log(z_k) \quad \text{and} \quad s_k = \sum_{i=1}^{k} y_k$$

are a sequence of iid standard exponential distributed random numbers and the failure times of an homogeneous Poisson process, respectively. The transformation

$$t_k = \alpha \cdot s_k^{1/\beta}$$

leads to the described Weibull process.

9.5.2 Transformation by Maintenance Actions

Now, we assume that at some time point, say $x = k \cdot \tau$ we make a preventive maintenance. Let t_{k-1} be the last failure time before x and v_{k-1} the virtual age at t_{k-1}. Then the next failure time t_k was simulated by the equation

$$H(t_k) - H(v_{k-1}) = -\log(z_k) .$$

Now, up to time x some part of this hazard is already used. Further, let be v_x the new virtual age after maintenance. Then the transformed failure time t_k^* is found to be

$$H(t_k) - H(v_{k-1}) = H(x) - H(v_{k-1}) + H(t_k^*) - H(v_x) .$$

For discrete lifetime distributions the simulation is the same. The only difference is that we must round the times.

9.6 Cost Optimal Maintenance

Now we consider optimal maintenance policies with respect to costs. Again, each failure is removed by an minimal repair and at fixed times τ, 2τ, ... preventive maintenance actions are undertaken. In classical policies, the maintenance renews the system, then we have a block replacement. We do not make the assumption that after maintenance the item is as good as new. Every maintenance action leads to an (constant) age v. In Nakagawa [Nak02] an incomplete maintenance model was considered under the following assumptions:

- The age after the k-th pm falls to $a_k t$
- $N-1$ pm's at $x_i, i=1,\dots,N-1$, renewal at x_N
- Minimal repairs at failures
- Each incomplete repair has the same costs

The problem was to find the times of pm's $x_i, i=1,\dots,$ and the number N of preventive maintenances with minimal costs per time.

We think that the assumption *each incomplete repair has the same costs* is very restrictive and want define a cost function which describes the costs of repair actions according to the degree of repair. Let c_F and c_M be the cost of a failure and the cost of a maintenance, respectively. We do not make the assumption that after maintenance the item is as good as new. Every maintenance action leads to an age v. Then the costs per time unit are given by

$$C(v,\tau) = \frac{c_F \cdot (H(\tau+v) - H(v)) + c_M(v,\tau)}{\tau} . \tag{9.5}$$

This function should be minimized with respect to v and τ.

9.6.1 Costs Proportional to the Removed Hazard

The first idea is that the costs are proportional to the cumulative hazard which is removed be the maintenance action, that is

$$c_M(v,\tau) \sim H(\tau+v) - H(v) .$$

In this case, the optimization problem has a trivial solution:

1. Since for distributions with increasing failure rates the cumulative hazard $H(t)$ is convex, $C(v,\tau)$ is increasing in v. Consequently, the pm should be a renewal.
2. If the pm renews the system, then the essential part of (9.5) is

$$\frac{H(\tau)}{\tau},$$

which is increasing in τ. For continuous lifetimes the problem doesn't have a solution. For discrete lifetimes the optimal policy is a renewal at each time.

9.6.2 Costs Proportional to the Degree of Repair

The next function, which can be good interpreted, is

$$c_R \cdot \left(\frac{\tau}{\tau+v}\right)^{\delta} ,$$

where c_R are the costs of a renewal. This function has the following properties:

- If the system is renewed ($v = 0$) then we get c_M.
- If $\tau = 0$ there are no costs.
- If $\delta < 1$ then the costs of a relatively small repair are closed to that of a renewal, if $\delta > 1$ then the costs of a large repair are relatively small.

Note that $\tau/(\tau + v)$ is $1/\xi$, where ξ is the degree of repair. Then the costs per time unit are given by

$$C(v, \tau) = \frac{c_F \cdot (H(\tau + v) - H(v)) + c_M \cdot \left(\frac{\tau}{\tau+v}\right)^{\delta}}{\tau}, \quad . \tag{9.6}$$

Again, $C(v, \tau)$ is increasing in v, because of the convexity of $H(t)$. And for $v = 0$ we get the classical problem of block replacement.

9.6.3 Costs Proportional to the Impact of Repair

Last, we consider a very simple cost function

$$c_M(v, \tau) = c_I \frac{1}{v^{\delta}} ,$$

that is, for a perfect repair the costs tend to infinity. In this case, we can find a unique optimum for both,v and τ. Figure 9.4 shows the contour lines for Weibull distributed lifetimes with shape $\beta = 4$, scale$\alpha = 20$, and $c_F/c_I = 2$. The optimal policy is $v = 3.1$, $\tau = 9.2$, and the resulting minimal costs per time unit are 0.093.

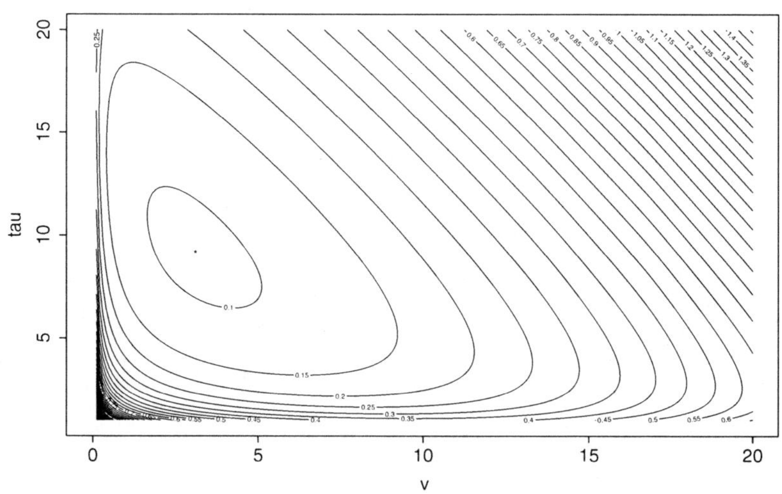

Figure 9.4. Contour lines of the cost function

9.7 Conclusion

We have considered failure rate optimal maintenance under the assumption that the maintenance action has an impact between the two extreme cases minimal repair and renewal. For finding cost optimal maintenance it was necessary to define a cost function which describes the costs of repair actions according to the degree of repair. There are many other possible cost functions, which can be considered. Further we can consider models where the repair after a failure is not minimal, but imperfect, too.

References

[Fin00] Finkelstein, M.S.: Modeling a Process of Non-Ideal Repair. In: Limnios, N., Nikulin, M. (eds.) Recent Advances in Reliability Theory, Statistics for Industry and Technology. Birkhhauser, Boston (2000)

[GLK03] Gasmi, S., Love, C.E., Kahle, W.: A General Repair/Proportional Hazard Framework to Model Complex Repairable Systems. IEEE Transactions on Reliability, **52**, 26–32 (2003)

[KL03] Kahle, W., Love, C.E.: Modeling the Influence of Maintenance Actions. In: Lindquist, B.H., Doksum, H.A. (eds.) Mathematical and Statistical Methods in Reliability, Quality, Reliability and Engeneering Statistics, World Scientific Publishing Co. (2003)

[Kij89] Kijima, M.: Some results for repairable systems with general repair. Journal of Applied Probability, **26**, 89–102 (1989)

[KMS88] Kijima, M., Morimura, H., Suzuki, Y.: Periodical Replacement Problem without Assuming Minimal Repair. European Journal of Operational Research, **37**, 194–203 (1988)

[LS98] Last, G., Szekli, R.: Stochastic Comparison of Repairable Systems by Coupling. Journal of Applied Probability, **35**, 348–70 (1998)

[LS78] Liptser, R.S., Shiryayev, A.N.: Statistics of Random Processes, vol II. Springer, New York (1978)

[NO75] Nakagawa,G., Osaki, S.: The Discrete Weibull Distribution. IEEE Transactions on Reliability **R-24**, 300–301 (1975)

[Nak02] Nakagawa, T.: Imperfect Preventive Maintenance Models. In: Osaki, S. (ed.) Stochastic Models in Reliability and Maintenance. Springer, Berlin (2002)

[PS85] Padgett, W.J., Spurrier, J.D.: Discrete Failure Models. IEEE Transactions on Reliability **R-34**, 253–256 (1985)

[XGB02] Xie, M., Gaudoin, O., Bracquemond, C.: Redefining Failure Rate Function for Discrete Distributions. International Journal of Reliability, Quality and Safety Engineering **9**, 275–285 (2002)

10

A Gini-Type Index for Aging/Rejuvenating Objects

M. P. Kaminskiy[1] and V. V. Krivtsov[2,*]

[1] Center of Technology and Systems Management, University of Maryland College Park, MD, USA
[2] Office of Technical Fellow in Quality Engineering, Ford Motor Company, Dearborn, MI, USA, vkrivtso@ford.com

Abstract: This paper introduces a simple index that helps to assess the degree of aging or rejuvenation of repairable systems and non-repairable systems (components). The index ranges from -1 to 1. It is negative for the class of decreasing failure rate distributions and point processes with decreasing ROCOF and is positive for the increasing failure rate distributions and point processes with increasing ROCOF. The introduced index is distribution free.

Keywords and phrases: Cumulative hazard function, ROCOF, Cumulative intensity function, Aging, Rejuvenation, Improving, Deteriorating, Stochastic point process

---

Acronyms

CDF	Cumulative distribution function
CFR	Constant failure rate
CIF	Cumulative intensity function
DFR	Decreasing failure rate
GPR	G-renewal process
HPP	Homogeneous Poison process
IFR	Increasing failure rate
NHPP	Non-homogeneous Poison process PP point process
ROCOF	Rate of occurrence of failures
RP	Renewal process
TTF	Time to failure

10.1 Introduction

In reliability and risk analysis, the terms *aging* and *rejuvenation* are used for describing reliability behavior of repairable as well as non-repairable systems (components).

V.V. Rykov et al. (eds.), *Mathematical and Statistical Models and Methods in Reliability: Applications to Medicine, Finance, and Quality Control*, Statistics for Industry and Technology, DOI 10.1007/978-0-8176-4971-5_10, © Springer Science+Business Media, LLC 2010

The *repairable systems* reliability is modeled by various point processes (PP), such as the homogeneous Poisson process (HPP), non-homogeneous Poisson process (NHPP), renewal process (RP), G-renewal process (GRP), to name a few. Among these PP, some special classes are introduced in order to model the so-called *improving* and *deteriorating* systems. An improving (deteriorating) system is defined as the system with decreasing (increasing) *rate of occurrence of failures* (ROCOF). It might be said that among the point processes used as models for repairable systems, the HPP (having a constant ROCOF) is a basic one.

Similarly, among the distributions used as models of time to failure (TTF) of *non-repairable* systems (components), the exponential distribution, which is the only distribution having a constant failure rate, plays a fundamental role. This distribution might be considered as the limiting between the class of *increasing failure rate* (IFR) distributions and the class of *decreasing failure rate* (DFR) distributions. The distribution is closely related to the above mentioned HPP. Indeed, in the framework of the HPP model, the distribution of the intervals between successive events observed during a time interval $[0, t]$ is the exponential one with parameter λ equal to parameter λ of the respective Poisson distribution with mean λt.

In many practical situations, it is important to make an assessment how far a given point process deviates from the HPP, which can be considered as a simple and, therefore, strong competing model. Note that if the HPP turns out to be an adequate model, the respective system is considered as non-aging, so that it does not need any preventive maintenance (as opposed to the case, when a repairable system reveals aging).

The statistical tools helping to find out if the HPP is an appropriate model are mainly limited to statistical hypothesis testing, in which the null hypothesis is

H_0 : "The times between successive events (interarrival times) are independent and identically exponentially distributed," and the alternative hypothesis is
H_1 : "The system is either aging or improving."

The most popular hypothesis testing procedures for the considered type of problems are the *Laplace test* [RH04] and the so-called *Military Handbook test* [AMSAA81]. It should be noted that these procedures do not provide a simple measure quantitatively indicating how different the ROCOF of a given point process is, compared to the respective constant ROCOF of the competing HPP model.

Analogously, for the non-repairable systems, some hypothesis testing procedures can be applied to help to determine if the exponential distribution is an appropriate TTF model. In such situations, in principle, any goodness-of-fit test procedure can be used. Some of these tests for the null-hypothesis: "the times to failure are independent and identically exponentially distributed" appear to have good power against the IFR or DFR alternatives [Law03].

Among such goodness-of-fit tests, one can mention the G-test, which is based on the so-called *Gini statistic* [GG78]. In turn, the Gini statistics originates from the so-called *Gini coefficient* used in macroeconomics for comparing an income distribution of a given country with the uniform distribution covering the same income interval. The Gini coefficient is used as a measure of income inequality [Sen97]. The coefficient takes on the values between 0 and 1. The closer the coefficient value to zero, the closer the distribution of interest is to the uniform one. The interested reader could find the index values sorted by countries in [LCIE07] that includes the UN and CIA data.

In the following sections, we first introduce a Gini-type coefficient for the repairable systems as in The coefficient takes on the values between -1 and 1. The closer it is to zero, the closer the PP of interest is to the HPP. A positive (negative) value of this coefficient will indicate whether a given repairable system is deteriorating (improving). Then, we introduce a similar coefficient for non-repairable systems. Again, the coefficient takes on the values between -1 and 1. The closer the coefficient's value is to zero, the closer the distribution of interest is to the exponential distribution. A positive (negative) value of the coefficient indicates an IFR (DFR) failure time distribution. For the sake of simplicity, in the following, this Gini-type coefficient will be referred to as *GT coefficient* and denoted as C.

10.2 GT Coefficient for Repairable Systems

10.2.1 Basic Definitions

A point process (PP) can be informally defined as a mathematical model for highly localized events distributed randomly in time. The major random variable of interest related to such processes is the number of events, $N(t)$, observed in time interval $[0, t]$. Using the nondecreasing integer-valued function $N(t)$, the point process $\{N(t), t \geq 0\}$ is introduced as the process satisfying the following conditions:

1. $N(t) \geq 0$
2. $N(0) = 0$
3. If $t_2 > t_1$, then $N(t_2) \geq N(t_1)$
4. If $t_2 > t_1$, then $[N(t_2) - N(t_1)]$ is the number of events occurred in the interval $(t_1, t_2]$

The mean value $E[N(t)]$ of the number of events $N(t)$ observed in time interval $[0, t]$ is called *cumulative intensity function* (CIF), *mean cumulative function* (MCF), or *renewal function*. In the following, term *cumulative intensity function* is used. The CIF is usually denoted by $\Lambda(t)$:

$$\Lambda(t) = E[N(t)].$$

Another important characteristic of point processes is the *rate of occurrence of events*. In reliability context, the *events* are *failures*, and the respective rate of occurrence is abbreviated to ROCOF. The ROCOF is defined as the derivative of CIF with respect to time, i.e.,

$$\lambda(t) = \frac{\mathrm{d}\Lambda(t)}{\mathrm{d}t}.$$

When an event is defined as a failure, the system modeled by a point process with an increasing ROCOF is called *aging* (*sad, unhappy,* or *deteriorating*) system. Analogously, the system modeled by a point process with a decreasing ROCOF is called *improving* (*happy* or *rejuvenating*) system.

The distribution of time to the first event (failure) of a point process is called the *underlying distribution.* For some point processes, this distribution coincides with the distribution of time between successive events; for others it does not.

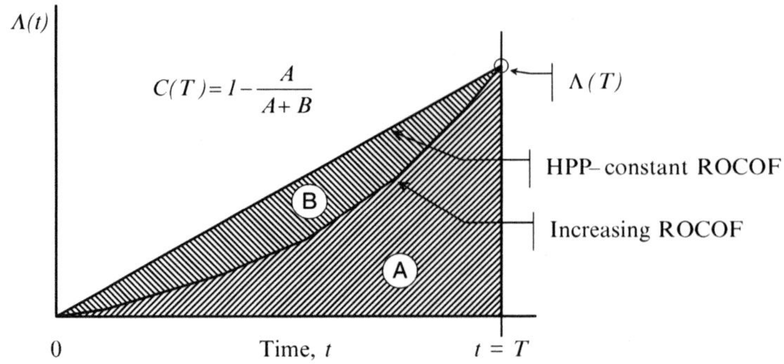

Figure 10.1. Graphical interpretation of GT coefficient for a point process with an increasing ROCOF

10.2.2 GT Coefficient

Consider a PP having an integrable over $[0, T]$ cumulative intensity function, $\Lambda(t)$. It is assumed that the respective ROCOF exists, and it is increasing function over the same interval $[0, T]$, so that $\Lambda(t)$ is concave upward, as illustrated by Fig. 10.1. Further consider the HPP with CIF $\Lambda_{\mathrm{HPP}}(t) = \lambda t$ that coincides with $\Lambda(t)$ at $t = T$, i.e., $\Lambda_{\mathrm{HPP}}(T) = \Lambda(T)$, – see Fig. 10.1.

Then, for a given time interval $[0, T]$ the GT coefficient is defined as

$$C(T) = 1 - \frac{\int_0^T \Lambda(t)\mathrm{d}t}{0.5T\Lambda(T)} = 1 - \frac{2\int_0^T \Lambda(t)\mathrm{d}t}{T\Lambda(T)}. \tag{10.1}$$

The smaller the absolute value of the GT coefficient, the closer the considered PP is to the HPP; clearly, for the HPP, $C(T) = 0$. GT coefficient satisfies the following inequality: $-1 < C(T) < 1$. It is obvious that for a PP with an increasing ROCOF, the GT coefficient is positive and for a PP with a decreasing ROCOF, the coefficient is negative. One can also show that the absolute value of GT coefficient $C(T)$ is proportional to the mean distance between the $\Lambda(t)$ curve and the CIF of the HPP.

For the most popular NHPP model – the *power law* model with the underlying Weibull CDF – the GT coefficient is expressed in a closed form:

$$C(T) = 1 - \frac{2}{\beta + 1}, \tag{10.2}$$

where β is the shape parameter of the underlying Weibull distribution.

Some examples of applying the GT coefficient to other PP commonly used in reliability and risk analysis are given in Table 10.1.

Repair effectiveness factor in Table 10.1 refers to the degree of restoration upon the failure of a repairable system; see [KS86], [KK98]. This factor equals zero for an RP: one, for an NHPP and is greater-or-equal-to zero, for a GRP (of which the RP and the NHPP are the particular cases).

Table 10.1. GT coefficients of some PP over time interval $[0, 2]$. Weibull with scale parameter $\alpha = 1$ is used as the underlying distribution

Stochastic point process	Shape parameter of underlying Weibull distribution	Repair effectiveness factor	GT coefficient
HPP	1	N/A	0
NHPP	1.1	1	0.05
NHPP	2	1	0.33
NHPP	3	1	0.50
RP	2	0	0.82
GRP	2	0.5	0.21

Note: the GT coefficient for RP and GRP was obtained using numerical techniques

---

10.3 GT Coefficient for Non-repairable Systems (Components)

Consider a non-repairable system (component) whose TTF distribution belongs to the class of the IFR distributions. Denote the *failure rate* or the *hazard function* associated with this distribution by $h(t)$. The respective *cumulative hazard function* is then

$$H(t) = \int_0^t h(\tau)\mathrm{d}\tau,$$

and is concave upward – see Fig. 10.2.

Consider time interval $[0, T]$. The cumulative hazard function at T is $H(T)$, the respective CDF is $F(T)$ and the reliability function is $R(T)$. Now, introduce h_{eff}, as the failure rate of the exponential distribution with the CDF equal to the CDF of interest at the time $t = T$, i.e.,

$$h_{\mathrm{eff}}(T) = -\frac{\ln(1 - F(T))}{T}.$$

In other words, the introduced exponential distribution with parameter h_{eff}, at $t = T$, has the same value of the cumulative hazard function as the IFR distribution of interest, see Fig. 10.2.

The GT coefficient, $C(T)$, is then introduced as

$$C(T) = 1 - \frac{\int_0^T H(t)\mathrm{d}t}{0.5Th_{\mathrm{eff}}(T)T} = 1 - \frac{2\int_0^T H(t)\mathrm{d}t}{TH(T)} = 1 - \frac{2\int_0^t \ln(R(t))\mathrm{d}t}{T\ln(R(T))}. \tag{10.3}$$

In terms of Fig. 10.2, $C(T)$, is defined as one minus the ratio of areas A and $A + B$. It is easy to check that the above expression also holds for the decreasing failure rate (DFR) distributions, for which $H(t)$ is concave downward.

It is clear that $C(T)$ satisfies the following inequality: $-1 < C(T) < 1$. The coefficient is positive for the IFR distributions, negative – for the DFR distributions and is equal to zero for the constant failure rate (CFR), i.e., exponential distribution. Note that the suggested coefficient is, in a sense, distribution-free.

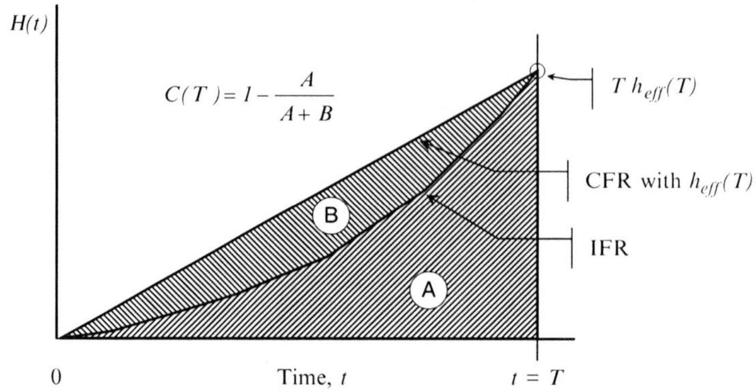

Figure 10.2. Graphical interpretation of the GT coefficient for an IFR distribution

Table 10.2. GT coefficient for Weibull Distribution as Function of Shape Parameter β

Shape parameter β	GT coefficient	TTF distribution
5	0.6(6)	IFR
4	0.6	IFR
3	0.5	IFR
2	0.3(3)	IFR
1	0	CFR
0.5	−0.3(3)	DFR
0.3	−0.5	DFR
0.25	−0.6	DFR
0.2	−0.6(6)	DFR

10.3.1 GT Coefficient for the Weibull Distribution

For some TTF distributions, the GT coefficient can be expressed in a closed form. For example, in the most important (in the reliability context) case of the Weibull distribution with the scale parameter α and the shape parameter β, and the CDF of the form:

$$F(t) = 1 - \exp\left(-\left(\frac{t}{\alpha}\right)^{\beta}\right),$$

the GT coefficient can be found as

$$C = 1 - \frac{2}{\beta + 1}. \qquad (10.4)$$

It's worth noting that in this case, the GT coefficient depends neither on the scale parameter α, nor on time interval T. Also note that (10.4) is exactly the same as (10.2). This is because NHPP's CIF is formally equal to the cumulative hazard function of the underlying failure time distribution; see, e.g., [Kri07].

Interestingly, $C(\beta) = -C\left(\frac{1}{\beta}\right)$, which is illustrated by Table 10.2.

Table 10.3. GT Coefficient for Gamma Distribution with $\lambda = 1$ and $T = 1$

Shape parameter k	GT coefficient	TTF distribution
5	0.623	IFR
4	0.543	IFR
3	0.428	IFR
2	0.258	IFR
1	0.000	CFR
0.5	−0.196	DFR
0.3	−0.285	DFR
0.25	−0.338	DFR
0.2	−0.375	DFR

10.3.2 GT Coefficient for the Gamma Distribution

Although not as popular as the Weibull distribution, the gamma distribution still has many important reliability applications. For example, it is used to model a standby system consisting of k identical components with exponentially distributed times to failure; the gamma distribution is also the conjugate prior distribution in Bayesian estimation of the exponential distribution.

Let's consider the gamma distribution with the CDF given by

$$F(t) = \frac{1}{\Gamma(k)} \int_0^{\lambda t} \tau^{k-1} e^{-\tau} \mathrm{d}\tau = I(k, \lambda t),$$

where $k > 0$ is the shape parameter, $\frac{1}{\lambda} > 0$ is the scale parameter, and $I(k, x) = \int_0^x y^{k-1} e^{-y} \mathrm{d}y$ is the incomplete gamma function. Similar to the Weibull distribution, the gamma distribution has the IFR, if the shape parameter $k > 1$; DFR, if $k < 1$, and CFR, if $k = 1$.

Using definition (10.3), the GT coefficient for the gamma distribution can be written as

$$C(T) = 1 - \frac{2 \int_0^T \ln(1 - I(k, \lambda\tau)) \mathrm{d}\tau}{T \ln(1 - I(k, \lambda T))}.$$

Table 10.3 displays $C(T)$ for the gamma distribution with $\lambda = 1$ evaluated at $T = 1$.

10.4 Conclusions

We have introduced a parsimonious index that helps to assess the degree of aging or rejuvenation of a (non)repairable system. The index ranges from −1 to 1. It is negative for the class of decreasing failure rate distributions and point processes with decreasing

ROCOF and is positive for the increasing failure rate distributions and point processes with increasing ROCOF. The index can also be found useful in hypothesis testing for exponentiality of the TTF or failure inter-arrival times.

Acknowledgments. We would like to acknowledge Professor Igor Ushakov for the valuable comments about the original version of the paper [KK08].

References

[GG78] Gail, M.H., Gastwirth, J.L.: A Scale-free Goodness-of-fit Test for the Exponential distribution Based on the Gini Statistic. *Journal of Royal Statistical Society*, Vol. 40, # 3, 350–357 (1978)

[KK98] Kaminskiy, M.P., Krivtsov, V.V.: A Monte Carlo Approach to Repairable System Reliability Analysis. In: *Probabilistic Safety Assessment and Management*, New York: Springer, 1061–1068 (1998)

[KK08] Kaminskiy, M.P., Krivtsov, V.V.: An Integral Measure of Aging/Rejuvenation for Repairable and Non-repairable Systems, *Reliability and Risk Analysis: Theory & Applications*, Vol. 1, 69–76 (2008)

[KS86] Kijima, M., Sumita, N.: A Useful Generalization of Renewal Theory: Counting Process Governed by Non-negative Markovian Increments. *Journal of Applied Probability*, Vol. 23, 71–88 (1986)

[Kri07] Krivtsov, V.V.: Practical Extensions to NHPP Application in Repairable System Reliability Analysis, *Reliability Engineering & System Safety*, Vol. 92, # 5, 560–562 (2007)

[LCIE07] List of Countries by Income Equality, Source: http://www.answers.com/topic/list-of-countries-by-income-equality (2007)

[Law03] Lawless, J.F.: Statistical models and methods for lifetime data, New Jersey: *Wiley* (2003)

[AMSAA81] MIL-HDBK-189, Reliability Growth Management, *AMSAA* (1981)

[Nel90] Nelson, W.: Accelerated Testing: Statistical Models, Test Plans, and Data Analysis. New York: *Wiley* (1990)

[RH04] Rausand, M., Hoyland A.: System Reliability Theory:Models, Statistical Methods, and Applications. New York: *Wiley* (2004)

[Sen97] Sen, A.: On Economic Inequality. Oxford, England: *Clarendon Press* (1997)

11

Redundancy Analysis for Multi-state System: Reliability and Financial Assessment

Anatoly Lisnianski[1,*] and Yi Ding[2]

[1] Israel Electric Corporation Ltd., Haifa, Israel, anatoly-l@iec.co.il
[2] School of EEE, Nanyang Technological University, Singapore

Abstract: This paper discusses a type of redundancy that is typical in a multi-state system. It considers two interconnected multi-state systems where one multi-state system can satisfy its own stochastic demand and also can provide abundant resource (performance) to another system in order to improve the assisted system reliability. The paper also considers the financial issue for such a system where satisfied demand is associated with financial benefit and unsatisfied demand is associated with penalty. Traditional methods are usually not effective enough for such multi-state systems because of the "dimensional curse" problem. This paper presents the method for evaluation of reliability and corresponding financial issue for multi-state system with such kind of redundancy. The proposed method is based on the special type of universal generating function – universal generating function associated with discrete-state continuous-time stochastic process. The numerical example is presented to illustrate the proposed method.

Keywords and phrases: Multi-state system, Redundancy, Markov stochastic process, Reliability, Universal generating function associated with stochastic process

11.1 Introduction

Redundancy problem in a multi-state system (MSS) is much more complex than that in a binary-state system. Some redundancy problems for MSSs were investigated in [LL03, KA03], where typical parallel connections of multi-state components, k-out-of-n multi-state system and corresponding extensions were discussed and summarized. Recent research has focused on the reliability evaluation and optimization of MSS [YEH06, HZF03, LLBE98, TZH08]. However, for the MSS there is an important type of redundancy that has not existed for binary-state systems and has also not been investigated till now in the framework of MSS reliability analysis.

For MSS it is typical that after satisfying its own demand one MSS can provide its abundant resource (performance) to another MSS directly or through the interconnection system (that can also be multi-state). In this case, the first MSS can be called as

V.V. Rykov et al. (eds.), *Mathematical and Statistical Models and Methods in Reliability: Applications to Medicine, Finance, and Quality Control*, Statistics for Industry and Technology, DOI 10.1007/978-0-8176-4971-5_11,

the reserve MSS and the second one as the main MSS. In general, case demand for the reserve and the main MSS can also be described by two different independent stochastic processes. Typical examples of such kinds of MSS include power generating systems where one power station can assist another power station to satisfy demands, oil and gas production and transportation systems, computing systems with distributed computation resources, etc. Such multi-state structure with redundancy may be treated as MSSs with mutual aid or structure with interconnected MSSs. This type of redundancy is common enough for MSS. However, by using existing methods it is very difficult to build the reliability model for a complex repairable MSS considering redundancy and to solve it for obtaining corresponding reliability indices.

In practice, each multi-state component in a MSS can have different numbers of performance levels. This number may be relatively large – up to ten and more [GOL06]. Even for relatively small MSSs consisting of 3-5 repairable components the number of the entire system states will be significantly great (ten thousands or more). In general, for a MSS consisting of n repairable components, where each component i has k_i different capacity levels, there are $K = \prod_{i=1}^{n} k_i$ system states. This number may be very large and increase dramatically with the increased number of components. For interconnected MSSs the problem can be more serious. For such MSS, enormous efforts have to be performed to develop a stochastic process model and solve it (in order to obtain corresponding reliability indices) by using traditional straightforward methods. However, it is difficult to develop the stochastic process model for such a complex interconnected MSS. Determining all system states and transitions correctly is an arduous job. Moreover, it can challenge the available computing resources. If the random process is identified as a Markov process, the system state probabilities can be obtained by solving $K = \prod_{i=1}^{n} k_i$ differential equations. Therefore, in practice only long-term reliability analysis is performed to assess reliability of such systems, which is based on steady-state probabilities. In such case instead of differential equations only algebraic equations will be solved. Therefore short-term transient dynamic behavior of a MSS is out of consideration. In general case, such an approach will lead to decreased accuracy.

In order to use multi-state models for all components and to avoid decreased accuracy for reliability analysis, a special technique is proposed in this paper. The technique is based on the special type of Universal Generating Function (UGF) – UGF associated with discrete-state continuous-time stochastic processes. The UGF was primarily introduced in [USH86] in order to reduce the MSS's computational complexity and it was associated with random variable (not with stochastic process!). The mathematical foundations of the ordinary UGF associated with random variable one can find in [GU95, USH00]. An updated comprehensive description of the ordinary UGF technique with many technical applications can be found in [LL03, LEV05]. The method suggested in the paper for reliability assessment of interconnected repairable MSSs is based on UGF extension – UGF associated with stochastic process. This type of UGF was introduced in [LIS04], corresponding definitions one can also find in [LFD10]. The proposed UGF is used to evaluate the financial issue for such a system where satisfied demand is associated with financial benefit and unsatisfied demand is associated with penalty.

11.2 Problem Formulation

According to the generic MSS model any system component i in MSS can have k_i different states corresponding to the performance levels, represented by the set

$$g_i = \{g_{i1}, \dots, g_{ik_i}\}.$$

The current state of the component i and the corresponding value of the component performance level $G_i(t)$ at any instant t, are random variables. $G_i(t)$ takes values from g_i: $G_i(t) \in g_i$. Therefore, for the time interval $[0, T]$, where T is the MSS operation period, the performance level of component i is defined as a discrete-state continuous-time stochastic process. In the paper, only Markov processes will be considered, where the process behavior at a future instant only depend on the current state. The general Markov model of a multi-state component was introduced in [LL03], which considered minor and major failures/repairs of components.

Minor failures are failures causing component transition from state j to the adjacent state $j-1$. In other words, minor failure causes minimal degradation of component performance. The major failure is a failure that causes the components transition from state j to state $l : l < j - 1$. The minor repair returns an component from state l to state $l+1$ while the major repair returns components transition from state l to state j, where $j < l + 1$. In this case for each component, its performance level $G_i(t)$ is a discrete-state continuous-time Markov stochastic process.

General redundancy scheme for a MSS is presented in Fig. 11.1. Main multi-state system MSSm should satisfy its demand that is presented as a discrete-state continuous time Markov stochastic process $W_m(t)$. MSSm consists of m multi-state components. Performance level of each component i in MSSm at any instant $t > 0$ is defined by its output Markov stochastic process $G_{mi}(t)$, $i = 1, \dots, m$. All m components in the main multi-state system are composed in the technical structure according to the given structure function ψ_m, which defines the main system output stochastic performance $G_m(t)$ over stochastic processes of the system components:

$$G_m(t) = \psi_m\{G_{m1}(t), \dots, G_{mm}(t)\}.$$

Analogously, the reserve multi-state system MSSr consists of r multi-state components arranged in the technical structure according its structure function ψ_r, which defines MSSr output stochastic performance $G_r(t)$ based on output stochastic processes $G_{ri}(t)$, $i = 1, \dots, r$ of its components:

$$G_r(t) = \psi_r\{G_{r1}(t), \dots, G_{rr}(t)\}.$$

Reserve multi-state system MSSr should also satisfy its own demand that can be presented as a stochastic process $W_r(t)$. If the output performance $G_r(t) > W_r(t)$, the abundant performance $G_r(t) - W_r(t)$ can be delivered to the main multi-state system MSSm through the connecting system. In this case the stochastic process $G_{\text{cinp}}(t)$ that represents an input of the connecting MSSc can be defined by the following structure function $\psi_{\text{cinp}} : G_{\text{cinp}}(t) = \psi_{\text{cinp}}\{G_r(t) - W_r(t), 0\} = \max\{G_r(t) - W_r(t), 0\}$. Structure function ψ_{cinp} defines the reserve system obligations concerning assistance to the main system.

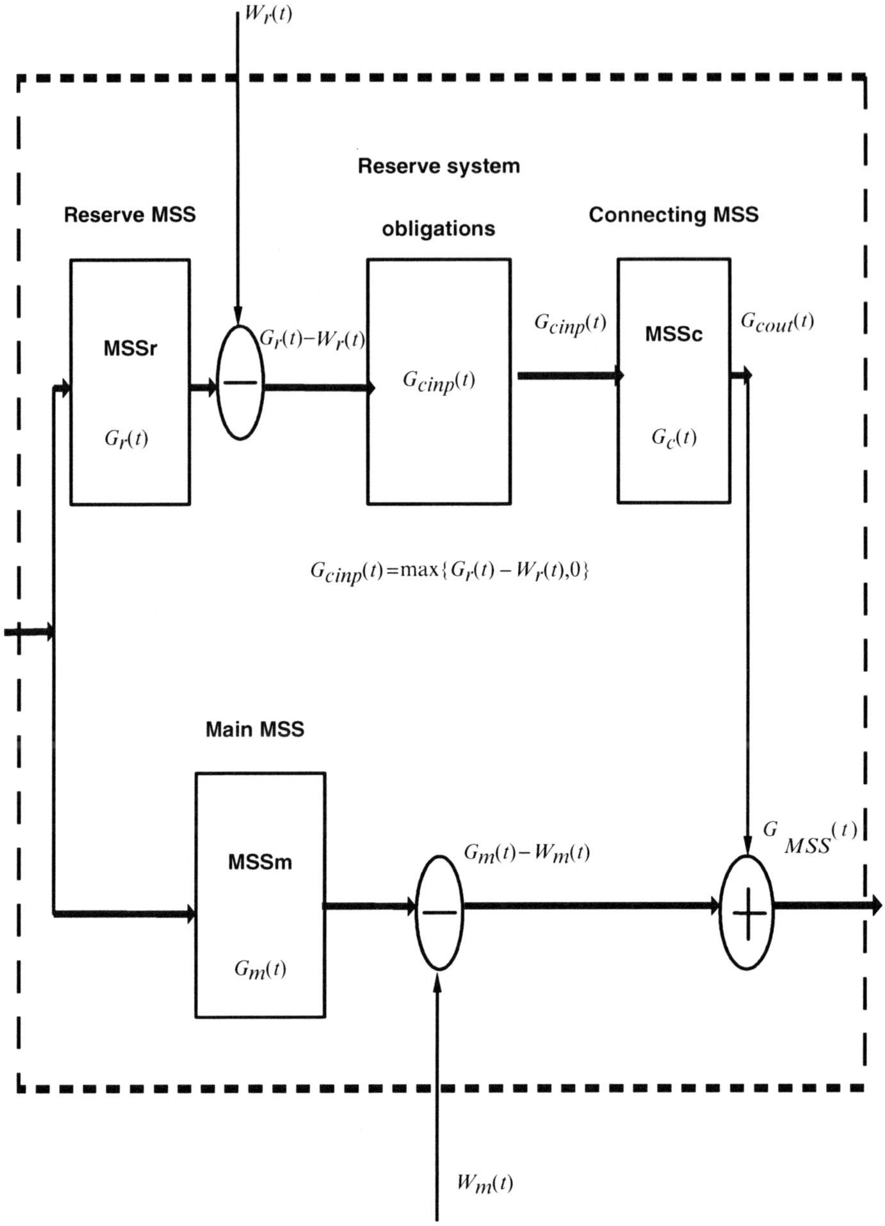

Figure 11.1. General redundancy scheme for the MSS

If the process $G_{\text{cinp}}(t)$ is defined by the above expression, it means that reserve MSSr will only send its abundant performance that remains after satisfying its own demand to the input of connecting MSSc. Generally speaking stochastic process $G_{\text{cinp}}(t)$ and function ψ_{cinp} can be defined by different ways. It will depend on the reserve system obligation agreement. For example, if, according to the agreement, even when $G_r(t) < W_r(t)$ it should be obligatory to send the specified performance g_s from the reserve system MSSr to the input of connecting system MSSc, then we have the following:

$$G_{\text{cinp}}(t) = \psi_{\text{cinp}}\{G_r(t) - W_r(t), g_s\} = \max\{G_r(t) - W_r(t), g_s\}.$$

The expression indicates that the reserve system according to its obligation agreement should send specified performance g_s to the connecting system even in the case when its own demand is not satisfied. When its demand is satisfied, the reserve system should send its abundant performance to the connecting system.

The connecting system can also be a multi-state system, which is designated as a MSSc. It consists of c multi-state components, which are composed in the technical structure with the given structure function ψ_c:

$$G_c(t) = \psi_c\{G_{c1}(t), \ldots, G_{cr}(t)\}.$$

In general case, such redundancy can be reversible. In other words, main MSSm can also be used as the redundant system in order to support the MSSr.

The problem is to evaluate reliability indices for the main MSSm that characterize the degree of satisfying demand $W_m(t)$, such as availability, expected instantaneous performance deficiency, expected accumulated performance deficiency, etc.

11.3 Model Descriptions

11.3.1 Main Definitions

The suggested method is based on special type of UGF–UGF associated with discrete-state continuous-time stochastic process. Here, we remind the following definitions [LFD10].

Definition 1. *Let individual z-transforms*

$$u_{X_j}(z) = \sum_{i=1}^{k_j} p_{ji} z^{x_{ji}}, \; j = 1, 2, \ldots, n$$

represent probability mass functions $\Pr\{X_j = x_{ji}\} = p_{ji},\; j = 1, 2, \ldots, n; i = 1, 2, \ldots, k_j$ *of* n *independent discrete random variables* X_j*. Universal Generating Operator (UGO) or composition operator* Ω_f *which produces z-transform for the random variable* $Y = f(X_1, X_2, \ldots, X_n)$ *is defined by the following*

$$\Omega_f\{u_{X_1}(z), u_{X_2}(z), \ldots, u_{X_n}(z)\} = \sum_{j_1=1}^{k_1} \sum_{j_2=1}^{k_2} \cdots \sum_{j_n=1}^{k_n} (p_{1j_1} p_{2j_2} \ldots p_{nj_n}) z^{f(x_{1j_1}, x_{2j_2}, \ldots, x_{nj_n})}.$$

Definition 2. *Let two vectors* $\{g_1, g_2, \ldots, g_K\}$ *and* $\{p_1(t_i), p_2(t_i), \ldots, p_K(t_i)\}$ *represent at any time* $t>0$ *states' performance levels and states' probability distribution respectively for discrete-state continuous-time stochastic process* $G(t) \in \{g_1, g_2, \ldots, g_K\}$*. z-transform*

$$u(z,t) = p_1(t)z^{g_1} + p_2(t)z^{g_2} + \cdots + p_K(t)z^{g_K}$$

is called as z-transform associated with stochastic process $G(t)$*.*

Definition 3. *If for* n *z-transforms* $u_j(z,t)$ *associated with independent stochastic processes* $G_j(t)$, $j = 1, 2, \ldots, n$, *Universal Generating Operator (UGO)* Ω_f *is defined, which produces resulting z-transform for the random variable* $G(t) = f(G_1(t), G_2(t), \ldots, G_n(t))$ *at any time instant* $t > 0$*, then these z-transforms are called as universal generating functions associated with corresponding discrete-states continuous-time stochastic processes* $G_j(t)$*.*

11.3.2 Model for Multi-State Component

In this subsection, when dealing with a single multi-state component, we will omit the index i for the designation of a set of the component's performance levels. This set is denoted as $\boldsymbol{g} = \{g_1, \ldots, g_k\}$. It is also assumed that this set is ordered so that $g_{j+1} \geq g_j$ for any j.

The state-space diagram for the general case of the repairable multi-state component with minor and major failures and repairs is presented in Fig. 11.2. Failures cause the component transition from the state l to the state j $(l > j)$ with corresponding transition intensity λ_{lj}. Repairs cause the component transition from the state j to the state l $(j < l)$ with corresponding transition intensity μ_{jl}.

Based on standard Markov technique the following system of differential equations (Kolmogorov equations) can be written for the state probabilities of the component

$$\begin{cases} \dfrac{dp_k(t)}{dt} = \sum\limits_{l=1}^{k-1} \mu_{l,k} p_l(t) - p_k(t) \sum\limits_{l=1}^{k-1} \lambda_{k,l} \\ \dfrac{dp_j(t)}{dt} = \sum\limits_{l=j+1}^{k} \lambda_{l,j} p_l(t) + \sum\limits_{l=1}^{j-1} \mu_{l,j} p_l(t) \\ \qquad - p_j(t) \Big(\sum\limits_{l=1}^{j-1} \lambda_{j,l} + \sum\limits_{l=j+1}^{k} \mu_{j,l} \Big), \text{for } 1 < j < k \\ \dfrac{dp_1(t)}{dt} = \sum\limits_{l=2}^{k} \lambda_{l,1} p_l(t) - p_1(t) \sum\limits_{l=2}^{k} \mu_{1,l} \end{cases} \tag{11.1}$$

with the initial conditions $p_k(0) = 1, p_{k-1}(0) = \cdots = p_1(0) = 0$.

Solving this system of differential equations one can obtain the state probabilities $p_j(t)$, $j = 1, \ldots, k$, which define probabilities that at instant $t > 0$ the component will be in state j.

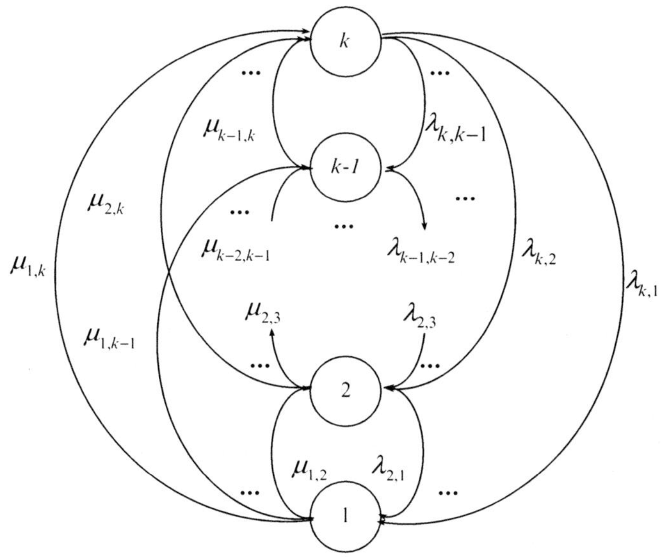

Figure 11.2. State-space diagram for repairable multi-state component

Based on these probabilities and given performance levels in every state j, one obtains universal generating function associated with component's output stochastic performance:

$$u(z,t) = p_1(t)z^{g_1} + p_2(t)z^{g_2} + \cdots + p_k(t)z^{g_k} \tag{11.2}$$

for any component at any instant $t > 0$.

The universal generating function (11.2) of a component is called as the individual universal generating function associated with discrete-state continuous-time Markov stochastic process $G(t) \in \boldsymbol{g}$, which describes the component evolution in its state space.

11.3.3 Model for Main Multi-State System and Its Demand

As stated in Sect. 11.2, main multi-state system MSSm consists of m multi-state components. Performance of each component i in MSSm at any instant $t > 0$ is defined by its output Markov stochastic process $G_{mi}(t)$, $i = 1, \ldots, m$. For any component i in MSSm we assume that its output performance stochastic process has $k_i^{(m)}$ different states with corresponding performance levels $g_{ij}^{(m)}$ and state probabilities $p_{ij}^{(m)}(t)$, $i = 1, \ldots, m; j = 1, \ldots, k_i^{(m)}$.

After solving corresponding system of differential equation (11.1) for component i, the following equation which defines individual UGF $u_{mi}(z,t)$ associated with output stochastic performance of the component i in MSSm can be written as:

$$u_{mi}(z,t) = \sum_{j=1}^{k_i^{(m)}} p_{ij}^{(m)}(t) z^{g_{ij}^{(m)}}, i = 1,\ldots,m. \tag{11.3}$$

All m components in the main multi-state system are composed in the technical structure according to the given structure function ψ_m, which defines the main system output stochastic performance $G_m(t)$:

$$G_m(t) = \psi_m\{G_{m1}(t),\ldots,G_{mm}(t)\},$$

where $G_m(t)$ is a discrete-state continuous-time Markov stochastic process with finite number of different performance levels.

We designate

K_m – number of output performance levels for main MSSm,

$p_j^{(m)}(t)$ – probability that stochastic output performance of main MSSm will be at level $g_j^{(m)}$, $j = 1,\ldots,K_m$ at time instant $t > 0$.

According to the definition of UGF, $U_m(z,t)$ associated with the stochastic output performance of the MSSm can be defined as the following format:

$$U_m(z,t) = \sum_{j=1}^{K_m} p_j^{(m)}(t) z^{g_j^{(m)}}. \tag{11.4}$$

By using composition operator Ω_{ψ_m} over individual UGFs representing output performance for each component, $U_m(z,t)$ for stochastic output performance of the main system can be obtained as:

$$U_m(z,t) = \sum_{j=1}^{K_m} p_j^{(m)}(t) z^{g_j^{(m)}} = \Omega_{\psi_m}\{u_{m1}(z,t),\ldots,u_{mm}(z,t)\}. \tag{11.5}$$

Taking into account (11.3) and using the general definition of composition operator, one can obtain the following expression:

$$\begin{aligned} U_m(z,t) &= \Omega_{\psi_m}\{u_{m1}(z,t),\ldots,u_{mm}(z,t)\} \\ &= \Omega_{\psi_m}\left\{\sum_{j_1=1}^{k_1^{(m)}} p_{1,j_1}^{(m)}(t) z^{g_{1,j_1}^{(m)}},\ldots,\sum_{j_m=1}^{k_m^{(m)}} p_{m,j_m}^{(m)}(t) z^{g_{m,j_m}^{(m)}}\right\} \\ &= \sum_{j_1=1}^{k_1^{(m)}}\sum_{j_2=1}^{k_2^{(m)}}\cdots\sum_{j_m=1}^{k_m^{(m)}}\left(\prod_{i=1}^{m} p_{i,j_i}^{(m)}(t) z^{\psi_m(g_{1,j_1}^{(m)},\ldots,g_{m,j_m}^{(m)})}\right). \end{aligned} \tag{11.6}$$

The procedures for computation of composition operators for major types of MSS (parallel, series, series-parallel or bridge configurations) have been well developed [LL03, LIS04] and provide a drastic reduction of the computational resources necessary to obtain the resulting UGF of a MSS. Based on these procedures the resulting UGF (11.4), corresponding states probabilities and performance levels for the output stochastic process of the MSSm can be obtained by using simple algebraic operations.

Demand $W_m(t)$ is assumed to be a discrete-state continuous time Markov stochastic process that at any instant $t > 0$ takes discrete values from the set $\boldsymbol{w_m} = \{w_{m1}, \ldots, w_{mM}\}$ with corresponding probabilities $p_1^{(w)}(t), \ldots, p_M^{(w)}(t)$. Therefore, the UGF $U_{W_m}(z,t)$ that corresponds to main system demand process has the following format:

$$U_{W_m}(z,t) = \sum_{l=1}^{M} p_l^{(w)}(t) z^{w_{ml}}. \tag{11.7}$$

We designate the UGF corresponding to Markov stochastic process $G_m(t) - W_m(t)$ as

$$U_{m-}(z,t) = \sum_{j=1}^{M_{m-}} p_j^{(m-)}(t) \cdot z^{g_j^{(m-)}}, \tag{11.8}$$

where
M_{m-} – number of possible performance levels for stochastic process $G_m(t) - W_m(t)$,
$p_j^{(m-)}(t)$ – probability that stochastic process $G_m(t) - W_m(t)$ will be at level $g_j^{(m-)}$, $j = 1, \ldots, M_{m-}$, at time instant $t > 0$.

By using the known structure function $\psi_{m-w} = G_m(t) - W_m(t)$, UGF $U_{m-}(z,t)$ can be obtained by using the following composition operator $\Omega_{\psi_{m-w}}$:

$$\begin{aligned} U_{m-}(z,t) &= \Omega_{\psi_{m-w}}\{U_m(z,t), U_{w_m}(z,t)\} \\ &= \Omega_{\psi_{m-w}}\left\{\sum_{j=1}^{K_m} p_j^{(m)}(t) z^{g_j^{(m)}}, \sum_{l=1}^{M} p_l^{(w)}(t) z^{w_{ml}}\right\} \\ &= \sum_{j=1}^{K_m}\sum_{l=1}^{M} p_j^{(m)} p_l^{(w)} z^{g_j^{(m)} - w_{ml}}. \end{aligned} \tag{11.9}$$

11.3.4 Model for Reserve Multi-State System and Its Demand

As stated in Sect. 11.2, reserve multi-state system MSSr consists of r multi-state components. Performance of each component i in MSSr is defined by its output Markov stochastic process $G_{ri}(t)$, $i = 1, \ldots, r$. For each component i in MSSr we assume that its stochastic process of output performance has $k_i^{(r)}$ different performance levels with corresponding performance levels $g_{ij}^{(r)}$ and state probabilities $p_{ij}^{(r)}(t)$, $i = 1, \ldots, r$; $j = 1, \ldots, k_i^{(r)}$.

After solving corresponding system of differential equation (11.1) for component i, the following equation which defines individual UGF $u_{ri}(z,t)$ for output stochastic performance of the component i in MSSr can be written as:

$$u_{ri}(z,t) = \sum_{j=1}^{k_i^{(r)}} p_{ij}^{(r)}(t) z^{g_{ij}^{(r)}}, i = 1, \ldots, r. \tag{11.10}$$

All r components in a reserve multi-state system are composed in the technical structure according to the given structure function ψ_r, which defines the reserve system output stochastic performance $G_r(t)$:

$$G_r(t) = \psi_r\{G_{r1}(t), \ldots, G_{rr}(t)\}.$$

$G_r(t)$ is a discrete-state continuous-time Markov stochastic process with finite number of different performance levels.

We designate

K_r – number of output performance levels for the reserve MSSr,

$p_j^{(r)}(t)$ – probability that stochastic output performance of the reserve MSSr will be at level $g_j^{(r)}$, $j = 1, \ldots, K_r$, at time instant $t > 0$.

According to the definition of UGF, $U_r(z,t)$ for stochastic output performance of MSSr can be defined as the following format:

$$U_r(z,t) = \sum_{j=1}^{K_r} p_j^{(r)}(t) z^{g_j^{(r)}}. \tag{11.11}$$

The resulting UGF $U_r(z,t)$ for the reserve system output stochastic performance $G_r(t)$ can be obtained by using composition operator Ω_{ψ_r} over individual UGFs representing output performance for each component in the reserve MSS:

$$U_r(z) = \sum_{j=1}^{K_r} p_j^{(r)}(t) z^{g_j^{(r)}} = \Omega_{\psi_r}\{u_{r1}(z,t), \ldots, u_{rr}(z,t)\}. \tag{11.12}$$

Taking into account expression (11.10) and using the general definition of composition operator [LL03], we obtain the following expressions:

$$\begin{aligned} U_r(z,t) &= \Omega_{\psi_r}\{u_{r1}(z,t), \ldots, u_{rr}(z,t)\} \\ &= \Omega_{\psi_r}\left\{\sum_{j_1=1}^{k_1^{(r)}} p_{1,j_1}^{(r)}(t) z^{g_{1,j_1}^{(r)}}, \ldots, \sum_{j_r=1}^{k_r^{(r)}} p_{r,j_r}^{(r)}(t) z^{g_{r,j_r}^{(r)}}\right\} \\ &= \sum_{j_1=1}^{k_1^{(r)}} \sum_{j_2=1}^{k_2^{(r)}} \cdots \sum_{j_r=1}^{k_r^{(r)}} \left(\prod_{i=1}^{r} p_{i,j_i}^{(r)}(t) z^{\psi_r(g_{1,j_1}^{(r)}, \ldots, g_{r,j_r}^{(r)})}\right). \end{aligned} \tag{11.13}$$

Demand $W_r(t)$ is also a discrete-state continuous time Markov stochastic process that at any instant $t > 0$ takes discrete values from the set $\boldsymbol{w_r} = \{w_{r1}, \ldots, w_{rN}\}$ with corresponding probabilities $p_1^{(w_r)}(t), \ldots, p_N^{(w_r)}(t)$. Therefore, the UGF $U_{w_r}(z,t)$ that corresponds to the demand process of the reserve system will be the following format:

$$U_{w_r}(z,t) = \sum_{l=1}^{N} p_l^{(w_r)}(t) z^{w_{rl}}. \tag{11.14}$$

We designate the UGF associated with Markov stochastic process $G_r(t) - W_r(t)$ as

$$U_{r-}(z,t) = \sum_{j=1}^{N_{r-}} p_j^{(r-)} z^{g_j^{(r-)}}. \tag{11.15}$$

The UGF $U_{r-}(z,t)$ can be obtained by the following way format:

$$\begin{aligned} U_{r-}(z,t) &= \Omega_{\psi_{r-w}}\{U_r(z,t), U_{w_r}(z,t)\} \\ &= \Omega_{\psi_{r-w}}\left\{\sum_{j=1}^{K_r} p_j^{(r)}(t) z^{g_j^{(r)}}, \sum_{l=1}^{N} p_l^{(w_r)}(t) z^{w_{rl}}\right\} \\ &= \sum_{j=1}^{K_r}\sum_{l=1}^{N} p_j^{(r)} p_l^{(w_r)} z^{g_j^{(r)} - w_{rl}}. \end{aligned} \tag{11.16}$$

11.3.5 Model for Reserve System Obligation and Connecting System

The reserve MSSr provides abundant resource (performance) to the main MSSm only after satisfying its own demand. Therefore, the stochastic process $G_{\text{cinp}}(t)$ that represents an input for the connecting MSSc can be defined by the following structure function ψ_{cinp}, which defines the reserve system obligation:

$$G_{\text{cinp}}(t) = \psi_{\text{cinp}}\{G_r(t) - W_r(t), 0\} = \max\{G_r(t) - W_r(t), 0\}. \tag{11.17}$$

If the process $G_{\text{cinp}}(t)$ is defined by expression (11.17), it indicates that reserve MSSr will only send its abundant performance that remains after satisfying its own demand to the input of connecting MSSc. As stated in Sect. 11.2, stochastic process $G_{\text{cinp}}(t)$ and function ψ_{cinp} are defined by the reserve system obligation agreement.

Based on (11.15), (11.16) and (11.17), UGF $U_{\text{cinp}}(z,t)$ associated with Markov stochastic process $G_{\text{cinp}}(t)$ can be obtained as:

$$\begin{aligned} U_{\text{cinp}}(z,t) &= \Omega_{\psi_{\text{cinp}}}\{U_{r-}(z,t), z^0\} = \Omega_{\psi_{\text{cinp}}}\left\{\sum_{j=1}^{N_{r-}} p_j^{(r-)} z^{g_j^{(r-)}}, z^0\right\} \\ &= \sum_{j=1}^{N_{r-}} p_j^{(r-)} z^{\max\{g_j^{(r-)},0\}}. \end{aligned} \tag{11.18}$$

In general case, connecting system MSSc can also be multi-state system. Its performance $G_c(t)$ is treated as the capability to transmit certain performance $g_l^{(c)}$, $l = 1,\ldots,c$ from the reserve system MSSr to the main system MSSm.

$$G_c(t) \in \{g_1^{(c)}, g_2^{(c)}, \ldots, g_c^{(c)}\}.$$

$p_l^{(c)}(t)$ is defined as the probability of state l corresponding to the performance level $g_l^{(c)}$ at instant $t > 0$. The UGF $U_c(z,t)$ of the MSSc corresponding to the underlying stochastic process $Gc(t)$ can be written as the following:

$$U_c(z,t) = \sum_{l=1}^{c} p_l^{(c)} z^{g_l^{(c)}}. \tag{11.19}$$

Output stochastic process $G_{\text{cout}}(t)$ of the connecting system MSSc can be obtained according to the following structure function:

$$G_{\text{cout}}(t) = \psi_{\text{cout}}\{G_c(t), G_{\text{cinp}}(t)\} = \min\{G_c(t), G_{\text{cinp}}(t)\}. \tag{11.20}$$

By using this structure function (11.20) and previously obtained UGFs $U_c(z,t)$ and $U_{\text{cinp}}(z,t)$ ((11.18) and (11.19), respectively) one can obtain UGF $U_{\text{cout}}(z,t)$ corresponding to the stochastic process of the output performance $G_{\text{cout}}(t)$. $G_{\text{cout}}(t)$ at any instant $t > 0$ is defined as the connecting system output performance that can be provided from the connecting system MSSc to the main system:

$$\begin{aligned} U_{\text{cout}}(z,t) &= \sum_{k=1}^{C_{\text{out}}} p_k^{(\text{cout})}(t) z^{g_k^{(\text{cout})}} \\ &= \Omega_{\psi_{\text{cout}}}\left\{\sum_{l=1}^{c} p_l^{(c)} z^{g_l^{(c)}}, \sum_{j=1}^{N_{r-}} p_j^{(r-)} z^{\max\{g_j^{(r-)},0\}}\right\} \\ &= \sum_{l=1}^{c}\sum_{j=1}^{N_{r-}} p_l^{(c)} p_j^{(r-)} z^{\min\{g_i^{(c)},\max(g_j^{(r-)},0)\}}, \end{aligned} \tag{11.21}$$

where C_{out} – number of output performance levels for discrete-state continuous-time stochastic process $G_{\text{cout}}(t)$,
$p_k^{(\text{cout})}(t)$ – probability that stochastic performance process $G_{\text{cout}}(t)$ will be at level $g_k^{(\text{cout})}$, $k = 1, \ldots, C_{\text{out}}$ at time instant $t > 0$.

11.3.6 Model for the Entire MSS

The output performance stochastic process $G_{\text{MSS}}(t)$ of the entire MSS considering redundancy is defined by the following structure function ψ_{MSS}:

$$G_{\text{MSS}}(t) = \psi_{\text{MSS}}\{G_m(t) - W_m(t), G_{\text{cout}}(t)\} = G_m(t) - W_m(t) + G_{\text{cout}}(t). \tag{11.22}$$

Based on this structure function and previously obtained $U_{m-}(z,t)$ and $U_{\text{cout}}(z,t)$ ((11.9) and (11.21), respectively), the UGF $U_{\text{MSS}}(z,t)$ for the entire MSS corresponding to the stochastic process $G_{\text{MSS}}(t)$ can be obtained as:

$$\begin{aligned} U_{\text{MSS}}(z,t) &= \sum_{j=1}^{M_{\text{MSS}}} p_j^{(\text{MSS})} z^{g_j^{(\text{MSS})}} \\ &= \Omega_{\psi_{\text{MSS}}}\{U_{m-}(z,t), U_{\text{cout}}(z,t)\} \\ &= \Omega_{\psi_{\text{MSS}}}\left\{\sum_{l=1}^{M_{m-}} p_l^{(m-)} z^{g_l^{(m-)}}, \sum_{k=1}^{C_{\text{out}}} p_k^{(\text{cout})}(t) z^{g_k^{(\text{cout})}}\right\} \\ &= \sum_{l=1}^{M_{m-}}\sum_{k=1}^{C_{\text{out}}} p_l^{(m-)} p_k^{(\text{cout})} z^{g_l^{(m-)}+g_k^{(\text{cout})}}, \end{aligned} \tag{11.23}$$

where M_{MSS} – number of output performance levels for discrete-state continuous-time stochastic process $G_{\text{MSS}}(t)$,
$p_j^{(\text{MSS})}(t)$ – probability that stochastic performance process $G_{\text{MSS}}(t)$ will be at level $g_j^{(\text{MSS})}$, $j = 1, \ldots, M_{\text{MSS}}$ at time instant $t > 0$.

11.4 Algorithm of UGF Computation for Entire MSS

The procedure of the UGF computation for the entire MSS considering redundancy is graphically presented in Fig. 11.3.

The procedure consists of the following steps.

1. Based on reliability data (failure and repair rates) for all components in MSSm and MSSr, individual UGFs (11.3) and (11.10) for all components can be obtained by solving the corresponding systems of differential equations (11.1).
2. Based on structure functions ψ_m, ψ_r and individual UGFs for all components in MSSm and MSSr UGFs $U_m(z,t), U_r(z,t)$ corresponding to performance stochastic processes $G_m(t), G_r(t)$ are evaluated according to (11.6) and (11.13), respectively.
3. In this step, UGFs $U_{m-}(z,t), U_{r-}(z,t)$ should be calculated corresponding to performance stochastic processes $G_m(t) - W_m(t)$, $G_r(t) - W_r(t)$ based on the (11.9) and (11.16), respectively.
4. UGF $U_{\text{cinp}}(z,t)$ corresponding to the stochastic processes $G_{\text{cinp}}(t)$ is evaluated according to expression (11.18), which is based on UGF $U_{r-}(z,t)$ and structure function ψ_{cinp}.
5. The UGF $U_c(z,t)$ corresponding to stochastic processes $G_c(t)$ is obtained according to the (11.19) for the connecting system.
6. The UGF $U_{\text{cout}}(z,t)$ corresponding to the performance stochastic process $G_{\text{cout}}(t)$, is evaluated by (11.21), which is based on UGFs $U_c(z,t), U_{\text{cinp}}(z,t)$ corresponding to performance stochastic processes $G_c(t)$, $G_{\text{cinp}}(t)$ respectively, and structure functions ψ_{cout} (expression (11.20)).

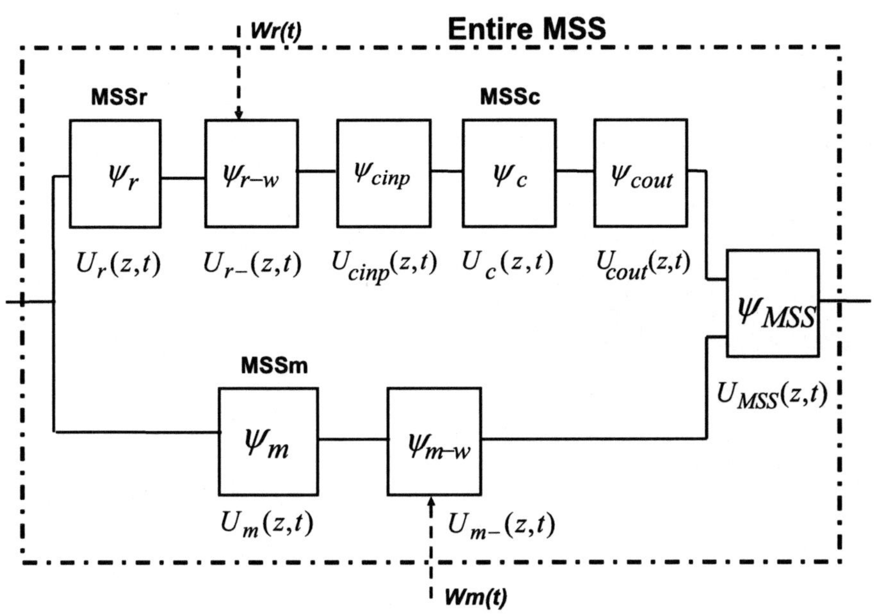

Figure 11.3. Recursive procedure for resulting UGF computation for entire MSS with redundancy

7. Based on the structure function ψ_{MSS} and previously obtained $U_{m-}(z,t)$ and $U_{\text{cout}}(z,t)$ (expressions (11.9) and (11.21) respectively), the resulting UGF $U_{\text{MSS}}(z,t)$ corresponding to the resulting output performance stochastic process $G_{\text{MSS}}(t)$ for the entire MSS is obtained according to (11.23).

11.5 Reliability Measures Computation for Entire MSS

When the UGF (expression (11.23)) of the entire interconnected MSS is obtained, the reliability measures for the system can be easily evaluated.

The entire MSS availability $A(t)$ at instant $t > 0$ can be evaluated as:

$$A(t) = \sum_{j=1}^{M_{\text{MSS}}} p_j^{(\text{MSS})}(t)\, \mathbf{1}(g_j^{(\text{MSS})} \geq 0), \tag{11.24}$$

where $\mathbf{1}(\mathit{True}) \equiv 1,\ \mathbf{1}(\mathit{False}) \equiv 0$.

The proposed UGF is used to evaluate the expected profit for the MSS, where satisfied demand is associated with financial benefit and unsatisfied demand is associated with penalty.

For a time period T, the system expected profit ($\mathit{EPROFIT}(T)$) can be calculated as:

$$\begin{aligned} \text{EPROFIT}(T) = \int_0^T \Bigg\{ \sum_{j=1}^{M_{\text{MSS}}} p_j^{(\text{MSS})}(t) \cdot \Big\{ \min\Big(\max(g_j^{(\text{MSS})}, 0), w_{mj} \Big) \cdot (\lambda - c) \\ - \max\Big(-g_j^{(\text{MSS})}, 0 \Big) \cdot \pi \Big\} \Bigg\} \cdot dt \end{aligned} \tag{11.25}$$

where w_{mj} is the demand of the main system for the system state j, λ is the electricity price to the customers, c is the production cost of electricity, and π is the incurred penalty cost.

For a time period T, the system expected interruption cost (ECOST(T)) can be calculated as:

$$\text{ECOST}(T) = \int_0^T \Bigg\{ \sum_{j=1}^{M_{\text{MSS}}} p_j^{(\text{MSS})}(t) \cdot \Big\{ \max\Big(-g_j^{(\text{MSS})}, 0 \Big) \cdot \pi \Big\} \Bigg\} \cdot dt. \tag{11.26}$$

11.6 Illustrative Example

The developed technique was used to evaluate two interconnected generation systems [ABA86]. System A consists of one 360 MW coal unit and one 220 MW gas unit. System B consists of one 360 MW coal units, two 220 MW gas units, and one 220 MW oil

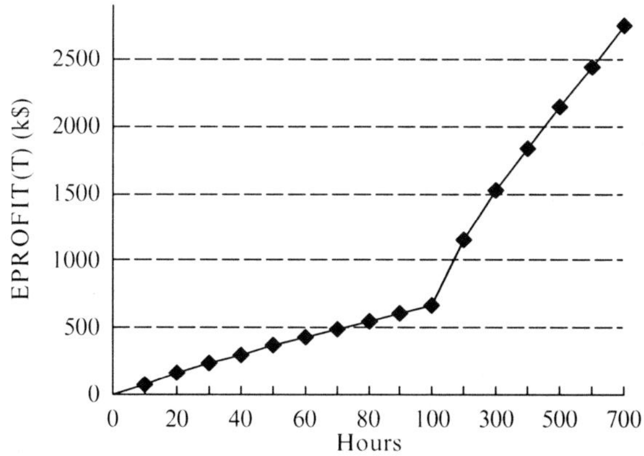

Figure 11.4. The system expected profit

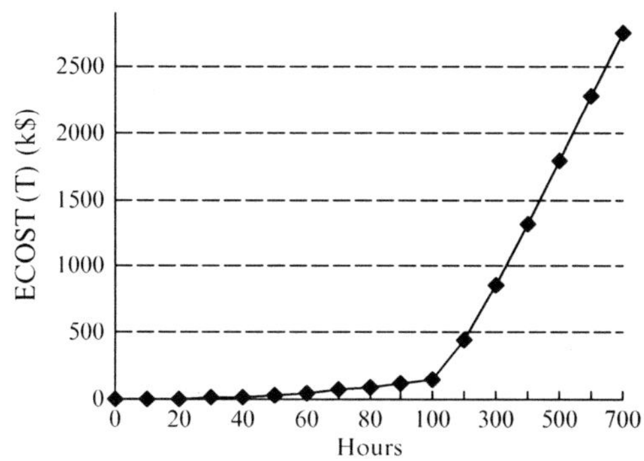

Figure 11.5. The system expected interruption cost

unit respectively. The coal unit, gas unit, and oil unit have 10, 10, and 11 states, respectively. The demands of system A and system B are 450 MW and 800 MW, respectively. Suppose that system A is the main system and system B is the corresponding reserve system. System B will provide a reserve to system A if system B can satisfy its own demand. Suppose the production cost of electricity and electricity price to the customers are set as 12$/MWh and 30$/MWh, respectively. A penalty cost (300$/MWh) will be incurred if the customer demand has been shed. Figure 11.4 illustrated the system expected profit of the example. It can be seen from the Fig. 11.4 that the system expected profit increases with the increasing of time. The system expected interruption cost is shown in Fig. 11.5, which also increases with the time. Because of the high penalty cost, the system expected interruption cost is almost the same as the system expected profit after about 700 h.

11.7 Conclusions

In the paper, an important type of redundancy is considered in MSS that has not existed in binary-state systems. Traditional methods applied to reliability computation for such systems are usually not effective enough because of dimension curse. A new approach to evaluate the dynamic reliability of MSS with such redundancy is suggested. The approach is based on using special type of UGF–UGF associated with random processes. The method presented in this paper is highly suitable for engineering applications since the procedure is well formalized and based on the natural decomposition of the entire interconnected systems. By using this method the short-term and long-term performance of multi-state systems with redundancy can be accurately predicted as well as financial issues.

References

[ABA86] Allan R.N., Billinton R., Abdel-Gawad N.M.K., IEEE reliability test system-extensions to and evaluation of the generating system. IEEE Transactions on Power Systems, **1**, No. 4, 1–7 (1986)

[GOL06] Goldner S., Markov model for a typical 360MW coal fired generation unit. Communications in Dependability and Quality Management, **9**, No. 1, 24–29 (2006)

[GU95] Gnedenko B., Ushakov I., Probabilitic Reliability Engineering. New York: Wiley (1995)

[HZF03] Huang J., Zuo M.J., Fang Z., Multi-state consecutive k-out-of-n systems. IIE Transactions **35**, 527–534 (2003)

[KA03] Kuo, W., Zuo M.J., Optimal Reliability Modeling Principles and Applications. New York: Wiley (2003)

[LEV05] Levitin G., Universal Generating Function and its Applications. Berlin: Springer (2005)

[LFD10] Lisnianski A., Frenkel I., Ding Y., Multi-state System Reliability Analysis and Optimization for Engineers and Industrial Managers. London, Heidelberg, New York: Springer (2010)

[LIS04] Lisnianski A., Universal generating function technique and random processes methods for multi-state system reliability analysis. Proceedings of 2nd International Workshop in Applied Probability (IWAP 2004), Piraeus, Greece, 237–242 (2004)

[LL03] Lisnianski, A., Levitin, G., Multi-State System Reliability Assessment, Optimization, Applications. Singapore: World Scientific (2003)

[LLBE98] Levitin G., Lisnianski A., Ben-Haim H., Elmakis D., Redundancy optimization for series-parallel multi-state systems. IEEE Transactions on Reliability, **47**, No. 2, 165–172 (1998)

[TZH08] Tian Z., Zuo M.J., Huang H., Reliability-redundancy allocation for multi-state series-parallel systems. IEEE Transactions on Reliability, **57**, No. 2, 303–310 (2008)

[USH00] Ushakov I., The method of generalized generating sequences. European Journal of Operating Research, **125**, No. 2, 316–323 (2000)

[USH86] Ushakov I., Universal generating function. Soviet Journal of Computer System Science, **24**, No. 5, 118–129 (1986)

[YEH06] Yeh W.C., The k-out-of-n acyclic multistate-node network reliability evaluation using the universal generating function method. Reliability Engineering and System Safety, **91**, 800–808 (2006)

12

On the Reliability Modeling of Hierarchical Systems

Vladimir Rykov* and Dmitry Kozyrev

Department of Probability Theory and Mathematical Statistics, Peoples' Friendship University of Russia, Moscow, Russia, vladimir_rykov@mail.ru

Abstract: For modeling and analysis of the reliability of complex hierarchical partially controllable systems the methods of decomposable semi-regenerative processes are used. A simple example illustrates our approach.

Keywords and phrases: Reliability of hierarchical systems, Decomposable semiregenerative processes

12.1 Introduction and Motivation

In terms of reliability, most of technical systems and biological objects with sufficiently high organization are complex hierarchical partially controllable systems. The failures in the systems of this type arise as a result of stress accumulation of the lowest (elementary) level, which passes several stages before the full failure. These faults lead to the decreasing efficiency of the system but do not lead to the full failure of the system. The system of control (SoC) fixes these fault stages of elements and gives the signal about the system "state of health" decreasing. According to these signals appropriate mechanisms of self-regulation are "switched on", and the system is self regenerated if the process disturbing is not too deep. In the last case, some external action is needed. It is supposed that this action being applied at time and in needed quality and quantity turns the system after some time to the normal functioning state. In the other case, the delay with maintenance of the system leads to the system degradation and as a result it leads to the full failure of the system. For biological systems, for instance, the neuron system plays a role of controlling system, and it possesses high reliability. This means that biological objects can be treated as a complex hierarchical partially controllable fault-tolerant reliability systems [DRS02, RD02, RDG04, RDGS04]. For different technical systems there exist analogous high reliable systems of control. These systems are special case of Multi-State Reliability Systems [LL03]. In some previous papers we considered such type of models under Markov assumptions [DRS02, RD02, RDG04, RDGS04].

In general case, the life times of units as well as the repair times of failed units, subsystem and the whole system should be considered as having general distributions.

V.V. Rykov et al. (eds.), *Mathematical and Statistical Models and Methods in Reliability: Applications to Medicine, Finance, and Quality Control*, Statistics for Industry and Technology, DOI 10.1007/978-0-8176-4971-5_12,

The respective distributions may depend on the type of the unit and also on the entire system state. The main characteristic for biological objects and complex technical systems are their survival functions. Thus, we focus in this paper on this function evaluation. Therefore, in this paper we propose a general mathematical model for the description and the survival function evaluation of complex hierarchical systems with general distributions of units life times as well as the repair times of failed units, subsystem and the whole system.

There are several approaches to model the reliability of systems with general life- and repair times distributions. However, anyhow all of them are reduced to markovization of the process that describes the system behavior. One of them was proposed by Yu. Belyaev [Be62], and consists in construction of so-called linear-wise Markov processes. Another approach was developed in the works of Buslenko et al. [Bu63, Bu71, BKK73, Ka78, Ka94, Ko65, Ko76, Kov76], who proposed and elaborated the mathematical technique for the study of so-called piecewise linear aggregative systems. The further development of this theory were done in the works [Ry75,Ry83,Ry84, Ry96,Ry97,Ry97,RYa71,RJ81], where the notion of Decomposable Semi-Regenerative Processes (DSRP) was proposed and methods for its investigation were developed. In this paper these methods are applied to the investigation of the reliability of complex hierarchical systems.

12.2 General Model

Consider a complex hierarchical multi-component system which is controlled and managed by a high-reliable system of control (SoC), shown in Fig. 12.1. Assume that the system is constructed from blocks and branches of several levels. Each block and the following after branches and blocks forms a hierarchical subsystem of the same type as the main one. The blocks of the lowest level will be referred to as units and they are subjected to gradual failures of their own type. They can take different values, depending on their types with some exhausted value. However, these multi-states units can be modified into the appropriate binary (i.e. with only two possible values) subsystems. Therefore, for the simplicity we limit ourselves to the binary systems only. We will denote by L the maximal level of units, and it is not necessary that any unit belongs to this level. Units of different levels are possible. The reliability of each unit is partially controllable. This means that after a fault it can be turned back into the normal (working) state after some repair time.

In order to specify the description of the system's behavior, we introduce a vector index $\mathbf{i} = (i_1, i_2, \ldots, i_{L_i})$ which determines each unit of the system as belonging to an appropriate chain of blocks at any level. Denote by $\mathcal{I}$ the set of these indices (and appropriate units). Then the state space of the system can be represented as $\mathcal{E} = \{\mathbf{j} = (j_\mathbf{i} : i \in \mathcal{I})\}$, where for any $\mathbf{i} \in \mathcal{I}$ the binary $j_\mathbf{i}$ represents the state of the $\mathbf{i}$th unit in sense of its reliability. To specify the subsystems of kth level we will refer to it as $\mathbf{i}_k$ and the states of the appropriate subsystem will be denoted by $j_\mathbf{i}^{(k)}$.

It is supposed that the life times of units as well as the repair times of failed units, subsystems and the whole system are independent random variables (r.v.) that have

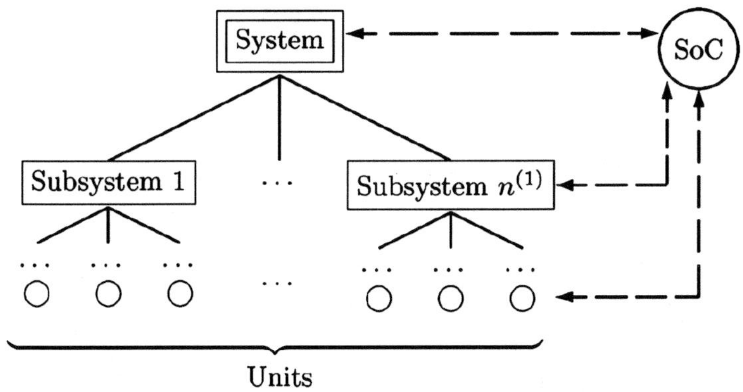

Figure 12.1. A complex multi-level hierarchical system

general distributions, which can depend on the type of unit $\mathbf{i}$ and appropriate subsystem $\mathbf{i}_k$. We will denote these r.v.'s by $A_{\mathbf{i}}$ and $B_{\mathbf{i}_k}$ and their cumulative distribution functions (c.d.f.) by

$$A_{\mathbf{i}}(x) = \mathbf{P}\{A_{\mathbf{i}} \leq x\}, \quad B_{\mathbf{i}_k}(x) = \mathbf{P}\{B_{\mathbf{i}_k} \leq x\}.$$

It is also supposed that these functions are differentiable, i.e. there exist appropriate probability density functions (p.d.f.). Some additional technique allows to omit this assumption, but we preserve them for the sake of simplicity.

12.3 Investigation of the Reliability of a General Model

According to the assumptions above we model the reliability of such a system by multi-dimensional process

$$\mathbf{J} = \{J_{\mathbf{i}}(t) : \mathbf{i} \in \mathcal{I}, \, t \geq 0\},$$

with the set of states $\mathcal{E}$, which should be concretized for every specific system. Moreover, for only the reliability investigation it is possible to consider the components $J_{\mathbf{i}}^{(k)}$ of the process $\mathbf{J}$ for subsystem of each level k as processes that takes only two states

$$J_{\mathbf{i}}^{(k)}(t) = \begin{cases} 0 & \text{if} \quad \mathbf{i}\text{th subsystem of } k\text{th level is working} \\ 1 & \text{if} \quad \mathbf{i}\text{th subsystem of } k\text{th level failed.} \end{cases}$$

According to the structure function of the system, the state of its units determines the state of appropriate subsystems and the whole system in sense of its reliability. After the repair of failed units, subsystems or the whole system they go to the initial states. Denote by $\mathcal{E}_{\mathbf{i}}^{(k)}$ and $\bar{\mathcal{E}}_{\mathbf{i}}^{(k)}$ the sets of the working and failure states for the $\mathbf{i}$th subsystem of the kth level. Then, the working period of the whole system is given by the relation

$$W^{(0)} = \inf\{t : \; J^{(0)}(t) \in \bar{\mathcal{E}}^{(0)}\} = \inf\{t : \; J_{\mathbf{i}_j}^{(1)}(t) \in \bar{\mathcal{E}}_{\mathbf{i}_j}^{(1)}, \; j = \overline{1, n_1}\}.$$

Analogously for each subsystem $\mathbf{i}$ of any kth level one has

$$W_{\mathbf{i}}^{(k)} = \inf\{t:\ t \le W^{(k)},\ J_{\mathbf{i}}^{(k)}(t) \in \bar{\mathcal{E}}_{\mathbf{i}}^{(k)}\} = \\ = \inf\left\{t:\ t \le W^{(k)},\ J_{\mathbf{i}_j}^{(k+1)}(t) \in \bar{\mathcal{E}}_{\mathbf{i}_j}^{(k+1)},\ j = \overline{1, n_{\mathbf{i}}^{(k)}}\right\}.$$

Therefore, considering the working period distributions of some subsystem as its life times one can investigate in the same way the reliability function of the subsystem of the next level. However, since the initial information about system is given for the lowest level only, the problem should be solved starting from the lowest level.

To calculate the cumulative distribution function (c.d.f.) of any subsystem $\mathbf{i}_k$ of any level k according to the definition of the working period as $W_{\mathbf{i}}^{(k)} = \inf\{t:\ J_{\mathbf{i}}^{(k)}(t) \in \bar{\mathcal{E}}_{\mathbf{i}}^{(k)}\}$ one has

$$W_{\mathbf{i}}^{(k)}(t) = \mathbf{P}\{W_{\mathbf{i}}^{(k)} \le t\} = 1 - \prod_{\mathbf{j} \in \bar{\mathcal{E}}_{\mathbf{i}}^{(k)}} [1 - \pi_{\mathbf{j}}(t)],$$

where $\pi_{\mathbf{j}}(t)$ is the probability distribution for the process $\mathbf{J}$ to be in failure state $\mathbf{j} \in \bar{\mathcal{E}}_{\mathbf{i}}^{(k)}$ at a time t. Therefore, the problem is divided into two parts:

- To investigate the process $J_{\mathbf{i}}^{(k)}(t)$ describing behavior of any subsystem of each level
- To find the working period $W_{\mathbf{i}}^{(k)}$ for any subsystem of each level

As the problems have identical solution for any subsystem they will be considered in some general construction.

12.4 The Subsystems Behavior Investigation

12.4.1 General Case

For investigation of a separate subsystem denote by $\mathbf{J}(t) = \{J_i(t),\ i = \overline{1,m}\}$ the binary process of its reliability, and by $A_i(x),\ B_i(x),\ (i = \overline{1,m})$ the c.d.f.'s of their life and repair times. In order to use the markovization approach, denote by W the working time of this subsystem (the time to the its first failure) and consider absorbing (with absorbtion in the set of failure states $\bar{\mathcal{E}}$) multi-dimensional Markov process $\mathbf{Z} = (\mathbf{J}(t),\ \mathbf{X}(t),\ t \le W)$ with general states space $\hat{\mathcal{E}} = \mathcal{E} \times R^m$, where components of the additional vector $\mathbf{X}$ describe the times elapsed after entering the appropriate states. Denote by (for simplicity index W will be omitted everywhere)

$$\pi_{\mathbf{j}}(t;\ \mathbf{dx}) = \pi_{(j_1,\dots j_m)}(t;\ \mathrm{d}x_1, \dots \mathrm{d}x_m) = \mathbf{P}\{J_i(t) = j_i,\ X_i(t) \in \mathrm{d}x_i,\ i = \overline{1,m},\ t \le W\} \tag{12.1}$$

the probability for the process $\mathbf{Z}$ to be in time t during its separate working period W at the state $(\mathbf{j},\ \mathbf{dx})$. Note that under some additional conditions if these functions are absolutely continuous with respect to Lebesgue measure they can be represented in terms of probability density functions (p.d.f.) $\pi_{\mathbf{j}}(t,\,\mathbf{dx}) = \pi_{\mathbf{j}}(t,\,\mathbf{x})\mathbf{dx}$. In order to apply the methods of Decomposable Semi-Regenerative Processes (DSRP) denote by:

- $S_1, S_2, \ldots S_n, \ldots$ the times of the process $\mathbf{J}$ jumps at separate working period W that coincide with times of failure or renovation of elements of the subsystem (the epochs, in which some of components of the process $\mathbf{Z}$ became equal zero). We will refer to S_n as times of the kth type renovation if $X_k(S_n + 0) = 0$
- $\mathbf{J}_n = \mathbf{J}(S_n + 0),\ \ \mathbf{X}_n = \mathbf{X}(S_n + 0),\ \ K_n = \#\{k:\ X_k(S_n + 0) = 0\}$
- $\mathbf{j}(k) = (j_1, \ldots, \bar{j}_k, \ldots j_m),\ \bar{i}_k = 1 - i_k$
- $\mathbf{x}(k) = (x_1, \ldots, x_{k-1}, 0, x_{k+1}, \ldots x_m)$
- $\mathbf{N} = \{N_{\mathbf{j}}^{(k)}(t;\,\mathbf{dx}(k)):\ 1 \le k \le m,\ \mathbf{j} \in \mathcal{E}\}$ embedded renewal process (ERP) with general renewal states space, which components $N_{\mathbf{j}}^{(k)}(t;\,\mathbf{dx}(k))$ are the kth type ERP with the set of renovation states $(\mathbf{j},\,\mathbf{dx}(k))$,

$$N_{\mathbf{j}}^{(k)}(t;\,\mathbf{dx}(k)) = \sum_{n \ge 0} \delta_{k,K_n} 1_{\{[0,t],\mathbf{j},\,\mathbf{dx}(k)\}}(S_n,\,\mathbf{J}_n,\,\mathbf{X}_n)$$

- $\mathbf{H} = \{H_{\mathbf{j}}^{(k)}(t;\,\mathbf{dx}(k)):\ 1 \le k \le m,\ \mathbf{j} \in \mathcal{E}\}$ embedded renewal function (ERF) with general states space, which components $H_{\mathbf{j}}^{(k)}(t;\,\mathbf{dx}(k)) = \mathbf{E}N_{\mathbf{j}}^{(k)}(t,\mathbf{dx}(k))$ are the kth type renewal function with the set of renovation states $(\mathbf{j},\,\mathbf{dx}(k))$

Denote also by $\Gamma_{j_k}(x)$ the c.d.f. of the time duration of the j_kth component of the process $\mathbf{Z}$ to stay at its state and by $\gamma_{j_k}(x)$ its hazard rate function (h.r.f.), which exists under our assumption,

$$\Gamma_{j_k}(x) = \delta_{j_k,0}A_k(x) + \delta_{j_k,1}B_k(x), \quad \gamma_{j_k}(x) = \frac{\Gamma'_{j_k}(x)}{1 - \Gamma_{j_k}(x)}.$$

Remind also that for p.d.f. $\Gamma(x)$ of any r.v. Γ and appropriate conditional distribution it holds

$$\Gamma(x) = 1 - \exp\left\{-\int_0^x \gamma(\xi)\mathrm{d}\xi\right\};\ \mathbf{P}\{\Gamma > y \,|\, \Gamma > x\} = \frac{1 - \Gamma(y)}{1 - \Gamma(x)} = \exp\left\{-\int_x^y \gamma(\xi)\mathrm{d}\xi\right\}.$$

In order to find the probabilities $\pi_{\mathbf{j}}(t;\,\mathbf{dx})$ in terms of ERP let us denote by $Q_{\mathbf{j}}^{(k,l)}(\mathrm{d}t;\,\mathbf{x}(k),\,\mathbf{dy}(l))$ the conditional probability measure of the following event: time between sequential renovations belongs to the interval $(t, t + \mathrm{d}t)$, and during this time a renovation of lth type with the state $(\mathbf{j}(l), \mathbf{dy}(l))$ occurs, being the previous one of the type k with the state $(\mathbf{j}, \mathbf{x}(k))$,

$$\begin{aligned} &Q_{\mathbf{j}}^{(k,\,l)}(\mathrm{d}t;\,\mathbf{x}(k),\,\mathbf{dy}(l)) \\ &= \mathbf{P}\{S_{n+1} - S_n \in \mathrm{d}t,\ \mathbf{J}_{n+1} = \mathbf{j}(l),\ \mathbf{X}_{n+1} \in \mathbf{dy}(l) \,|\, \mathbf{J}_n = \mathbf{j},\ \mathbf{X}_n = \mathbf{x}(k)\}, \end{aligned}$$

and by $\bar{Q}_{\mathbf{j}}^{(k,\,l)}(\mathrm{d}t;\,\mathbf{x}(k),\,\mathbf{dy})$ appropriate measure at the time directly before jump. For the first renovation (for the initial state $\mathbf{0}$), the appropriate measure is denoted by $Q_{\mathbf{0}}^{(l)}(\mathrm{d}t;\,\mathbf{0},\,\mathbf{dy}(l))$. Denote also by $\bar{Q}_{\mathbf{j}}(t;\,\mathbf{x},\,\mathbf{dy})$ the conditional probability measure that beginning from the state $(\mathbf{j}, \mathbf{x})$ the process does not leave it during the time t, and therefore it occurs at the state $(\mathbf{j}, \mathbf{y} = t\mathbf{1} + \mathbf{x})$,

$$\bar{Q}_{\mathbf{j}}(t;\, \mathbf{x},\, \mathbf{dy}) = \mathbf{P}\{S_{n+1} - S_n > t,\, \mathbf{X}(S_n + t) \in \mathbf{dy} \,|\, \mathbf{J}_n = \mathbf{j},\, \mathbf{X}_n = \mathbf{x}\}.$$

With these notations the following theorem holds.

Theorem 1. *The probabilities $\pi_{\mathbf{j}}(t;\, \mathbf{dx})$ $(\mathbf{j} \in E)$ satisfy to the equalities*

$$\pi_{\mathbf{0}}(t;\, \mathbf{dy}) = \bar{Q}_{\mathbf{0}}(t;\, \mathbf{0}, \mathbf{dy}) + \sum_{1 \le k \le m} \int_0^t \int_{R^{m-1}} H_{\mathbf{0}}^{(k)}(\mathrm{d}u;\, \mathbf{dx}(k)) \bar{Q}_{\mathbf{0}}(t-u;\, \mathbf{x}(k),\, \mathbf{dy}),$$

$$\pi_{\mathbf{j}}(t; \mathbf{dy}) = \sum_{1 \le k \le m} \int_0^t \int_{R^{m-1}} H_{\mathbf{j}}^{(k)}(\mathrm{d}u;\, \mathbf{dx}(k)) \bar{Q}_{\mathbf{j}}(t-u;\, \mathbf{x}(k),\, \mathbf{dy}), \tag{12.2}$$

Proof. Immediately follows from the Complete Probability Formulae if only remark that in kth type renovation epoch the kth component of the vector $\mathbf{x}$ became equal zero, $X_k(S_n + 0) = 0$.

On the other hand the theory of DSRP gives the following.

Theorem 2. *The family of ERF's $H_{\mathbf{k}}^{(l)}(t;\, \mathbf{dx}(k))$ satisfies to the equations:*

- *for $\mathbf{j} \in \mathcal{E}$ such that $\mathbf{j}(l) \in \mathcal{E}$*

$$H_{\mathbf{e}_l}^{(l)}(\mathrm{d}t; \mathbf{dy}(l)) = Q_{\mathbf{0}}^{(l)}(\mathrm{d}t;\, \mathbf{0},\, \mathbf{dy}(l)) + \sum_{1 \le k \le m} \int_0^t \int_{R^{m-1}} \mathrm{d}H_{\mathbf{0}}^{(k)}(\mathrm{d}u,\, \mathbf{dx}(k)) Q_{\mathbf{0}}^{(k,l)}(\mathrm{d}t-u,\, \mathbf{x}(k), \mathbf{dy}(l)), \tag{12.3}$$

$$H_{\mathbf{j}}^{(l)}(\mathrm{d}t;\, \mathbf{dy}(l)) = \sum_{1 \le k \le m} \int_0^t \int_{R^{m-1}} H_{\mathbf{j}}^{(k)}(\mathrm{d}u;\, \mathbf{dx}(k)) Q_{\mathbf{j}}^{(k,\,l)}(\mathrm{d}t-u;\, \mathbf{x}(k),\, \mathbf{dy}(l)) \tag{12.4}$$

- *and for $\mathbf{j} \in \mathcal{E}$ such that $\mathbf{j}(l) \in \bar{\mathcal{E}}$*

$$\pi_{\mathbf{j}(l)}(\mathrm{d}t;\, \mathbf{dy}(l)) = \sum_{1 \le k \le m} \int_0^t \int_{R^{m-1}} H_{\mathbf{j}}^{(k)}(\mathrm{d}u;\, \mathbf{dx}(k)) Q_{\mathbf{j}}^{(k,\,l)}(\mathrm{d}t-u;\, \mathbf{x},\, \mathbf{dy}) \tag{12.5}$$

The representation of the functions $\bar{Q}_{\mathbf{j}}^{(k,\,l)}(\mathrm{d}t;\, \mathbf{x}(k),\, \mathbf{dy})$ and $\bar{Q}_{\mathbf{j}}(t;\, \mathbf{x},\, \mathbf{dy})$ in terms of the model characteristics is given in the following lemma.

Lemma 1. *For the functions* $\bar{Q}_{\mathbf{j}}^{(k,\,l)}(\mathrm{d}t;\,\mathbf{x}(k),\,\mathbf{dy})$ *and* $\bar{Q}_{\mathbf{j}}(t;\,\mathbf{x},\,\mathbf{dy})$, *the following representations holds*

$$\bar{Q}_{\mathbf{0}}^{(l)}(\mathrm{d}t;\,\mathbf{0},\,\mathbf{dy}) = \beta_l(y_l)\prod_{1\le i\le m}\exp\left\{-\int_0^{y_i}\alpha_i(\xi)\mathrm{d}\xi\right\}\delta(t-y_i)\mathrm{d}y_i\,\mathrm{d}t,$$

$$\bar{Q}_{\mathbf{j}}^{(k,\,l)}(\mathrm{d}t;\,\mathbf{x}(k),\,\mathbf{dy}) = \gamma_{j_l}(y_l)\prod_{i\ne l}\exp\left\{-\int_{x_i}^{y_i}\gamma_{j_i}(\xi)\mathrm{d}\xi\right\}\delta(t-(y_i-x_i))\mathrm{d}y_i\,\mathrm{d}t,$$

$$\bar{Q}_{\mathbf{j}}(t;\,\mathbf{x},\,\mathbf{dy}) = \prod_{i}\exp\left\{-\int_{x_i}^{y_i}\gamma_i(\xi)\mathrm{d}\xi\right\}\delta(t-(y_l-x_l))\mathrm{d}y_i. \tag{12.6}$$

Proof. Indeed, for the functions $\bar{Q}_{\mathbf{j}}^{(k,\,l)}(\mathrm{d}t;\,\mathbf{x}(k),\,\mathbf{dy})$ one has

$$\begin{aligned}
&\bar{Q}_{\mathbf{j}}^{(k,\,l)}(\mathrm{d}t;\,\mathbf{x}(k),\,\mathbf{dy})\\
&= \mathbf{P}\{S_{n+1}-S_n\in\mathrm{d}t,\,\mathbf{J}_{n+1}=\mathbf{j}(l),\,\mathbf{X}_{n+1}\in\mathbf{dy}\,|\,\mathbf{J}_n=\mathbf{j},\,\mathbf{X}_n=\mathbf{x}(k)\}\\
&= \mathbf{P}\left\{\Gamma_{j_l}\in\mathrm{d}y_{j_l}\,|\,\Gamma_{j_l}>x_{j_l}\right\}\prod_{i\ne l}\mathbf{P}\left\{\Gamma_i>y_i\,|\,\Gamma_i>x_i\right\}\delta(t-(y_i-x_i))\mathrm{d}y_i\,\mathrm{d}t\\
&= \frac{\Gamma'_{j_l}(y_l)\mathrm{d}y_l}{1-\Gamma_{j_l}(x_l)}\prod_{i\ne l}\frac{1-\Gamma_i(y_i)}{1-\Gamma_i(x_i)}\delta(t-(y_i-x_i))\mathrm{d}y_i\,\mathrm{d}t\\
&= \frac{\Gamma'_{j_l}(y_l)}{1-\Gamma_{j_l}(y_l)}\prod_{1\le i\le m}\frac{1-\Gamma_i(y_i)}{1-\Gamma_i(x_i)}\delta(t-(y_i-x_i))\mathrm{d}y_i\,\mathrm{d}t\\
&= \gamma_{j_l}(y_l)\prod_{1\le i\le m}\exp\left\{-\int_{x_i}^{y_i}\gamma_{j_i}(\xi)\mathrm{d}\xi\right\}\delta(t-(y_i-x_i))\mathrm{d}y_i\,\mathrm{d}t.
\end{aligned}$$

Analogously, the next formulas can be obtained.

To get appropriate expressions in terms of the Laplace transform (LT) denote by $v=\sum_{1\le i\le m}v_i$ and $v(k)=\sum_{i\ne k}v_i$ and introduce the following functions

$$\phi_{\mathbf{0}}(s,\,v) = \int_0^\infty\exp\left\{-(s+v)t-\sum_{1\le i\le m}\int_0^t\alpha_i(\xi)\mathrm{d}\xi\right\}\beta_l(t)\mathrm{d}t;$$

$$\phi_{\mathbf{j}}(s,\,\mathbf{x},\,v) = \int_0^\infty\exp\left\{-(s+v)t-\sum_{1\le i\le m}\int_{x_i}^{x_i+t}\gamma_{j_i}(\xi)\mathrm{d}\xi\right\}\gamma_{j_l}(x_l+t)\mathrm{d}t;$$

$$\psi_{\mathbf{j}}(s,\,\mathbf{x},\,v) = \int_0^\infty\exp\left\{-(s+v)t-\sum_{1\le i\le m}\int_{x_i}^{x_i+t}\gamma_{j_i}(\xi)\mathrm{d}\xi\right\}\mathrm{d}t.$$

With these notations consider the LT $q_{\mathbf{j}}^{(k,l)}(s;\mathbf{x}(k),\mathbf{v}(l))$ and $\bar{q}_{\mathbf{j}}(s;\,\mathbf{x},\mathbf{v})$ of the functions $Q_{\mathbf{j}}^{(k,l)}(t;\,\mathbf{x}(k),\mathbf{dy}(l))$ and $\bar{Q}_{\mathbf{j}}(t;\,\mathbf{x},\mathbf{dy})$,

$$q_{\mathbf{j}}^{(k,l)}(s;\mathbf{x}(k),\,\mathbf{v}(l)) = \int_0^\infty \mathrm{e}^{-st} \int_{R^{m-1}} \mathrm{e}^{-\mathbf{v}'(l)\mathbf{y}(l)} Q_{\mathbf{j}}^{(k,l)}(t;\,\mathbf{x}(k),\,\mathbf{dy}(l));$$

$$\bar{q}_{\mathbf{j}}(s;\mathbf{x},\,\mathbf{v}) = \int_0^\infty \mathrm{e}^{-st} \int_{R^m} \mathrm{e}^{-\mathbf{v}'\mathbf{y}} \bar{Q}_{\mathbf{j}}(t;\,\mathbf{x},\,\mathbf{dy})$$

Lemma 2. *The following expressions hold*

$$\begin{aligned} q_{\mathbf{0}}(s;\,\mathbf{0},\,\mathbf{v}(l)) &= \phi_{\mathbf{0}}(s,\,v(l)); \\ q_{\mathbf{j}}^{(k,l)}(s;\,\mathbf{x}(k),\,\mathbf{v}(l)) &= \mathrm{e}^{-\mathbf{v}'\mathbf{x}}\phi_{\mathbf{j}}(s,\,\mathbf{x},\,v(l)); \\ \bar{q}_{\mathbf{j}}(s;\,\mathbf{v}) &= \mathrm{e}^{-\mathbf{v}'\mathbf{x}}\psi_{\mathbf{j}}(s,\,\mathbf{x},\,v); \end{aligned} \tag{12.7}$$

Proof. Taking into account that $q_{\mathbf{j}}^{(k,l)}(s;\,\mathbf{x}(k),\,\mathbf{v}(l)) = \bar{q}_{\mathbf{j}}^{(k,l)}(s;\,\mathbf{x}(k),\,\mathbf{v})|_{v_l=0}$ calculate first the LT for $\bar{Q}_{\mathbf{j}}^{(k,l)}(t;\,\mathbf{x}(k),\,\mathbf{dy})$.

$$\begin{aligned} \bar{q}_{\mathbf{0}}^{(l)}(s;\,\mathbf{0},\,\mathbf{v}) &\equiv \int_0^\infty \int_{R^m} \mathrm{e}^{-st-\mathbf{v}'\mathbf{y}} \bar{Q}_{\mathbf{0}}^{(l)}(\mathrm{d}t;\,\mathbf{0},\,\mathbf{dy}) \\ &= \int_0^\infty \int_{R^m} \mathrm{e}^{-st-\mathbf{v}'\mathbf{y}} \beta_l(y_l) \prod_{1\le i\le m} \exp\left\{-\int_0^{y_i} \alpha_i(\xi)\mathrm{d}\xi\right\} \delta(t-y_i)\mathrm{d}y_i\,\mathrm{d}t \\ &= \int_0^\infty \exp\left\{-(s+v)t - \sum_{1\le i\le m} \int_0^t \alpha_i(\xi)\mathrm{d}\xi\right\} \beta_l(t)\mathrm{d}t \equiv \phi_{\mathbf{0}}(s,\,v); \end{aligned}$$

$$\begin{aligned} \bar{q}_{\mathbf{j}}^{(k,l)}(s;\,\mathbf{x}(k),\,\mathbf{v}) &\equiv \int_0^\infty \int_{R^m} \mathrm{e}^{-st-\mathbf{v}'\mathbf{y}} \bar{Q}_{\mathbf{j}}^{(k,l)}(\mathrm{d}t;\,\mathbf{x}(k),\,\mathbf{dy}) \\ &= \int_0^\infty \int_{R^m} \mathrm{e}^{-st-\mathbf{v}'\mathbf{y}} \gamma_{j_l}(y_l) \prod_{1\le i\le m} \exp\left\{-\int_{x_i}^{y_i} \gamma_i(\xi)\mathrm{d}\xi\right\} \\ &\quad \times\delta(t-(y_i-x_i))\mathrm{d}y_i\,\mathrm{d}t, \\ &= \mathrm{e}^{-\mathbf{v}'\mathbf{x}} \int_0^\infty \exp\left\{-(s+v)t - \sum_{1\le i\le m} \int_{x_i}^{x_i+t} \gamma_{j_i}(\xi)\mathrm{d}\xi\right\} \gamma_{j_l}(x_l+t)\mathrm{d}t \\ &= \mathrm{e}^{-\mathbf{v}'\mathbf{x}}\phi_{\mathbf{j}}(s,\,\mathbf{x},\,v); \end{aligned}$$

$$\bar{q}_{\mathbf{j}}(s;\,\mathbf{v}) \equiv \int\limits_0^\infty \int\limits_{R^m} \mathrm{e}^{-st-\mathbf{v}'\mathbf{y}} \bar{Q}_{\mathbf{j}}(t;\,\mathbf{dy})\,\mathrm{d}t$$

$$= \int\limits_0^\infty \int\limits_{R^m} \mathrm{e}^{-st-\mathbf{v}'\mathbf{y}} \prod_i \exp\left\{-\int\limits_{x_i}^{y_i} \gamma_i(\xi)\mathrm{d}\xi\right\} \delta(t-(y_i-x_i))\mathrm{d}y_i\,\mathrm{d}t$$

$$= \mathrm{e}^{-\mathbf{v}'\mathbf{x}} \int\limits_0^\infty \exp\left\{-(s+v)t - \sum_i \int\limits_{x_i}^{x_i+t} \gamma_{j_i}(\xi)\mathrm{d}\xi\right\} \mathrm{d}t = \mathrm{e}^{-\mathbf{v}'\mathbf{x}} \psi_{\mathbf{j}}(s,\,\mathbf{x},\,v).$$

Substituting $v_l = 0$ into these relations proves the lemma.

Using these results for the Laplace Transforms (LT) $\tilde{\pi}$ of the probabilities π

$$\tilde{\pi}_{\mathbf{j}}(s;\,\mathbf{v}) \equiv \int\limits_0^\infty \int\limits_{R^m} \mathrm{e}^{-st-\mathbf{v}'\mathbf{y}} \pi_{\mathbf{j}}(t;\,\mathbf{dy})\mathrm{d}t$$

the following theorem can be proved.

Theorem 3. *The LT $\tilde{\pi}$ of the probabilities π can be expressed in forms*

$$\tilde{\pi}_{\mathbf{0}}(s;\,\mathbf{v}) = \psi_{\mathbf{0}}(s,\,v) + \sum_{1\le k\le m} \int\limits_0^\infty \int\limits_{R^{m-1}} \mathrm{e}^{-su-\mathbf{v}'\mathbf{x}(k)} H_{\mathbf{0}}^{(k)}(\mathrm{d}u,\,\mathbf{dx}(k))\psi_{\mathbf{0}}(s,\,\mathbf{x}(k),\,v),$$

$$\tilde{\pi}_{\mathbf{j}}(s;\,\mathbf{v}) = \sum_{1\le k\le m} \int\limits_0^\infty \int\limits_{R^{m-1}} \mathrm{e}^{-su-\mathbf{v}'\mathbf{x}(k)} H_{\mathbf{j}}^{(k)}(\mathrm{d}u,\,\mathbf{dx}(k))\psi_{\mathbf{j}}(s,\,\mathbf{x}(k),\,v). \tag{12.8}$$

Proof. The proof could be also done by the following calculations:

$$\tilde{\pi}_{\mathbf{0}}(s;\,\mathbf{v}) \equiv \int\limits_0^\infty \int\limits_{R^m} \mathrm{e}^{-st-\mathbf{v}'\mathbf{y}} \pi_{\mathbf{0}}(t;\,\mathbf{dy})\mathrm{d}t = \int\limits_0^\infty \int\limits_{R^m} \mathrm{e}^{-st-\mathbf{v}'\mathbf{y}} \bar{Q}_{\mathbf{0}}(t;\,\mathbf{0},\,\mathbf{dy})\mathrm{d}t$$

$$+ \sum_{1\le k\le m} \int\limits_0^\infty \int\limits_{R^m} \mathrm{e}^{-st-\mathbf{v}'\mathbf{y}} \int\limits_0^\infty \int\limits_{R^{m-1}} H_{\mathbf{0}}^{(k)}(\mathrm{d}u,\,\mathbf{dx}(k))\bar{Q}_{\mathbf{0}}(t-u,\,\mathbf{x}(k),\,\mathbf{dy})\mathrm{d}t$$

$$= \psi_{\mathbf{0}}(s,\,v) + \sum_{1\le k\le m} \int\limits_0^\infty \mathrm{e}^{-su} \int\limits_{R^{m-1}} H_{\mathbf{0}}(\mathrm{d}u,\,\mathbf{dx}(k)) \int\limits_{t\ge u} \mathrm{e}^{-s(t-u)-\mathbf{v}'\mathbf{y}}$$

$$\times \int\limits_{R^m} \bar{Q}_{\mathbf{0}}(t-u,\,\mathbf{x}(k),\,\mathbf{dy})\,\mathrm{d}t$$

$$= \psi_{\mathbf{0}}(s,\,v) + \sum_{1\le k\le m} \int\limits_0^\infty \int\limits_{R^{m-1}} \mathrm{e}^{-su-\mathbf{v}'\mathbf{x}(k)} H_{\mathbf{0}}^{(k)}(\mathrm{d}u,\,\mathbf{dx}(k))\psi_{\mathbf{0}}(s,\,\mathbf{x}(k),\,v);$$

$$\tilde{\pi}_{\mathbf{j}}(s;\,\mathbf{v}) \equiv \int_0^\infty \int_{R^m} \mathrm{e}^{-st-\mathbf{v}'\mathbf{y}} \pi_{\mathbf{j}}(t;\,\mathbf{dy})\mathrm{d}t$$

$$= \sum_{1\le k\le m} \int_0^\infty \int_{R^m} \mathrm{e}^{-st-\mathbf{v}'\mathbf{y}} \int_0^\infty \int_{R^{m-1}} H_{\mathbf{j}}^{(k)}(\mathrm{d}u, \mathbf{dx}(k))\bar{Q}_{\mathbf{j}}(t-u, \mathbf{x}(k),\, \mathbf{dy})\mathrm{d}t$$

$$= \sum_{1\le k\le m} \int_0^\infty \mathrm{e}^{-su} \int_{R^{m-1}} H_{\mathbf{j}}^{(k)}(\mathrm{d}u, \mathbf{dx}(k)) \int_{t\ge u} \mathrm{e}^{-s(t-u)-\mathbf{v}'\mathbf{y}}$$

$$\times \int_{R^m} \bar{Q}_{\mathbf{j}}(t-u, \mathbf{x}(k), \mathbf{dy})\mathrm{d}t$$

$$= \sum_{1\le k\le m} \int_0^\infty \int_{R^{m-1}} \mathrm{e}^{-su-\mathbf{v}'\mathbf{x}(k)} H_{\mathbf{j}}^{(k)}(\mathrm{d}u, \mathbf{dx}(k))\psi_{\mathbf{j}}(s, \mathbf{x}(k), v).$$

Analogously for the LT

$$h_{\mathbf{j}}^{(l)}(s;\,\mathbf{v}(l)) \equiv \int_0^\infty \int_{R^{m-1}} \mathrm{e}^{-st-\mathbf{v}'\mathbf{y}(l)} H_{\mathbf{j}}^{(l)}(\mathrm{d}t; \mathbf{dy}(l))$$

of the functions $H_{\mathbf{j}}^{(l)}(\mathrm{d}t;\,\mathbf{dy}(l))$ the following theorem holds.

Theorem 4. *The LT $h_{\mathbf{j}}^{(k)}(s;\,\mathbf{v}(k))$ of the embedded renewal functions $H_{\mathbf{j}}^{(k)}(\mathrm{d}t;\,\mathbf{dy}(k))$ can be expressed in the following forms. For $\mathbf{j}$, such that $\mathbf{j}(l) \in \mathcal{E}$*

$$h_{\mathbf{e}_l}^{(l)}(s;\,\mathbf{v}(l)) = \phi_{\mathbf{0}}(s;\,v(l)) + \sum_{1\le k\le m} \int_0^\infty \int_{R^{m-1}} \mathrm{e}^{-su}\mathrm{d}H_{\mathbf{0}}^{(k)}(\mathrm{d}u,\,\mathbf{dx}(k))\mathrm{e}^{-\mathbf{v}'\mathbf{x}(k)}\phi_{\mathbf{0}}(s;\,\mathbf{x}(k),\,v(l)),$$

$$h_{\mathbf{j}}^{(l)}(s;\mathbf{v}(l)) = \sum_{1\le k\le m} \int_0^\infty \int_{R^{m-1}} \mathrm{e}^{-su}\mathrm{d}H_{\mathbf{j}}^{(k)}(\mathrm{d}u,\,\mathbf{dx}(k))\mathrm{e}^{-\mathbf{v}'\mathbf{x}(k)}\phi_{\mathbf{j}}(s;\,\mathbf{x}(k),\,v(l)), \tag{12.9}$$

and for $\mathbf{j} \in \mathcal{E}$ such that $\mathbf{j}(l) \in \bar{\mathcal{E}}$

$$\tilde{\pi}_{\mathbf{j}}(s;\,\mathbf{v}(l)) = \sum_{1\le k\le m} \int_0^\infty \int_{R^{m-1}} \mathrm{e}^{-su}\mathrm{d}H_{\mathbf{j}}^{(k)}(\mathrm{d}u,\,\mathbf{dx}(k))\mathrm{e}^{-\mathbf{v}'\mathbf{x}(k)}\phi_{\mathbf{j}}(s;\,\mathbf{x}(k),\,v(l)). \tag{12.10}$$

Proof. Using the formulas (12.3) one can find for $\mathbf{j}$ with $\mathbf{j}(l) \in \mathcal{E}$

$$h^{(l)}_{\mathbf{e}_l}(s; \mathbf{v}(l)) \equiv \int_0^\infty \int_{R^{m-1}} \mathrm{e}^{-st-\mathbf{v}'\mathbf{y}(l)} H^{(l)}_{\mathbf{e}_l}(\mathrm{d}t; \mathbf{dy}(l))$$

$$= \int_0^\infty \int_{R^{m-1}} \mathrm{e}^{-st-\mathbf{v}'\mathbf{y}(l)} Q^{(l)}_{\mathbf{0}}(\mathrm{d}t;\, \mathbf{0},\, \mathbf{dy}(l))$$

$$+ \int_0^\infty \int_{R^{m-1}} \mathrm{e}^{-st-\mathbf{v}'\mathbf{y}(l)} \sum_{1\le k\le m} \int_0^t \int_{R^{m-1}} \mathrm{d}H^{(k)}_{\mathbf{0}}(\mathrm{d}u,\, \mathbf{dx}(k))$$

$$\times Q^{(l)}_{\mathbf{0}}(\mathrm{d}t-u,\, \mathbf{x}(k),\, \mathbf{dy}(l))$$

$$= \phi_{\mathbf{0}}(s;\, v(l)) + \sum_{1\le k\le m} \int_0^\infty \int_{R^{m-1}} \mathrm{e}^{-su}\mathrm{d}H^{(k)}_{\mathbf{0}}(\mathrm{d}u, \mathbf{dx}(k)) \int_{t\ge u} \mathrm{e}^{-s(t-u)-\mathbf{v}'\mathbf{y}(l)}$$

$$\times \int_{R^{m-1}} Q^{(+)}_{\mathbf{0}}(\mathrm{d}t-u, \mathbf{x}(k), \mathbf{dy}(l))$$

$$= \phi_{\mathbf{0}}(s; v(l)) + \sum_{1\le k\le m} \int_0^\infty \int_{R^{m-1}} \mathrm{e}^{-su}\mathrm{d}H^{(k)}_{\mathbf{0}}(\mathrm{d}u, \mathbf{dx}(k))\mathrm{e}^{-\mathbf{v}'\mathbf{x}(k)}$$

$$\times \phi_{\mathbf{0}}(s; \mathbf{x}(k), v(l));$$

$$h^{(l)}_{\mathbf{j}}(s; \mathbf{v}(l)) \equiv \int_0^\infty \int_{R^{m-1}} \mathrm{e}^{-st-\mathbf{v}'\mathbf{y}(l)} H^{(l)}_{\mathbf{j}}(\mathrm{d}t; \mathbf{dy}(l))$$

$$= \int_0^\infty \int_{R^{m-1}} \mathrm{e}^{-st-\mathbf{v}'\mathbf{y}(l)} \sum_{1\le k\le m} \int_0^t \int_{R^{m-1}} \mathrm{d}H^{(k)}_{\mathbf{j}}(\mathrm{d}u,\, \mathbf{dx}(k))$$

$$\times Q^{(k,l)}_{\mathbf{j}}(\mathrm{d}t-u,\, \mathbf{x}(k),\, \mathbf{dy}(l))$$

$$= \sum_{1\le k\le m} \int_0^\infty \int_{R^{m-1}} \mathrm{e}^{-su}\mathrm{d}H^{(k)}_{\mathbf{j}}(\mathrm{d}u,\, \mathbf{dx}(k)) \int_{t\ge u} \mathrm{e}^{-s(t-u)-\mathbf{v}'\mathbf{y}(l)}$$

$$\times \int_{R^{m-1}} Q^{(k,l)}_{\mathbf{j}}(\mathrm{d}t-u,\, \mathbf{x}(k),\, \mathbf{dy}(l))$$

$$= \sum_{1\le k\le m} \int_0^\infty \int_{R^{m-1}} \mathrm{e}^{-su}\mathrm{d}H^{(k)}_{\mathbf{j}}(\mathrm{d}u,\, \mathbf{dx}(k))\mathrm{e}^{-\mathbf{v}'\mathbf{x}(k)}\phi_{\mathbf{j}}(s;\, \mathbf{x}(k),\, v(l)).$$

At least for $\mathbf{j} \in \mathcal{E}$ with $\mathbf{j}(l) \in \bar{\mathcal{E}}$

$$\begin{aligned}
\tilde{\pi}_{\mathbf{j}(l)}(s;\, \mathbf{v}(l)) &\equiv \int_0^\infty \int_{R^{m-1}} \mathrm{e}^{-st-\mathbf{v}'\mathbf{y}(l)} \pi_{\mathbf{j}(l)}(\mathrm{d}t;\, \mathbf{dy}(l)) \\
&= \int_0^\infty \int_{R^{m-1}} \mathrm{e}^{-st-\mathbf{v}'\mathbf{y}(l)} \sum_{1\le k\le m} \int_0^t \int_{R^{m-1}} H_{\mathbf{j}}^{(k)}(\mathrm{d}u;\, \mathbf{dx}(k)) \\
&\quad \times Q_{\mathbf{j}(l)}^{(k,l)}(\mathrm{d}t-u;\, \mathbf{x}(k),\, \mathbf{dy}(l)) \\
&= \sum_{1\le k\le m} \int_0^\infty \int_{R^{m-1}} \mathrm{e}^{-su}\mathrm{d}H_{\mathbf{j}}^{(k)}(\mathrm{d}u, \mathbf{dx}(k)) \int_{t\ge u} \mathrm{e}^{-s(t-u)-\mathbf{v}'\mathbf{y}(l)} \\
&\quad \times \int_{R^{m-1}} Q_{\mathbf{j}}^{(k,l)}(\mathrm{d}t-u, \mathbf{x}(k), \mathbf{dy}(l)) \\
&= \sum_{1\le k\le m} \int_0^\infty \int_{R^{m-1}} \mathrm{e}^{-su}\mathrm{d}H_{\mathbf{j}}^{(k)}(\mathrm{d}u,\, \mathbf{dx}(k))\mathrm{e}^{-\mathbf{v}'\mathbf{x}(k)}\phi_{\mathbf{j}}(s;\, \mathbf{x}(k),\, v(l)).
\end{aligned}$$

Using working period of any subsystem as its life time one can calculate the working period of subsystems of the next level etc. up to the whole (entire) system.

The results above show that to investigate the problem in general case the solution of complex integral equations is needed. On the other hand they show that the problem could be reduced to the investigation of the functions ψ and ϕ. For this numerical analysis is needed.

Under the quick restoration condition in order to get an approximate solution it is possible to use a phase states enlarging method. Consider the Exponential case.

12.4.2 Exponential Case

Note that under exponential distributions of the life and repair times the functions ϕ and ψ do not depend on the additional variables and with $\gamma_{\mathbf{j}} = \sum_{1\le k\le m} \gamma_{j_k}$ take the following form

$$\phi_{\mathbf{j}}^{(l)}(s;\, \mathbf{x}, v(l)) = \frac{\gamma_{j_l}}{s+v(l)+\gamma_{\mathbf{j}}}, \qquad \psi_{\mathbf{j}}(s;\, \mathbf{x}, v) = \frac{1}{s+v+\gamma_{\mathbf{j}}}.$$

This remark gives the possibility to simplify the general equations. Indeed, in this case the (12.8) for the time dependent probabilities in terms of their LT take the form

$$\begin{aligned}
\tilde{\pi}_{\mathbf{0}}(s;\, \mathbf{v}) &= \frac{1}{s+v+\gamma_{\mathbf{0}}} + \sum_{1\le k\le m} h_{\mathbf{0}}^{(k)}(s;\, \mathbf{v}(k))\frac{1}{s+v+\gamma_{\mathbf{0}}}; \\
\tilde{\pi}_{\mathbf{j}}(s;\, \mathbf{v}) &= \sum_{1\le k\le m} h_{\mathbf{j}}^{(k)}(s;\, \mathbf{v}(k))\frac{1}{s+v+\gamma_{\mathbf{j}}}.
\end{aligned} \tag{12.11}$$

The (12.9) for functions h take the form: for $\mathbf{j} \in \mathcal{E}$

$$h_{\mathbf{e}_l}^{(l)}(s; \mathbf{v}(l)) = \frac{\gamma_l}{s + v(l) + \gamma_{\mathbf{0}}} + \sum_{1 \le k \le m} h_{\mathbf{0}}^{(k)}(s; \mathbf{v}(k)) \frac{\gamma_k}{s + v(l) + \gamma_{\mathbf{0}}},$$

$$h_{\mathbf{j}}^{(l)}(s; \mathbf{v}(l)) = \sum_{1 \le k \le m} h_{\mathbf{j}}^{(k)}(s; \mathbf{v}(k)) \frac{\gamma_{j_k}}{s + v(k) + \gamma_{\mathbf{j}(l)}}, \tag{12.12}$$

and for $\mathbf{j} \in \mathcal{E}$ with $\mathbf{j}(l) \in \bar{\mathcal{E}}$

$$\tilde{\pi}_{\mathbf{j}(l)}(s; \mathbf{v}(l)) = \sum_{1 \le k \le m} h_{\mathbf{j}}^{(k)}(s; \mathbf{v}) \frac{\gamma_{j_k}}{s + v(k) + \gamma_{\mathbf{j}(l)}}. \tag{12.13}$$

To find the appropriate functions for macro-states (independent from $\mathbf{x}$) one should set in these relations $\mathbf{v} = \mathbf{0}$. In this case, the (12.8) for the time dependent probabilities in terms of their LT take the form

$$\tilde{\pi}_{\mathbf{0}}(s) = \frac{1}{s + \alpha_{\mathbf{0}}} + \sum_{1 \le k \le m} h_{\mathbf{0}}^{(k)}(s) \frac{1}{s + \alpha_{\mathbf{0}}} \equiv (1 + h_{\mathbf{0}}(s)) \frac{1}{s + \alpha_{\mathbf{0}}}$$

$$\tilde{\pi}_{\mathbf{j}}(s) = \sum_{1 \le k \le m} h_{\mathbf{j}}^{(k)}(s) \frac{1}{s + \gamma_{\mathbf{j}}} \equiv h_{\mathbf{j}}(s) \frac{1}{s + \gamma_{\mathbf{j}}}. \tag{12.14}$$

On the other hand, for the functions $h_{\mathbf{j}}^{(k)}(s)$ from (12.12) the following system of equations holds

$$h_{\mathbf{e}_l}^{(l)}(s) = \frac{\beta_l}{s + \alpha(l)} + \sum_{1 \le k \le m} h_{\mathbf{0}}^{(k)}(s) \frac{\beta_l}{s + \alpha(l)} = (1 + h_{\mathbf{0}}(s)) \frac{\beta_l}{s + \alpha(l)}$$

$$h_{\mathbf{j}}^{(l)}(s) = \sum_{1 \le k \le m} h_{\mathbf{j}}^{(k)}(s) \frac{\gamma_{j_l}}{s + \gamma_{\mathbf{j}(l)}} = h_{\mathbf{j}}(s) \frac{\gamma_{j_l}}{s + \gamma_{\mathbf{j}(l)}}. \tag{12.15}$$

and for $\mathbf{j} \in \mathcal{E}$ with $\mathbf{j}(l) \in \bar{\mathcal{E}}$

$$\tilde{\pi}_{\mathbf{j}(l)}(s) = \sum_{1 \le k \le m} h_{\mathbf{j}}^{(k)}(s) \frac{\gamma_{j_l}}{s + \gamma_{\mathbf{j}(l)}} = h_{\mathbf{j}}(s) \frac{\gamma_{j_l}}{s + \gamma_{\mathbf{j}(l)}}. \tag{12.16}$$

To concretize this methodic let us consider a simple example.

12.5 An Example

Consider an example with only two elements in the system. Denote: $\mathbf{0} = (0,0)$, $\mathbf{1} = (0,1)$, $\mathbf{2} = (1,0)$, $\mathbf{3} = (1,1)$. With these notations one has

$$\tilde{\pi}_{\mathbf{0}}(s) = \frac{1}{s + \gamma_{\mathbf{0}}} + h_{\mathbf{0}}^{(1)}(s) \frac{1}{s + \gamma_{\mathbf{0}}} + h_{\mathbf{0}}^{(2)}(s) \frac{1}{s + \gamma_{\mathbf{0}}};$$

$$\tilde{\pi}_{\mathbf{k}}(s) = h_{\mathbf{k}}^{(k)}(s) \frac{1}{s + \gamma_{\mathbf{k}}} \qquad (k = 1, 2);$$

$$\tilde{\pi}_{\mathbf{3}}(s) = h_{\mathbf{1}}^{(1)}(s) \frac{\alpha_2}{s + \gamma_1} + h_{\mathbf{2}}^{(2)}(s) \frac{\alpha_1}{s + \gamma_{\mathbf{2}}},$$

where the functions $h_{\mathbf{j}}^{(k)}(s)$ satisfy to the equations

$$h_{\mathbf{0}}^{(k)}(s) = h_{\mathbf{k}}^{(k)}(s)\frac{\beta_k}{s+\gamma_{\mathbf{k}}} \qquad (\text{for } \mathbf{k}=\mathbf{1},\ k=1, \text{ and } \mathbf{k}=\mathbf{2},\ k=2),$$

$$h_{\mathbf{k}}^{(k)}(s) = \frac{\alpha_k}{s+\gamma_{\mathbf{0}}}(1+h_{\mathbf{0}}^{(1)}(s)+h_{\mathbf{0}}^{(2)}(s)) \qquad (\text{for } \mathbf{k}=\mathbf{1}, \text{ and } k=1,\ \mathbf{k}=\mathbf{2},\ k=2).$$

The solutions of these equations are (where for $k=1$, $\bar{\mathbf{k}}=\mathbf{2}=(1,0)$ and for $k=2$, $\bar{\mathbf{k}}=\mathbf{1}=(0,1)$)

$$h_{\mathbf{0}}^{(k)}(s) = \frac{\beta_k(s+\gamma_{\bar{\mathbf{k}}})}{(s+\gamma_{\mathbf{0}})(s+\gamma_{\mathbf{1}})(s+\gamma_{\mathbf{2}})-\alpha_1\beta_1(s+\gamma_{\mathbf{2}})-\alpha_2\beta_2(s+\gamma_{\mathbf{1}})} \qquad (k=1,2),$$

$$h_{\mathbf{k}}^{(k)}(s) = \frac{\alpha_k(s+\gamma_{\mathbf{1}})(s+\gamma_{\mathbf{2}})}{(s+\gamma_{\mathbf{0}})(s+\gamma_{\mathbf{1}})(s+\gamma_{\mathbf{2}})-\alpha_1\beta_1(s+\gamma_{\mathbf{2}})-\alpha_2\beta_2(s+\gamma_{\mathbf{1}})} \qquad (k=1,2).$$

Therefore, for the LST of the working period $\tilde{w}(s)=\int_0^\infty \mathrm{e}^{-st}W(\mathrm{d}t)=\tilde{\pi}_{\mathbf{3}}(s)$ one has

$$\tilde{\pi}_{\mathbf{3}}(s) = \frac{\alpha_1\alpha_2(2s+\gamma_{\mathbf{1}}+\gamma_{\mathbf{2}})}{(s+\gamma_{\mathbf{0}})(s+\gamma_{\mathbf{1}})(s+\gamma_{\mathbf{2}})-\alpha_1\beta_1(s+\gamma_{\mathbf{2}})-\alpha_2\beta_2(s+\gamma_{\mathbf{1}})},$$

that coincides with the results obtained by classic Markov approach.

In the case of homogeneous system, when $\alpha_1=\alpha_2=\alpha$ and $\beta_1=\beta_2=\beta$ we have $\gamma_{\mathbf{0}}=2\alpha$, $\gamma_{\mathbf{1}}=\gamma_{\mathbf{2}}=\alpha+\beta$, and the expression for $\tilde{\pi}_{\mathbf{3}}(s)$ takes the following form

$$\tilde{\pi}_{\mathbf{3}}(s) = \frac{2\alpha^2}{s^2+(3\alpha+\beta)s+2\alpha^2}.$$

By splitting this expression into partial fractions and applying the inverse Laplace transform, it's easy to obtain the distribution function of time to first system failure. The mean time to system failure can be obtained by differentiating $\tilde{\pi}_{\mathbf{3}}(s)$ at point $s=0$:

$$\mathbf{E}W = -\tilde{\pi}_{\mathbf{3}}'(0) = \left(3+\frac{\beta}{\alpha}\right)\frac{1}{2\alpha},$$

that coincides with the results, obtained by analyzing the appropriate Markov process.

We omit the result for the case of non-homogeneous system because of its awkwardness.

12.6 Further Problems

In a particular case, when the life times are approximately of the same magnitude and are large enough compared to the repair times (the case of quick restoration), it is possible to use the idea of the system phase states enlargement, proposed by Korolyuk and Turbin [KT78] (see also [KK99]). It is well known that for the system in parallel in the case of quick restoration the time to failure is approximately exponential

[So70, So71] (see also [Ko80]). Therefore, in the case when all elements have the same magnitude of life times all working states could be joined into one working state and the probability characteristic of the system with one working state and one failure state will be close to the initial system.

This approach could be recursively applied to the system under consideration in order to evaluate the life time distribution of the whole system in terms of its entire characteristics.

References

[DRS02] Dimitrov, B., Rykov, V., Stanchev, P.: On multi-state reliability systems. In: Proceedings MMR-2002. Trondheim (Norway), June 17-21 (2002)

[RD02] Rykov, V., Dimitrov, B.: On multi-state reliability systems. In: Applied Stochastic Models and Information Processes, 128–135. Proceedings of the International Seminar. Petrozavodsk, Sept. 8-13. Petrozavodsk (2002)

[RDG04] Rykov, V., Dimitrov, B., Green Jr., D., Stanchev, P.: Reliability of complex hierarchical systems with fault tolerance units. In: Proceedings MMR-2004. Santa Fe (U.S.A.), June (2004)

[RDGS04] Rykov, V., Dimitrov, B., Green Jr., D., Stanchev, P.: Reliability model for biological objects. In: The Model of Durability, Aging and Degradation in Medicine, Biology and Technique, 230–240. Proceedings of the International Conference. St.Petersburg, Russia, June 7-9, (2004)

[LL03] Lisniansky, A., Levitin, G.: Multi-state system reliability. Assessment, optimization and application. World Scientific. New Jersey, London, Singapore, Hong Kong, 358p. (2003)

[Be62] Belyaev, Yu.K.: Line Markov processes and their application to problems of the theory of reliability. (English) Select. Translat. Math. Statist. Probab. 9, 233–250 (1971); translation from Trudy VI vsesojuzn. Sovescan. po Teor. Verojatn. Mat. Statist. Vil'njus-Palanga 1960, 309–323 (1962)

[Bu63] Buslenko, N.P.: To the theory of complex systems. Izv. AN SSSR. Tachn. Cybern., Moscow, **5** (In Russian) (1963)

[Bu71] Buslenko, N.P.: On a class of complex systems. Problems of applied mathematics and mechanics. "Nauka", Moscow. (In Russian) (1971)

[BKK73] Buslenko, N.P., Kalashnikov, V.V., Kovalenko, I.N.: Lectures on the theory of complex systems. "Sovetyskoe Radio", Moscow, 440p. (In Russian) (1973)

[Ka78] Kalashnikov, V.V.: Qualitative analysis of complex systems behavior with the method of testing function. "Nauka", Moscow, 248p. (In Russian) (1978)

[Ka94] Kalashnikov, V.V.: Topics on regenerative processes, CRC, Boca Raton (1994)

[Ko65] Kovalenko, I.N.: On some classes of complex systems. Izv. AN SSSR, Techn. Cybern., Moscow, **6** (1964), **1**,**3** (In Russian) (1965)

[Ko76] Kovalenko, I.N.: Analitico-statistical method for settlement of high-reliable systems characteristics. Cybernetics, Kiev, **6** (In Russian) (1976)

[Kov76] Kovalenko, I.N.: Investigations on complex systems reliability. "Naukova dumka", Kiev, 212p. (In Russian) (1976)

[Ko80] Kovalenko, I.N.: Analysis of rare events for the systems efficiency and reliability evaluation. "Sovetskoe Radio", Moscow, 209p. (In Russian) (1980)

[Ry75] Rykov, V.V.: Regenerative processes with embedded regeneration periods and their aplication for priority queueing systems investigation. Cibernetics., Kiev, **6** (In Russian) (1975)

[Ry83] Rykov, V.V.: An investigation of an one-server general queueing system by regenerative methods. I. Izv. AN SSSR, Techn. Cybern., **6**, 13–20 (In Russian) (1983)

[Ry84] Rykov, V.V.: An investigation of an one-server general queueing system by regenerative methods. II. An investigation of the main processes on separate regeneration period. Izv. AN SSSR, Techn. Cybern., **1**, 126–132 (In Russian) (1984)

[Ry96] Rykov, V.V.: On decomposition of hierarchical computer communication networks. Proceedings of The International Conference on Distributed Computer Communication Networks. 77–85. Inst. for Information Transmission Problems RAS, Moscow (1996)

[Ry97] Rykov, V.V.: Two approaches to decomposition of complex hierarchical systems. Continuously interacting subsytems. Autom. and Romote Control, **10**, 91–104 (1997)

[Ry97] Rykov, V.V.: Two approaches to decomposition of complex hierarchical systems. Aggregative sytems. Autom. and Romote Control, **12**, 140–149 (1997)

[RYa71] Rykov, V.V., Yastrebenetsky, M.A.: On regenerative processes with several types of regeneration points. Cibernetics, Kiev. **3**, 82–86 (In Russian) (1971)

[RJ81] Rykov, V.V., Jolkoff, S.Yu.: Generalized regeneranive processes with embedded regeneranion periods and their applications. MOS. Ser. Optimization. V.12, **4**, 575–591 (1981)

[KT78] Korolyuk, V.S., Turbin, A.F.: Mathematical basis for phase enlarging of complex systems. "Naukova dumka", Kiev (1978)

[KK99] Korolyuk, V.S., Korolyuk, V.V.: Stochastic models of systems. Kluwer, Boston (1999)

[So70] Solov'ev, A.D.: Reservation with quick restoration. Izv. AN SSSR, Techn. Cybern., **1**, 56–71 (In Russian) (1970)

[So71] Solov'ev, A.D.: Asymptotic behavior of the first rare events occurence epoch in regenerative process. Izv. AN SSSR, Techn. Cybern., **6**, 79–89 (In Russian) (1971)

Part II

Statistical Methods in Reliability

13

Parametric Estimation of Redundant System Reliability From Censored Data

Vilijandas Bagdonavičius[1], Inga Masiulaitytė[1], and Mikhail Nikulin[2,*]

[1] Vilnius University, Vilnius, Lithuania
[2] IMB, Victor Segalen University, Bordeaux, France, mikhail.nikouline@u-bordeaux2.fr

Abstract: Parametric point and interval estimators of the cumulative distribution function of redundant systems with several "warm" stand-by units are proposed. The data are supposed to be right censored. Accuracy of interval estimators obtained from complete data was investigated by simulation.

Keywords and phrases: AFT model; Asymptotic properties of estimators; Censored failure data; Redundant system; Sedyakin model; Warm operating conditions

13.1 Introduction

In literature on reliability (see Rausand and Hoyland [RH04], Shah and Dhillon [SD07], Srinivasan and Subramanian [SS06], Veklerov [Vek87], Wasson [Was05]), various characteristics of redundant systems with stand-by units functioning in various operating conditions are studied from probabilistic point of view. If expressions of reliability characteristics of redundant system are written as time functions depending on parameters of the components and parameters relating the distributions of units functioning in "warm" and "hot" operating conditions, then reliability of the system may be estimated using estimators of the reliability characteristics of the components. A question of accuracy of system reliability estimators rises.

In this paper, we consider redundant systems with one main unit and $m-1$ stand-by units operating in "warm" conditions, i.e. under lower stress than the main one. The problem is to obtain confidence intervals for the cumulative distribution functions of redundant systems using censored failure data of two units tested in "hot" and in "warm" conditions.

Denote by T_1, F_1, S_1, f_1 and λ_1 the failure time, cumulative distribution , survival, probability density and hazard function, respectively, of an unit functioning in "hot" conditions. In "warm" conditions, we use the same notation using index 2 instead of 1. We suppose that switching from "warm" to "hot" conditions does not do any damage to units. Bagdonavičius et al. [BMN08,BMN09] give mathematical formulation

V.V. Rykov et al. (eds.), *Mathematical and Statistical Models and Methods in Reliability: Applications to Medicine, Finance, and Quality Control*, Statistics for Industry and Technology, DOI 10.1007/978-0-8176-4971-5_13,

of "fluent switch on" based on AFT and Sedyakin's models (see Bagdonavičius [BAG78, BAG90], Bagdonavičius and Nikulin [BN02], Sedyakin [Sed66]) and propose tests for verification of this hypothesis.

In this paper, we suppose that the distribution of units functioning in "warm" and "hot" conditions differ only in scale, i.e. $F_2(t) = F_1(rt)$ for all $t \geq 0$ and some $r > 0$. In such a case, the c.d.f. of units functioning in "hot" and "warm" conditions mostly belong to the same parametric classes of distributions, for example, exponential, Weibull, loglogistic, gamma.

Denote by $K_j(t)$ the c.d.f. of the system with one main and $j-1$ stand-by units. If the hypothesis of "fluent switch on" of stand-by units is verified (see Bagdonavičius et al.), then this function can be commuted recurrently:

$$K_j(t) = \int_0^t F_1(t + ry - y) \mathrm{d}K_{j-1}(y). \tag{13.1}$$

In this paper, we consider parametric point and interval estimators of the cumulative distribution function $K_j(t)$ using right censored data obtained from reliability experiments of units in "hot" and "warm" conditions. Accuracy of interval estimators is investigated by simulation.

13.2 Parametric Point Estimators of the c.d.f. $K_j(t)$

Suppose that the following data are available:

(a) Right censored sample

$$(X_{11}, \delta_{11})^T, \ldots, (X_{1n_1}, \delta_{1n_1})^T$$

of size n_1; here

$$X_{1i} = T_{1i} \wedge C_{1i}, \quad \delta_{1i} = \mathbf{1}_{\{T_{1i} \leq C_{1i}\}},$$

T_{1i} are failure times of units tested in "hot" conditions, C_{1i} – censoring times;

(b) Right censored sample

$$(X_{21}, \delta_{21})^T, \ldots, (X_{2n_2}, \delta_{2n_2})^T$$

of size n_2; here

$$X_{2j} = T_{2j} \wedge C_{2j}, \quad \delta_{2j} = \mathbf{1}_{\{T_{2j} \leq C_{2j}\}},$$

T_{2j} are failure times of units tested in "warm" conditions, C_{2j} – censoring times.

Suppose that in hot conditions the c.d.f $F_1(t;\theta)$ is absolutely continuous and depends on finite dimensional parameter $\theta \in \Theta \subset \mathbf{R}^k$. Set $\gamma = (r, \theta^T)^T$.

The maximum likelihood estimator $\gamma^* = (r^*, (\theta^*)^T)^T$ of the parameter γ maximizes the loglikelihood function

$$\ell(\gamma) = \sum_{i=1}^{n_1} \delta_{1i} \ln \lambda_1(X_{1i};\theta) + \sum_{i=1}^{n_1} \ln S_1(X_{1i};\theta) + m_2 \ln r +$$

$$\sum_{j=1}^{n_2} \delta_{2j} \ln \lambda_1(rX_{2j};\theta) + \sum_{j=1}^{n_2} \ln S_1(rX_{2j};\theta); \quad (13.2)$$

here $m_2 = \sum_{j=1}^{n_2} \delta_{2j}$.

The c.d.f. $K_j(t)$ is estimated recurrently:

$$\hat{K}_j(t) = \int_0^t F_1(t + r^* y - y; \theta^*) \mathrm{d}\hat{K}_{j-1}(y), \quad \hat{K}_1(t) = F_1(t;\theta^*). \quad (13.3)$$

13.3 Asymptotic Confidence Intervals for $K_j(t)$

Suppose that

$$\frac{n_i}{n} = l_i + O(\frac{1}{n}), \quad l_i \in (0,1), \quad \text{as} \quad n = n_1 + n_2 \to \infty.$$

Under classical assumptions on the family of distributions $F_1(t,\theta)$and the censoring mechanism there exists positively definite matrix $i(\gamma)$ such that

$$\frac{1}{\sqrt{n}} \dot{\ell}(\gamma) \xrightarrow{d} V = (V_1, V_2)^T \sim N_{k+1}(0, i(\gamma)), \quad -\frac{1}{n}\ddot{\ell}(\gamma^*) \xrightarrow{P} i(\gamma),$$

$$\sqrt{n}(\gamma^* - \gamma) \xrightarrow{d} Y = (Y_1, Y_2^T)^T \sim N_{k+1}(0, i^{-1}(\gamma)).$$

V_1 and Y_1 are one-dimensional, V_2 and Y_2 – k-dimensional.

Set $\hat{I} = -\ddot{\ell}(\hat{\gamma}^*)$. We have $\frac{1}{n}\hat{I} \xrightarrow{P} i(\gamma)$.

Using delta method we obtain:

$$\sqrt{n}(\hat{K}_2(t) - K_2(t)) \xrightarrow{\mathcal{D}} W_2(t) = C_2(t;\gamma)^T Y,$$

where

$$C_2(t;\gamma) = (C_{21}(t;\gamma), C_{22}^T(t;\gamma))^T, \quad C_{21}(t;\gamma) = \int_0^t \frac{\partial}{\partial r} F_1(t + ry - y;\theta) \mathrm{d}F_1(y;\theta),$$

$$C_{22}(t;\gamma) = \int_0^t \frac{\partial}{\partial \theta} F_1(t + ry - y;\theta) \mathrm{d}F_1(y;\theta) + F_1(t + ry - y;\theta) \mathrm{d}(\frac{\partial}{\partial \theta} F_1(y;\theta)).$$

The random variable $W_2(t)$ is linear function of Y.

If $j \geq 2$ then

$$\sqrt{n}(\hat{K}_j(t) - K_j(t)) \xrightarrow{\mathcal{D}} W_j(t),$$

where the random variable $W_j(t)$, $j \geq 2$, is linear function of Y:

$$W_j(t) = Y^T C_j(t;\gamma), \quad C_j(t;\gamma) \in (C[0,t])^{k+1}.$$

It follows by induction:

$$W_j(t) = Y^T\Big(\int_0^t \frac{\partial}{\partial\gamma} F_1(t + ry - y;\theta)\mathrm{d}K_{j-1}(y;\gamma) + F_1(t + ry - y;\theta)\mathrm{d}C_{j-1}(t;\gamma)\Big).$$

So the variance

$$\mathbf{Var}(W_j(t)) = \mathbf{Var}(C_j(t;\gamma)^T Y) = C_j^T(t;\gamma)i^{-1}(\gamma)C_j(t;\gamma)$$

is estimated by $nC_2^T(t;\hat{\gamma})\hat{I}^{-1})C_j(t;\hat{\gamma})$, and the variance $\sigma^2_{\hat{K}_j(t)}$ of the estimator $\hat{K}_2(t)$ is estimated by

$$\hat{\sigma}^2_{\hat{K}_j(t)} = C_j^T(t;\hat{\gamma})I^{-1}(\hat{\gamma})C_j(t;\hat{\gamma}).$$

The asymptotic $1-\alpha$ confidence interval for $K_j(t)$ is

$$\hat{K}_j(t) \pm \hat{\sigma}_{\hat{K}_j(t)} z_{1-\alpha/2}, \tag{13.4}$$

or, alternatively, $(\underline{K}_j(t), \overline{K}_j(t))$, where

$$\underline{K}_j(t) = \left(1 + \frac{1-\hat{K}_j(t)}{\hat{K}_j(t)} \exp\left\{\frac{\hat{\sigma}_{\hat{K}_j} z_{1-\alpha/2}}{\sqrt{\hat{K}_j(t)(1-\hat{K}_j(t))}}\right\}\right)^{-1},$$

$$\overline{K}_j(t) = \left(1 + \frac{1-\hat{K}_j(t)}{\hat{K}_j(t)} \exp\left\{-\frac{\hat{\sigma}_{\hat{K}_j} z_{1-\alpha/2}}{\sqrt{\hat{K}_j(t)(1-\hat{K}_j(t))}}\right\}\right)^{-1}; \tag{13.5}$$

here $z_{1-\alpha/2}$ is $(1-\alpha/2)$ quantile of the standard normal law.

13.3.1 Exponential Distribution

Suppose that the distribution of failure times in "hot" and "warm" conditions is exponential, i.e. $S_1(t) = e^{-\lambda t}$.
The loglikelihood function has the form

$$l(r;\lambda) = m\ln\lambda + m_2\ln r - \lambda\Big(\sum_{i=1}^{n_1} X_{1i} + r\sum_{j=1}^{n_2} X_{2j}\Big);$$

here $m_k = \sum_{i=1}^{n_k}\delta_{ki}$, $k = 1,2$, $m = m_1 + m_2$. Equating the score function to zero we obtain the system of equations

$$\dot{\ell}_r = \frac{\partial l}{\partial r} = \frac{m_2}{r} - \lambda\sum_{j=1}^{n_2} X_{2j} = 0, \quad \dot{\ell}_\lambda = \frac{\partial l}{\partial\lambda} = \frac{m}{\lambda} - \sum_{i=1}^{n_1} X_{1i} - r\sum_{j=1}^{n_2} X_{2j} = 0.$$

So the estimators of the parameters r and λ are:

$$\hat{r} = \frac{\sum_{i=1}^{n_1} X_{1i}/m_1}{\sum_{j=1}^{n_2} X_{2j}/m_2}; \quad \hat{\lambda} = \frac{m_1}{\sum_{i=1}^{n_1} X_{1i}}.$$

Second partial derivatives are

$$\ddot{\ell}_{r^2} = -\frac{m_2}{r^2}; \quad \ddot{\ell}_{\lambda^2} = -\frac{m}{\lambda^2}; \quad \ddot{\ell}_{\lambda r} = -\sum_{j=1}^{n_2} X_{2j}.$$

So

$$\hat{I}^{-1} = \begin{pmatrix} \frac{m\hat{r}^2}{m_1 m_2} & -\frac{\hat{\lambda}\hat{r}}{m_1} \\ -\frac{\hat{\lambda}\hat{r}}{m_1} & \frac{\hat{\lambda}^2}{m_1} \end{pmatrix}.$$

In the case of exponential distribution, the c.d.f. of the redundant system has explicit form

$$K_j(t) = \prod_{i=1}^{j-1} \left(1 + \frac{1}{ir}\right) \sum_{i=0}^{j-1} (-1)^i C_{j-1}^i \frac{1 - e^{-\lambda(1+ir)t}}{1 + ir}.$$

So the weights $C_j = (C_{j1}, C_{j2})^T$ are:

$$C_{j1}(t; r, \lambda) = \frac{\partial K_j(t)}{\partial r} = \prod_{i=1}^{j-1} \left(1 + \frac{1}{ir}\right) \left[-\frac{1}{r} \sum_{i=1}^{j-1} \frac{1}{1+ir} \sum_{i=0}^{j-1} (-1)^i C_{j-1}^i \frac{1 - e^{-\lambda(1+ir)t}}{1+ir} \right.$$

$$\left. + \sum_{i=0}^{j-1} (-1)^i C_{j-1}^i \frac{e^{-\lambda(1+ir)t}[i\lambda t(1+ir) + i] - i}{(1+ir)^2} \right],$$

$$C_{j2}(t; r, \lambda) = \frac{\partial K_j(t)}{\partial \lambda} = t \prod_{i=1}^{j-1} \left(1 + \frac{1}{ir}\right) \sum_{i=0}^{j-1} (-1)^i C_{j-1}^i e^{-\lambda(1+ir)t}.$$

The estimator of the variance $\sigma^2_{\hat{K}_j(t)}$ of the estimator $\hat{K}_j(t)$ is

$$\hat{\sigma}^2_{\hat{K}_j(t)} = C_j^T(t; \hat{r}, \hat{\lambda}) \hat{I}^{-1} C_j(t; \hat{r}, \hat{\lambda}),$$

and the asymptotic $1 - \alpha$ confidence interval for $K_j(t)$ has the form (13.4) or, alternatively, (13.5).

In the case $j = 2$,

$$C_2(t; r, \lambda) = (C_{21}(t; r, \lambda), C_{22}(t; r, \lambda))^T$$

$$= \left(\frac{S_1(t)}{r^2}(F_2(t) - r\lambda t S_2(t)), \quad \frac{(1+r)t}{r} S_1(t) F_2(t) \right),$$

$$\hat{\sigma}^2_{\hat{K}_2(t)} = \frac{\hat{S}_1^2(t)}{m \hat{l}_1 \hat{l}_2 r^2} \left(\hat{l}_1 \left[\hat{F}_2(t) - \hat{\lambda}\hat{r}t\hat{S}_2(t) \right]^2 + \hat{l}_2 \left[(1 - \hat{\lambda}t)\hat{F}_2(t) - \hat{\lambda}\hat{r}t \right]^2 \right);$$

here $\hat{l}_i = m_i/m$, $\hat{S}_1(t) = e{-}\hat{\lambda}t$, $\hat{S}_2(t) = e{-}\hat{r}\hat{\lambda}t$, $\hat{F}_i(t) = 1 - \hat{S}_i(t)$.

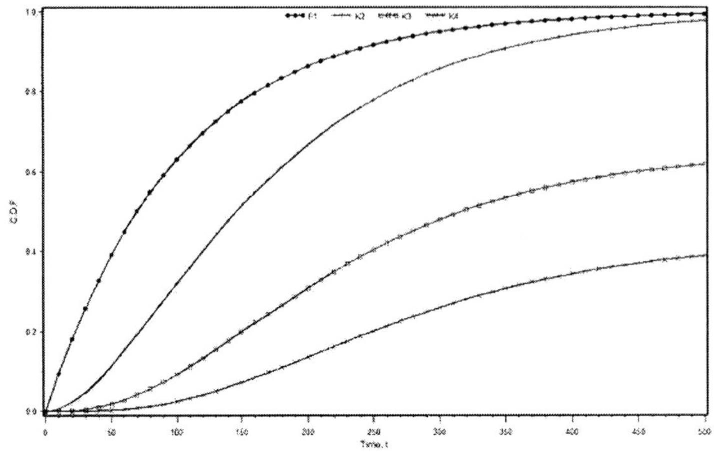

Figure 13.1. Graphs of the trajectories of the parametric estimators $\hat{F}_1, \hat{K}_2$ (exponential distribution)

Table 13.1. Confidence level for finite samples ($n_1 = n_2 = 100$)

Time, t	50	100	200	300	400	500
$K_2(t)$	0.114	0.319	0.667	0.856	0.941	0.977
Confidence level (%)	89.9	89.4	88.9	90.2	90.0	91.5

Table 13.2. Confidence level for finite samples ($n_1 = n_2 = 100$)

Time, t	50	100	200	300	400	500
$K_3(t)$	0.018	0.092	0.309	0.479	0.573	0.617
Confidence level (%)	88.6	91.8	92.8	90.8	89.9	90.3

In the case of complete samples of size $n_1 = n_2 = 100$, we found by simulation finite sample confidence levels of the intervals obtained using asymptotic formulas with $1 - \alpha = 0.9$. We simulated failure times T_{1j} and T_{2j} from exponential distribution with following parameters:

$$T_{1j} \sim \mathcal{E}(\lambda_1), \quad T_{2j} \sim \mathcal{E}(\lambda_2), \quad \lambda_1 = \frac{1}{100}, \quad \lambda_2 = \frac{1}{300}.$$

The number of replications was 2,000. For various values of t the proportions of confidence interval realizations covering the true value of the distributional function $K_2(t)$ are given in Table 13.1.

For various values of t the proportions of confidence interval realizations covering the true value of the distributional function $K_3(t)$ are given in Table 13.2.

13.3.2 Weibull Distribution

Suppose that the distribution of failure times in "hot" and "warm" conditions is Weibull, i.e. $S_1(t) = e^{-(t/\mu)^{\nu}}$.

The loglikelihood function is

$$l(r;\theta) = m(\ln\nu - \nu\ln\mu) + \nu m_2 \ln r + (\nu-1)(\sum_{i=1}^{n_1}\delta_{1i}\ln X_{1i} + \sum_{j=1}^{n_2}\delta_{2j}\ln X_{2j})$$

$$-\frac{1}{\mu^\nu}(\sum_{i=1}^{n_1}X_{1i}^\nu + r^\nu\sum_{j=1}^{n_2}X_{2j}^\nu).$$

Equating the score function to zero the following system of equations is obtained:

$$\dot{\ell}_r = \frac{m_2\nu}{r} - \frac{\nu r^{\nu-1}}{\mu^\nu}\sum_{j=1}^{n_2}X_{2j}^\nu = 0;$$

$$\dot{\ell}_\nu = \frac{m}{\nu} - m\ln\mu + m_2\ln r + \sum_{i=1}^{n_1}\delta_{1i}\ln X_{1i} + \sum_{j=1}^{n_2}\delta_{2j}\ln X_{2j}$$

$$+\frac{1}{\mu^\nu}[\ln\mu(\sum_{i=1}^{n_1}X_{1i}^\nu + r^\nu\sum_{j=1}^{n_2}X_{2j}^\nu) - \sum_{i=1}^{n_1}X_{1i}^\nu\ln X_{1i} - r^\nu\ln r\sum_{j=1}^{n_2}X_{2j}^\nu - r^\nu\sum_{j=1}^{n_2}X_{2j}^\nu\ln X_{2j}] = 0;$$

$$\dot{\ell}_\mu = -\frac{m\nu}{\mu} + \frac{\nu}{\mu^{\nu+1}}(\sum_{i=1}^{n_1}X_{1i}^\nu + r^\nu\sum_{j=1}^{n_2}X_{2j}^\nu) = 0.$$

Resolving this system of equations we obtain that the estimators $\hat{\mu}$ and $\hat{r}$ are explicit functions of the estimator $\hat{\nu}$:

$$\hat{\mu} = \left(\frac{1}{m_1}\sum_{i=1}^{m_1}X_{1i}^{\hat{\nu}}\right)^{1/\hat{\nu}}, \quad \hat{r} = \left(\frac{m_2}{m_1}\frac{\sum_{i=1}^{m_1}X_{1i}^{\hat{\nu}}}{\sum_{j=1}^{m_2}X_{2j}^{\hat{\nu}}}\right)^{1/\hat{\nu}}.$$

The estimator $\hat{\nu}$ satisfies the equation

$$\frac{m}{\hat{\nu}} + \sum_{i=1}^{n_1}\delta_{1i}\ln X_{1i} + \sum_{j=1}^{n_2}\delta_{2j}\ln X_{2j} - m_1\frac{\sum_{i=1}^{n_1}X_{1i}^\nu\ln X_{1i}}{\sum_{i=1}^{n_1}X_{1i}} - n_2\frac{\sum_{j=1}^{n_2}X_{2j}^\nu\ln X_{2j}}{\sum_{j=1}^{n_2}X_{2j}} = 0.$$

Second partial derivatives of the loglikelihood function are

$$\ddot{\ell}_{r^2} = -\frac{m_2\nu}{r^2} - \frac{\nu(\nu-1)r^{\nu-2}}{\mu^\nu}\sum_{j=1}^{n_2}X_{2j}^\nu; \quad \ddot{\ell}_{r\mu} = \frac{\nu^2 r^{\nu-1}}{\mu^{\nu+1}}\sum_{j=1}^{n_2}X_{2j}^\nu;$$

$$\ddot{\ell}_{r\nu} = \frac{m_2}{r} - \frac{r^{\nu-1} + \nu r^{\nu-1}(\ln r - \ln\mu)}{\mu^\nu}\sum_{j=1}^{n_2}X_{2j}^\nu - \frac{\nu r^{\nu-1}}{\mu^\nu}\sum_{j=1}^{n_2}X_{2j}^\nu\ln X_{2j};$$

$$\ddot{\ell}_{\nu^2} = -\frac{m}{\nu^2} - \frac{1}{\mu^\nu}\ln\mu[\ln\mu(\sum_{i=1}^{n_1}X_{1i}^\nu + r^\nu\sum_{j=1}^{n_2}X_{2j}^\nu t) - \sum_{i=1}^{n_1}X_{1i}^\nu\ln X_{1i}$$

$$-r^\nu \ln r \sum_{j=1}^{n_2} X_{2j}^\nu - r^\nu \sum_{j=1}^{n_2} X_{2j}^\nu \ln X_{2j}]$$

$$+\frac{1}{\mu^\nu}[\ln \mu t(\sum_{i=1}^{n_1} X_{1i}^\nu \ln X_{1i} + r^\nu \ln r \sum_{j=1}^{n_2} X_{2j}^\nu + r^\nu \sum_{j=1}^{n_2} X_{2j}^\nu \ln X_{2j})$$

$$-\sum_{i=1}^{n_1} X_{1i}^\nu \ln^2 X_{1i} - r^\nu \ln^2 r \sum_{j=1}^{n_2} X_{2j}^\nu - r^\nu \ln r \sum_{j=1}^{n_2} X_{2j}^\nu \ln X_{2j}$$

$$-r^\nu \ln r \sum_{j=1}^{n_2} X_{2j}^\nu \ln X_{2j} - r^\nu \sum_{j=1}^{n_2} X_{2j}^\nu \ln^2 X_{2j}];$$

$$\ddot{\ell}_{\mu\nu} = -\frac{m}{\mu} + \frac{\mu^{\nu+1} - \nu\mu^{\nu+1}\ln\mu}{\mu^{2(\nu+1)}}(\sum_{i=1}^{n_1} X_{1i}^\nu + r^\nu \sum_{j=1}^{n_2} X_{2j}^\nu)c$$

$$+\frac{\nu}{\mu^{\nu+1}}(\sum_{i=1}^{n_1} X_{1i}^\nu \ln X_{1i} + r^\nu \ln r \sum_{j=1}^{n_2} X_{2j}^\nu + r^\nu \sum_{j=1}^{n_2} X_{2j}^\nu \ln X_{2j});$$

$$\ddot{\ell}_{\mu^2} = \frac{n\nu}{\mu^2} - \frac{\nu(\nu+1)}{\mu^{\nu+2}}(\sum_{i=1}^{n_1} X_{1i}^\nu + r^\nu \sum_{j=1}^{n_2} X_{2j}^\nu).$$

The estimator of the c.d.f. K_2 is

$$\hat{K}_2(t) = 1 - e^{-(\frac{t}{\hat{\mu}})^{\hat{\nu}}} - \frac{\hat{\nu}}{\hat{\mu}} \int_0^t \left(\frac{y}{\hat{\mu}}\right)^{\hat{\nu}-1} e^{-(\frac{t+\hat{r}y-y}{\hat{\mu}})^{\hat{\nu}} - (\frac{y}{\hat{\mu}})^{\hat{\nu}}} dy,$$

and the functions C_{2i} are

$$C_{21}(t) = \frac{\partial K_2(t)}{\partial r} = \frac{\nu^2}{\mu} \int_0^t \left(\frac{y}{\mu}\right)^\nu \left(\frac{t+ry-y}{\mu}\right)^{\nu-1} e^{-(\frac{t+ry-y}{\mu})^\nu - (\frac{y}{\mu})^\nu} dy,$$

$$C_{22}(t) = \frac{\partial K_2(t)}{\partial \mu} = -\frac{\nu}{\mu}\left(\frac{t}{\mu}\right)^\nu e^{-(\frac{t}{\mu})^\nu}$$

$$+\frac{\nu^2}{\mu^2} \int_0^t \left(\frac{y}{\mu}\right)^{\nu-1} \left[1 - \left(\frac{y}{\mu}\right)^\nu - \left(\frac{t+ry-y}{\mu}\right)^\nu\right] e^{-(\frac{t+ry-y}{\mu})^\nu - (\frac{y}{\mu})^\nu} dy,$$

$$C_{23}(t) = \frac{\partial K_2(t)}{\partial \nu} = \left(\frac{t}{\mu}\right)^\nu e^{-(\frac{t}{\mu})^\nu} \ln\left(\frac{t}{\mu}\right) - \frac{1}{\mu}\int_0^t \left(\frac{y}{\mu}\right)^{\nu-1} e^{-(\frac{t+ry-y}{\mu})^\nu - (\frac{y}{\mu})^\nu} dy$$

$$+\frac{\nu}{\mu}\int_0^t \left(\frac{y}{\mu}\right)^{\nu-1} \left[\left(\frac{y}{\mu}\right)^\nu \ln\left(\frac{y}{\mu}\right) + \left(\frac{t+ry-y}{\mu}\right)^\nu \ln\left(\frac{t+ry-y}{\mu}\right)\right] e^{-(\frac{t+ry-y}{\mu})^\nu - (\frac{y}{\mu})^\nu} dy$$

$$-\frac{\nu}{\mu}\int_0^t \left(\frac{y}{\mu}\right)^{\nu-1} \ln\left(\frac{y}{\mu}\right) e^{-(\frac{t+ry-y}{\mu})^\nu - (\frac{y}{\mu})^\nu} dy,$$

The estimator of the variance $\sigma^2_{\hat{K}_2(t)}$ of the estimator $\hat{K}_2(t)$ is

$$\hat{\sigma}^2_{\hat{K}_2(t)} = C_2^T(t;\hat{r},\hat{\mu},\hat{\nu})(-\ddot{\ell}(\hat{r},\hat{\mu},\hat{\nu}))^{-1}C_2(t;\hat{r},\hat{\mu},\hat{\nu}), \tag{13.6}$$

$$C_2(t;\hat{r},\hat{\mu},\hat{\nu}) = (C_{21}(t;\hat{r},\hat{\mu},\hat{\nu}), C_{22}(t;\hat{r},\hat{\mu},\hat{\nu}), C_{23}(t;\hat{r},\hat{\mu},\hat{\nu}))^T.$$

The asymptotic $1-\alpha$ confidence interval for $K_2(t)$ is of the form (13.4) or, alternatively, (13.5) with $j = 2$.

The formulas simplify in the case of complete samples. Noting that the random variables $Z_{1i} = \left(\frac{T_{1i}}{\mu}\right)^\nu$ have the standard exponential distribution, i.e. $Z_{1i} \sim \mathcal{E}(1)$ we obtain

$$T_{1i} = \mu Z_{1i}^{1/\nu}, \quad T_{1i}^\nu = \mu^\nu Z_{1i}, \quad T_{1i}^\nu \ln T_{1i} = \mu^\nu \ln\mu\, Z_{1i} + \frac{\mu^\nu}{\nu} Z_{1i}\ln Z_{1i};$$

$$T_{1i}^\nu \ln^2 T_{1i} = \mu^\nu \ln^2\mu\, Z_{1i} + \frac{2\mu^\nu}{\nu}\ln\mu\, Z_{1i}\ln Z_{1i} + \frac{\mu^\nu}{\nu^2} Z_{1i}\ln^2 Z_{1i}.$$

We have

$$\mathbf{E}Z_{1i} = 1, \quad \mathbf{E}Z_{1i}\ln Z_{1i} = \int_0^\infty x\ln x e^{-x}\mathrm{d}x = \Gamma'(2),$$

$$\mathbf{E}Z_{1i}^\nu \ln^2 Z_{1i} = \int_0^\infty x e^{-x}\ln^2 x\mathrm{d}x = \Gamma''(2).$$

So

$$\mathbf{E}T_{1i}^\nu = \mu^\nu,$$

$$\mathbf{E}T_{1i}^\nu \ln T_{1i} = \mathbf{E}\left(\frac{\mu^\nu}{\nu} Z_{1i}\ln Z_{1i} + \mu^\nu \ln\mu Z_{1i}\right) = \frac{\mu^\nu}{\nu}\left[\Gamma'(2) + \nu\ln\mu\right];$$

$$\mathbf{E}T_{1i}^\nu \ln^2 T_{1i} = \frac{\mu^\nu}{\nu^2}\left[\Gamma''(2) + 2\nu\Gamma'(2)\ln\mu + \nu^2\ln^2\mu\right].$$

The random variables rT_{2j} and T_{1i} have the same distribution, so

$$\mathbf{E}T_{2i}^\nu = \mathbf{E}\frac{1}{r^\nu}(rT_{2i})^\nu = \left(\frac{\mu}{r}\right)^\nu,$$

$$T_{2j}^\nu \ln T_{2j} = \frac{1}{r^\nu}(rT_{2j})^\nu[\ln(rT_{2j}) - \ln r] = \frac{1}{r^\nu}\left((rT_{2j})^\nu \ln(rT_{2j}) - \ln r(rT_{2j})^\nu\right);$$

$$\mathbf{E}T_{2j}^\nu \ln T_{2j} = \frac{1}{r^\nu}\left(\frac{\mu^\nu}{\nu}(\Gamma'(2) + \nu\ln\mu) - \mu^\nu\ln r\right) = \frac{\mu^\nu}{r^\nu\nu}\left(\Gamma'(2) + \nu\ln\frac{\mu}{r}\right).$$

$$\mathbf{E}T_{2j}^\nu \ln^2 T_{2j} = \frac{1}{\nu^2}\left(\frac{\mu}{r}\right)^\nu\left[\Gamma''(2) + 2\nu\Gamma'(2)\ln\frac{\mu}{r} + \nu^2\ln^2\frac{\mu}{r}\right].$$

Using the obtained means we compute the means of the second partial derivatives of the loglikelihood function:

$$\mathbf{E}\ddot{\ell}_{r^2} = -\frac{\nu^2 n_2}{r^2}; \quad \mathbf{E}\ddot{\ell}_{r\mu} = \frac{n_2\nu^2}{r\mu}; \quad \mathbf{E}\ddot{\ell}_{r\nu} = -\frac{n_2}{r}\Gamma'(2);$$

$$\mathbf{E}\ddot{\ell}_{\mu^2} = -\frac{n\nu^2}{\mu^2}; \quad \mathbf{E}\ddot{\ell}_{\mu\nu} = \frac{n\Gamma'(2)}{\mu}; \quad \mathbf{E}\ddot{\ell}_{\nu^2} = \frac{n}{\nu^2}(1 + \Gamma''(2)).$$

So the Fisher information matrix is

$$I(r, \mu, \nu) = \begin{pmatrix} \frac{n_2\nu^2}{r^2} & -\frac{n_2\nu^2}{r\mu} & \frac{n_2}{r}\Gamma'(2) \\ -\frac{n_2\nu^2}{r\mu} & \frac{n\nu^2}{\mu^2} & -\frac{n}{\mu}\Gamma'(2) \\ \frac{n_2}{r}\Gamma'(2) & -\frac{n}{\mu}\Gamma'(2) & \frac{n}{\nu^2}(1 + \Gamma''(2)) \end{pmatrix}.$$

The inverse of the Fisher information matrix is

$$I^{-1}(r, \mu, \nu) = \begin{pmatrix} \frac{nr^2}{n_1 n_2 \nu^2} & \frac{r\mu}{n_1\nu^2} & 0 \\ \frac{r\mu}{n_1\nu^2} & \frac{\mu^2 n[1+\Gamma''(2)] - \mu^2 n_2[\Gamma'(2)]^2}{n_1 n\nu^2(1+\Gamma''(2)-[\Gamma'(2)]^2)} & \frac{\mu\Gamma'(2)}{n(1+\Gamma''(2)-[\Gamma'(2)]^2)} \\ 0 & \frac{\mu\Gamma'(2)}{n(1+\Gamma''(2)-[\Gamma'(2)]^2)} & \frac{\nu^2}{n(1+\Gamma''(2)-[\Gamma'(2)]^2)} \end{pmatrix}.$$

The estimator of the variance $\sigma^2_{\hat{K}_2(t)}$ of the estimator $\hat{K}_2(t)$ is

$$\hat{\sigma}^2_{\hat{K}_2(t)} = C_2^T(t; \hat{r}, \hat{\mu}, \hat{\nu}) I^{-1}(\hat{r}, \hat{\mu}, \hat{\nu}) C_2(t; \hat{r}, \hat{\mu}, \hat{\nu}). \quad (13.7)$$

The asymptotic $1-\alpha$ confidence interval for $K_2(t)$ is of the form (13.4) or, alternatively, (13.5) with $j = 2$.

In the case of complete samples of size $n_1 = n_2 = 100$, we found by simulation finite sample confidence levels of the intervals obtained using asymptotic formulas with

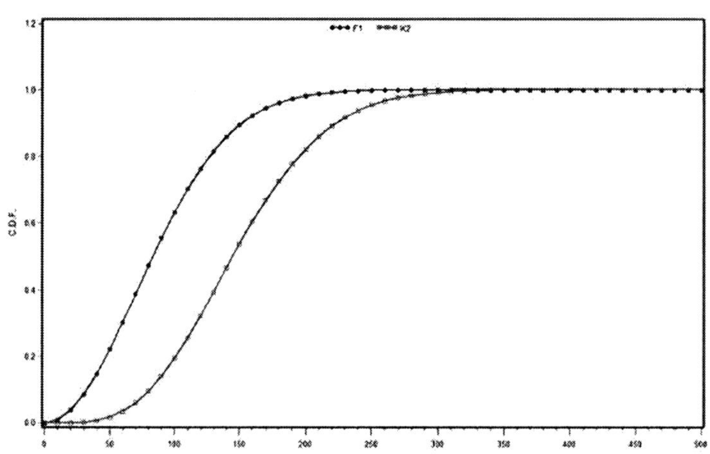

Figure 13.2. Graphs of the trajectories of the parametric estimators $\hat{F}_1, \hat{K}_2$ (Weibull distribution)

Table 13.3. Confidence level for finite samples ($n_1 = n_2 = 100$)

Time, t	50	100	200	300	400	500
$K_2(t)$	0.018	0.194	0.822	0.992	0.999	1.000
Confidence level (%)	89.4	89.2	89.2	89.5	89.5	90.2

$1-\alpha = 0.9$. We simulated failure times T_{1j} and T_{2j} from the Weibull distribution with following parameters:

$$T_{1j} \sim W(\alpha_1, \beta_1), \quad T_{2j} \sim W(\alpha_2, \beta_2),$$

$$\alpha_1 = \alpha_2 = 2, \quad \beta_1 = 100, \quad \beta_2 = 300.$$

The number of replications was 2,000. For various values of t the proportions of confidence interval realizations covering the true value of the distributional function $K_2(t)$ are given below:

13.3.3 Loglogistic Distribution

Suppose that the distribution of failure times in "hot" and "warm" conditions is loglogistic. So $S_1(t) = \frac{1}{1+(t/\mu)^\nu}$. The loglikelihood function has the form

$$l(r;\mu,\nu) = m\ln\nu - \nu m\ln\mu + \nu m_2\ln r + (\nu-1)(\sum_{i=1}^{n_1}\delta_{1i}\ln X_{1i} + \sum_{j=1}^{n_2}\delta_{2j}\ln X_{2j})$$

$$-\sum_{i=1}^{n_1}(1+\delta_{1i})\ln\left(1+\left(\frac{X_{1i}}{\mu}\right)^\nu\right) - \sum_{j=1}^{n_2}(1+\delta_{2j})\ln\left(1+\left(\frac{rX_{2j}}{\mu}\right)^\nu\right).$$

Partial derivatives are

$$\dot{\ell}_r = \frac{\nu m_2}{r} - \frac{\nu}{r}\sum_{j=1}^{n_2}(1+\delta_{2j})\frac{\left(\frac{rX_{2j}}{\mu}\right)^\nu}{1+\left(\frac{rX_{2j}}{\mu}\right)^\nu};$$

$$\dot{\ell}_\mu = -\frac{\nu m}{\mu} + \frac{\nu}{\mu}\sum_{i=1}^{n_1}(1+\delta_{1i})\frac{\left(\frac{X_{1i}}{\mu}\right)^\nu}{1+\left(\frac{X_{1i}}{\mu}\right)^\nu} + \frac{\nu}{\mu}\sum_{j=1}^{n_2}(1+\delta_{2j})\frac{\left(\frac{rX_{2j}}{\mu}\right)^\nu}{1+\left(\frac{rX_{2j}}{\mu}\right)^\nu};$$

$$\dot{\ell}_\nu = \frac{m}{\nu} - m\ln\mu + m_2\ln r + \sum_{i=1}^{n_1}\delta_{1i}\ln X_{1i} + \sum_{j=1}^{n_2}\delta_{2j}\ln X_{2j}$$

$$-\sum_{i=1}^{n_1}(1+\delta_{1i})\frac{\left(\frac{X_{1i}}{\mu}\right)^\nu\log\left(\frac{X_{1i}}{\mu}\right)}{1+\left(\frac{X_{1i}}{\mu}\right)^\nu} - \sum_{j=1}^{n_2}(1+\delta_{2j})\frac{\left(\frac{rX_{2j}}{\mu}\right)^\nu\log\left(\frac{rX_{2j}}{\mu}\right)}{1+\left(\frac{rX_{2j}}{\mu}\right)^\nu}.$$

Set

$$U_i = \left(\frac{X_{1i}}{\mu}\right)^\nu, \quad V_j = \left(\frac{rX_{2j}}{\mu}\right)^\nu.$$

Second partial derivatives of the loglikelihood function are

$$\ddot{\ell}_{r^2} = -\frac{\nu m_2}{r^2} + \frac{\nu}{r^2}\sum_{j=1}^{n_2}(1+\delta_{2j})\frac{V_j}{1+V_j} - \frac{\nu^2}{r^2}\sum_{i=1}^{n_2}(1+\delta_{2j})\frac{V_j}{(1+V_j)^2},$$

$$\ddot{\ell}_{r\mu} = \frac{\nu^2}{r\mu}\sum_{j=1}^{m_2}(1+\delta_{2j})\frac{V_j}{(1+V_j)^2},$$

$$\ddot{\ell}_{r\nu} = \frac{m_2}{r} - \frac{1}{r}\sum_{j=1}^{n_2}(1+\delta_{2j})\frac{V_j}{1+V_j} - \frac{1}{r}\sum_{j=1}^{n_2}(1+\delta_{2j})\frac{V_j \ln V_j}{(1+V_j)^2},$$

$$\ddot{\ell}_{\mu^2} = \frac{m\nu}{\mu^2} - \frac{2\nu}{\mu^2}\left(\sum_{i=1}^{n_1}(1+\delta_{1i})\frac{U_i}{1+U_i} + \sum_{j=1}^{n_2}(1+\delta_{2j})\frac{V_i}{1+V_i}\right)$$

$$-\frac{2\nu^2}{\mu^2}\left(\sum_{i=1}^{n_1}(1+\delta_{1i})\frac{U_i}{(1+U_i)^2} + \sum_{j=1}^{n_2}(1+\delta_{2j})\frac{V_i}{(1+V_i)^2};\right),$$

$$\ddot{\ell}_{\mu\nu} = -\frac{m}{\mu} + \frac{1}{\mu}\left(\sum_{i=1}^{n_1}(1+\delta_{1i})\frac{U_i}{1+U_i} + \sum_{j=1}^{n_2}(1+\delta_{2j})\frac{V_j}{1+V_j}\right)$$

$$+\frac{2}{\mu}\left(\sum_{i=1}^{n_1}(1+\delta_{1i})\frac{U_i \ln U_i}{(1+U_i)^2} + \sum_{j=1}^{n_2}(1+\delta_{2j})\frac{V_i \ln V_i}{(1+V_i)^2}\right),$$

$$\ddot{\ell}_{\nu^2} = -\frac{m}{\nu^2} - \frac{1}{\nu^2}\left(\sum_{i=1}^{n_1}(1+\delta_{1i})\frac{U_i \ln^2 U_i}{(1+U_i)^2} + \sum_{j=1}^{n_2}(1+\delta_{2j})\frac{V_i \ln^2 V_i}{(1+V_i)^2}\right).$$

The estimator of the c.d.f. $K_2(t)$ is

$$\hat{K}_2(t) = 1 - \frac{1}{1+\left(\frac{t}{\hat{\mu}}\right)^{\hat{\nu}}} - \frac{\hat{\nu}}{\hat{\mu}}\int_0^t \left(\frac{y}{\hat{\mu}}\right)^{\hat{\nu}-1} \frac{1}{1+\left(\frac{t+\hat{r}y-y}{\hat{\mu}}\right)^{\hat{\nu}}} \frac{1}{\left(1+\left(\frac{y}{\hat{\mu}}\right)^{\hat{\nu}}\right)^2} dy,$$

and the functions C_{2i} are

$$C_{21}(t) = \nu^2 \int_0^t \left(\frac{y}{\mu}\right)^{\nu}\left(\frac{t+ry-y}{\mu}\right)^{\nu-1}\left(1+\left(\frac{y}{\mu}\right)^{\nu}\right)^{-2}\left(1+\left(\frac{t+ry-y}{\mu}\right)^{\nu}\right)^{-2} dy,$$

$$C_{22}(t) = -\frac{\nu}{\mu}\left(\frac{t}{\mu}\right)^{\nu}\left(1+\left(\frac{t}{\mu}\right)^{\nu}\right)^{-2} - \int_0^t \frac{\nu^2\left(\frac{y}{\mu}\right)^{\nu-1}}{\mu^2\left(1+\left(\frac{t+ry-y}{\mu}\right)^{\nu}\right)\left(1+\left(\frac{y}{\mu}\right)^{\nu}\right)^2}$$

$$\times \frac{1-\left(\frac{y}{\mu}\right)^{\nu} - 2\left(\frac{y}{\mu}\right)^{\nu}\left(\frac{t+ry-y}{\mu}\right)^{\nu}}{\left(1+\left(\frac{t+ry-y}{\mu}\right)^{\nu}\right)\left(1+\left(\frac{y}{\mu}\right)^{\nu}\right)} dy,$$

$$C_{23}(t) = \left(\frac{t}{\mu}\right)^{\nu}\left(1+\left(\frac{t}{\mu}\right)^{\nu}\right)^{-2}\ln\left(\frac{t}{\mu}\right) - \frac{1}{\mu}\int_0^t \left(\frac{y}{\mu}\right)^{\nu-1}\left(1+\left(\frac{y}{\mu}\right)^{\nu}\right)^{-2}$$

$$\times\left(1+\left(\frac{t+ry-y}{\mu}\right)^{\nu}\right)^{-1}\left[\left(1+\left(\frac{y}{\mu}\right)^{\nu} - \nu\ln\left(\frac{y}{\mu}\right)\left(1-\left(\frac{y}{\mu}\right)^{\nu}\right)\right)\left(1+\left(\frac{y}{\mu}\right)\right)^{-1}\right.$$

$$\left. -\nu\left(\frac{t+ry-y}{\mu}\right)^{\nu}\left(1+\left(\frac{t+ry-y}{\mu}\right)^{\nu}\right)^{-1}\ln\left(\frac{t+ry-y}{\mu}\right)\right]dy,$$

The estimator of the variance $\sigma^2_{\hat{K}_2(t)}$ of the estimator $\hat{K}_2(t)$ has the form (13.6) and the asymptotic $1-\alpha$ confidence interval for $K_2(t)$ is of the form (1) or, alternatively, (2) taking $j = 2$.

The formulas simplify in the case of complete samples. In such a case the random variables U_i and V_j are identically distributed with the probability density function of the standard loglogistic distribution: $f_{U_i}(x) = f_{V_j}(x) = 1/(1+x)^2$. It implies that for any $k > -2$ and $a \in (-1, k+1)$

$$g(a) = \mathbf{E}\frac{U_i^a}{(1+U_i)^k} = \int_0^{\infty}\frac{x^a}{(1+x)^{k+2}}dx = \frac{\Gamma(k-a+1)\Gamma(a+1)}{\Gamma(k+2)}.$$

$$g'(a) = \mathbf{E}\frac{U_i^a \ln U_i}{(1+U_i)^k} = \int_0^{\infty}\frac{x^a}{(1+x)^{k+2}}\ln x dx$$

$$= \frac{(\Gamma(k-a+1)\Gamma(a+1))'_a}{\Gamma(k+2)} = \frac{-\Gamma'(k-a+1)\Gamma(a+1)+\Gamma(k-a+1)\Gamma'(a+1)}{\Gamma(k+2)};$$

$$g''(a) = \mathbf{E}\frac{X_i^a \ln^2 X_i}{(1+X_i)^k} = \int_0^{\infty}\frac{x^a}{(1+x)^{k+2}}\ln^2 x dx$$

$$= \frac{\Gamma''(k-a+1)\Gamma(a+1) - 2\Gamma'(k-a+1)\Gamma'(a+1) + \Gamma(k-a+1)\Gamma''(a+1)}{\Gamma(k+2)}.$$

If $a = 1$ and $k = 1$, then

$$\mathbf{E}\frac{U_i}{1+U_i} = \frac{1}{2}, \quad \mathbf{E}\frac{U_i \ln U_i}{1+U_i} = \frac{-\Gamma'(1)+\Gamma'(2)}{2};$$

if $a = 1$ and $k = 2$, then

$$\mathbf{E}\frac{U_i}{(1+U_i)^2} = \frac{1}{6}, \quad \mathbf{E}\frac{U_i\ln U_i}{(1+U_i)^2} = 0, \quad \mathbf{E}\frac{U_i\ln^2 U_i}{(1+U_i)^2} = \frac{\Gamma''(2)-[\Gamma'(2)]^2}{3},$$

if $a = 2$ and $k = 2$, then

$$\mathbf{E}\frac{U_i^2}{(1+U_i)^2} = \frac{-2\Gamma'(1)+\Gamma'(3)}{6}.$$

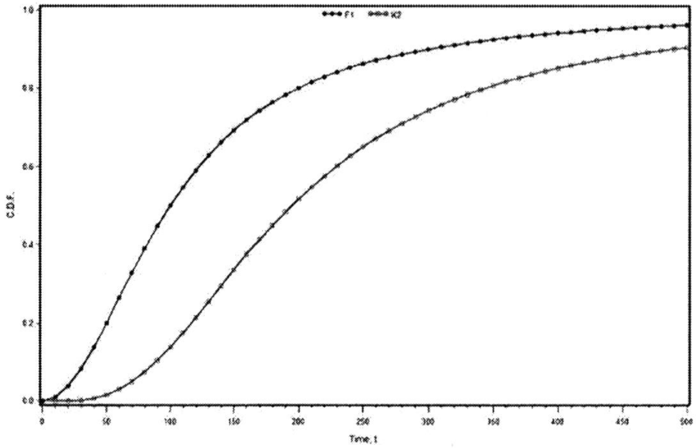

Figure 13.3. Graphs of the trajectories of the parametric estimators $\hat{F}_1, \hat{K}_2$ (loglogistic distribution)

Now we are able to compute the means of the second partial derivatives of the loglikelihood function:

$$\mathbf{E}\ddot{\ell}_{r^2} = -\frac{n_2\nu^2}{3r^2}, \quad \mathbf{E}\ddot{\ell}_{r\mu} = \frac{n_2\nu^2}{3r\mu}, \quad -\mathbf{E}\ddot{\ell}_{r\nu} = 0;$$

$$\mathbf{E}\ddot{\ell}_{\mu^2} = -\frac{n\nu^2}{3\mu^2}, \quad \mathbf{E}\ddot{\ell}_{\mu\nu} = 0, \quad -\mathbf{E}\ddot{\ell}_{\nu^2} = \frac{n}{3\nu^2}(3 + 2\Gamma''(2) - 2(\Gamma'(2))^2);$$

So the Fisher information matrix is

$$I(r,\mu,\nu) = \begin{pmatrix} \frac{n_2\nu^2}{3r^2} & -\frac{n_2\nu^2}{3r\mu} & 0 \\ -\frac{n_2\nu^2}{3r\mu} & \frac{n\nu^2}{3\mu^2} & 0 \\ 0 & 0 & \frac{n\{3+2\Gamma''(2)-2[\Gamma'(2)]^2\}}{3\nu^2} \end{pmatrix}$$

The inverse of the Fisher information matrix is

$$I^{-1}(r,\mu,\nu) = \begin{pmatrix} \frac{3nr^2}{n_1n_2\nu^2} & \frac{3r\mu}{n_1\nu^2} & 0 \\ \frac{3r\mu}{n_1\nu^2} & \frac{3\mu^2}{n_1\nu^2} & 0 \\ 0 & 0 & \frac{3\nu^2}{n\{3+2\Gamma''(2)-2[\Gamma'(2)]^2\}} \end{pmatrix}.$$

The estimator of the variance $\sigma^2_{\hat{K}_2(t)}$ of the estimator $\hat{K}_2(t)$ is of the form (13.7).

The asymptotic $1-\alpha$ confidence interval for $K_2(t)$ is of the form (13.4) or, alternatively, (13.5) with $j = 2$.

In the case of complete samples of size $n_1 = n_2 = 100$ we found by simulation finite sample confidence levels of the intervals obtained using asymptotic formulas with $1 - \alpha = 0.9$. We simulated failure times T_{1j} and T_{2j} from loglogistic distribution with following parameters:

$$T_{1j} \sim L(\alpha_1, \beta_1), \quad T_{2j} \sim L(\alpha_1, \beta_1),$$
$$\alpha_1 = \alpha_2 = 2, \quad \beta_1 = 100, \quad \beta_2 = 300.$$

Table 13.4. Confidence level for finite samples ($n_1 = n_2 = 100$).

Time, t	50	100	200	300	400	500
$K_2(t)$	0.016	0.138	0.517	0.743	0.851	0.905
Confidence level (%)	89.0	88.8	90.4	89.6	89.5	90.5

The number of replications was 2000. For various values of t the proportions of confidence interval realizations covering the true value of the distributional function $K_2(t)$ are given below:

References

[BAG78] Bagdonavičius, V., Testing the hyphothesis of the additive accumulation of damages. Probab. Theor. Appl., **23**, 2, 403–408 (1978)

[BAG90] Bagdonavičius, V., Accelerated life models when the stress is not constant. Kybernetika, **26**, 289–295, (1990)

[BN02] Bagdonavičius, V., Nikulin, M., Accelerated Life Models. Chapman and Hall, Boca Raton (2002)

[BMN08] Bagdonavicius, V., Masiulaityte, I., Nikulin, M., Statistical analysis of redundant systems with "warm" stand-by units. Stochast. Int. J. Probab. Stochast. Process., **80**, 2–3, 115–128 (2008)

[BMN09] Bagdonavicius, V., Masiulaityte, I., Nikulin, M., Statistical analysis of redundant systems with one stand-by unit. In: (Eds. C. Huber, N. Limnios, M. Mesbah, M. Nikulin) Maathematical Methods in Survival Analysis, Reliability and Quality of Life. Wiley, London, 189–202

[RH04] Rausand, A., Hoyland A., System Reliability Theory Models. Statistical Methods and Applications. Wiley, New Jersey (2004)

[Sed66] Sedyakin, N.M., On one physical principle in reliability theory (in russian). Techn. Cybern., **3**, 80–87 (1966)

[SD07] Shah, A., Dhillon, B.S., Reliability and availability analysis of three-state device redundant systems with human errors and common-cause failures. Int. J. Perform. Eng., **3**, 411–418 (2007)

[SS06] Srinivasan, S. K., Subramanian, R., Reliability analysis of a three unit warm standby redundant system with repair. Ann. Oper. Res., **143**, 227–235 (2006)

[Vek87] Veklerov, E., Reliability of redundant systems with unreliable switches. IEEE Trans. Reliab., **36**, 470–472 (1987)

[Was05] Wasson Ch.S., System Analysis, Design, and Development: Concepts, Principles, and Practices. Wiley, New York (2005)

14

Assessing Accuracy of Statistical Inferences by Resamplings

Yuri K. Belyaev

Department of Mathematics and Mathematical Statistics, Umeå University, Umeå, Sweden and Department of Forest Economics, Swedish University of Agricultural Sciences, Umeå, Sweden, `yuri.belyaev@math.umu.se`, `yuri.belyaev@sekon.se`

Abstract: Suppose that a list of explanatory variables and corresponding random responses was obtained during a series of regression experiments. The characteristic of interest is the mean value of responses considered as a regression function of corresponding values of explanatory variables. For example, if responses are failure times of tested elements, then the conditional mean value of life time given the value of explanatory variable is one of the important reliability characteristics of the tested elements. The analysis of this type of data can be realized in the framework of linear heteroscedastic regression models. Here, one of the central problems is a consistent estimation of the unknown regression function when the size of data grows unboundedly. The problems related to analysis of regression data attracted many researches, see Wu [Ann. Statist. **14**, 1261–1350 (1986)]. We give an approach to consistent solution of the problems under the assumption that values of explanatory variables are real numbers and the regression function is a polynomial with unknown degree and coefficients. The selection of regression function is based on resamplings from terms in the sum of the residuals estimated by the ordinary least squares method with various values of polynomial degree. In a similar way, resamplings from the weighted estimated residuals are used for consistent estimation of the deviations distributions of estimated coefficients from their true unknown values. The consistency of applied resamplings methods holds under certain assumptions, e.g. it is assumed that the residuals distributions have uniformly integrable second moments (assumption $\mathbf{AW}_2$). Given in Appendix a variant of the Central Limit Resampling Theorem is used in the proofs of Theorems 1 and 2.

Keywords and phrases: Asymptotic normality, Distributions of deviations, Least squares estimators, Linear heteroscedastic regression, Overparametrisation, Resampled sums of weighted estimated residuals, Selection of regression function

14.1 Introduction

Usually, the analysis of statistical data includes three steps. In the first step, after a pilot analysis of a given statistical data, an appropriate statistical model is suggested. In the second step, statistical inferences, e.g. point estimation, p-values, classification,

V.V. Rykov et al. (eds.), *Mathematical and Statistical Models and Methods in Reliability: Applications to Medicine, Finance, and Quality Control*, Statistics for Industry and Technology, DOI 10.1007/978-0-8176-4971-5_14,

etc., are realized. In the third step, the accuracy of the obtained statistical inferences are evaluated. In some statistical models, the last step can be done by using computer intensive methods Efron [E79], Davison and Hinkley [DH97], Belyaev [Bel07].

We characterize accuracy of an estimator as follows. Suppose that there are n researchers (n is large). Each researcher realizes an experiment and has obtained data. Conditions of experiments are the same. Then the obtained data can be considered as realizations of independent and identically distributed random sets. Each researcher knows only its own data set. All researches used the same software. They had the same recommendation to estimate a parameter θ_0 of interest. The ith researcher obtains only one value $\hat{\theta}_i$ of an estimator $\hat{\Theta}_i$ and he knows nothing about values of estimates $\hat{\theta}_j, j \neq i$, of others researches. The accuracy of such experiments we characterize by the deviations distribution of the estimates $\hat{\theta}_i, i = 1, ..., n$, from the parameter of interest θ_0.

If we could know θ_0 and all values of estimates $\hat{\theta}_i, i = 1, ..., n$, then we could consistently estimate the distribution of deviations $F_{\theta_0}[x] = \mathrm{P}[\hat{\Theta}_i - \theta_0 \leq x]$ as $n \to \infty$ by the *empirical distribution function* (e.d.f.)

$$\hat{F}_n[x] = \frac{1}{n} \sum_{i=1}^{n} \mathrm{I}[\hat{\theta}_i - \theta_0 \leq x]. \tag{14.1}$$

We will call $F_{\theta_0}[\cdot]$ the *total distribution of deviations* (total d.d.).

We underline that $F_{\theta_0}[\cdot]$ is the same for all researches. It would be useful for every researcher to know the total d.d. which characterizes the accuracy of the used estimator. If it would be possible to find a consistent estimator of the total d.d. then each researcher would have the useful information on possible deviations of its estimate value $\hat{\theta}_i$ from the true unknown value θ_0. This will be useful addition to the common confidence intervals which are different for different researches. To find a consistent estimator for $F_{\theta_0}[\cdot]$, is not a trivial problem because as usual only one data set is known. Under certain assumptions resampling methods can help to obtain consistent estimator of the total d.d. $F_{\theta_0}[\cdot]$ if the size n of the known data set grows unboundedly. A variant of the Central Limit Resampling Theorem (CLRT), which we apply in assessing accuracy of considered estimators, is given in Appendix.

The main aim of this work is an illustration of the proper resampling methods in evaluation of accuracy of statistical inferences in the case of regression heteroscedastic data.

In the analysis of reliability data, it is common to estimate the mean failure time of tested elements. It is one of important reliability characteristics. Suppose that before the life testing for each element it is known a value of an explanatory variable. Then the mean failure should be considered as a function of this explanatory variable. We suggest estimation of this function and its accuracy in the framework of the linear heteroscedastic regression model.

We use the following notation. Capital letters denote random variables (r.v.s) and the corresponding ordinary letters their values, E and P expectation and probability, respectively, $\xrightarrow{\mathrm{P}}$ means convergence in probability. The normal distribution function (d.f.) with mean μ and variance σ^2 is $\Phi\left[\frac{x-\mu}{\sigma}\right]$, $\Phi[x] = \frac{1}{\sqrt{2\pi}} \int_{-\infty}^{x} e^{-z^2/2} \mathrm{d}z$, $\mathrm{I}[\mathcal{A}]$ is the indicator of an event $\mathcal{A}$. For vectors we use bold characters and for matrices double-struck characters. The definition of resampling is given in Appendix. The mark "$*$" is used for resampled r.v.s and their characteristics.

14.2 Linear Heteroscedastic Regression Model

Suppose that we have statistical data $\mathbf{d}_n = \{\{x_1, y_1\}, ..., \{x_n, y_n\}\}$, $x_1, ..., x_n$ are values of *explanatory variable* and $y_1, ..., y_n$ are values of *responses.* One may interpret the responses as failure times of tested elements. We consider the simplest case of data: $x_1, ..., x_n$, are values of independent identically distributed real r.v.s and responses $y_1, ..., y_n$ are values of independent real r.v.s $Y_1, ..., Y_n$. We also assume that the distribution of Y_i may vary for different i and it only depends on the value of related explanatory variable x_i, $i = 1, ..., n$. We assume that $\mathrm{E}[Y_i^2] < \infty$, then, the variances of $Y_i, i = 1, ..., n$, are finite. The considered type of statistical data is called *heteroscedastic.* We suppose that there are no other explanatory variables with influence on the response distributions. It is a strong and restrictive assumption.

We can write

$$Y_i = \mathrm{E}_{x_i}[Y_i] + (Y_i - \mathrm{E}_{x_i}[Y_i]) = f[x_i] + W_i, i = 1, ..., n.$$

W_i is called *residual,* $\mathrm{E}_{x_i}[W_i] = 0$. We do not know the function $f[x] = \mathrm{E}_x[Y_i]$ and do not observe the true values $W_i = w_i, i = 1, ..., n$. If all Y_i are failure times then $f[x]$ is the mean failure time. The function $f[x]$ is called a regression function if in the relations

$$y_i = f[x_i] + w_i, \quad i = 1, ..., n, \tag{14.2}$$

w_i are values of independent r.v.s W_i with zero expectations. It is desirable to know how it is possible consistently select an appropriate regression function $f[x]$ and to evaluate accuracy of corresponding estimators. Our approach can be generalized to data with vector explanatory variables and vector responses $\{\{\mathbf{x}_1, \mathbf{y}_1\}, ..., \{\mathbf{x}_n, \mathbf{y}_n\}\}$, Belyaev [Bel04]. The generalization is based on usage of the Cramér–Wold device, Cramér and Wold [CW36], Belyaev and Sjöstedt de-Luna [BS00].

We simplify the stated problem to find an appropriate regression function. Suppose that the explanatory variables are values in a finite interval $[x_-, x_+], -\infty < x_- < x_+ < +\infty$, and the true regression function $f_{tr}[x]$ is continuous. Then $f_{tr}[x]$ can be uniformly approximated on $[x_-, x_+]$ by polynomials $f_k[x] = b_0 + b_1 x + \cdots + b_k x^k, k = 0, 1, 2, ...$ Similarly, we could use for approximation $f[x]$ by other systems of functions, e.g. trigonometric functions. We further simplify the problem. We suppose that the true regression function $f_{tr}[x]$ is a polynomial $f_{tr}[x] = b_0(0) + b_1(0)x + \cdots + b_{k_0}(0)x^{k_0}$, $b_{k_0}(0) \neq 0$. Its degree k_0 and its $k_0 + 1$ parameters are not known. The regression data can be written as the following system of relations

$$\begin{aligned} y_1 &= b_0(0) + b_1(0)x_1 + \cdots + b_{k_0}(0)x_1^{k_0} + w_1, \\ \cdots & \quad \cdots \quad \cdots \quad \cdots \\ y_n &= b_0(0) + b_1(0)x_n + \cdots + b_{k_0}(0)x_n^{k_0} + w_n, \end{aligned} \tag{14.3}$$

$b_{k_0}(0) \neq 0$. Recall that the vector $\{b_0(0), ..., b_{k_0}(0)\}$, and the true degree k_0 are unknown parameters.

Let $\mathcal{F}_k$ be the set of all polynomials of degree k. We try to find $f_{tr}[x]$ among functions

$$f_k[x] = b_0 + b_1 x + \cdots + b_k x^k \in \mathcal{F}_k.$$

If $k > k_0$, then we have the case with *overparametrisation.* If $k < k_0$, then we have the case with *underparametrisation.* We will show how it is possible consistently recognize as $n \to \infty$ which of these two cases we have for any given k. We can do that by a special investigation of estimated residuals.

We say that a polynomial $f_k[x] = b_0 + b_1 x + b_2 x^2 + \cdots + b_k x^k$ belongs to the set of selected models sMI$_0$I$_1$...I$_k$ where I$_j$ = 1 if we do not know b_j and we want to estimate it, otherwise if we know that $b_j = 0$ then we let I$_j = 0$, $j < k$, I$_k = 1$. For example, $f_2(x) = \frac{4}{9} + \frac{4}{3}x - x^2$ belongs to sM111, $f_2(x) = \frac{4}{3}x - x^2$ belongs to sM011, $f_1(x) = \frac{9}{4} - x^2$ belongs to $sM101$ $f_2(x) = \frac{9}{4} + \frac{4}{3}x$ belongs to sM11, etc. In the overparametrisation case $b_k(0) = 0$ if $k > k_0$. We will use vectors' notation $\mathbf{y}_n = \{y_1, ..., y_n\}^T$, ("$T$" – transposed), i.e. $\mathbf{y}_n$ is a column vector, $\tilde{\mathbf{x}}_h = \{1, x_h, x_h^2, ..., x_h^k\}^T, h = 1, ..., n$

$\mathbf{b}_k(0) = \{b_0(0), ..., b_k(0)\}^T$, $\mathbf{w}_n = \{w_1, ..., w_n\}^T$, and the matrices

$$\mathbb{X}_{nk} = \begin{pmatrix} 1 & x_1 & x_1^2 & \dots & x_1^k \\ \dots & \dots & \dots & \dots & \dots \\ 1 & x_n & x_n^2 & \dots & x_n^k \end{pmatrix}, \mathbb{X}_{nk}^T = \begin{pmatrix} 1 & 1 & \cdots & 1 \\ x_1 & x_2 & \dots & x_n \\ \dots & \dots & \dots & \dots \\ x_1^n & x_2^n & \dots & x_n^n \end{pmatrix}.$$

The regression equations can be written as follows,

$$\mathbf{Y}_n = \mathbb{X}_{nk}\mathbf{b}_k(0) + \mathbf{W}_n. \tag{14.4}$$

$\mathbb{X}_{nk}^T\mathbb{X}_{nk}$ is a symmetric $(k+1) \times (k+1)$-matrix. Let the matrix $\mathbb{X}_{nk(j)}$, the vectors $\mathbf{b}_{k(j)}$, and $\tilde{\mathbf{x}}_{h(j)}$ be obtained by exclusion the jth column in $\mathbb{X}_{nk}$, the jth component in $\mathbf{b}_k$, and x_h^j in $\tilde{\mathbf{x}}_h$, respectively. The notation $f_{k(j)}[x]$ is used if the term $b_j(0)x^j$ is excluded from the regression polynomial $f_k[x]$.

In general case, when rank $(\mathbb{X}_{nk}^T\mathbb{X}_{nk}) < k+1$ then the Moore–Penrose inverse matrix $(\mathbb{X}_{nk}^T\mathbb{X}_{nk})^+$ to $(\mathbb{X}_{nk}^T\mathbb{X}_{nk})$ should be used, Searly [S71]. In this case $(\mathbb{X}_{nk}^T\mathbb{X}_{nk})^+$ should be used.

By $\mathcal{M}(\mathbb{X}_{nk}^T)$ we denote the linear subspace in $\mathbb{R}^{k+1}$ generated by the column vectors.

Let $\mathbf{c} = \{c_0, c_1, ..., c_k\}^T$, $\mathbf{c} \in \mathcal{M}(\mathbb{X}_{nk}^T)$, $\mathbb{X}_{nk} = \begin{bmatrix} \tilde{\mathbf{x}}_1^T \\ \cdots \\ \tilde{\mathbf{x}}_n^T \end{bmatrix}$, $\mathbb{X}_{nk}^T = [\tilde{\mathbf{x}}_1 ... \tilde{\mathbf{x}}_n]$.

The following relations hold

$$\mathbb{X}_{nk}^T\mathbb{X}_{nk} = \sum_{h=1}^{n} \tilde{\mathbf{x}}_h\tilde{\mathbf{x}}_h^T, \quad (\mathbb{X}_{nk}^T\mathbb{X}_{nk})^+(\mathbb{X}_{nk}^T\mathbb{X}_{nk})(\mathbb{X}_{nk}^T\mathbb{X}_{nk})^+ = (\mathbb{X}_{nk}^T\mathbb{X}_{nk})^+. \tag{14.5}$$

The symmetric matrix $(\mathbb{X}_{nk}^T\mathbb{X}_{nk})^+\mathbb{X}_{nk}^T\mathbb{X}_{nk}$ is a projector on $\mathcal{M}(\mathbb{X}_{nk}^T) = \mathcal{M}(\mathbb{X}_{nk}^T\mathbb{X}_{nk})$. Hence, we have

$$\tilde{\mathbf{x}}_h^T(\mathbb{X}_{nk}^T\mathbb{X}_{nk})^+\mathbb{X}_{nk}^T\mathbb{X}_{nk} = \tilde{\mathbf{x}}_h^T. \tag{14.6}$$

We assume that for any given k and all sufficiently large n the rank $(\mathbb{X}_{nk}^T\mathbb{X}_{nk}) = (k+1)$ then $(\mathbb{X}_{nk}^T\mathbb{X}_{nk})^+ = (\mathbb{X}_{nk}^T\mathbb{X}_{nk})^{-1}$, $\mathcal{M}(\mathbb{X}_{nk}^T) = \mathbb{R}^{k+1}$, and

$$\hat{\mathbf{b}}(k,n) = \{\hat{b}_0(k,n), ..., \hat{b}_k(k,n)\}^T = (\mathbb{X}_{nk}^T\mathbb{X}_{nk})^{-1}\mathbb{X}_{nk}^T\mathbf{Y}_n \tag{14.7}$$

is the *ordinary least squares* (OLS-) estimate of $\mathbf{b}_k(0) = \{b_0(0), ..., b_k(0)\}^T$. The OLS-estimate for the true residuals is

$$\hat{\mathbf{W}}_n(k) = \{\hat{W}_{1n}(k), ..., \hat{W}_{nn}(k)\}^T = \mathbf{Y}_n - \mathbb{X}_{nk}\hat{\mathbf{b}}_k(n). \tag{14.8}$$

We introduce the following assumptions on residuals and explanatory variables:

AW$_1$: (i) *True residuals* $W_1, ..., W_n$ *are independent real-valued random variables; distribution of* W_h *may depend on the value of* $x_h, h \geq 1, -\infty < x_- \leq x_h \leq x_+ < \infty$;

(ii) $\mathrm{E}[W_h] = 0, h \geq 1$;

(iii) *There exist two* $0 < \sigma_-^2 \leq \sigma_+^2 < \infty$ *such that*

$$\sigma_-^2 \leq \sigma_h^2[x_h] = \mathrm{E}[W_h^2] \leq \sigma_+^2;$$

AW$_2$: For each $\varepsilon > 0$ there is an $a_\varepsilon > 0$ such that

$$\sup_{h \geq 1} \mathrm{E}[W_h^2 \mathrm{I}[| W_h |> a_\varepsilon]] < \varepsilon;$$

AX$_1$: Trace $\mathrm{tr}\,(\mathbb{X}_{nk}^T \mathbb{X}_{nk})^{-1} = O\left(\frac{1}{n}\right)$, and $\mathrm{tr}(\mathbb{X}_{nk(j)}^T \mathbb{X}_{nk(j)})^{-1} = O\left(\frac{1}{n}\right)$, $j \leq k$, $n \longrightarrow \infty$;

AX$_2$: $t(\mathbf{c}, n) = \max_{1 \leq h \leq n} \frac{|\mathbf{c}^T (\mathbb{X}_{nk}^T \mathbb{X}_{nk})^{-1} \tilde{\mathbf{x}}_h|}{\sqrt{\mathbf{c}^T (\mathbb{X}_{nk}^T \mathbb{X}_{nk})^{-1} \mathbf{c}}} \longrightarrow 0, \ n \longrightarrow \infty$;

AX$_3$: $\mathrm{s}(\mathbf{c}, n) = \sum_{h=1}^n \frac{(\mathbf{c}^T (\mathbb{X}_{nk}^T \mathbb{X}_{nk})^{-1} \tilde{\mathbf{x}}_h)^2 \tilde{\mathbf{x}}_h^T (\mathbb{X}_{nk}^T \mathbb{X}_{nk})^{-1} \tilde{\mathbf{x}}_h}{\mathbf{c}^T (\mathbb{X}_{nk}^T \mathbb{X}_{nk})^{-1} \mathbf{c}} \longrightarrow 0, n \longrightarrow \infty.$

Numerical experiments show that if explanatory variables "uniformly dense", e.g. they are values of independent r.v.s with positive probability density on $[x_-, x_+]$, then the sequences of values in **AX**$_1$–**AX**$_3$ approach to zero as $n \to \infty$.

14.3 Resampling from Estimates of Residuals

Recall that k is the degree of the regression polynomial $f_k[x]$. From (14.7) and (14.8), the OLS-estimators $\hat{W}_{hn}(k)$ of the residuals $W_h, h = 1, ..., n$, satisfy the following relations

$$\hat{W}_{hn}(k) = W_h - \sum_{h'=1}^{n} \tilde{\mathbf{x}}_h^T (\mathbb{X}_{nk}^T \mathbb{X}_{nk})^{-1} \tilde{\mathbf{x}}_{h'} W_{h'}, \quad h = 1, ..., n, \tag{14.9}$$

if $k \geq k_0$. From **AW**$_1$ and (14.28) we have

$$\mathrm{E}\left[\mathrm{E}^\star\left[\left(\sum_{h=1}^{n}\sum_{h'=1}^{n} (N_{hn}^\star - 1) \tilde{\mathbf{x}}_h^T (\mathbb{X}_{nk}^T \mathbb{X}_{nk})^{-1} \tilde{\mathbf{x}}_{h'} W_{h'}\right)^2\right]\right]$$

$$\leq \sigma_+^2 \sup_{x_- \leq x \leq x_+} \|\tilde{\mathbf{x}}_h\|^2 \mathrm{tr}(\mathbb{X}_{nk}^T \mathbb{X}_{nk})^{-1} = (k+1)^2 \sigma_+^2 x_0^2 \mathrm{tr}(\mathbb{X}_{nk}^T \mathbb{X}_{nk})^{-1},$$

$$x_0 = \max_{j \leq k}\{| x_- |^j, | x_+ |^j\}. \tag{14.10}$$

If the coefficient $b_j(0) \neq 0$ is erroneously excluded from the regression polynomial $f_{tr}[x]$, then instead of (14.4) we have that the vector of responses

$$\mathbf{Y}_n = \mathbb{X}_{nk(j)}\mathbf{b}_{k(j)}(0) + b_j(0)\mathbf{x}_n^j + \mathbf{W}_n,$$

where $\mathbf{x}_n^j = \{x_1^j, x_2^j, ..., x_n^j\}^T$. The estimators of corresponding residuals, obtained by the least squares method, satisfy the following relation

$$\hat{\mathbf{W}}_{n(j)}(k) = \mathbf{W}_n + b_j(0)\mathbf{x}_n^j - \mathbb{X}_{nk(j)}(\mathbb{X}_{nk(j)}^T\mathbb{X}_{nk(j)})^{-1}\mathbb{X}_{nk(j)}^T(\mathbf{W}_n + b_j(0)\mathbf{x}_n^j). \quad (14.11)$$

Let $\hat{w}_{hn(j)}(k), h = 1, ..., n,$ be the estimated components values of random vector (14.11), then $\hat{w}_{hn(j)} = w_{hn} + b_j(0)u[x_h]$, where

$$u_j[x_h] = x_h^j - \tilde{\mathbf{x}}_h(\mathbb{X}_{nk(j)}^T\mathbb{X}_{nk(j)})^{-1}\mathbb{X}_{nk(j)}^T\mathbf{x}_n^j. \quad (14.12)$$

We can calculate values $u_j[x_h], h = 1, ..., n, 1 \leq j \leq k$.

Then, the two sequences of resampled sums

$$\frac{1}{\sqrt{n}}\sum_{h=1}^{n}(N_{hn}^\star - 1)\hat{w}_{hn}(k), \quad (14.13)$$

$$\frac{1}{\sqrt{n}}\sum_{h=1}^{n}(N_{hn}^\star - 1)\hat{w}_{hn(j)}(k), \quad n = 1, 2, ... \quad (14.14)$$

are defined and we can simulate their values via simulated copies $\mathbf{j}_n^\star, r = 1, ..., R$ for any given large R.

Let $u[x]$ be a real bounded function on $[x_-, x_+] \subset \mathbb{R}^1, u_+ = \sup_{x_- \leq x \leq x_+} \mid u[x] \mid$, $\{x_h\}_{h \geq 1}$ be a sequence, $x_h \in [x_-, x_+], u_h = u[x_h], h \geq 1$. Residuals $\{W_h\}_{h\geq 1}$ are independent r.v.s, $E[W_h] = 0, \sigma_-^2 \leq \sigma_h^2[x_h] = E[W_h^2] \leq \sigma_+^2 < \infty$. We introduce the following notation: $u_h^o = u_h - \overline{u_{\cdot n}}$, $\overline{u_{\cdot n}} = \frac{1}{n}\sum_{h=1}^n u_h$, $\overline{u_{\cdot n}^{o2}} = \frac{1}{n}\sum_{h=1}^n (u_h^o)^2$, $\overline{\sigma_{\cdot n}^2} = \frac{1}{n}\sum_{h=1}^n \sigma_h^2[x_h]$. From (14.28) and $\mathbf{AW}_1$ it follows that

$$\begin{aligned}\sigma_{W+u,n}^2 &= \mathrm{E}\left[\mathrm{E}^\star\left[\left(\frac{1}{\sqrt{n}}\sum_{h=1}^{n}(N_{hn}^\star - 1)(W_h + u_h)\right)^2\right]\right] \\ &= \left(1 - \frac{1}{n}\right)\overline{\sigma_{\cdot n}^2} + \overline{u_{\cdot n}^{o2}} - (\overline{u_{\cdot n}})^2. \end{aligned} \quad (14.15)$$

We call the sequence of explanatory variables $x_h, h = 1, 2, ...,$, and the function u_h *essential* if $\underline{\lim}_{h\to\infty}(\overline{u_{\cdot n}^{o2}} - (\overline{u_{\cdot n}})^2) > 0$. Let $\{Z_h'\}_{h\geq 1}$ and $\{Z_h''\}_{h\geq 1}$ be two sequences of r.v.s with uniformly bounded variances $\{\sigma_h'^2\}_{h\geq 1}$ and $\{\sigma_h''^2\}_{h\geq 1}$. We say that $\{Z_h'\}_{h\geq 1}$ has essentially larger variances than $\{Z_h''\}_{h\geq 1}$ if $\underline{\lim}_{h\to\infty}(\sigma_h'^2 - \sigma_h''^2) > 0$. If $\lim_{h\to\infty}(\sigma_h'^2 - \sigma_h''^2) = 0$, then we say that both sequences of r.v.s have equivalent variances. The following statement follows from (14.15).

Lemma 1. *If assumptions* $\mathbf{AW}_1, \mathbf{AW}_2$, *and* $\mathbf{AX}_1$ *hold,* $W_h, h \geq 1$, *are the residuals,* $\mathbf{x}_h, h = 1, 2, ...,$ *and a real function* $u[x]$ *are essential then*

$$\underline{\lim}_{n\to\infty}(\overline{\sigma_{W+u,n}^2} - (\overline{\sigma_{\cdot n}^2}) = \underline{\lim}_{n\to\infty}(\overline{u_{\cdot n}^{o2}} - (\overline{u_{\cdot n}})^2) > 0, \quad (14.16)$$

i.e. the sequence r.v.s $\left\{\frac{1}{\sqrt{n}}\sum_{h=1}^{n}(N^{\star}_{hn}-1)(W_h+u_h)\right\}$ *has essentially larger variances than the sequence* $\left\{\frac{1}{\sqrt{n}}\sum_{h=1}^{n}(N^{\star}_{hn}-1)W_h\right\}$.

Theorem 1. *If* $\mathbf{AW}_1, \mathbf{AW}_2, \mathbf{AX}_1$ *and* $\mathbf{AX}_2$ *hold, then*

(*i*) *For each fixed* h *r.v.s* $\hat{W}_{hn}(k)$, $k \geq k_0$, *are unbiased and consistent OLS-estimators of* W_h $n \to \infty$;
(*ii*) $\left|\mathrm{P}\left[\frac{1}{\sqrt{n}}\sum_{h=1}^{n} W_n \leq z\right] - \Phi\left[\frac{z}{\sigma_{\cdot n}}\right]\right| \to 0,\ n \to \infty,\ \sigma_{\cdot n} = \left(\frac{1}{n}\sum_{h=1}^{n}\sigma_h^2[x_h]\right)^{1/2}$;
(*iii*) $\sup_z\left|\mathrm{P}\left[\frac{1}{\sqrt{n}}\sum_{h=1}^{n} W_h \leq z\right] - \frac{1}{\sqrt{n}}\sum_{h=1}^{n}(N^{\star}_{hn}-1)\mathrm{I}[\hat{w}_{hn}(k) \leq z]\right| \xrightarrow{\mathrm{P}} 0,\ n \to \infty$;
(*iv*) *If* $b_j(0) = 0$ *is excluded from the list of coefficients* $\mathbf{b}_k(0) = \{b_0(0), ..., b_k(0)\}^T$ *then the sequence of r.v.s (14.13)and (14.14) have equivalent variances.*

Proof. Part (i) follows from (14.9) and $\mathbf{AW}_1$. All assumptions in Theorem 3 hold for r.v.s $Z_{hn} = W_h/\sqrt{n}$. Therefore, (ii) holds. The distributions of the resampled sums of $W_h/\sqrt{n}$ and $\hat{W}_{hn}(k)/\sqrt{n}, h = 1, ..., n$, are asymptotically normal. From (14.10) it follows that the distributions' variances approach each other as $n \to \infty$, and (iii) also holds. To check (iv), we use relation (14.10) (14.11) with $b_j(0) = 0$, and $\mathbf{AX}_1$.

Suppose that for the sequence of explanatory variables $x_h, h = 1, 2, ...,$ holds

$$\underline{\lim}_{n\to\infty}\mathrm{E}^{\star}\left[\left(\sum_{h=1}^{n}(N^{\star}_{hn}-1)u_j[x_h]\right)^2\right] > 0. \tag{14.17}$$

Then, the sequence of resampled sums (14.14) has essentially larger variances than the sequence of resampled sums (14.13). In this case we have to estimate $b_j(0)x^j$.

The suggested approach shows possibility to apply resamplings in consistent identification of the degree of the true regression polynomial and exclude from estimation zero coefficients.

14.4 OLS-Estimators of Regression Coefficients and Their Accuracies

The resampling methods can be used in the consistent assessing distributions of deviations OLS-estimators $\hat{b}_j(k,n)$ from the true values $b_j(0)$. If we erroneously include $b_j(0) = 0$ into the list of unknown coefficients and find their OLS-estimators, then $\mathbb{X}_{nk}$ does not been reduced to $\mathbb{X}_{nk(j)}$. Then, accuracy of others OLS-estimators will be worsened, i.e. their variances will be essentially larger. Therefore, it is important before estimation exclude the zero coefficients from the estimated regression polynomial. This worsening was observed in a series of numerical experiments.

It is convenient to consider assessing accuracies of the OLS-estimators of linear forms $\mathbf{c}^T\mathbf{b}_k(0)$, $\mathbf{c} = \{c_0, ..., c_k\}^T, k \geq k_0$. If $\mathbf{c}_j = \{\delta_{0j}, \delta_{1j}, ..., \delta_{kj}\}^T$, then all regression coefficients $b_j(0), j = 0, 1, ..., k, k \geq k_0$ are unbiasly estimable as in (14.7).

The OLS-estimator of $\mathbf{c}^T\mathbf{b}_k(0)$ is $\mathbf{c}^T(\mathbb{X}_{nk}^T\mathbb{X}_{nk})^{-1}\mathbb{X}^T\mathbf{Y}_n$. The normed deviation of this estimate from the linear form $\mathbf{c}^T\mathbf{b}_k(0)$ is

$$\frac{\mathbf{c}^T(\mathbb{X}_{nk}^T\mathbb{X}_{nk})^{-1}\mathbb{X}^T\mathbf{Y}_n - \mathbf{c}^T\mathbf{b}_k(0)}{\sqrt{\mathbf{c}^T(\mathbb{X}_{nk}^T\mathbb{X}_{nk})^{-1}\mathbf{c}}} = \sum_{h=1}^{n} U_{hn}(\mathbf{c}, k), \tag{14.18}$$

where

$$U_{hn}(\mathbf{c}, k) = \frac{\mathbf{c}^T(\mathbb{X}_{nk}^T\mathbb{X}_{nk})^{-1}\tilde{\mathbf{x}}_h}{\sqrt{\mathbf{c}^T(\mathbb{X}_{nk}^T\mathbb{X}_{nk})^{-1}\mathbf{c}}} W_h. \tag{14.19}$$

The mean of (14.18) is zero and the variance is

$$\sigma_n^2(\mathbf{c}, k) = \sum_{h=1}^{n} \frac{\mathbf{c}^T(\mathbb{X}_{nk}^T\mathbb{X}_{nk})^{-1}\tilde{\mathbf{x}}_h\tilde{\mathbf{x}}_h^T(\mathbb{X}_{nk}^T\mathbb{X}_{nk})^{-1}\mathbf{c}\sigma_h^2[x_h]}{\mathbf{c}^T(\mathbb{X}_{nk}^T\mathbb{X}_{nk})^{-1}\mathbf{c}}, \quad \sigma_-^2 \le \sigma_n^2(\mathbf{c}, k) \le \sigma_+^2.$$

Theorem 2. *If assumptions* $\mathbf{AW}_1, \mathbf{AW}_2$, $\mathbf{AX}_1$ *and* $\mathbf{AX}_2$ *hold, and* $\mathbf{c} = \{c_0, ..., c_k\}^T$, $k \ge k_0$, *then*

(i) $\mathbf{c}^T(\mathbb{X}_{nk}^T\mathbb{X}_{nk})^{-1}\mathbb{X}^T\mathbf{Y}_n$ *is an unbiased and consistent estimator of* $\mathbf{c}^T\mathbf{b}_k(0)$, *and for each* $z \in \mathbb{R}^1$

$$\sup_z \left| \mathrm{P}\left[\sum_{h=1}^{n} U_{hn}(\mathbf{c}, k) \le z\right] - \Phi\left[\frac{z}{\sigma_n(\mathbf{c}, k)}\right] \right| \to 0, \quad n \to \infty; \tag{14.20}$$

(ii) If the assumption $\mathbf{AX}_3$ *also holds then for each* $z \in \mathbb{R}^1$

$$\sup_z \left| P^\star\left[\sum_{h=1}^{n} (N_{hn}^\star - 1)\hat{U}_{hn}(\mathbf{c}, k) \le z \,\middle|\, \mathbb{D}_n = \{\{x_1, y_1\}, ..., \{x_n, y_n\}\}\right]\right.$$
$$\left. - \mathrm{P}\left[\sum_{h=1}^{n} U_{hn}(\mathbf{c}, k) \le z\right] \right| \xrightarrow{\mathrm{P}} 0, \quad n \to \infty. \tag{14.21}$$

Proof. We check validity of Assumptions (i)–(v) in the Theorem 3 in Appendix. From (14.18) and (14.19) we have that the terms $Z_{hn} = U_{hn}(\mathbf{c}, k), 1 \le h \le n$, are independent and $E[U_{hn}(\mathbf{c}, k)] = 0$. Assumptions (ii)–(iv) follow from $\mathbf{AW}_1$. From $\mathbf{AW}_2$, (14.5) and the Chebyshev inequality for any small $\varepsilon > 0$ and all sufficiently large n such that $t(\mathbf{c}, n) < \varepsilon/a_\varepsilon$ it follows

$$\begin{aligned}
&\sum_{h=1}^{n} E[U_{hn}^2(\mathbf{c}, k)\mathrm{I}[\mid U_{hn}(\mathbf{c}, k) \mid > \varepsilon]] \\
&\le \sum_{h=1}^{n} \left(\frac{\mathbf{c}^T(\mathbb{X}_{nk}^T\mathbb{X}_{nk})^{-1}\tilde{\mathbf{x}}_h}{\sqrt{\mathbf{c}^T(\mathbb{X}_{nk}^T\mathbb{X}_{nk})^{-1}\mathbf{c}}}\right)^2 \mathrm{E}\left[W_h^2 I\left[\mid W_h \mid > \frac{\varepsilon}{t(\mathbf{c}, n)}\right]\right] \\
&\le \frac{\mathbf{c}^T(\mathbb{X}_{nk}^T\mathbb{X}_{nk})^{-1}\left(\sum_{h=1}^{n}\tilde{\mathbf{x}}_h\tilde{\mathbf{x}}_h^T\right)(\mathbb{X}_{nk}^T\mathbb{X}_{nk})^{-1}\mathbf{c}\varepsilon}{\mathbf{c}^T(\mathbb{X}_{nk}^T\mathbb{X}_{nk})^{-1}\mathbf{c}} < \varepsilon, \quad n \ge 1.
\end{aligned} \tag{14.22}$$

The inequalities (14.22) and $\mathbf{AX}_2$ imply that (i) and (v) in Theorem 3 hold. Hence, the stated relations (14.20) hold. The idea to prove (14.21) is similar to those used in the proof of Theorem 1. From (14.9) we can write

$$\begin{aligned}
&\sum_{h=1}^{n}(N_{hn}^{\star}-1)\hat{U}_{hn}(\mathbf{c},k)-\sum_{h=1}^{n}(N_{hn}^{\star}-1)U_{hn}(\mathbf{c},k)\\
&=-\sum_{h=1}^{n}(N_{hn}^{\star}-1)\frac{\mathbf{c}^{T}(\mathbb{X}_{nk}^{T}\mathbb{X}_{nk})^{-1}\tilde{\mathbf{x}}_{h}}{\sqrt{\mathbf{c}^{T}(\mathbb{X}_{nk}^{T}\mathbb{X}_{nk})^{-1}\mathbf{c}}}(\hat{W}_{hn}(k)-W_{h})\\
&=-\sum_{h=1}^{n}\sum_{h'=1}^{n}(N_{hn}^{\star}-1)\frac{\mathbf{c}^{T}(\mathbb{X}_{nk}^{T}\mathbb{X}_{nk})^{-1}\mathbf{x}_{h}\tilde{\mathbf{x}}_{h}^{T}(\mathbb{X}_{nk}^{T}\mathbb{X}_{nk})^{T}\tilde{\mathbf{x}}_{h'}}{\sqrt{\mathbf{c}^{T}(\mathbb{X}_{nk}^{T}\mathbb{X}_{nk})^{-1}\mathbf{c}}}W_{h'}. \qquad (14.23)
\end{aligned}$$

From $\mathbf{AX}_3$ and (14.28), it follows that the variance of right hand side in (14.23) tends to zero as $n\to\infty$. Hence, part (ii) in Theorem 2 holds. We have

$$\begin{aligned}
&P^{\star}\left[\sum_{h=1}^{n}(N_{hn}^{\star}-1)U_{hn}(\mathbf{c},k)\le z|\mathbb{D}_n=\{\{x_1,y_1\},...,\{x_n,y_n\}\}\right]\\
&=\lim_{n\to\infty,R\to\infty}\frac{1}{R}\sum_{r=1}^{R}\mathrm{I}\left[\sum_{h=1}^{n}(n_{hn}^{\star r}-1)\hat{u}_{hn}(\mathbf{c},k)\le z\right], \qquad (14.24)
\end{aligned}$$

where $n_{hn}^{\star r}=\sum_{i=1}^{n}\mathrm{I}[j_{in}^{\star r}=h]$, and

$$\hat{u}_{hn}(\mathbf{c},k)=\frac{\mathbf{c}^{T}(\mathbb{X}_{nk}^{T}\mathbb{X}_{nk})^{-1}\tilde{\mathbf{x}}_{h}}{\sqrt{\mathbf{c}^{T}(\mathbb{X}_{nk}^{T}\mathbb{X}_{nk})^{-1}\mathbf{c}}}\hat{w}_{hn}(k).$$

From (14.20)–(14.24) it follows

$$\sup_{z}\left|\mathrm{P}\left[\frac{\mathbf{c}^{T}\hat{\mathbf{b}}_{k}(n)-\mathbf{c}^{T}\mathbf{b}_{k}(0)}{\sqrt{\mathbf{c}^{T}(\mathbb{X}_{nk}^{T}\mathbb{X}_{nk})^{-1}\mathbf{c}^{T}}}\le z\right]-\frac{1}{R}\sum_{r=1}^{R}I\left[\sum_{h=1}^{n}(n_{hn}^{\star r}-1)\hat{u}_{hn}(\mathbf{c},k)\le z\right]\right|\xrightarrow{\mathrm{P}}0, \qquad (14.25)$$

$R\to\infty, n\to\infty$.

Under the same assumptions by Cramèr–Wold device one can generalize (14.25) to consistent simultaneous estimation the deviations distribution of two or more linear forms $\mathbf{c}_i^T\hat{\mathbf{b}}_k(n)-\mathbf{c}_i^T\mathbf{b}_k(0)$, $i=1,2,....$ Then it is possible to estimate accuracy of estimated maximum $\hat{f}_m=\max_x f_k[x]=f[\hat{x}_m]$ and its position $\hat{x}_m$, Belyaev [Bel04]. This possibility can be used in searching of optimal designs of regression experiments.

If rank$(\mathbb{X}_{nk}^T\mathbb{X}_{nk})<k+1$, then the most of relations in Sects. 14.3 and 14.4 hold with $(\mathbb{X}_{nk}^T\mathbb{X}_{nk})^{+}$ instead of $(\mathbb{X}_{nk}^T\mathbb{X}_{nk})^{-1}$, and $\mathbf{c}\in\mathcal{M}(\mathbb{X}_{nk}^T)$.

14.5 Conclusion

Usage of resamplings methods in the analysis of statistical data with explanatory variables gives new possibilities of consistent identification corresponding statistical models and consistent evaluation accuracies of related statistical inferences. In the considered approach to the analysis data of regression experiments the decision, to add or exclude $b_j(0)x^j$ from estimated regression polynomial, can be done by comparison the related distributions of sums of resampled estimated residuals. If $b_j(0) \neq 0$ then from lemma 1 it follows that the two related distributions, of sums of resampled estimated residuals, with and without $b_j(0)$ will be essentially different for all sufficiently large n. Then $\hat{b}_j(k,n)$ should be included in the list with coefficients of estimated regression function. Estimates of components in (14.7) are calculated simultaneously for each copy of $\mathbf{J}_n^{\star}$. The erroneous inclusion $\hat{b}_j(k,n)$ in the list of components (14.7) if $b_j(0) = 0$ essentially decreases accuracies of others estimates $\hat{b}_{j'}(k,n), j' \neq j$. Hence, the overparametrisation is not desirable. Our numerical experiments show that the resampling methods from sums of estimated residuals (14.13) and from sums of weighted deviations (14.18) are efficient methods in assessing accuracies of inferences in the analysis data of regression experiments. One may detect relationship of the resamplings methods with the theoretical problem of consistent estimation distribution of a sum of n independent r.v.s if we know only one value for each of these r.v.s. Then usage of the CLRT immediately follows from presentation deviations of the considered OLS-estimators (14.18)–(14.19) as sums of weighted residuals values.

In several applications, we can also observe such possibility to use the CLRT but in a more complicated form. In paper Belyaev [Bel05], the resamplings methods are used in assessing accuracy of the cross-classification probabilities estimated by the cross-validation method. Here, the estimators are sums of values of locally dependent random variables and resamplings are taken from independent subsets which are directed graphs. The considered in Belyaev [Bel05] $1NN$- and kNN-classifiers were applied in the analysis of digital images transmitted from satellites. The classifiers can also be applied for rejection as defective the elements with non-admissible probability of appearance failures after short working time, depending on the values of elements explanatory variables.

Evaluation the accuracy of estimated reliability characteristics is actual in constructing some mechanical systems with redistributions of applied loads between non-failed systems' elements. Cables with many fibres are examples of such systems, Crowder et al. [CKSS88]. Reliability of cables can be characterised by the distribution of tensile strengths destroying pieces of tested cables. Each cable consists of m fibers. If a fiber breaks in a tested cable, then the force applied to the piece of cable will be immediately equally redistributed between the not yet unbroken fibers. The redistributed forces may simultaneously destroy one or several new fibers. The applied force is growing up to the value when all fibers, of the tested piece of cable, will be broken. The numbers of simultaneously broken fibers and corresponding forces are registered for each of n tested cable. The martingale theory can be used to obtain a non-parametric estimator similar to Aalen–Nelson one for the distribution of tensile forces destroying cables. The resampling methods can be used in assessing accuracy of such non-parametric estimators, Belyaev and Rydén [BR97], Rydén [Ryd00], where a variant of the Functional CLRT is given.

There are several economic problems related to the cost-benefit analysis of improvement reliability characteristics. An appropriate increasing of price, which consumers will be ready to pay for elements with improved reliability characteristics, is essentially to know for future development and production of such elements. Necessary information may be collected by a contingent valuation (CV) where randomly sampled consumers should freely state intervals containing the most reasonable for them future price. It is not known where inside the stated intervals exact values of reasonable prices are placed. The ends of stated intervals depend on the unknown suggested values of prices, and besides that the ends of the stated intervals may be essentially rounded. Is it possible to find the distribution of exact prices hidden inside the stated intervals? It is a rather difficult statistical problem. Its solution depends on developing a corresponding statistical model which is essentially different from the models with censored data considered in the Survival Analysis, Turnbull [T76], where usually censoring intervals are independent from positions of exact values.The author of this paper participates in a research program on evaluation acceptable costs needed for improvement of environment. The initial stage of research is presented in report, Belyaev, Håkansson, and Kriström [BHK09], where we have not yet a developed appropriate statistical model. The suggested statistical model will be given in forthcoming research report, Belyaev and Kriström [BK10]. It seems that there are many similarities with the above statistical problem with prices for elements with improved reliability characteristics.

Acknowledgments. The research results presented in this paper have been obtained as a part of research programs at the Department of Mathematics and Mathematical Statistics (Umeå University) and the Department of Forest Economics (SLU) financed by Elforsk, the Swedish Energy Agency, the National Board of Fisheries, the Swedish Environmental Agency, PlusMinus program (SLU).

References

[Bel03] Belyaev, Yu.: Necessary and sufficient conditions for consistency of resampling. Research report, Center of Biostochastics. Swedish University of Agricultural Sciences (2003)

[Bel04] Belyaev, Yu.K.: Application of Resampling Methods to Linear Heteroscedastic Regression with Vector Responses. Research Report, Department of Mathematical Statistics, Umeå University (2004)

[Bel05] Belyaev, Yu.K.: Application of Clustered Resampling Methods in Assessing Accuracy of Cross-Validated Estimators of Cross-Classification Probabilities of Nearest-Neighbor Classifiers. Research Report, Centre of Biostochastics, Swedish University of Agricultural Sciences (2005)

[Bel07] Belyaev, Yu.: Resampling for Lifetime Distributions. Encyclopedia of Statistics in Quality and Reliability. Wiley, Chichester, 1653–1657 (2007)

[BHK09] Belyaev, Yu.K., Håkansson, C. and Kriström, B.: Rounding it up: Interval and Open Ended Valuation questions. Working paper 373, Department of Forest Economics, Swedish University of Agricultural Sciences (2009)

[BK10] Belyaev, Yu.K. and Kriström, B.: Approach to Analysis of Self-Selected Interval Data. Working paper CERE, Umeå University. #2/2010, `http://www.cere.se/section.asp?id=home`

[BR97] Belyaev, Yu.K. and Rydén, P.: Non-Parametric Estimators of the Distribution of Tensile Strengths for Wires. Research Report, Department of Mathematical Statistics, Umeå Uviverslty (1997)

[BS00] Belyaev, Yu. K. and Sjöstedt-de Luna, S.: Weakly approaching sequences of random distributions. J. Appl. Probab., **37**, No 3, 807–822 (2000)

[CW36] Cramèr, H. and Wold, H: Some theorems on distribution function. J. Lond. Math. Soc., **11**, 290–295 (1936)

[CKSS88] Crowder, M., Kimber, A., Smith, R. and Sweeting, T.: Extreme Value Theory in Engineering. Academic, San Diego (1988)

[DH97] Davison, A.C. and Hinkley, D.V.: Bootstrap Methods and their Application. Cambridge University Press, London (1997)

[E79] Efron, B.: Bootstrap methods: another look at the jacknife. Ann. Statist., **7**, 1–26 (1979)

[L77] Loéve, M.: Probability Theory, vI, 4th ed., Springer, New York (1977)

[M92] Mammen, E.: Bootstrap, wild bootstrap and asymptotic normality. Probab. Theor. Relat. Field, **93**, 439–455 (1992)

[Ryd00] Rydén, P.: Statistical Analysis and Simulation Methods Related to Load-Sharing Models. Doctoral Dissertation, Department of Mathematical Statistics, Umeå University (2000)

[S71] Searly, S.R.: Linear Models. Wiley, New York (1971)

[T76] Turnbull, B.W.: The empirical distribution function with arbitrarily grouped, censored and truncated data. J. Roy. Statist. Soc. Ser. B, **38**, 290–295 (1976)

[W86] Wu, C.-F.J.: Jackknife, bootstrap and other resampling methods in regression analysis (with discussion). Ann. Statist., **14**, 1261–1350 (1986)

Appendix

One approach to evaluation accuracies of statistical inferences is closely related to a solution of the following problem. Let $\mathbf{z}_n = \{z_{1n}, ..., z_{rn}\}^T$ be observed values of r.v.s $\mathbf{Z}_n = \{Z_{1n}, ..., Z_{nn}\}^T$ with independent components. The unknown distributions of the r.v.s Z_{hn} can be different. It is useful to estimate the distribution of $Z_{\cdot n} = \sum_{h=1}^{n}(Z_{hn} - \mathrm{E}[Z_{hn}])$ if only one value z_{hn} of $Z_{hn}, h = 1, ..., n$, is observed. Then the only one value $z_{\cdot n} = \sum_{h=1}^{n} z_{hn}$ is known. Under rather general assumptions, sums of n values, resampled from the list $\mathbf{z}_n$, can be considered as if they have the desirous distributions of $\sum_{h=1}^{n}(Z_{hn} - \mathrm{E}[Z_{hn}])$ as $n \to \infty$.

Let $\{X_{hn} : \{h, n\} \in \mathcal{T}\}$ be a triangular array of real r.v.s $\{X_{hn}\}$ which are independent for each $n = 2, 3, ...$, $\mathcal{T} = \{\{h, n\} : n = 1, 2, ..., h = 1, 2, ..., n\}$, $\mathbf{X}_n = \{X_{1n}, ..., X_{nn}\}$. The r.v.s X_{hn}, $h = 1, ..., n$, may be non-identically distributed. By $\{\rho_n\}_{n \geq 1}$ we denote a sequence of positive non-random values which will be used to rescale r.v.s. We consider the rescaled r.v.s $Z_{hn} = X_{hn}/\rho_n$. Let $\bar{Z}_{\cdot n} = Z_{\cdot n}/n$, and

$$\mathbf{Z}_n^{\odot} = \{Z_{1n} - \bar{Z}_{\cdot n}, Z_{2n} - \bar{Z}_{\cdot n}, ..., Z_{nn} - \bar{Z}_{\cdot n}\}. \tag{14.26}$$

Let $\mathbf{J}_n^{\star} = \{J_{1n}^{\star}, ..., J_{nn}^{\star}\}$, be n independent r.v.s uniformly distributed on $\{1, 2, ..., n\}$, i.e. $\mathrm{P}^{\star}[J_{in}^{\star} = h] = 1/n$, $h = 1, ..., n$. We use the mark "$\star$" to show that r.v.s' probabilities and expectations are related to $\mathbf{J}_n^{\star}$. Values of r.v.s $J_{in}^{\star}$ can be obtained on a personal computer by simulation. Resampling copies, based on the list $\mathbf{Z}_n^{\odot}$, obtained via simulation of the r.v.s $\mathbf{J}_n^{\star}$, will be

$$\mathbf{Z}_n^{\odot\star} = \{Z_{1n}^{\odot\star}, ..., Z_{nn}^{\odot\star}\}, \tag{14.27}$$

where $Z^{\odot\star}_{hn} = Z^{\star}_{hn} - \bar{Z}_{\cdot n}$, $Z^{\star}_{hn} = Z_{J^{\star}_{hn},n}$. Let $N^{\star}_{hn} = \sum_{i=1}^{n} \mathrm{I}(J^{\star}_{in} = h)$. The mean values, the variances and the covariances of $N^{\star}_{hn}$ are

$$E^{\star}[N^{\star}_{hn}] = 1,\ E^{\star}[(N^{\star}_{hn} - 1)^2] = 1 - \frac{1}{n},\ E^{\star}[(N^{\star}_{h_1 n} - 1)(N^{\star}_{h_2 n} - 1)] = -\frac{1}{n},\ h_1 \neq h_2. \tag{14.28}$$

The resampled sums of components in $\mathbf{Z}^{\odot\star}_{n}$ can be written as follows

$$Z^{\odot\star}_{\cdot n} = \sum_{h=1}^{n} (N^{\star}_{hn} - 1) Z_{hn}. \tag{14.29}$$

The variance of $Z_{\cdot n}$ is $\sigma^2_{\cdot n} = \sum_{h=1}^{n} \mathrm{E}[(Z_{hn} - \mathrm{E}[Z_{hn}])^2] < \infty$, if $\mathrm{E}[Z^2_{hn}] < \infty, h \geq 1$.

Theorem 3. *(CLRT with sufficient assumptions) Suppose that*

(*i*) $\max_{1 \leq h \leq n} \mid Z_{hn} \mid \xrightarrow{\mathrm{P}} 0,\ n \to \infty,$

(*ii*) *All second order moments exists,* $\mathrm{E}[Z^2_{hn}] < \infty,$

(*iii*) *There are constants* $0 < \sigma^2_{-} \leq \sigma^2_{+} < \infty$ *such that* $\sigma^2_{-} \leq \sigma^2_{\cdot n} \leq \sigma^2_{+},$

(*iv*) $\sum_{h=1}^{n} (\mathrm{E}[Z_{hn}])^2 \to 0,\ n \to \infty,$

(*v*) *(Lindeberg assumption) for every* $\tau > 0$

$$\sum_{h=1}^{n} \mathrm{E}[(Z_{hn})^2 \mathrm{I}(\mid Z_{hn} \mid > \tau)] \to 0, \quad n \to \infty. \tag{14.30}$$

Then for any z

$$\mathrm{P}\left[\sum_{h=1}^{n} (Z_{hn} - \mathrm{E}[Z_{hn}]) \leq z\right] - \Phi\left[\frac{z}{\sigma_{\cdot n}}\right] \to 0,\ n \to \infty, \tag{14.31}$$

and

$$\mathrm{P}[Z^{\odot\star}_{\cdot n} \leq z \mid Z_{1n}, ..., Z_{nn}] - \mathrm{P}\left[\sum_{h=1}^{n} (Z_{hn} - \mathrm{E}[Z_{hn}]) \leq z\right] \xrightarrow{\mathrm{P}} 0,\ n \to \infty. \tag{14.32}$$

Note that (14.32) implies that both sequences of the d.f.s converge mutually to each other in the uniform metric. For proof of Theorem 1, see e.g. Belyaev [Bel03] and Belyaev and Sjösted-de-Luna [BS00].

Assumptions (i)–(iii), and (v) imply the sufficient part of the Central Limit Theorem (CLT), see Loeve [L77]. Assumptions (i)–(v) imply the sufficient part of the Central Limit Resampling Theorem (CLRT). See also Mammen [M92], and Belyaev [Bel03] were variants of CLRTs with necessary and sufficient assumptions are given.

This Theorem 3 can be used in the justification consistency of estimators of total d.d.s in several statistical models. The main idea to use resamplings is follows.

Suppose that we are interested in finding a consistent estimator for the distribution of $\sum_{h=1}^{n} (Z_{hn} - \mathrm{E}[Z_{hn}])$ knowing only values $\{z_{1n}, ..., z_{nn}\}$ of r.v.s $\{Z_{1n}, ..., Z_{nn}\}$.

Let $\bar{z}_{\cdot n} = \frac{1}{n}\sum_{h=1}^{n} z_{hn}$, and $\{j_{1n}^{\star r}, ..., j_{nn}^{\star r}\}$, $r = 1, ..., R$, be R independently simulated copies of the r.v. $\mathbf{J}_n^{\star}$. We find values $z_{\cdot n}^{\odot\star r} = \sum_{h=1}^{n}(z_{j_{hn}^{\star r},n} - \bar{z}_{\cdot n}) = \sum_{h=1}^{n}(n_{hn}^{\star r} - 1)$ z_{hn}, $r = 1, ..., R$. The mean $\frac{1}{R}\sum_{r=1}^{R} \mathrm{I}[z_{\cdot n}^{\odot\star r} \le z]$ is a stepwise function of z. This stepwise function can be applied to consistent estimation of $\mathrm{P}[Z_{\cdot n}^{\odot\star} \le z \mid z_{1n}, ..., z_{nn}]$ as $R \to \infty$ and $n \to \infty$. From (i)–(v) and (14.32) we have for any z

$$\sup_{z}\left|\frac{1}{R}\sum_{r=1}^{R}\mathrm{I}[z_{\cdot n}^{\odot\star r} \le z\,] - \mathrm{P}\left[\sum_{h=1}^{n}(Z_{hn} - \mathrm{E}[Z_{hn}]) \le z\right]\right| \xrightarrow{\mathrm{P}} 0, \tag{14.33}$$

as $R \to \infty$ and $n \to \infty$. In addition, relation (14.31) implies that both terms in (14.33) approach to the sequence of normal d.f.s $\Phi\left[\frac{z}{\sigma_{\cdot n}}\right]$ included in (14.31). Note that (14.33) is the convergence in the uniform metric in probability.

15

Change Point Estimation in Regression Models with Fixed Design

Maik Döring and Uwe Jensen*

Institut für Angewandte Mathematik und Statistik, Universität Hohenheim, Stuttgart, Germany, jensen@uni-hohenheim.de

Abstract: In this paper, we consider a simple regression model with change points in the regression function which can be one of two types: A so called smooth bent-line change point or a discontinuity point of a regression function. In both cases we investigate the consistency of the M-estimates of the change points. It turns out that the rates of convergence are $n^{1/2}$ or n, respectively, where n denotes the sample size in a fixed design. In addition, the asymptotic distributions of the change point estimators are investigated.

Keywords and phrases: Regression, Change-points, M-estimates, Rate of Consistency

15.1 Introduction

To motivate our study we begin to look at change point models in Survival Analysis. Methods of Survival Analysis were developed to estimate lifetime or survival distributions in medicine or reliability. Sometimes additional variables, so called covariates, are observed, which characterize the objects under consideration in more detail. The link between these covariates and the lifetime distribution is given in regression models by the intensities of corresponding lifetime models (see [ABG08, MS06]). One of the most popular regression models is the Cox model:

If for each object i, $i = 1, \ldots, n$, k covariates $Z_{i,1}, \ldots, Z_{i,k}$ are observed, then the hazard rate process is given by

$$\lambda_i (t) = \lambda_0 (t) \cdot R_i (t) \cdot \exp \{\beta_1 Z_{i,1} + \ldots + \beta_k Z_{i,k}\},$$

where λ_0 is the deterministic baseline hazard rate. R_i is the risk indicator, equal to one as long as object i is observed (at risk). $\beta^T = (\beta_1, \ldots, \beta_k)$ is the vector of the unknown regression parameters. The analysis of several datasets has shown that the basic model should in some cases be extended to a Cox model with a change point. An overview over change point models can be found in [JL07].

V.V. Rykov et al. (eds.), *Mathematical and Statistical Models and Methods in Reliability: Applications to Medicine, Finance, and Quality Control*, Statistics for Industry and Technology, DOI 10.1007/978-0-8176-4971-5_15,

In the following, we briefly describe the simplest version of such a model with a change point. More complex versions are investigated in detail in [JL08]. In the simplest version, we assume that for one covariate with index, say k, we have a change in the regression parameter from β_k to $\beta_k + \beta_{k+1}$, when the covariate $Z_{i,k}$ hits the threshold (change point) ξ. The corresponding intensity is

$$\lambda_i(t) = \lambda_0(t) \cdot R_i(t) \cdot \exp\left\{\beta_1 Z_{i,1} + \dots + \beta_{k-1} Z_{i,k-1} + \beta_k Z_{i,k} + \beta_{k+1}(Z_{i,k} - \xi)^+\right\}, \tag{15.1}$$

where $a^+ = \max\{a, 0\}$. The change point ξ has to be estimated. The asymptotic behavior of the (partial) maximum likelihood estimates $\hat{\xi}_n$ has been investigated in detail in [JL08]. It turned out that this sequence is consistent at a rate of $n^{\frac{1}{2}}$ and follows asymptotically a normal distribution. Estimates of change points are often known to converge at a rate of n. For instance Pons [Pon03] and Kosorok and Song [KS07] considered extensions of the Cox model where the covariates jump at a certain threshold. They derived n-consistent estimates of the change points. It seems that the smoothness of the regression function or the criterion function, here the partial likelihood, determines the rate of convergence of change point estimators. In (15.1) the regression function is continuous in the change point ξ in contrast to the models of Pons and Kosorok and Song with jump change points. The question arises whether this difference also can be found in simple regression models. At first sight this question looks easier to answer as it is. It is quite challenging and includes the use of methods of M-estimation theory. In this paper, simple regression models in a fixed design with smooth (continuous case) or jump change points are investigated.

We assume that the observations $Y_{1,n}, \dots, Y_{n,n}$ for $n \in \mathbb{N}$ and $1 \leq i \leq n$ are given by

$$Y_{i,n} = f_{\theta_0}\left(\frac{i}{n}\right) + \epsilon_{i,n}, \quad 1 \leq i \leq n, \ n \in \mathbb{N}.$$

For $\theta \in [0,1]$ the regression function $f_\theta : [0,1] \to \mathbb{R}$ is given by

$$f_\theta(x) := g(x) + h_\theta(x) \cdot 1\!\!1_{[\theta,1]}(x),$$

where $g : [0,1] \to \mathbb{R}$, $h_\theta : [0,1] \to \mathbb{R}$ and $1\!\!1_A$ is the indicator function of a set A. For $n \in \mathbb{N}$ and $1 \leq i \leq n$ let $\epsilon_{i,n}$ be i.i.d. random variables with $\mathbf{E}\epsilon_{1,1} = 0$ and suitably integrable, i.e. there exists a real number $1 < p < \infty$ such that $\mathbf{E}|\epsilon_{1,1}|^p < \infty$. In the following the focus will be on estimating the change point θ_0 by the least squares method. We assume that the regression function is known except the change point $\theta_0 \in [0,1]$. Therefore, we can without loss of generality set $g = 0$.

The problem to estimate the location of a change point in a regression model has been studied in the literature to some extent, see among others Müller [Mue92], Müller and Song [MSo97], Müller and Stadtmüller [MSt99], Bai [Bai97], Dempfle and Stute [DS02], Koul et al. [KQS03], and Lan et al. [LBM09] and the cited references therein. In most cases locating a jump discontinuity is considered and properties of the estimators are studied. Müller [Mue92] investigates the problem of estimating a jump change point in the derivative of some order $\nu \geq 0$ of the regression function. His change point estimators are based on one-sided kernels. This includes the case of continuous regression functions with a change in the derivative at some point which we call smooth change point. In a number of applications one would rather model a smooth change point than a jump in the regression function. in particular, in the recently published article by Lan et al. [LBM09] the plotted dataset would suggest to fit a regression function with a smooth change instead of the proposed jump model. In this paper, we investigate both types of regression models, the jump case and the continuous case, for a fixed design.

The jump case: Here, we assume that the regression function has a jump at θ:

$$h_\theta(x) = h(x) \text{ with } h: \ [0,1] \to \mathbb{R}.$$

In the special case of $h = 1$ we have $f_\theta(x) = \mathbb{1}_{[\theta,1]}(x)$. Further, we assume that h is continuous and that there exist a $\delta > 0$ and a constant $\tilde{C} = \tilde{C}(\delta) > 0$ such that for all $\theta \in [\theta_0 - \delta, \theta_0 + \delta] \cap [0,1]$

$$\int_{\theta_0 \wedge \theta}^{\theta_0 \vee \theta} h^2(x)\, \mathrm{d}x \geq \tilde{C}\,|\theta - \theta_0|, \tag{15.2}$$

where $a \wedge b = \min\{a,b\}$ and $a \vee b = \max\{a,b\}$. For example, (15.2) obviously holds true, if

$$\inf\left\{h^2(x): \ x \in [\theta_0 - \delta, \theta_0 + \delta] \cap [0,1]\right\} \geq \tilde{C} > 0.$$

The continuous case: Here we assume that the regression function has a smooth change point θ:

$$h_\theta(x) = (x - \theta)^q, \ q \geq 1.$$

The following conclusion can also be drawn for the more complex model $h_\theta(x) = h(x) \cdot (x-\theta)^q$, where $h: \ [0,1] \to \mathbb{R}$ is continuous. In the special case of $q = 1$ we have $f_\theta(x) = (x-\theta)^+$.

We consider the least squares error for any possible change-point. For $\theta \in [0,1]$ and $n \in \mathbb{N}$ we define

$$M_n(\theta) := -\frac{1}{n}\sum_{i=1}^{n}\left(Y_{i,n} - f_\theta\left(\frac{i}{n}\right)\right)^2.$$

For $n \in \mathbb{N}$ our estimator is defined as the maximizing point of M_n:

$$\hat{\theta}_n := \underset{\theta \in [0,1]}{\operatorname{argmax}}\, M_n(\theta).$$

Observe that

$$M_n(\theta) = \tilde{M}_n(\theta) - \frac{1}{n}\sum_{i=1}^{n}\epsilon_{i,n}^2, \text{ with}$$

$$\tilde{M}_n(\theta) := -\frac{1}{n}\sum_{i=1}^{n} 2\epsilon_{i,n} \cdot \left(f_{\theta_0}\left(\frac{i}{n}\right) - f_\theta\left(\frac{i}{n}\right)\right) + \left(f_{\theta_0}\left(\frac{i}{n}\right) - f_\theta\left(\frac{i}{n}\right)\right)^2. \tag{15.3}$$

It follows that M_n and $\tilde{M}_n$ have the same maximizers. The $\epsilon_{i,n}$'s are centered random variables, hence

$$\mathbf{E}\tilde{M}_n(\theta) = -\sum_{i=1}^{n} \frac{1}{n}\left(f_{\theta_0}\left(\frac{i}{n}\right) - f_\theta\left(\frac{i}{n}\right)\right)^2.$$

Note that for fixed $\theta \in [0,1]$ the expectation of $\tilde{M}_n(\theta)$ is a Riemann sum. If the function f_θ is integrable, then this sum converges for $n \to \infty$ to the corresponding integral $\tilde{M}(\theta)$, where the deterministic function $\tilde{M} : [0,1] \to \mathbb{R}$ is given by

$$\tilde{M}(\theta) := -\int_0^1 (f_{\theta_0}(x) - f_\theta(x))^2 \, dt.$$

If the function f is continuous, i.e. h is continuous with $h_\theta(\theta) = 0$, then this convergence holds uniformly.

Lemma 1. *If $f_{(\cdot)}(\cdot) : [0,1]^2 \to \mathbb{R}$ is continuous, then*

$$\lim_{n\to\infty} \sup_{\theta\in[0,1]} \left|\mathbf{E}\tilde{M}_n(\theta) - \tilde{M}(\theta)\right| = 0.$$

Proof. By the continuity of f on the compact set $[0,1]^2$ we have that f is uniformly continuous. Thus for every $\epsilon > 0$ a $k \in \mathbb{N}$ exists such that

$$|f_\theta(x) - f_{\tilde{\theta}}(x)| < \frac{1}{3}\epsilon \text{ for all } \theta, \tilde{\theta} \in \left[\frac{i-1}{k}, \frac{i}{k}\right] \text{ for } 1 \le i \le k.$$

Furthermore, for arbitrary $\theta_i \in \left[\frac{i-1}{k}, \frac{i}{k}\right]$ there exists an $n_0(\theta_i)$, such that for all $n \ge n_0(\theta_i)$ by the convergence of the Riemann sums

$$\left|\mathbf{E}\tilde{M}_n(\theta_i) - \tilde{M}(\theta_i)\right| < \frac{1}{3}\epsilon.$$

Hence, for all $n \ge n_0 = \max_{1\le i\le k} n_0(\theta_i)$ and $1 \le i \le k$

$$\sup_{\theta\in\left[\frac{i-1}{k}, \frac{i}{k}\right]} \left|\mathbf{E}\tilde{M}_n(\theta) - \tilde{M}(\theta)\right| \le \epsilon.$$

Since $\bigcup_{i=1}^k \left[\frac{i-1}{k}, \frac{i}{k}\right] = [0,1]$ the statement follows. □

15.2 The Jump Case

The jump case in a simple regression model has been already investigated by Kosorok (Chap. 14.5.1 in [Kos08]), under different conditions for a random design. Here, we consider the jump case for a fixed design. We assume that the regression function has a jump at an unknown point in time θ. The easiest case would be to set $h_\theta(x) = c$, where $c \neq 0$ is a constant. Here, we consider a slightly more general model, namely $h_\theta(x) = h(x)$ with $h: [0,1] \to \mathbb{R}$ continuous, which fulfills in addition (15.2) to ensure that we have a jump at θ_0. Then we have for $\theta \in [0,1]$ by (15.3) that

$$\tilde{M}_n(\theta) = \frac{1}{n} \sum_{i=[n(\theta\wedge\theta_0)]+1}^{[n(\theta\vee\theta_0)]} 2\epsilon_{i,n} \cdot h\left(\frac{i}{n}\right) \cdot \operatorname{sgn}(\theta_0 - \theta) - h^2\left(\frac{i}{n}\right)$$

$$\tilde{M}(\theta) = -\int_{\theta\wedge\theta_0}^{\theta\vee\theta_0} h^2(x)\, \mathrm{d}x,$$

where sgn denotes the signum function. We get a similar result as in Lemma 1 in the jump case. Since the function h does not depend on θ one can show the uniform convergence directly. Therefore, we state the following lemma without proof.

Lemma 2. *Let $h: [0,1] \to \mathbb{R}$ be continuous. Then*

$$\lim_{n\to\infty} \sup_{\theta\in[0,1]} \left| \mathbf{E}\tilde{M}_n(\theta) - \tilde{M}(\theta) \right| = 0.$$

Next, we prove that our estimators $\hat{\theta}_n$ are (strongly) consistent. By (15.2) we have that θ is a unique maximizer of the deterministic function $\tilde{M}$. Our estimator $\hat{\theta}_n$ is a maximizer of $\tilde{M}_n$. We show that the sequence $\tilde{M}_n$ converges uniformly for $n \to \infty$ to $\tilde{M}$. By a corresponding argmax theorem we can transfer this convergence to the maximizing points.

Lemma 3. *Let $h: [0,1] \to \mathbb{R}$ be continuous and let $\mathbf{E}|\epsilon_{1,1}|^p < \infty$ for some $p > 1$. Then*

$$\sup_{\theta\in[0,1]} \left| \tilde{M}_n(\theta) - \tilde{M}(\theta) \right| \xrightarrow{n\to\infty} 0 \begin{cases} P\text{-stochastically, if } 1 < p < \infty \\ P\text{-a.s., if } 2 < p < \infty. \end{cases}$$

Proof. First we prove an upper bound of $\sup_{\theta\in[0,1]} \left| \tilde{M}_n(\theta) - \mathbf{E}\tilde{M}_n(\theta) \right|$. Let $\rho > 0$ and C be a generic constant. Then,

$$P\left(\sup_{\theta\in[0,1]} \left| \tilde{M}_n(\theta) - \mathbf{E}\tilde{M}_n(\theta) \right| > \rho \right) = P\left(\sup_{\theta\in[0,1]} \left| \frac{1}{n} \sum_{i=[n(\theta\wedge\theta_0)]+1}^{[n(\theta\vee\theta_0)]} 2\epsilon_{i,n} \cdot h\left(\frac{i}{n}\right) \right| > \rho \right)$$

$$\leq P\left(\max_{[n\theta_0]+1\leq k\leq n} \left| \frac{1}{n} \sum_{i=[n\theta_0]+1}^{k} \epsilon_{i,n} \cdot h\left(\frac{i}{n}\right) \right|^p > C\rho^p \right)$$

$$+ P\left(\max_{0\leq k\leq [n\theta_0]} \left| \frac{1}{n} \sum_{i=k+1}^{[n\theta_0]} \epsilon_{i,n} \cdot h\left(\frac{i}{n}\right) \right|^p > C\rho^p \right) = P_1 + P_2.$$

For $1 \leq i \leq n$ the random variables $\epsilon_{i,n} \cdot h\left(\frac{i}{n}\right)$ are centered and independent. So the sums are martingales. By Doob's inequality it follows that

$$P_1 \leq Cn^{-p}\rho^{-p}\mathbf{E} \left| \sum_{i=[n\theta_0]+1}^{n} \epsilon_{i,n} \cdot h\left(\frac{i}{n}\right) \right|^p.$$

By Lemma A.2 in [Fer01] we can change the sum with the expectation. This is similar to the inequality of Burkholder, see for example Theorem 1 on page 396 in [CT88].

$$P_1 \leq Cn^{-p}\rho^{-p} \sum_{i=[n\theta_0]+1}^{n} \mathbf{E}\left|\epsilon_{i,n} \cdot h\left(\frac{i}{n}\right)\right|^p \begin{cases} 2 & 1 \leq p \leq 2 \\ (n-[n\theta_0])^{\frac{p}{2}-1} & 2 < p < \infty \end{cases} \tag{15.4}$$

By the boundedness of the function h on the compact interval $[0,1]$ and the identical distribution of the $\epsilon_{i,n}$'s we get

$$P_1 \leq C \cdot \mathbf{E}\left|\epsilon_{1,1}\right|^p \rho^{-p} \begin{cases} n^{-(p-1)} & 1 \leq p \leq 2 \\ n^{-\frac{p}{2}} & 2 < p < \infty, \end{cases}$$

The same conclusion can be drawn for P_2. Hence we have

$$\lim_{n\to\infty} P\left(\sup_{\theta\in[0,1]}\left|\tilde{M}_n(\theta) - \mathbf{E}\tilde{M}_n(\theta)\right| > \rho\right) = 0 \text{ if } 1 < p < \infty,$$

$$\sum_{n=1}^{\infty} P\left(\sup_{\theta\in[0,1]}\left|\tilde{M}_n(\theta) - \mathbf{E}\tilde{M}_n(\theta)\right| > \rho\right) < \infty \text{ if } 2 < p < \infty.$$

By Lemma 2 it follows that $\mathbf{E}\tilde{M}_n$ converges uniformly to $\tilde{M}$. Finally the assertion is proved by applying the Borel–Cantelli Lemma. □

In the next step, an argmax theorem is used to carry over the convergence to the maximizer.

Theorem 1. *Let $h_\theta(x) = h(x)$ with $h: [0,1] \to \mathbb{R}$ continuous and let h satisfy* (15.2). *Further, let $\mathbf{E}\left|\epsilon_{1,1}\right|^p < \infty$ for some $p > 1$. Then*

$$\hat{\theta}_n \xrightarrow{n\to\infty} \theta_0 \begin{cases} P\text{-stochastically, if } 1 < p < \infty \\ P\text{-a.s., if } 2 < p < \infty. \end{cases}$$

Proof. In the case of $2 < p < \infty$ it follows by Lemma 3 that

$$\sup_{\theta\in[0,1]}\left|\tilde{M}_n(\theta) - \tilde{M}(\theta)\right| \xrightarrow{n\to\infty} 0 \text{ P-a.s.}$$

This convergence may be transferred to the maximizing points by a corresponding argmax theorem, which can be found in the appendix. It remains to prove that the assumptions (1) – (4) of Theorem 7 hold. By (15.2) θ_0 is the unique maximizing point of $\tilde{M}$. Hence the assumption (1) and (2) are satisfied. The definition of our estimator yields (3) directly and by Lemma 3 we get (4). By a subsequences argument the required stochastic convergence follows. □

We have that the estimator $\hat{\theta}_n$ is consistent at a rate of n.

Theorem 2. *Let $h: [0,1] \to \mathbb{R}$ be continuous satisfying* (15.2). *Further, let $\mathbf{E}\left|\epsilon_{1,1}\right|^p < \infty$ for some $p > 1$. Then*

$$n\left(\hat{\theta}_n - \theta_0\right) = O_P(1).$$

Proof. For $\delta > 0$ there exists a constant $\tilde{C} > 0$ from (15.2) such that for all $\theta \in [\theta_0 - \delta, \theta_0 + \delta] \cap [0,1]$

$$-\tilde{M}(\theta) \geq \tilde{C} |\theta - \theta_0| .$$

It follows by Lemma 2, that for eventually all $n \in \mathbb{N}$

$$-\mathbf{E}\tilde{M}_n(\theta) \geq \frac{\tilde{C}}{2} \frac{|[n\theta] - [n\theta_0]|}{n}. \tag{15.5}$$

Let $x > 0$ and C be a positive generic constant. By the definition of $\hat{\theta}_n$ it follows that

$$\begin{aligned} P\left(x < \left|n\hat{\theta}_n - [n\theta_0]\right| < n\delta\right) &\leq P\left(\sup_{x<|n\theta-[n\theta_0]|<n\delta} \tilde{M}_n(\theta) \geq \tilde{M}_n(\theta_0)\right) \\ &\leq P\left(\max_{x<|k-[n\theta_0]|<n\delta} \tilde{M}_n\left(\frac{k}{n}\right) - \mathbf{E}\tilde{M}_n\left(\frac{k}{n}\right) + \mathbf{E}\tilde{M}_n\left(\frac{k}{n}\right) \geq 0\right). \end{aligned}$$

By (15.5) we have

$$\begin{aligned} &\leq P\left(\max_{x<|k-[n\theta_0]|<n\delta} \tilde{M}_n\left(\frac{k}{n}\right) - \mathbf{E}\tilde{M}_n\left(\frac{k}{n}\right) - \frac{\tilde{C}}{2}\frac{|k-[n\theta_0]|}{n} \geq 0\right) \\ &\leq P\left(\max_{x<|k-[n\theta_0]|<n\delta} \frac{1}{|k-[n\theta_0]|^p} \left|\sum_{i=([n\theta_0]\wedge k)+1}^{[n\theta_0]\vee k} \epsilon_{i,n} \cdot h\left(\frac{i}{n}\right)\right|^p \geq C\right) \\ &\leq P\left(\max_{x<k-[n\theta_0]<n\delta} \frac{1}{(k-[n\theta_0])^p} \left|\sum_{i=[n\theta_0]+1}^{k} \epsilon_{i,n} \cdot h\left(\frac{i}{n}\right)\right|^p \geq C\right) \\ &\quad + P\left(\max_{x<[n\theta_0]-k<n\delta} \frac{1}{([n\theta_0]-k)^p} \left|\sum_{i=k+1}^{[n\theta_0]} \epsilon_{i,n} \cdot h\left(\frac{i}{n}\right)\right|^p \geq C\right) := P_1 + P_2. \end{aligned}$$

For $1 \leq i \leq n$ the random variables $\epsilon_{i,n} \cdot h\left(\frac{i}{n}\right)$ are centered and independent. So the sums are martingales. The term $\frac{1}{(k-[n\theta_0])^p}$ is decreasing in k, hence by the inequality of Chow it follows that

$$\begin{aligned} P_1 &\leq C \sum_{k=[x]+[n\theta_0]+1}^{[n\theta_0]+[n\delta]-1} \left(\frac{1}{(k-[n\theta_0])^p} - \frac{1}{(k+1-[n\theta_0])^p}\right) \mathbf{E}\left|\sum_{i=[n\theta_0]+1}^{k} \epsilon_{i,n} \cdot h\left(\frac{i}{n}\right)\right|^p \\ &\quad + C\frac{1}{[n\delta]^p}\mathbf{E}\left|\sum_{i=[n\theta_0]+1}^{[n\theta_0]+[n\delta]} \epsilon_{i,n} \cdot h\left(\frac{i}{n}\right)\right|^p =: P_{1,1} + P_{1,2}. \end{aligned}$$

As in the proof of Lemma 3, (15.4) gives

$$P_{1,2} = C\frac{1}{[n\delta]^p}\mathbf{E}\left|\sum_{i=[n\theta_0]+1}^{[n\theta_0]+[n\delta]} \epsilon_{i,n} \cdot h\left(\frac{i}{n}\right)\right|^p \leq C \cdot \mathbf{E}|\epsilon_{1,1}|^p \begin{cases} n^{-(p-1)} & 1 < p \leq 2 \\ n^{-\frac{1}{2}p} & 2 < p < \infty. \end{cases}$$

By Lemma A.2 in [Fer01] we claim that

$$P_{1,1} \le C \sum_{k=[x]+[n\theta_0]+1}^{[n\theta_0]+[n\delta]-1} \frac{1}{(k-[n\theta_0])^{p+1}} \mathbf{E} \left| \sum_{i=[n\theta_0]+1}^{k} \epsilon_{i,n} \cdot h\left(\frac{i}{n}\right) \right|^p$$

$$\le C \sum_{k=[x]+[n\theta_0]+1}^{[n\theta_0]+[n\delta]-1} \frac{1}{(k-[n\theta_0])^{p+1}} \sum_{i=[n\theta_0]+1}^{k} \mathbf{E} \left| \epsilon_{i,n} \cdot h\left(\frac{i}{n}\right) \right|^p \begin{cases} 2 & 1<p\le 2 \\ (k-[n\theta_0])^{\frac{p}{2}-1} & 2<p<\infty \end{cases}$$

By the boundedness of the function h on the compact interval $[0,1]$ and the identical distribution of the $\epsilon_{i,n}$'s we get

$$\le C \cdot \mathbf{E} \left|\epsilon_{1,1}\right|^p \sum_{k=[x]+[n\theta_0]+1}^{[n\theta_0]+[n\delta]-1} \begin{Bmatrix} (k-[n\theta_0])^{-p} \\ (k-[n\theta_0])^{-\frac{1}{2}p-1} \end{Bmatrix} \le C \cdot \mathbf{E} \left|\epsilon_{1,1}\right|^p \begin{cases} x^{-(p-1)} & 1<p\le 2 \\ x^{-\frac{1}{2}p} & 2<p<\infty. \end{cases}$$

A corresponding upper bound for P_2 can be derived in the same manner. So we have for $x>0$, $\delta>0$ and eventually all $n \in \mathbb{N}$ that

$$P\left(x < \left|n\hat{\theta}_n - [n\theta_0]\right| < n\delta\right) \le C \cdot \mathbf{E} \left|\epsilon_{1,1}\right|^p \begin{cases} n^{-(p-1)} + x^{-(p-1)} & 1<p\le 2 \\ n^{-\frac{1}{2}p} + x^{-\frac{1}{2}p} & 2<p<\infty. \end{cases}$$

By Theorem 1 it follows that $\hat{\theta}_n \xrightarrow{n\to\infty} \theta_0$ P-stochastically for $1<p<\infty$. Therefore, we have

$$\lim_{x\to\infty} \limsup_{n\to\infty} P\left(x < n\left|\hat{\theta}_n - \theta_0\right|\right)$$

$$\le \lim_{x\to\infty} \limsup_{n\to\infty} \left(P\left(\frac{1}{2}x < \left|n\theta_0 - [n\theta_0]\right|\right) + C \cdot \mathbf{E} \left|\epsilon_{1,1}\right|^p \begin{Bmatrix} n^{-(p-1)} + x^{-(p-1)} \\ n^{-\frac{1}{2}p} + x^{-\frac{1}{2}p} \end{Bmatrix} \right.$$

$$\left. + P\left(\frac{1}{4}n\delta \le \left|n\hat{\theta}_n - n\theta_0\right|\right) + P\left(\frac{1}{4}n\delta \le \left|n\theta_0 - [n\theta_0]\right|\right)\right) = 0. \qquad \square$$

We next show that $n\hat{\theta}_n - [n\theta_0]$ converges in distribution to a maximizing point of a random walk with drift. For that purpose, let $(\epsilon_k)_{k\in\mathbb{Z}}$ be i.i.d. random variables with the same distribution as $\epsilon_{1,1}$. Further, we define a sequence of stochastic processes $(Z_n)_{n\in\mathbb{N}}$ with $Z_n = \{Z_n(k) : k \in \mathbb{Z}\}$ and a random walk $Z = \{Z(k) : k \in \mathbb{Z}\}$ by

$$Z_n(k) := \begin{cases} n \cdot \tilde{M}_n\left(\theta_0 + \frac{k}{n}\right) & -n\theta_0 \le k \le n(1-\theta_0) \\ -7 & \mathit{otherwise}. \end{cases}$$

$$Z(k) := -h^2(\theta_0) \cdot |k| - 2h(\theta_0) \cdot \operatorname{sgn}(k) \cdot \sum_{j=1}^{|k|} \epsilon_{\operatorname{sgn}(k)\cdot j}$$

Observe that $[n\hat{\theta}_n] - [n\theta_0]$ is a maximizer of the process Z_n for $n \in \mathbb{N}$. If the sequence Z_n converges for $n \to \infty$ to Z in some sense, then we can transfer this convergence to the maximizing points by a continuous mapping theorem for the argmax functional of Ferger (Theorem 2 in [Fer04]).

Lemma 4. *If $h: [0,1] \to \mathbb{R}$ is continuous, then*

$$\{Z_n(k): -d \le k \le d\} \xrightarrow[n\to\infty]{L} \{Z(k): -d \le k \le d\} \quad \forall d \in \mathbb{N}.$$

Proof. Let $d \in \mathbb{N}$ and $n_0 = \frac{d}{\min\{\theta_0, 1-\theta_0\}}$, then we have for all $n > n_0$ that $-n\theta_0 \le -d \le k \le d \le n(1-\theta_0)$. It follows that

$$\begin{aligned}
Z_n(k) &= \sum_{i=([n\theta_0]\wedge[n\theta_0]+k)+1}^{[n\theta_0]\vee[n\theta_0]+k} 2\epsilon_{i,n}\cdot \operatorname{sgn}(-k)\cdot h\left(\frac{i}{n}\right) - h^2\left(\frac{i}{n}\right)\\
&= \sum_{j=1}^{|k|} 2\epsilon_{[n\theta_0]+\operatorname{sgn}(k)\cdot j,n}\cdot sgn(-k)\cdot h\left(\frac{[n\theta_0]+\operatorname{sgn}(k)\cdot j}{n}\right)\\
&\quad - h^2\left(\frac{[n\theta_0]+\operatorname{sgn}(k)\cdot j}{n}\right)\\
&\overset{L}{=} \sum_{j=1}^{|k|} 2\epsilon_{\operatorname{sgn}(k)\cdot j}\cdot sgn(-k)\cdot h\left(\frac{[n\theta_0]+\operatorname{sgn}(k)\cdot j}{n}\right) - h^2\left(\frac{[n\theta_0]+\operatorname{sgn}(k)\cdot j}{n}\right)\\
&\xrightarrow{n\to\infty} \sum_{j=1}^{|k|} 2\epsilon_{\operatorname{sgn}(k)\cdot j}\operatorname{sgn}(-k)\cdot h(\theta_0) - h^2(\theta_0) = Z(k),
\end{aligned}$$

where the convergence follows by the continuity of the function h. □

Theorem 3. *Let $h: [0,1] \to \mathbb{R}$ be continuous satisfying (15.2). Further let $\mathbf{E}|\epsilon_{1,1}|^p < \infty$ for some $p > 1$, then:*

(*i*) *The trajectories of Z possess a smallest and a largest maximizer τ_s and τ_l.*

(*ii*) $\limsup_{n\to\infty} P\left(n\hat{\theta}_n - [n\theta_0] \le z\right) \le P(\tau_s \le z)$

(*iii*) $\liminf_{n\to\infty} P\left(n\hat{\theta}_n - [n\theta_0] \le z\right) \ge P(\tau_l \le z)$.

Proof. We apply a corresponding argmax theorem of Ferger, which can be found in the appendix, Theorem 8. We have $\mathbf{E}Z(k) = -h^2(\theta_0)\cdot|k| \xrightarrow{|k|\to\infty} \infty$ and by the strong law of large numbers it follows that $Z(k) \xrightarrow[P-a.s.]{|k|\to\infty} -\infty$, hence (i) holds true. Observe that $[n\hat{\theta}_n] - [n\theta_0]$ is a maximizer of the process Z_n for $n \in \mathbb{N}$. Together with Lemma 4, statement (i) and Theorem 2 the assumptions (1)–(4) of Theorem 8 follow, hence (ii) and (iii) hold true by $P\left(n\hat{\theta}_n - [n\theta_0] \le z\right) = P\left([n\hat{\theta}_n] - [n\theta_0] \le z\right)$. □

If the trajectories of Z possess a unique maximizer almost surely, then we have convergence in distribution.

Theorem 4. *Let $h: [0,1] \to \mathbb{R}$ be continuous satisfying (15.2) and let $\mathbf{E}|\epsilon_{1,1}|^p < \infty$ for some $p > 1$. Further, let $\epsilon_{1,1}$ have an absolutely continuous distribution, then:*

(*i*) *The trajectories of Z possess a unique maximizer τ almost surely.*

(*ii*) $n\hat{\theta}_n - [n\theta_0] \xrightarrow[n\to\infty]{L} \tau.$

Proof. As $\epsilon_{1,1}$ has an absolutely continuous distribution we conclude that for $k \neq l \in \mathbb{Z}$ $Z(k) - Z(l)$ has also an absolutely continuous distribution. Hence,

$$\begin{aligned}
&P(Z \text{ has a unique maximizer}) \\
&= 1 - P\big(\{Z \text{ has no maximizer}\} \cup \{Z \text{ has at least two maximizer } \tau \text{ and } \sigma\}\big) \\
&\geq 1 - P\big(Z(\tau) = Z(\sigma)\big) \\
&= 1 - P\left(\bigcup_{k \neq l \in \mathbb{Z}} \{\tau = k, \sigma = l, Z(k) = Z(l)\}\right) \\
&\geq 1 - \sum_{k \neq l \in \mathbb{Z}} P\big(Z(k) - Z(l) = 0\big) \\
&= 1.
\end{aligned}$$

So we get (i). Now by Theorem 3 it follows that for $z \in \mathbb{Z}$ we have

$$\limsup_{n\to\infty} P\left(n\hat{\theta}_n - [n\theta_0] \leq z\right) \leq P(\tau \leq z) \leq \liminf_{n\to\infty} P\left(n\hat{\theta}_n - [n\theta_0] \leq z\right),$$

which yields statement (ii). □

15.3 The Continuous Case

In this case, we assume that the regression function f_θ has a smooth change point in the sense that there is no jump but a continuous change at θ:

$$f_\theta(x) = h_\theta(x) \cdot 1\!\!1_{[\theta,1]}(x),$$

where

$$h_\theta(x) = (x - \theta)^q, \ q \geq 1.$$

First we prove the consistence of the least squares estimators $\hat{\theta}_n$ described in Chap. 15.1. By Lemma 1 we know that $\mathbf{E}\tilde{M}_n$ converges for $n \to \infty$ uniformly to $\tilde{M}$. We use Andrews' [And87] uniform law of large numbers to show that $\tilde{M}_n$ converge for $n \to \infty$ uniformly to $\tilde{M}$.

Lemma 5. *Let* $\mathbf{E}\,|\epsilon_{1,1}|^p < \infty$ *for some* $p > 1$. *Then*

$$\sup_{\theta\in[0,1]} \left|\tilde{M}_n(\theta) - \tilde{M}(\theta)\right| \xrightarrow{n\to\infty} 0 \begin{cases} P\text{-stochastically, if } 1 < p < \infty \\ P\text{-a.s., if } 2 < p < \infty. \end{cases}$$

Proof. Let ϵ_i for $i \in \mathbb{N}$ be i.i.d. random variables with distribution of $\epsilon_{1,1}$. We define for $\theta \in [0,1]$, $\delta > 0$, $n \in \mathbb{N}$ and $1 \le i \le n$ the following functions

$$q_i(\epsilon_i, \theta) = -2\epsilon_i \cdot \left(f_\theta\left(\frac{i}{n}\right) - f_{\theta_0}\left(\frac{i}{n}\right) \right)$$
$$q_i^*(\epsilon_i, \theta, \delta) = \sup\left\{ q_i\left(\epsilon_i, \tilde{\theta}\right) : \tilde{\theta} \in [\theta-\delta, \theta+\delta] \cap [0,1] \right\}$$
$$q_{*,i}(\epsilon_i, \theta, \delta) = \inf\left\{ q_i\left(\epsilon_i, \tilde{\theta}\right) : \tilde{\theta} \in [\theta-\delta, \theta+\delta] \cap [0,1] \right\}.$$

We apply Theorem 9 to $q_i(\epsilon_i, \theta)$. Since $[0,1]$ is compact and $q_i(\epsilon_i, \theta)$, $q_i^*(\epsilon_i, \theta, \delta)$ and $q_{*,i}(\epsilon_i, \theta, \delta)$ are random variables, the assumptions A1 and A2(a) of Theorem 9 hold. $f_\theta(x) = (x-\theta)^q \cdot \mathbb{1}_{[\theta,1]}(x)$ is a decreasing function in θ, which gives

$$q_i^*(\epsilon_i, \theta, \delta) = 2\epsilon_i \cdot \left(f_{\theta_0}\left(\frac{i}{n}\right) - \mathbb{1}_{\{\epsilon_i > 0\}} \cdot f_{\theta+\delta}\left(\frac{i}{n}\right) - \mathbb{1}_{\{\epsilon_i < 0\}} \cdot f_{\theta-\delta}\left(\frac{i}{n}\right) \right)$$
$$q_{*,i}(\epsilon_i, \theta, \delta) = 2\epsilon_i \cdot \left(f_{\theta_0}\left(\frac{i}{n}\right) - \mathbb{1}_{\{\epsilon_i > 0\}} \cdot f_{\theta-\delta}\left(\frac{i}{n}\right) - \mathbb{1}_{\{\epsilon_i < 0\}} \cdot f_{\theta+\delta}\left(\frac{i}{n}\right) \right),$$

where $\theta + \delta$ has to be understood as $\min\{\theta+\delta, 1\}$ and $\theta - \delta$ as $\max\{\theta-\delta, 0\}$. The brackets on the right hand side are bounded. If $\mathbf{E}|\epsilon_{1,1}|^p < \infty$ for some $1 < p \le 2$, then we have by Theorem 1 on page 124 of [CT88] that $q_i^*(\epsilon_i, \theta, \delta)$ and $q_{*,i}(\epsilon_i, \theta, \delta)$ satisfy pointwise a weak law of large numbers. If $\mathbf{E}|\epsilon_{1,1}|^p < \infty$ for some $p \ge 2$, then $q_i^*(\epsilon_i, \theta, \delta)$ and $q_{*,i}(\epsilon_i, \theta, \delta)$ satisfy a strong law of large numbers. Hence, Assumption A2(b) of Theorem 9 holds true. Further, we have for all $\theta \in [0,1]$

$$\lim_{\delta \to 0} \sup_{n \in \mathbb{N}} \frac{1}{n} \sum_{i=1}^n \mathbf{E} \sup_{\theta-\delta \le \tilde{\theta} \le \theta+\delta} \left| q_i\left(\epsilon_i, \tilde{\theta}\right) - q_i(\epsilon_i, \theta) \right| = 0,$$

since

$$\sup_{\theta-\delta \le \tilde{\theta} \le \theta+\delta} \left| q_i\left(\epsilon_i, \tilde{\theta}\right) - q_i(\epsilon_i, \theta) \right| = \sup_{\theta-\delta \le \tilde{\theta} \le \theta+\delta} 2|\epsilon_i| \cdot \left| f_\theta\left(\frac{i}{n}\right) - f_{\tilde{\theta}}\left(\frac{i}{n}\right) \right|$$
$$\le 2|\epsilon_i| \cdot \left(f_{\theta-\delta}\left(\frac{i}{n}\right) - f_{\theta+\delta}\left(\frac{i}{n}\right) \right) \le 2|\epsilon_{1,1}| \cdot 2q\delta.$$

The last inequality follows for $q \ge 1$ from $\left|\frac{\partial}{\partial\theta} f_\theta(x)\right| \le q$. Hence, all assumptions of Theorem 9 hold true and a uniform law of large numbers for the sequences $q_i(\epsilon_{i,n}, \theta)$ can be applied. By (15.3) we have

$$\tilde{M}_n(\theta) - \mathbf{E}\tilde{M}_n(\theta) = \frac{1}{n} \sum_{i=1}^n q_i(\epsilon_{i,n}, \theta) \stackrel{\mathcal{L}}{=} \frac{1}{n} \sum_{i=1}^n q_i(\epsilon_i, \theta),$$

hence by Lemma 1 the statement follows. □

Theorem 5. *Let* $\mathbf{E}|\epsilon_{1,1}|^p < \infty$ *for some* $p > 1$. *Then*

$$\hat{\theta}_n \xrightarrow{n \to \infty} \theta_0 \begin{cases} P\text{-stochastically, if } 1 < p < \infty \\ P\text{-a.s., if } 2 < p < \infty. \end{cases}$$

Proof. In the case of $2 < p < \infty$, Lemma 5 yields

$$\sup_{\theta \in [0,1]} \left| \tilde{M}_n(\theta) - \tilde{M}(\theta) \right| \xrightarrow{n \to \infty} 0 \text{ P-a.s.}$$

This can be shown by the same method as in Theorem 1. Here, a simple calculation proves that θ_0 is the unique maximizing point of $\tilde{M}$. □

We are going to show that the least squares estimators $\hat{\theta}_n$ are consistent at a rate of $n^{\frac{1}{2}}$. We adopt an approach which can be found in [VdV98]. We define the measurable function

$$m_\theta(x,y) := -\big(y - f_\theta(x)\big)^2$$

which fulfills a Lipschitz condition for every θ_1, θ_2 in $[0,1]$:

$$\begin{aligned}|m_{\theta_1}(x,y) - m_{\theta_2}(x,y)| &= |f_{\theta_1}(x) - f_{\theta_2}(x)| \cdot |2y - f_{\theta_1}(x) - f_{\theta_2}(x)| \\ &\leq 2q\big(1 + |y|\big) \cdot |\theta_1 - \theta_2| =: \dot{m}(x,y) \cdot |\theta_1 - \theta_2|,\end{aligned}$$

since $\left|\frac{\partial}{\partial\theta} f_\theta(x)\right| \leq q$ for $q \geq 1$. Furthermore,

$$\mathbf{E} m_\theta(x, Y_{1,1}) = -\mathbf{E}\epsilon_{1,1}^2 - \big(f_\theta(x) - f_{\theta_0}(x)\big)^2$$

admits a second-order Taylor expansion at θ_0:

$$\begin{aligned}\mathbf{E} m_\theta(x, Y_{1,1}) &= \mathbf{E} m_{\theta_0}(x, Y_{1,1}) - (\theta - \theta_0)^2 \cdot \frac{\partial}{\partial\theta} f_{\theta_0}(x) + o\left(|\theta - \theta_0|^2\right) \\ &= \mathbf{E}\epsilon_{1,1}^2 - q^2 (x - \theta_0)^{2(q-1)} \cdot \mathbb{1}_{[\theta_0,1]}(x) \cdot (\theta - \theta_0)^2 + o\left(|\theta - \theta_0|^2\right).\end{aligned}$$

Since $\hat{\theta}_n$ is the least squares estimator

$$\hat{\theta}_n := \operatorname*{argmax}_{\theta \in [0,1]} \frac{1}{n} \sum_{i=1}^n m_\theta\left(\frac{i}{n}, Y_{i,n}\right)$$

we conclude that

$$\frac{1}{n} \sum_{i=1}^n m_{\hat{\theta}_n}\left(\frac{i}{n}, Y_{i,n}\right) \geq \frac{1}{n} \sum_{i=1}^n m_{\theta_0}\left(\frac{i}{n}, Y_{i,n}\right) - O_P\left(n^{-1}\right).$$

Therefore, all condition of Corollary 5.53 in [VdV98] are satisfied, which gives us the following result.

Theorem 6. *Let* $\mathbf{E}\epsilon_{1,1}^2 < \infty$, *then*

$$\sqrt{n}\left(\hat{\theta}_n - \theta_0\right) = O_p(1) \quad \textit{for all } q \geq 1.$$

References

[ABG08] Aalen, O.O., Borgan, O., Gjessing, S.: Survival and Event History: A Process Point of View. Springer, New York (2008)

[And87] Andrews, D.W.K.: Consistency in nonlinear econometric models: A generic uniform law of large numbers. Econometrica, **55**, 1465–1471 (1987)

[Bai97] Bai, J.: Estimation of a change point in multiple regression models. Review of Economics and Statistics, **79**, 551–560 (1997)

[CT88] Chow, Y.S., Teicher, H.: Probability Theorey. Springer, New York (1988)

[DS02] Dempfle, A., Stute, W.: Nonparametric estimation of a discontinuity in regression. Statistica Neerlandica, **56**, 233–242 (2002)

[Fer01] Ferger, D.: Exponential and polynomial tailbounds for change-point estimators. Journal of Statistical Planning and Inference, **92**, 73–109 (2001)

[Fer04] Ferger, D.: A continuous mapping theorem for the argmax-functional in the non-unique case. Statistica Neerlandica, **58**, 83–96 (2004)

[Fer09] Ferger, D.: Stochastische Prozesse mit Strukturbrüchen. Manuskript [in German], Dresdner Schriften zur Mathematischen Stochastik (2009)

[JL07] Jensen, U., Lütkebohmert C.: Change-Point Models. Review in Ruggeri, F., Kenett, R., Faltin, W. (ed) Encyclopedia of Statistics in Quality and Reliability, Vol. 1, 306–312. Wiley, New York (2007)

[JL08] Jensen, U., Lütkebohmert C.: A Cox-type regression model with change-points in the covariates. Lifetime Data Analysis, **14**, 267–285 (2008)

[Kos08] Kosorok, M.R.: Introduction to Empirical Processes and Semiparametric Inference. Springer, New York (2008)

[KS07] Kosorok, M., Song, R.: Inference under right censoring for transformation models with a change-point based on a covariate threshold. The Annals of Statistics, **35**, 957–989 (2007)

[KQS03] Koul, H.L., Qian, L., Surgailis, D.: Asymptotics of M-estimators in two-phase linear regression models. Stochastic Process and their Applications, **103**, 123–154 (2003)

[MS06] Martinussen, T., Scheike, T.H.: Dynamic Regression Models for Survival Data. Springer, New York (2006)

[LBM09] Lan, Y., Banerjee, M., Michailidis: Change-point estimation under adaptive sampling. The Annals of Statistics, **37**, 1752–1791 (2009)

[Mue92] Müller, H.G.: Change-point in nonparametric regression analysis. The Annals of Statistics, **20**, 737–761 (1992)

[MSo97] Müller, H.G., Song, K.S.: Two-stage change-point estimators in smooth regression models. Statistics and Probability Letters, **34**, 323–335 (1997)

[MSt99] Müller, H.G., Stadtmüller, U.: Discontinuous versus smooth regression. The Annals of Statistics, **27**, 299–337 (1999)

[Pon03] Pons, O.: Estimation in a Cox regression model with a change-point according to a threshold in a covariate. The Annals of Statistics, **31**, 442–463 (2003)

[VdV98] Van der Vaart, A.W.: Asymptotic Statistics. Cambridge Series in Statistical and Probabilistic Mathematics (1998)

Appendix

Theorem 7. *Let X and X_n for $n \in \mathbb{N}$ be stochastic processes defined on a common probability space $(\Omega, \mathfrak{A}, P)$ with trajectories in $D[0,1]$. Let $\tau_n \in [0,1]$ for $n \in \mathbb{N}$ be a sequence and $\tau \in [0,1]$ satisfying the following properties:*

(1) $X(\tau) \vee X(\tau-) = sup_{t\in[0,1]}X(t)$ *a.s.*

(2) $sup_{t\in[0,1]}X(t) > sup\{X(t) : |t-\tau| > \epsilon\}$ *a.s. for all* $\epsilon > 0$.

(3) $X_n(\tau_n) \vee X(\tau_n-) = sup_{t\in[0,1]}X_n(t)$ *a.s. with* τ_n *measurable for all* $n \in \mathbb{N}$.

(4) $sup_{t\in[0,1]}|X_n(t) - X(t)| \xrightarrow{n\to\infty} 0$ *a.s. for* $n \to \infty$.

Then, it follows that $\tau_n \xrightarrow{n\to\infty} \tau$ *a.s.*

Proof. See [Fer09] (p 25, Theorem 4.6.) □

Theorem 8. *Let* X *and* X_n *for* $n \in \mathbb{N}$ *be stochastic processes defined on a common probability space* $(\Omega, \mathfrak{A}, P)$ *with index set* $\mathbb{Z}$. *Let* $\tau_n \in [0,1]$ *for* $n \in \mathbb{N}$ *be a sequence satisfying the following properties:*

(1) $X_n(\tau_n) \geq sup_{k\in\mathbb{Z}}X_n(k) - \alpha_n, \quad \alpha_n = o_P(1)$.

(2) $\{X_n(k) : \; -d \leq k \leq d\} \xrightarrow[n\to\infty]{L} \{X(k) : \; -d \leq k \leq d\} \quad \forall d \in \mathbb{N}$.

(3) *The trajectories of* X *posses a smallest and a largest maximizer* τ_s *and* τ_l, *respectively, which are Borel measurable.*

(4) $\tau_n = O_P(1)$.

Then for every $z \in \mathbb{Z}$ *it follows that*

$$\limsup_{n\to\infty} P(\tau_n \leq z) \leq P(\tau_s \leq z) \qquad \text{and} \qquad \liminf_{n\to\infty} P(\tau_n \leq z) \geq P(\tau_l \leq z).$$

Proof. This follows from Theorem 2 in [Fer04] (p 85) □

To state Andrews [And87] uniform law of large numbers we introduce some notation. Let $(W_i)_{i\in\mathbb{N}}$ be a sequence of W-valued random variables defined on a common probability space. For $i \in \mathbb{N}$ and some metric space (Θ, d) we define the functions $q_i : W \times \Theta \to \mathbb{R}$. For $\delta > 0$ define

$$q_i^*(W_i, \theta, \delta) := \sup\left\{q_i\left(W_i, \tilde{\theta}\right) : \; d\left(\theta, \tilde{\theta}\right) < \delta\right\}$$
$$q_{*,i}(W_i, \theta, \delta) := \inf\left\{q_i\left(W_i, \tilde{\theta}\right) : \; d\left(\theta, \tilde{\theta}\right) < \delta\right\}$$

The theorem is based on the following assumptions:

Assumption A1: (Θ, d) is a compact metric space.

Assumption A2: (a) $q_i(W_i, \theta)$, $q_i^*(W_i, \theta, \delta)$ and $q_{*,i}(W_i, \theta, \delta)$ are random variables for all $\theta \in \Theta$, $i \in \mathbb{N}$ and all δ sufficiently small, where δ may depend on θ.

(b) $\{q_i^*(W_i, \theta, \delta)\}$ and $\{q_{*,i}(W_i, \theta, \delta)\}$ satisfy pointwise strong (weak) laws of large numbers, i.e.

$$\frac{1}{n}\sum_{i=1}^{n}\{q_{*,i}(W_i, \theta, \delta)\} - \mathbf{E}\{q_{*,i}(W_i, \theta, \delta)\} \xrightarrow{n\to\infty} 0 \text{ P-a.s. (in Probability)}$$
$$\frac{1}{n}\sum_{i=1}^{n}\{q_i^*(W_i, \theta, \delta)\} - \mathbf{E}\{q_i^*(W_i, \theta, \delta)\} \xrightarrow{n\to\infty} 0 \text{ P-a.s. (in Probability)},$$

for all $\theta \in \Theta$, and all δ sufficiently small, where δ may depend on θ.

Assumption A3: For all $\theta \in \Theta$,

$$\lim_{\delta\to 0}\sup_{n\in\mathbb{N}}\left|\frac{1}{n}\sum_{i=1}^{n}\mathbf{E}q_{*,i}\left(W_i,\theta,\delta\right)-\mathbf{E}q_i\left(W_i,\theta\right)\right|=0$$

$$\lim_{\delta\to 0}\sup_{n\in\mathbb{N}}\left|\frac{1}{n}\sum_{i=1}^{n}\mathbf{E}q_i^*\left(W_i,\theta,\delta\right)-\mathbf{E}q_i\left(W_i,\theta\right)\right|=0.$$

Theorem 9. *If Assumptions A1-A3 hold, then*

$$\sup_{\theta\in\Theta}\left|\frac{1}{n}\sum_{i=1}^{n}q_i\left(W_i,\theta\right)-\mathbf{E}q_i\left(W_i,\theta\right)\right|\xrightarrow{n\to\infty}0\ P\text{-}a.s.\ (in\ Probability).$$

Proof. See [And87]. □

16

A Model for Field Failure Prediction Using Dynamic Environmental Data

Yili Hong[1] and William Q. Meeker[2,*]

[1] Department of Statistics, Virginia Tech, Blacksburg, VA, USA, yilihong@vt.edu
[2] Department of Statistics, Iowa State University, Ames, IA, USA, wqmeeker@iastate.edu

Abstract: Due to the dynamics of the environment and the variability in product usage, product units in the field are usually exposed to varying failure-causing stresses. Some products are equipped with sensors and smart chips that measure and record usage/environmental information over the life of the product. For some products, it is possible to track environmental variables dynamically, even in real time, providing useful information for field-failure prediction. In many applications, predictions are needed for individual units, giving the remaining life of individuals, and for the population, giving the cumulative number of failures at a future time. It is always desirable to obtain more accurate predictions for both the population and the individuals. This paper outlines a model and methods that can be used for field-failure prediction using dynamic environmental data. Multivariate time series models are also used to describe the dynamic covariate information. The cumulative exposure model is used to link the explanatory variables which are recorded as a multivariate time series, and the failure-time model.

Keywords and phrases: Cumulative exposure model, Covariate process, Failure time data, Multivariate time series, Reliability, Usage history

16.1 Introduction

16.1.1 Background

Laboratory tests that are conducted to obtain product reliability information are often done under a constant stress. Product units in the field, however, are usually exposed to varying failure-causing stresses due to the dynamics of the environment and the variability in product usage. These variations include environmental variables such as temperature, humidity, vibration, UV intensity and spectrum, and usage variables such as loading and use rate, which vary from unit to unit and over time within each unit.

For some products, it is possible to track environmental variables dynamically, even in real time. This usage/environmental information can be obtained from sensors and

V.V. Rykov et al. (eds.), *Mathematical and Statistical Models and Methods in Reliability: Applications to Medicine, Finance, and Quality Control*, Statistics for Industry and Technology, DOI 10.1007/978-0-8176-4971-5_16,

smart chips that are installed in a product to measure and record such information over the life of the product. For products that are connected to a network or installed with a wireless transmission device, such information is available dynamically or periodically. For products that are not connected to a network, this information is available at the time of product inspection, return, or repair.

Several examples of products/systems that provide dynamic information are as follows.

- OnStar™ [OnS09] is an in-vehicle safety and security system created to help protect automobile occupants. The system consists of various sensors and has the ability to communicate vehicle information to the driver as well as to a central location, via a satellite wireless connection. The system also collects usage and environmental information and, with the vehicle owner's permission, transmits this information periodically to the central location.
- Large medical systems, such as CT scanners, have sensors and devices that can provide real-time system information to those who do system maintenance.
- Hahn and Doganaksoy [HD08, Sect. 9.9] describe an application involving modern locomotive engines installed with sensors that indicate operating status variables such as oil pressure, oil temperature, and water temperature. Such information is automatically recorded and transmitted to a central location and can be used to shutdown an engine, should a dangerous condition arise. Aircraft subsystems also have similar sensors.
- High-voltage power transformers can be monitored by an automatic dissolved gas analyzer (DGA) system (e.g., [STW+05]). DGA automatically performs periodic analyses (typically every hour) to indicate the presence of different kinds of dissolved gases in the transformer insulating oil and moisture content. Certain combinations of gas mixtures are known to be a precursor of a failure event. In addition, the DGA system reports real-time dynamic loading and thermal information. This information is automatically transmitted to a control center for monitoring and analyses.
- Computers and high-end printers with smart chips can record the usage history and the environmental condition such as operating temperature. This information is available dynamically through the network or other communications channels and can, in cooperation with the owner, be downloaded periodically.

16.1.2 Applications in Prediction

One important reason for outfitting products with sensors, smart chips, and communications channels is to assist in the delivery of timely maintenance actions and to increase system availability. It can be expected, however, that using dynamic usage/environmental information in modeling and data analysis will also provide stronger statistical methods and more accurate inferences or predictions of field failures. These improvements can be realized when one or more of the important sources of variability the field data can be explained by the additional information. In applications, predictions for the number of field failures or warranty returns for the population is important for financial planning decisions, such as setting warranty reserves for a manufactured product or capital budgeting for a company's fleet of assets. For example, after a product has been introduced into the field for a certain period of time (e.g., 1 year), the finance department is often interested to know what will be the total number of returns for some future period of time (i.e., the next 3 years), based on early warranty returns of the product. By taking advantage of the dynamic information available from the product, one can expect to get more accurate predictions than what would be obtained by using only the traditional failure-time data.

The prediction of the remaining life of individual units sometimes is also of interest, especially for fleets of assets for a company (e.g., locomotives in Sect. 9.9 of [HD08], and high-voltage power transformers in [HMM09]). [HMM09] give the prediction intervals for the remaining life of high-voltage power transformers based only on currently available failure-time data. The prediction intervals given there are wide for individual units. The dynamic usage/environmental information can be expected to improve the accuracy of prediction intervals for individuals.

16.1.3 Related Literature

[Nel90, Chap. 10] describes the cumulative exposure model in the context of life tests. The cumulative exposure model is equivalent to the time scale accelerated failure time model with time-dependent covariates used in [RT92]. [RT92], however, used a nonparametric estimation method that does not require specification of the baseline failure-time distribution.

[Nel01] describes prediction for field reliability of units under dynamic stresses using the cumulative exposure model. The problem considered in our paper, however, is different. We consider predictions and prediction intervals (PIs) for both the population and individual units based on the distribution of remaining life. The uncertainty in the covariate process is also considered.

In the area of warranty prediction involving dynamic stress, [GMMO09] consider warranty prediction with stress that is random from unit to unit but constant within a unit. Stress information, however, is not available for individual units. [HM10] consider a warranty prediction problem where the average use-rate is available for both failed and censored units. However, warranty prediction procedures using the dynamic information need to be developed.

16.2 Data and Model

16.2.1 Notation

Let T be the time to failure random variable. The usage/environmental information at time t is denoted by a random vector $X(t) = [X_1(t), \cdots, X_p(t)]'$ where p is the number of covariates. The history of the covariate process is denoted by $\boldsymbol{X}(t) = \{X(s) : 0 \leq s \leq t\}$ which records the dynamic information from time 0 to time t. Because of the dynamic information on usage and environmental conditions, observations of $X(t)$ are available for each small time interval with length Δ. $X(t)$ is assumed to be constant over these intervals. Thus, the covariate history is recorded as a multivariate time series.

The data are denoted by $\{ t_i, \delta_i, \boldsymbol{x}_i(t_i) \}$ for $i = 1, 2 \cdots, n$ where n is the number of observations in the dataset. Here, t_i is the failure time (time in service) for unit i if it failed (did not fail). The censoring indicator $\delta_i = 1$ if unit i failed and $\delta_i = 0$ otherwise. Let $x(t)$ be the observed covariate information at time t. Then $\boldsymbol{x}_i(t_i) = \{x(s) : 0 \leq s \leq t_i\}$ is the observed covariate history from the time origin to t_i for unit i.

16.2.2 Cumulative Exposure Model

We use the cumulative exposure model, as described in [Nel90], to model the failure-time data with covariates which were recorded as a multivariate time series. In particular, the cumulative exposure U for a unit with failure time T is defined as

$$U = \int_0^T g[X(s); \boldsymbol{\beta}] \mathrm{d}s \sim F_0(u, \boldsymbol{\theta}_0) \tag{16.1}$$

given the covariate entire history $\boldsymbol{X}(\infty) = \boldsymbol{x}(\infty)$. Here, $g[X(t); \boldsymbol{\beta}]$ is the time scale acceleration rate which is a function of the covariate process history with parameter $\boldsymbol{\beta}$, and $F_0(u, \boldsymbol{\theta}_0)$ is the baseline cumulative distribution function (cdf). $g[X(t); \boldsymbol{\beta}]$ gives the instantaneous effect of the stress/exposure on the product life from both the usage and the environment at time t. If a unit is operated under harsh environmental conditions and/or has a large use rate, then $g[X(t); \boldsymbol{\beta}] > 1$. That is, the calender time scale is accelerated. The unit would be expected to fail sooner than those used under mild conditions.

There needs to be a restriction on $g[X(t); \boldsymbol{\beta}]$ in order for the parameters to be estimable. In particular, the function needs to have $g[X(t); \boldsymbol{\beta}] = 1$ when $\boldsymbol{\beta} = \mathbf{0}$. Given the covariate history $\boldsymbol{X}(\infty) = \boldsymbol{x}(\infty)$ and $\boldsymbol{\beta} = \mathbf{0}$, the cumulative exposure is $U = \int_0^T g[x(s); \boldsymbol{\beta}]\mathrm{d}s = \int_0^T 1\mathrm{d}s = T$. That is, the cumulative exposure has the same scale as the calender time scale.

We assume that lifetimes of product units that are all used at the same constant use rate and environmental conditions can be adequately described by the same distribution. This is because the failure mechanisms of those units are similar. The cumulative exposure model converts units under different usage and environmental conditions into a comparable scale which is called the cumulative exposure. We assume that failure time of the population under the cumulative exposure time scale can be adequately described by a single distribution.

16.2.3 Modeling the Time Scale Acceleration Rate

The following log-linear relationship is widely used as an acceleration factor

$$g[X(t); \boldsymbol{\beta}] = \exp[\boldsymbol{\beta}' X(t)]. \tag{16.2}$$

This model assumes the effect on the time scale acceleration is proportional for different values of $X(t)$. This model is sometimes called the proportional quantiles (PQ) model or the scale accelerated failure-time (SAFT) model (e.g., [ME98, Chap. 17]). Here, $X(t)$ might be transformed values of the original explanatory variable.

- If use-rate information is available, one possible relationship is $g[X(t); \boldsymbol{\beta}] = \exp[\beta_0 + \beta_1 X(t)]$ where $X(t) = \log(\text{use rate})$, which is the inverse power law (e.g., [ME98, Page 480]).

- If information on temperature is available, the Arrhenius relationship (e.g., [ME98, Page 472]) can be used, in which case $g[X(t);\boldsymbol{\beta}] = \exp[\beta_0 + \beta_1 X(t)]$ where $X(t) = -11605/(\texttt{temp} + 273.6)$ is a transformation of Celsius temperature `temp`. β_1 can be interpreted as the effective activation energy in electron volts. Both β_0 and β_1 are product or material characteristics.
- If information on both temperature and use rate is available, the following relationship can be used,

$$g[X(t);\boldsymbol{\beta}] = \exp[\beta_0 + \beta_1 X_1(t) + \beta_2 X_2(t)],$$

where $X_1(t)$ is the transformed temperature and $X_2(t)$ is the transformed use rate.

16.2.4 Modeling the Baseline Distribution

The baseline cdf is defined as the cdf of the failure-time distribution for a unit which is used at typical fixed conditions. These baseline conditions are similar to the use conditions in a standard life test or accelerated life test. We will model the baseline cdf $F_0(u;\boldsymbol{\theta}_0)$ as a log-location-scale distribution. The general log-location-scale cdf is

$$F_0(u;\boldsymbol{\theta}_0) = \Phi\left[\frac{\log(u)-\mu}{\sigma}\right], \quad u > 0. \tag{16.3}$$

Here, $\boldsymbol{\theta}_0 = (\mu,\sigma)'$, μ is the location parameter, σ is the scale parameter, and $\Phi(z)$ is the standard cdf for the location-scale family of distributions (location 0 and scale 1). The corresponding probability density function (pdf) is

$$f_0(u;\boldsymbol{\theta}_0) = \frac{\mathrm{d}F_0(u)}{\mathrm{d}u} = \frac{1}{\sigma u}\phi\left[\frac{\log(u)-\mu}{\sigma}\right],$$

where $\phi(z) = \mathrm{d}\Phi(z)/\mathrm{d}z$. The Weibull and lognormal distributions are the most commonly used distributions for modeling of failure-time data from this family of distributions. The cdf and pdf of T, given the entire history $\boldsymbol{X}(\infty) = \boldsymbol{x}(\infty)$, is

$$F(t;\boldsymbol{\beta},\boldsymbol{\theta}_0) = F_0\left(\int_0^t g[x(s);\boldsymbol{\beta}]\mathrm{d}s;\boldsymbol{\theta}_0\right)$$

and

$$f(t;\boldsymbol{\beta},\boldsymbol{\theta}_0) = g[x(t);\boldsymbol{\beta}]f_0\left(\int_0^t g[x(s);\boldsymbol{\beta}]\mathrm{d}s;\boldsymbol{\theta}_0\right)$$

respectively.

16.2.5 Modeling the Covariate Process

For the purpose of prediction of failure times, a parametric model for $X(t)$ is needed. This allows prediction of the future covariate vector for an individual unit. $X(t)$ is modeled as

$$X(t) = m(t;\boldsymbol{\eta}) + a(t)\,. \tag{16.4}$$

Here, $m(t;\boldsymbol{\eta})$ is the mean function with parameter $\boldsymbol{\eta}$ and $a(t)$ is the error term which is assumed to be a stationary process. The parametric form for the mean function $m(t;\boldsymbol{\eta})$ needs to be specified according to the particular application. Some components of $\boldsymbol{\eta}$ can be random to allow for population nonhomogeneity of the covariate process. Also, depending on the application, the following gives possible models for the distribution of $a(t)$.

- In a simple case, $a(t)$ for different values of t can be modeled as independently and identically distributed (iid) with $N(0,\Sigma)$ where Σ is the covariance matrix.
- To allow for more complicated models for $a(t)$, the vector autoregressive (VAR) moving average time series models in [Rei03] can be used. For example, the VAR(1) model is represented as

$$a(t) = \Psi\, a(t-1) + \varepsilon(t), \tag{16.5}$$

where Ψ is an unknown coefficient matrix. For the purpose of prediction, a parametric distribution assumption is needed for the noise term $\varepsilon(t)$. One common choice is that $\varepsilon(t)$ are iid with $N(0,\nu^2 I)$ where I is the identity matrix and ν^2 is the variance factor.

16.3 Parameter Estimation

In this section, we use the method of maximum likelihood (ML) to obtain estimates for unknown model parameters. The ML estimates for the failure-time distribution parameters are obtained by conditioning on the observed covariate history. The ML estimates for the parameters of the covariate history can also be obtained by assuming that the data were generated from a specific class of multivariate time series models.

16.3.1 ML Estimate for Parameters of the Failure-Time Distribution

The likelihood of the failure-time data, conditional on the observed covariate history, is

$$L(\boldsymbol{\beta},\boldsymbol{\theta}_0|DATA) = \\ \prod_{i=1}^{n}\left\{g[x_i(t_i);\boldsymbol{\beta}]f_0\left(\int_0^{t_i} g[x_i(s);\boldsymbol{\beta}]\mathrm{d}s;\boldsymbol{\theta}_0\right)\right\}^{\delta_i}\left\{1-F_0\left(\int_0^{t_i} g[x_i(s);\boldsymbol{\beta}]\mathrm{d}s;\boldsymbol{\theta}_0\right)\right\}^{1-\delta_i}. \tag{16.6}$$

The maximum likelihood (ML) estimate $\left(\widehat{\boldsymbol{\beta}}',\widehat{\boldsymbol{\theta}}'\right)'$ is obtained by finding those values of $(\boldsymbol{\beta}',\boldsymbol{\theta}')'$ that maximize (16.6).

16.3.2 ML Estimate for Parameters of the Covariate Process

The likelihood for the covariate history, assuming the parameter $\boldsymbol{\eta}$ in the mean function $m(t;\boldsymbol{\eta})$ is non-random and the error term $a(t)$ is distributed with $N(0,\Sigma)$, is

$$L(\boldsymbol{\eta}, \Sigma|DATA) = \prod_{i=1}^{n}\prod_{s\le t_i}\frac{1}{(2\pi)^{p/2}|\Sigma|^{1/2}}\exp\left\{-\frac{1}{2}[x_i(s)-m(s;\boldsymbol{\eta})]'\Sigma^{-1}[x_i(s)-m(s;\boldsymbol{\eta})]\right\}. \quad (16.7)$$

The ML estimates are denoted by $\widehat{\boldsymbol{\eta}}$ and $\widehat{\Sigma}$. For more complicated models describing $a(t)$, the parameters can also be estimated using the method of ML. The ML estimation techniques described, for example, in [Rei03, Chap. 5] can be used.

16.4 Predictions

As described in Sect. 16.1.2, predictions are needed for both the cumulative number of field failures and the remaining life of individual units. There is need for accurate predictions on both the population and individuals in business and industry. The availability of dynamic environmental information can be expected to improve the accuracy of predictions, especially for individuals. In applications, predictions need to correspond to the real time scale, after the data-freeze date (DFD). These predictions will be based on the distribution of remaining life of units that have survived until the DFD.

16.4.1 Distribution of Remaining Life

The distribution of remaining life provides the basis for calculating the prediction for the population and the individuals. The distribution of T_i, given $T_i > t_i$ and the covariate history $\boldsymbol{X}_i(t_i)$, is

$$\rho_i(t_w;\boldsymbol{\theta}) = \Pr[t_i < T_i \le t_w | T_i > t_i, \boldsymbol{X}_i(t_i)], \quad t_w > t_i. \quad (16.8)$$

Here, $\boldsymbol{\theta}$ is the collection of parameters including $\boldsymbol{\beta}, \boldsymbol{\theta}_0$, and the parameters for the covariate process. In particular,

$$\begin{aligned}\rho_i(t_w;\boldsymbol{\theta}) &= \mathbf{E}_{\boldsymbol{X}_i(t_i,t_w)|\boldsymbol{X}_i(t_i)}\left\{\Pr[t_i < T_i \le t_w | T_i > t_i, \boldsymbol{X}_i(t_i), \boldsymbol{X}_i(t_i,t_w)]\right\} \\ &= \frac{\mathbf{E}_{\boldsymbol{X}_i(t_i,t_w)|\boldsymbol{X}_i(t_i)}\left\{F_0\left(\int_0^{t_w} g[X_i(u);\boldsymbol{\beta}]\mathrm{d}u;\boldsymbol{\theta}_0\right)\right\} - F_0\left(\int_0^{t_i} g[x_i(u);\boldsymbol{\beta}]\mathrm{d}u;\boldsymbol{\theta}_0\right)}{1 - F_0\left(\int_0^{t_i} g[x_i(u);\boldsymbol{\beta}]\mathrm{d}u;\boldsymbol{\theta}_0\right)}\end{aligned} \quad (16.9)$$

where $\boldsymbol{X}_i(t_1,t_2) = \{X_i(u) : t_1 < u \le t_2\}$. When the model for $X(t)$ is complicated, the distribution of $\boldsymbol{X}_i(t_i,t_w)|\boldsymbol{X}_i(t_i)$ may be mathematically intractable. Numerical methods can, however, be applied to evaluate $\rho_i(t_w;\boldsymbol{\theta})$.

16.4.2 Prediction for the Population

When focusing on the overall population, we need to generate predictions for the cumulative number of failures for the units in the field. The prediction for the warranty returns can also be obtained in a similar way but there is a need to adjust the risk set for the length of the warranty period. Prediction intervals are also needed for quantifying the statistical uncertainties.

Let $N(s)$ be the number of field failures at s time units after the DFD. $N(s) = \sum_{i \in RS} I_i(s)$ where RS is the risk set and $I_i(s) \sim \text{Bernoulli}[\rho_i(t_i + s; \boldsymbol{\theta})]$. The point prediction for $N(s)$ is $\widehat{N}(s) = \sum_{i \in RS} \rho_i(t_i + s; \widehat{\boldsymbol{\theta}})$. A prediction interval (PI) for $N(s)$ is denoted by $[\underline{N}, \overline{N}]$. The naive (plug-in) PI is obtained by solving

$$F_N(\underline{N}; \widehat{\boldsymbol{\theta}}) = \frac{\alpha}{2}, \quad \text{and} \quad F_N(\overline{N}; \widehat{\boldsymbol{\theta}}) = 1 - \frac{\alpha}{2}. \tag{16.10}$$

Here, $F_N(n_k; \boldsymbol{\theta}), n_k = 0, 1, \cdots, n^*$ is the cdf of $N(s)$ where n^* is the number of units in the RS at the DFD. $1 - \alpha$ is the desired coverage probability. Note that $N(s)$ is a sum of non-identically distributed Bernoulli random variables. The cdf of N_k does not have a simple closed-form expression. An approximation is usually needed in applications. The Volkova approximation ([Vol96]), which is based on a refined normal approximation with correction for the skewness of $N(s)$, is used by [HMM09] for a prediction problem. The Poisson approximation is also used in the literature (e.g., [EM99, Sect. A.3]) when the expected number of failure (after the DFD in setting of this paper) is small (e.g., less than 10).

16.4.3 Prediction for Individuals

When focusing on individuals, we will compute prediction intervals for each individual. The naive prediction interval for the individual remaining life is denoted by $[\underline{T}_i, \overline{T}_i]$ and can be obtained by solving

$$\rho_i(\underline{T}_i; \widehat{\boldsymbol{\theta}}) = \frac{\alpha}{2}, \text{ and } \rho_i(\overline{T}_i; \widehat{\boldsymbol{\theta}}) = 1 - \frac{\alpha}{2}, \tag{16.11}$$

where $\rho_i(\cdot, \boldsymbol{\theta})$ is given in (16.9). Note here the PI of remaining life is conditional on the individual's current time in service t_i and its observed covariate process $\boldsymbol{x}_i(t_i)$. Thus each individual will have a distinct PI.

16.5 Calibration of Prediction Intervals

The PIs in (16.10) and (16.11) ignore the uncertainty in $\widehat{\boldsymbol{\theta}}$. Thus, the coverage probability is generally smaller than the nominal $1 - \alpha$ level. These PIs can be calibrated to improve the coverage probability property. We will use simulations to do the calibration.

16.5.1 Bootstrapping the Distribution of $\widehat{\theta}$

To account for the uncertainty in $\widehat{\boldsymbol{\theta}}$, we use a parametric bootstrap simulation to approximate the distribution of $\widehat{\boldsymbol{\theta}}$. The calibration has two parts: first, we use bootstrap to generate the bootstrap version of the covariate process $\boldsymbol{x}_i^*(t_i), i = 1, 2, \cdots, n$. Because we assume a parametric model for the covariate process as in (16.4), parametric simulation methods can be used here to generate $\boldsymbol{x}_i^*(t_i)$. Repeating the ML estimation procedure in Sect. 16.3.2, one obtains the bootstrap version of estimates of the parameters for the covariate process.

The second part of the calibration process is to obtain the bootstrap version estimates of parameters for the failure-time distribution. The traditional bootstrap method that uses simple random sampling with replacement can be problematic with heavy censoring, as it can result in bootstrap samples without enough failures for the estimation of the parameters. Here, we use the random weighted bootstrap method (e.g., [NR94], [JYW01]) to obtain the bootstrap version estimates of the parameters. See [HMM09] for another application of random weighted bootstrap in calibration PIs. In particular, with a set of random weights Z_i generated from any positive continuous distribution with $\mathbf{E}(Z_i) = \sqrt{\mathrm{Var}(Z_i)}$, the random weighted likelihood is

$$
\begin{aligned}
&L^*(\boldsymbol{\beta}, \boldsymbol{\theta}_0 | DATA) \\
&= \prod_{i=1}^{n} \left\{ g[x_i^*(t_i); \boldsymbol{\beta}] f_0 \left(\int_0^{t_i} g[x_i^*(s); \boldsymbol{\beta}] \mathrm{d}s; \boldsymbol{\theta}_0 \right) \right\}^{\delta_i Z_i} \\
&\quad \times \left\{ 1 - F_0 \left(\int_0^{t_i} g[x_i^*(s); \boldsymbol{\beta}] \mathrm{d}s; \boldsymbol{\theta}_0 \right) \right\}^{(1-\delta_i) Z_i} .
\end{aligned}
$$

Here, $\boldsymbol{x}_i^*(t_i)$ is the bootstrap sample generated in the first part. The bootstrap versions of the parameter estimates for the failure-time distribution can be obtained by maximizing the random weighted likelihood. Combining with the bootstrap version estimates of the parameter for the covariates, we obtain the bootstrap version of $\widehat{\boldsymbol{\theta}}$, which is denoted by $\widehat{\boldsymbol{\theta}}^*$.

16.5.2 Calibration for Prediction Intervals

With B bootstrap samples of $\widehat{\boldsymbol{\theta}}^*$, the calibration of PIs for the population can be done by using a procedure similar to the procedure described in Sect. 6.2 of [HMM09]. Here, B is usually chosen to be a large number (e.g., B = 10,000). The calibration of PIs for individuals can be done by using a procedure similar to the procedure described in Sect. 5.4 of [HMM09].

16.6 Conclusions and Areas for Future Research

In this paper, we outline a model and methods that can be used for field-failure prediction using dynamic environmental data. We also describe predictions of the cumulative number of field failures and the remaining life for individuals. Prediction intervals are also given and the associated calibration procedures are also described.

In future work, we will consider more general modeling of effects of covariate processes on the failure-time distribution. For example, one can model the time scale acceleration rate function g in (16.1) as $g[\boldsymbol{X}(t), \boldsymbol{\beta}]$. This means the time scale acceleration rate function depends on the history from the time origin to time t. For example, some environmental variables may have a delayed effect on the failure-time distribution. Alternative models, such as the proportional hazards model with time dependent covariates could also be considered. Parametric models for the baseline hazard function and the covariate process will be needed if prediction is the main goal of the application.

Modern sensor technology also allows us to obtain dynamic degradation measurements (or indirect measurements) for products or components of products on the field. Prediction and intervals using dynamic degradation can be expected to have some advantages and provide more useful information. Some research has been done in this direction. [GP08] used dynamic environmental data to update the distribution of remaining life under a Bayesian frame work. [VTM09] developed a statistical model for linking field and laboratory exposure data that measure the chemical degradation processes of a coating system, where environmental variables such as UV spectrum and intensity, temperature, and relative humidity were also measured repeatedly. More general models and methods for prediction and prediction intervals, however, need to be developed for these situations.

Acknowledgment. We would like to thank Katherine Meeker for helpful comments on an earlier version of this paper.

References

[EM99] L. A. Escobar and W. Q. Meeker. Statistical prediction based on censored life data. *Technometrics*, 41:113–124, 1999

[GMMO09] Huairui Guo, Arai Monteforte, Adamantios Mettas, and Doug Ogden. Warranty prediction for products with random stresses and usages. In *Proceedings Annual Reliability and Mantainability Symposium*, pages 72–77, Fort Worth, TX, 2009

[GP08] Nagi Gebraeel and Jing Pan. Prognostic degradation models for computing and updating residual life distributions in a time-varing environment. *IEEE Transactions on Reliability*, 57:539–550, 2008

[HD08] Gerald J. Hahn and Necip Doganaksoy. *The Role of Statistics in Business and Industry*. Wiley, New Jersey, 2008

[HM10] Y. Hong and W. Q. Meeker. Field-failure and warranty prediction based on auxiliary use-rate information. *Technometrics*, 52:148–159, 2010

[HMM09] Y. Hong, W. Q. Meeker, and J. D. McCalley. Prediction of remaining life of power transformers based on left truncated and right censored lifetime data. *The Annals of Applied Statistics*, 3:857–879, 2009

[JYW01] Z. Jin, Z. Ying, and L. J. Wei. A simple resampling method by perturbing the minimand. *Biometrika*, 88(2):381–390, 2001

[ME98] W. Q. Meeker and L. A. Escobar. *Statistical Methods for Reliability Data*. Wiley, New York, 1998

[Nel90] W. Nelson. *Accelerated Testing: Statistical Models, Test Plans, and Data Analyses, (Republished in a paperback in Wiley Series in Probability and Statistics, 2004)*. Wiley, New York, 1990

[Nel01] W. Nelson. Prediction of field reliability of units, each under differing dynamic stresses, from accelerated test data. In: N. Balakrishnan and C. R. Rao, editors, *Handbook of Statistics 20: Advances in Reliability, Chapter IX*. North-Holland, Amsterdam, 2001

[NR94] Michael A. Newton and Adrian E. Raftery. Approximate Bayesian inference with the weighted likelihood bootstrap. *Journal of the Royal Statistical Society: Series B*, 56:3–48, 1994

[OnS09] OnStar. *The OnStar Official site.* `http://www.onstar.com/us_english/jsp/index.jsp`, 2009

[Rei03] G. C. Reinsel. *Elements of Multivariate Time Series Analysis*, 2nd edition. Springer, New York, 2003

[RT92] J. Robins and A. A. Tsiatis. Semiparametric estimation of an accelerated failure time model with time-dependent covariates. *Biometrika*, 79:311–319, 1992

[STW+05] K. Spurgeon, W. H. Tang, Q. H. Wu, Z. J. Richardson, and G. Moss. Dissolved gas analysis using evidential reasoning. *IEE Proceedings – Science, Measurement and Technology*, 152:110–117, 2005

[Vol96] A. Y. Volkova. A refinement of the central limit theorem for sums of independent random indicators. *Theory of Probability and its Applications*, 40(4):791–794, 1996

[VTM09] I. Vaca-Trigo and W. Q. Meeker. A statistical model for linking field and laboratory exposure results for a model coating. In: Jonathan Martin, Rose A. Ryntz, Joannie Chin, and Ray A. Dickie, editors, *Service Life Prediction of Polymeric Materials*. Springer, Berlin, 2009

17

Efficient Regression Estimation Under General Censoring and Truncation

Catherine Huber

Université Paris Descartes, Paris, France and U1018 INSERM, Villejuif, France,
catherine.huber@parisdescartes.fr

Abstract: Much attention has been paid to the semi-parametric approach for right censored data in survival analysis while the case of data that are also truncated has been considered less frequently. We consider here the case of data that are censored and truncated in the most general way and obtain an efficient estimator of a regression model relying on a basic hazard that is non parametric. The model is defined in Sect. 17.1, the censorship and truncation scheme is described in Sect. 17.2. The resulting model for the really observed data is derived in Sect. 17.3 and the efficiency of the proposed estimator proved in Sect. 17.4. The last section deals with commentaries and perspectives of this work which was conducted in collaboration with Valentin Solev of the Steklov Institute in Saint Petersbourg (Russia) and Filia Vonta of the National Technical University in Athens (Greece).

Keywords and phrases: AIDS, Censored data, Kullback-Leibler distance, Least favorite parametric sub-model, Semiparametric approach, Seropositive patient, Truncation by interval

17.1 Framework

We want to estimate the impact of some factors on the time to onset X of a specific event among patients of a certain type, without assuming a parametric model for the baseline law of X, while X itself is not observed.

1. X is the random time elapsed until some event takes place. Example: time to onset of AIDS among seropositive patients.
2. We aim at estimating the impact of some factors, named covariates, on the probability of X.
   Example: age, sex, and transmission mode.
3. Each patient is observed at certain random times $Y_1, \ldots, Y_k$, so that X is known to lie inside an interval: it is interval censored. Moreover, some patients are skipped from the sample: they are truncated.
   Example: observation of seropositive patients is scheduled 3 or 4 times a year to check for possible onset of AIDS.

V.V. Rykov et al. (eds.), *Mathematical and Statistical Models and Methods in Reliability: Applications to Medicine, Finance, and Quality Control*, Statistics for Industry and Technology, DOI 10.1007/978-0-8176-4971-5_17,

17.2 Data Description

Time X to an event that changes permanently the state of subject i under study (0 before X, 1 afterwards) is a random variable whose distribution is influenced by certain factors. This influence is to be estimated under the following observation scheme:

- Censorship: observation of each subject i does not take place continuously but is scheduled at a random number $r(i)$ of random times
$$a < Y_{i,1} <, \cdots, < Y_{i,r(i)} < b.$$
- Truncation: only those scheduled times that are inside a given random window $[Z_{i,1}\ Z_{i,2}]$ give rise to an observation of subject i.

Thus, if subject i is observed in state 0 at time $y_{i,j}$ and in state 1 at time $y_{i,j+1}$, inside its window $z_{i,1}, z_{i,2}$, one observes two embedded intervals bracketing the unobserved $X = x$:

$$z_{i,1} \leq y_{i,k_1} \leq y_{i,j} < x? \leq y_{i,j+1} \leq y_{i,k_2} \leq z_{i,2}$$

17.2.1 Censorship

The random partition

$$\tau = \{Y_0 = a < Y_1 <, \cdots, < Y_j, \cdots, < Y_r < Y_{r+1} = b\}$$

defines, for each $x \in (a, b)$, through $j(x) = \inf\{j : x \leq Y_{j+1}\}$ a bracketing interval $L(x) := Y_{j(x)}$; $R(x) := Y_{j(x)+1}$:

$$L(x) < x \leq R(x)$$

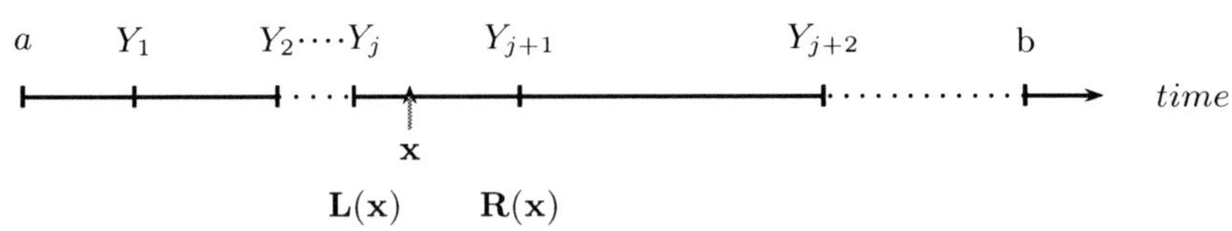

17.2.2 Truncation by Interval $[z_1, z_2]$.

Observations can take place only inside an interval $\triangle = [z_1, z_2]$, for some $z_1 < z_2$. Then, for a fixed value of the censoring partition $\tau = t$, all observations take place inside interval

$$[R(z_1)\ ;\ L(z_2)], \quad \text{provided } R(z_1) < L(z_2).$$

Otherwise, there is no observation at all: truncated subject.

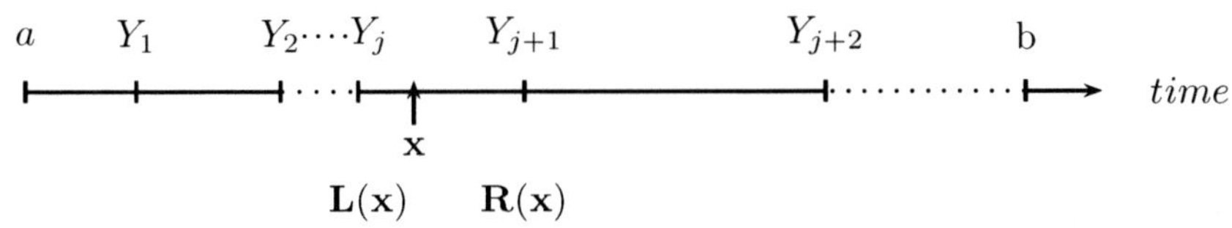

Special case of right truncation: $z_1 = a$.

17.2.3 The Data

The data is a 4-dimensional vector embedding the unobserved x:

$$R(z_1) \leq L(x) < R(x) \leq L(z_2)$$

and in the special case of right truncation by $Z = z$, a 3-dimensional vector:

$$L(x) < R(x) \leq L(z)$$

As the data include also the covariate $\Xi_i, i = 1, \ldots, n$ whose influence on the distribution of the unobserved X is to be estimated, the data are finally

$$(L(X), R(X), L(Z), \Xi)$$

for each subject i, $i = 1, \ldots, n$.

17.3 The Semi-Parametric Model

We denote ν^k the Lebesgue measure on $\mathbf{R}^k$, while ν_k is any measure on $\mathbf{R}^k$.

17.3.1 Model for X

X has density f with respect to ν^k and survival function S conditional on $\Xi = \xi$ defined as

$$S(t|\Xi = \xi) = P(X > t|\Xi = \xi) = e^{-e^{\beta'\xi}\Lambda(t)} \tag{17.1}$$

- $\beta \in R^d$ is the parameter of interest
- Ξ a d-dimensional vector of covariates, with distribution $dP_\Xi = \phi(\xi)d\nu_d$ for some $\sigma-$finite measure ν_d on $\mathbf{R}^d$ possibly equal to ν^d.
- Λ the cumulative hazard function, infinite-dimensional nuisance parameter.

17.3.2 Resulting Model for the Observations Q

Observed data are $Q_1, \ldots, Q_n$, i.i.d. random vectors,

$$Q = (L(X), R(X), L(Z), \Xi)$$

with density $p(u,v,w,\xi)$ with respect to a measure ν^* [HCSV07] given as

$$p(u,v,w,\xi) = r(u,v,w) \cdot \frac{\int_u^v f(t|\xi)\,\mathrm{d}t}{\int_0^w f(t|\xi)\,\mathrm{d}t} \cdot \phi(\xi) \tag{17.2}$$

Two components for r, density of the joint law of censoring and truncation

$$r = r_3 + r_2$$

- r_3 absolutely continuous with respect to ν^3 (when $u < v < w$),
- r_2 absolutely continuous with respect to ν^2 (when $u < v = w$).

- r is the known density with respect to ν^{**}, the marginal of ν^* integrated over ξ, of the joint law of censoring and truncation.
- ν^* is the measure on $\mathbf{R}^{3+d}$ which is defined for continuous nonnegative functions $\psi(s) = \psi(u,v,w,\xi)$ by the relation

$$\iiiint \psi(s)\,\mathrm{d}\nu^* = \iiiint \psi(u,v,w,\xi)\,\mathrm{d}(\nu^3 \otimes \nu_d)(u,v,w,\xi) + \\ + \iiint \psi(u,v,v,\xi)\,\mathrm{d}(\nu^2 \otimes \nu_d)(u,v,\xi).$$

For details and an example of such a law r, see [HCSV07]
From model (17.1) p is equal to

$$r(u,v,w) \cdot \frac{\int_u^v e^{-e^{\beta'\xi}\Lambda(t)} e^{\beta'\xi}\lambda(t)\,\mathrm{d}t}{\int_0^w e^{-e^{\beta'\xi}\Lambda(t)} e^{\beta'\xi}\lambda(t)\,\mathrm{d}t} \cdot \phi(\xi) := r(u,v,w) \cdot \varphi(u,v,w,\xi|\beta,\lambda) \cdot \phi(\xi) \tag{17.3}$$

by definition of φ which is the likelihood of each unobserved survival conditional on the censoring, truncation and covariate.

We prove efficiency by following the methodology given in [SV05].
The data-space $\mathbf{D} = \mathbf{R} \times \mathbf{R} \times \mathbf{R} \times \mathbf{R}^d$ consists of vectors $s = (u,v,w,\xi)$.

- True parameters: (β_0, λ_0).
- True law of the observed data: μ_0

$$d\mu_0(s) := p_0(s)\mathrm{d}\nu^*(s) \;\; ; \;\; p_0(s) := p(s|\beta_0,\lambda_0)$$

Densities $f_Q(s|\beta,\lambda)$ with respect to μ_0 are

$$f_Q(s|\beta,\lambda) = \frac{\int_u^v e^{-e^{\beta'\xi}\Lambda(t)} e^{\beta'\xi}\lambda(t)\,\mathrm{d}t}{\int_0^w e^{-e^{\beta'\xi}\Lambda(t)} e^{\beta'\xi}\lambda(t)\,\mathrm{d}t} \times \frac{\int_0^w e^{-e^{\beta_0'\xi}\Lambda_0(t)} e^{\beta_0'\xi}\lambda_0(t)\,\mathrm{d}t}{\int_u^v e^{-e^{\beta_0'\xi}\Lambda_0(t)} e^{\beta_0'\xi}\lambda_0(t)\,\mathrm{d}t}. \tag{17.4}$$

17.3.3 Least Favorable Parametric Sub-Model (LFPS)

For fixed β, find the least favorable parametric sub-model (Λ_β, β) by minimizing the Kullback–Leibler distance between $P(\cdot|\beta_0, \lambda_0)$ and $P(\cdot|\beta, \lambda(\beta))$, denoted $\mathcal{K}(\beta, \lambda)$:

$$\mathcal{K}(\beta,\lambda) = -\int \log\left(\frac{p(u,v,w,\xi|\beta,\lambda)}{p(u,v,w,\xi|\beta_0,\lambda_0)}\right) p(u,v,w,\xi|\beta_0,\lambda_0)\,\mathrm{d}\nu^*.$$

Due to the form of the law $r(u, v, w)$, written equivalently as

$$= -\int \log\left(\frac{p(u,v,w,\xi|\beta,\lambda)}{p(u,v,w,\xi|\beta_0,\lambda_0)}\right) p(u,v,w,\xi|\beta_0,\lambda_0)\,\mathrm{d}(\nu^3\otimes\nu_d)$$

$$-\int \log\left(\frac{p(u,v,v,\xi|\beta,\lambda)}{p(u,v,v,\xi|\beta_0,\lambda_0)}\right) p(u,v,v,\xi|\beta_0,\lambda_0)\,\mathrm{d}(\nu^2\otimes\nu_d)$$

which because of model (17.1) is equal to

$$-\int\left\{\log\left(\int_u^v e^{-e^{\beta'\xi}\Lambda(t)}e^{\beta'\xi}\lambda(t)\mathrm{d}t\right) - \log\left(\int_0^w e^{-e^{\beta'\xi}\Lambda(t)}e^{\beta'\xi}\lambda(t)\mathrm{d}t\right)\right\}$$
$$p(u,v,w,\xi|\beta_0,\lambda_0)d(\nu^3\otimes\nu_d)(u,v,w,\xi)$$
$$-\int\left\{\log\left(\int_u^v e^{-e^{\beta'\xi}\Lambda(t)}e^{\beta'\xi}\lambda(t)\mathrm{d}t\right) - \log\left(\int_0^v e^{-e^{\beta'\xi}\Lambda(t)}e^{\beta'\xi}\lambda(t)\mathrm{d}t\right)\right\}$$
$$p(u,v,v,\xi|\beta_0,\lambda_0)\mathrm{d}(\nu^2\otimes\nu_d)(u,v,\xi) + C', \tag{17.5}$$

where C' denotes a term that does not depend on (β, λ).

Keeping β **fixed**, differentiate $\mathcal{K}(\beta, \lambda)$ (Gâteaux differentiation) with respect to the functional parameter λ. We consider perturbations of functions $\lambda \in \mathcal{V}$ by small multiples of functions γ in $\mathcal{G}_0$:

Hazard λ belongs to space $\mathcal{V}$:

$$\mathcal{V} = \{\lambda \text{ measurable } : 0 < c_1 \le \lambda(t) \le c_2 = C\cdot c_1\ \forall t\in[0,b]\}.$$

Perturbations γ belong to space $\mathcal{G}_0$:

$$\mathcal{G}_0 = \{\gamma\in\mathcal{C}^2\cap L^\infty([0,b],\nu)\ :\ \|\gamma(t)\|_\infty \le C\cdot c_1\}$$

such that:

$$\{\frac{\gamma}{c_1}\ :\ \gamma\in\mathcal{G}_0\} \text{ is } \|\cdot\|_\infty \text{ dense in } \{g\in L^\infty([0,b],\nu)\ :\ \|g\|_\infty\le C\}.$$

Differentiation operator D_λ for all functionals $\Phi : \mathcal{V} \to \mathbf{R}$, and all $\gamma \in \mathcal{G}_0$:
Let θ be in a small neighborhood of 0. Then

$$\lambda_\theta(t) := \lambda(t) + \theta\gamma(t).$$

$$(D_\lambda\,\Phi(\lambda))(\gamma) := \frac{d}{d\theta}\Phi(\lambda+\theta\gamma)\Big|_{\theta=0} := \frac{d}{d\theta}\Phi(\lambda_\theta)\Big|_{\theta=0} \tag{17.6}$$

Lemma 1 (Least favorable Parametric Sub-model (LFPS)). *For the observational scheme defined in Sect. 17.2 under which* $Q = (L(X), R(X), L(Z), \Xi)$ *is the observed random vector of variables with values* $s = (u, v, w, \xi)$, μ_0 *the true law such that* $d\mu_0 = p_0 d\nu^*$ *and under model (17.1) assumed for the survival time* X, *for fixed* β, *a least favorable nuisance parametrization* Λ_β *is defined recursively through the equation*

$$\Lambda_\beta(t) = \frac{I_1(t,b,\beta)}{I_2(t,b,\beta)} \tag{17.7}$$

where

$$I_1(t,b,\beta) = E_{\mu_0}\left(\frac{\Lambda_\beta(R(X))S_\beta(R(X))e^{\beta'\Xi}}{S_\beta(L(X)) - S_\beta(R(X))}\middle| R(X) > t\right)$$

$$-E_{\mu_0}\left(\frac{\Lambda_\beta(L(X))S_\beta(L(X))e^{\beta'\Xi}}{S_\beta(L(X)) - S_\beta(R(X))}\middle| L(X) > t\right) - E_{\mu_0}\left(\frac{\Lambda_\beta(L(Z))S_\beta(L(Z))e^{\beta'\Xi}}{1 - S_\beta(L(Z))}\middle| L(Z) > t\right)$$

and

$$I_2(t,b,\beta) = E_{\mu_0}\left(e^{\beta'\Xi}\middle| L(X) \geq t\right) + E_{\mu_0}\left(\frac{\Lambda_\beta(L(Z))S_\beta(L(Z))e^{\beta'\Xi}}{1 - S_\beta(L(Z))}\middle| L(Z) \geq t\right)$$

$$-E_{\mu_0}\left(\frac{\Lambda_\beta(R(X))S_\beta(R(X))e^{\beta'\Xi}}{S_\beta(L(X)) - S_\beta(R(X))}\middle| L(X) < t \leq R(X)\right).$$

The least favorable sub-model is defined implicitly through (17.7).

Proof. Proof of Lemma 1 is rather technical.

17.4 Inference About β

17.4.1 Main Result

We impose regularity assumptions on the baseline hazard λ (bounded), on β, the parameter of interest (which has to lie in the interior of a compact set), on the laws of the censoring, the truncation and the covariate Ξ.

Lemma 2 (Efficiency). *Under the assumed regularity assumptions the profile likelihood estimator* $\tilde{\beta}$, *for model (17.1) and interval censored and truncated data, is efficient for* β.
Moreover, the semi-parametric information bound about β_0 *is given at the least favorable sub-model and is defined implicitly from (17.7).*

Hint on proof. We prove that under the assumed regularity assumptions, all requirements for Theorem 3 in Slud and Vonta (2005) are fulfilled. Then, in the current situation we have that

$$\sqrt{n}(\tilde{\beta} - \beta_0) \xrightarrow{\mathcal{D}} \mathcal{N}(0, (I^0_\beta)^{-1})$$

where, in our setting, the semi-parametric information bound is given below as

$$\int \Big(\Big(- a'\xi \{ \frac{(\Lambda(u)S(u) - \Lambda(v)S(v))e^{\beta'\xi}}{S(u) - S(v)} + \frac{\Lambda(w)S(w)e^{\beta'\xi}}{1 - S(w)} \} \Big)$$

$$\times \Big(- \{ \frac{(\Gamma(u)S(u) - \Gamma(v)S(v))e^{\beta'\xi}}{S(u) - S(v)} + \frac{\Gamma(w)S(w)e^{\beta'\xi}}{1 - S(w)} \} \Big) \Big) \Big|_{\beta_0, \lambda_0} p_0 \mathrm{d}\nu^*$$

where γ is chosen as the least favorable direction defined implicitly from the recursive equation (17.7). ■

17.5 Perpectives and References

This is work in progress so that there is the need for simulations and real examples. This can be done along the lines of an example for the joint law of censoring and truncation that we treated in our paper in JSPI [HCSV09].

In a more theoretical direction, we hope to extend the method to other semi-parametric models like for example accelerated models [BN02] and transformation models [HCV04].

References

[BN02] VilijandasBagdonavicius and Mikhail Nikulin: Accelerated Life Models. Chapman & Hall/CRC (2002)

[HCV04] Huber-Carol, C. and Vonta, F.: Frailty models for arbitrarily censored and truncated data. Lifetime Data Analysis, **10**, 369–388 (2004)

[HCSV07] Huber-Carol, C., Solev, V., and Vonta, F.: Estimation of density for arbitrarily censored and truncated data. In: Probability, Statistics and Modelling in Public Health. Springer, New York, 246–265 (2007)

[HCSV09] Huber-Carol, C., Solev,V., and Vonta, F.: Interval censored and truncated data: rate of convergence of NPMLE of the density. Journal of Statistical Planning and Inference, **139**, 1734–1749 (2009)

[SV05] Slud, E. and Vonta, F.: Efficient semiparametric estimators via modified profile likelihood. Journal of Statistical Planning and Inference, **129**, 339–367 (2005)

18

On Generalized Tests of Fit for Multinomial Populations

Alex Karagrigoriou*, Kyriacos Mattheou and Panayiotis Panayiotou

Department of Mathematics and Statistics, University of Cyprus, Cyprus, alex@ucy.ac.cy

Abstract: The aim of this work is the investigation of tests of fit for multinomial distributions based on the Φ-family of test statistics for goodness of fit (gof) tests [Mattheou and Karagrigoriou, Aust. New Zeal. J. Statist. **52**(2), 187–200 (2010)]. For comparing the various goodness of fit tests the asymptotic distribution, which is known to be chi-square, [Cressie and Read, JRSSB, **5**, 440–454 (1984); Zografos et al. Comm. Statist. Theor. Meth., **19(5)**, 1785–1802 (1990); Mattheou and Karagrigoriou, Australian and New Zealand Journal of Statistics **52**(2), 187–200 (2010)] and the empirical distribution of all test statistics under investigation are obtained and at the same time the appropriate critical values are evaluated. For the comparison, samples from trinomial distributions are used and both the size and the power of each test for various alternative hypotheses are calculated.

Keywords and phrases: Generalized goodness of fit tests, Φ-family of test statistics, Multinomial populations, Asymptotic and empirical distribution

18.1 Introduction

The problem of goodness-of-fit to a distribution, $H_0 : F = F_0$, is frequently treated by partitioning the range of data in disjoint intervals and by testing the hypothesis $H_0 : \mathbf{p} = \mathbf{p}_0$ about the vector of parameters of a multinomial distribution.

Let $P = \{E_i\}_{i=1,\ldots,m}$ be a partition of the real line $\mathbb{R}$ in m intervals. Let $\mathbf{p} = (p_1, \ldots, p_m)'$ and $\mathbf{p}_0 = (p_{10}, \ldots, p_{m0})'$ be the true and the hypothesized probabilities of the intervals E_i, $i = 1, \ldots, m$, respectively, in such a way that $p_i = P_F(E_i)$, $i = 1, \ldots, m$ and $p_{i0} = P_{F_0}(E_i) = \int\limits_{E_i} dF_0$, $i = 1, \ldots, m$.

Let $Y_1, \ldots, Y_N$ be a random sample from F, let $n_i = \sum\limits_{j=1}^{n} I_{E_i}(Y_j)$, where

$$I_{E_i}(Y_j) = \begin{cases} 1 & \text{if } Y_j \in E_i \\ 0 & \text{otherwise} \end{cases},$$

V.V. Rykov et al. (eds.), *Mathematical and Statistical Models and Methods in Reliability: Applications to Medicine, Finance, and Quality Control*, Statistics for Industry and Technology, DOI 10.1007/978-0-8176-4971-5_18,

$\widehat{\mathbf{p}} = (\widehat{p}_1, \widehat{p}_2, ..., \widehat{p}_m)'$ with $\widehat{p}_i = n_i/N, i = 1, \ldots, m$ be the maximum likelihood estimator (MLE) of p_i, the true probability of the E_i interval, and $\sum_{i=1}^{m} n_i = N$. For testing the simple null hypothesis

$$H_0 : \mathbf{p} = \mathbf{p}_0, \tag{18.1}$$

the most commonly used test statistics are Pearson's or chi-squared test statistic, given by

$$X^2 = \sum_{i=1}^{m} \frac{(n_i - Np_{i0})^2}{Np_{i0}} \tag{18.2}$$

and the likelihood ratio test statistic given by

$$G^2 = 2\sum_{i=1}^{m} n_i \log\left(\frac{n_i}{Np_{i0}}\right). \tag{18.3}$$

Both these test statistics are particular cases of the family of power-divergence test statistics (CR test) which has been introduced by Cressie and Read [CR84] and is given by

$$I_n^{\lambda}(\widehat{\mathbf{p}}, \mathbf{p}_0) = \frac{2N}{\lambda(\lambda+1)} \sum_{i=1}^{m} \widehat{p}_i \left(\left(\frac{\widehat{p}_i}{p_{i0}}\right)^{\lambda} - 1\right), \qquad \lambda \neq -1, 0 \tag{18.4}$$

where $-\infty < \lambda < \infty$. Particular values of λ in (18.4) correspond to well known test statistics: Chi-squared test statistic X^2 $(\lambda = 1)$, likelihood ratio test statistic G^2 $(\lambda \to 0)$, Freeman–Tukey test statistic $(\lambda = -1/2)$, minimum discrimination information statistic [GK78] $(\lambda \to -1)$, modified chi-squared test statistic $(\lambda = -2)$ and Cressie–Read test statistic $(\lambda = 2/3)$.

Although the power-divergence test statistics yield to an important family of gof tests, it is possible to consider the more general family of $\phi-$ divergence test statistics for testing (18.1) which contains (18.4) as a particular case and is defined by

$$I_n^{\phi}(\widehat{\mathbf{p}}, \mathbf{p}_0) = \frac{2n}{\phi''(1)} \sum_{i=1}^{m} p_{i0}\phi\left(\frac{\widehat{p}_i}{p_{i0}}\right), \tag{18.5}$$

with $\phi(x)$ a twice continuously differentiable function for $x > 0$ such that $\phi''(1) \neq 0$. The above family of tests is based on the well known ϕ-divergence family of measures defined simultaneously by [CS63] and [AS66]:

Definition 1. *The ϕ-divergence family of measures is defined by*

$$D_{\phi}(\widehat{\mathbf{p}}, \mathbf{p}_0) = \sum_{i=1}^{m} p_{i0}\phi\left(\frac{\widehat{p}_i}{p_{i0}}\right), \quad \phi \in \Phi \tag{18.6}$$

where $\hat{p}_i = n_i/N$ the MLE of p_i, $i = 1, \ldots, m$ and Φ is the class of all convex functions $\phi(x)$, $x \geq 0$, such that at $x = 1$, $\phi(1) = 0$, at $x = 0$, $0\phi(0/0) = 0$ and $0\phi(u/0) = \lim_{u\to\infty} \phi(u)/u$.

Note that if $\phi \in \Phi$ is differentiable at $x = 1$, then $\psi(x) \equiv \phi(x) - \phi'(1)(x-1) \in \Phi$ and $\psi'(1) = 0$. This property, together with the convexity, implies that $\psi(x) \geq 0$, $\forall\, x \geq 0$ and $D_\psi(\widehat{\mathbf{p}}, \mathbf{p}_0) = D_\phi(\widehat{\mathbf{p}}, \mathbf{p}_0)$. Since the two divergence measures coincide, we can consider the sets Φ and $\Phi^* \equiv \Phi \cap \{\phi : \phi'(1) = 0\}$ to be equivalent.

Note that the well known Kullback–Leibler (KL) divergence measure [KL51] is obtained for $\psi(x) = x\log(x) - x + 1 \in \Phi$ or $\phi(x) = x\log(x) \in \Phi^*$. If $\varphi(u) = 1/2(1-u)^2$, $\varphi(u) = \left(u^{\lambda+1} - u - \lambda(u-1)\right)/(\lambda(\lambda+1))$, $\lambda \neq 0, -1$, or $\varphi(u) = (1-\sqrt{u})^2$, Csiszar's measure yields the Pearson's chi-squared divergence, the Cressie and Read power divergence [CR84], and the Matusita's divergence [MAT67], respectively. More examples can be found in Arndt [AR01], Pardo [PA06], and Vajda [VA89, VA95].

Csiszar's family of measures was recently generalized by Mattheou and Karagrigoriou [MK10] to the Φ-family of measures given by

$$d_a = \sum_{i=1}^{m} p_{i0}^{1+a} \Phi\left(\frac{\hat{p}_i}{p_{i0}}\right). \tag{18.7}$$

For Φ having the special form

$$\Phi_1(u) = u^{1+a} - \left(1 + \frac{1}{a}\right) u^a + \frac{1}{a} \tag{18.8}$$

we obtain the BHHJ measure of Basu et al. [BHHJ98] which was proposed for the development of a minimum divergence estimating method for robust parameter estimation. Observe that for $\Phi(u) = \phi(u)$ and $a = 0$ we obtain the Csiszár's ϕ-divergence family of measures while for $a = 0$ and for

$$\Phi(u) = \Phi_{2,\lambda}(u) = \frac{1}{\lambda(\lambda+1)}\left(u^{\lambda+1} - u - \lambda(u-1)\right)$$

it reduces to the Cressie and Read power divergence measure. Other important special cases of the Φ–divergence family are the ones for which the function $\Phi(u)$ takes the form

$$\Phi_2(u) = (1+\lambda)\Phi_{2,\lambda}(u)$$

and

$$\Phi_{1,\alpha}(u) = \frac{1}{1+a}\Phi_1(u) = \frac{1}{1+a}\left(u^{1+a} - \left(1 + \frac{1}{a}\right)u^a + \frac{1}{a}\right). \tag{18.9}$$

It is easy to see that for $a \to 0$ the measure $\Phi_2(\cdot)$ reduces to the KL measure.

A general test of fit based on the above Φ-family of measures was recently proposed by Mattheou and Karagrigoriou [MK10]. In this work, we focus on the same test and establish its empirical distribution which then is compared with the asymptotic one. The comparison is based on graphical illustrations as well as on power evaluations through extensive simulations. In Sect. 18.2, we present the Φ-family of tests based on (18.7) and state its asymptotic distribution. In Sect. 18.3, we establish the empirical distribution not only of the Φ-test but also of all the well known test statistics used in such problems, namely the KL, Matusita, CR, and Pearson test statistics. Finally, we provide a comparative study between the asymptotic and the empirical distributions.

18.2 The Φ-Family of Test Statistics

Mattheou and Karagrigoriou [MK10] have recently defined for any function Φ such that $\Phi(1) = \Phi'(1) = 0$ and $\Phi''(1) \neq 0$, the Φ-divergence test statistics for tests of fit discussed in the previous section, which is given by:

$$I_n^{\Phi}(\widehat{\mathbf{p}}, \mathbf{p}_0) = \frac{2Nd_a}{\Phi''(1)}, \quad d_a = \sum_{i=1}^{m} p_{i0}^{1+a} \Phi\left(\frac{\widehat{p}_i}{p_{i0}}\right), \quad \Phi \in \Phi^*. \tag{18.10}$$

Cressie and Read [CR84] obtained the asymptotic distribution of the power-divergence test statistic $I_n^{\Phi}(\widehat{\mathbf{p}}, \mathbf{p}_0)$ for $\Phi(u) = \Phi_{2,a}(u)$, Zografos et al. [ZFP90] extended the result to the family $I_n^{\Phi}(\widehat{\mathbf{p}}, \mathbf{p}_0)$ for $a = 0$ and $\Phi = \phi \in \Phi^*$ and Mattheou and Karagrigoriou [MK10] extended the result to cover any function $\Phi \in \Phi^*$:

Theorem 1. *Under the null hypothesis $H_0 : \mathbf{p} = \mathbf{p}_0 = (p_{10}, \ldots, p_{m0})'$, the asymptotic distribution of the Φ-divergence test statistic, $I_n^{\Phi}(\widehat{\mathbf{p}}, \mathbf{p}_0)$, is a chi-square with $m-1$ degrees of freedom times a constant c:*

$$I_n^{\Phi}(\widehat{\mathbf{p}}, \mathbf{p}_0) \xrightarrow[n\to\infty]{L} c\chi^2_{m-1},$$

where $c = 0.5\left(\min_i p_{i0}^a + \max_i p_{i0}^a\right)$.

The following two theorems for the CR test statistic and the ϕ-family of test statistics are special cases of the above theorem for $c = 1$, $a = 0$ and for the appropriate forms of the function $\Phi(\cdot)$.

Theorem 2. *Under the null hypothesis $H_0 : \mathbf{p} = \mathbf{p}_0 = (p_{10}, \ldots, p_{m0})'$, the asymptotic distribution of the divergence test statistic $I_n^{\Phi}(\widehat{\mathbf{p}}, \mathbf{p}_0)$ with $\Phi = \Phi_{2,\lambda}$ given in (18.4), is chi-square with $m-1$ degrees of freedom:*

$$I_n^{\Phi_{2,\lambda}}(\widehat{\mathbf{p}}, \mathbf{p}_0) \xrightarrow[n\to\infty]{L} \chi^2_{m-1}.$$

Theorem 3. *Under the null hypothesis $H_0 : \mathbf{p} = \mathbf{p}_0 = (p_{10}, \ldots, p_{m0})'$, the asymptotic distribution of the $\phi-$ divergence test statistic, $I_n^{\phi}(\widehat{\mathbf{p}}, \mathbf{p}_0)$ given in (18.5), is chi-square with $m-1$ degrees of freedom:*

$$I_n^{\phi}(\widehat{\mathbf{p}}, \mathbf{p}_0) \xrightarrow[n\to\infty]{L} \chi^2_{m-1}.$$

The following theorem by Mattheou and Karagrigoriou [MK10] provides the asymptotic power of the proposed test statistic.

Theorem 4. *Let $(n_1, \ldots, n_m)' \sim M(N, \mathbf{p})$ with $\mathbf{p} = (p_1, \ldots, p_m)'$, p_i, $i = 1, \ldots, m-1$ unknown parameters and $p_m = 1 - \sum_{i=1}^{m-1} p_i$. Under the alternative hypothesis $H_a : \mathbf{p} = \mathbf{p}_b = (p_{1b}, \ldots, p_{mb})'$ the power of the test given in (18.10) is asymptotically equal to*

$$\gamma_a = P\left(Z \geq (2\sqrt{n}\sigma_a)^{-1}\left(\Phi''(1)\, c\chi^2_{m-1,\alpha} - 2nd_a^*\right)\right), \tag{18.11}$$

where d_a^* is given by (18.7) with p_{ib} in place of $\hat{p}_i$, Z is a standard Normal random variable, $\chi^2_{m-1,\alpha}$ is the αth percentile of the χ^2_{m-1} distribution, and

$$\sigma_a^2 = \sum_{i=1}^{m} p_{ib} \left[p_{i0}^a \Phi' \left(\frac{p_{ib}}{p_{i0}} \right) \right]^2 - \left[\sum_{j=1}^{m} p_{ib} p_{i0}^a \Phi' \left(\frac{p_{ib}}{p_{i0}} \right) \right]^2 .$$

Note that from the statistical point of view, the Cressie and Read family of measures with $\lambda = 2/3$ is very important since as a by-product, the power divergence statistic CR that emerged for goodness of fit purposes for multinomial distributions has received widespread attention [RC88]. Other related goodness of fit tests have been investigated among others by Aguirre and Nikulin [AN94], Bagdonavičius and Nikulin [BN02], Huber and Vonta [HV04], Marsh [MAR06], Menéndez et al. [MPPP97] and Zhang [ZH02].

18.3 Empirical Distribution and Power Simulations

For checking the accuracy of the asymptotic distribution of I_n^Φ test statistic as well as the asymptotic distribution of the other four popular tests, namely the tests of fit based on the Kullback measure (KL), the Pearson's chi-squared measure (X^2), the Matusita measure (Mat) and the Cressie and Read measure (CR), we evaluate first their empirical cumulative distribution functions (cdfs).

The proposed Φ test of fit is applied for $\Phi = \Phi_1$ and $\Phi = \Phi_2$ and for two different values of the index a, namely for $a = 0.01$ and 0.05 while the CR test is applied for $\lambda = 2/3$. For the empirical distributions we focus on the trinomial distribution $\mathbf{p}_0 = (p_{10}, p_{20}, p_{30})' = (0.2, 0.6, 0.2)'$ with sample size equal to 150 and 25,000 simulations. For each simulation, the value of each test statistic is evaluated. The 25,000 values of each test are ranked and the graph of the empirical distribution is compared with the corresponding asymptotic distribution presented in the appropriate theorem of Sect. 14.2. The graphs for the Φ-test for the function Φ_1 with $a = 0.01$ and $a = 0.05$ are given in Figs. 18.1 and 18.2, respectively, while the graphs for the same values of the index a for the function Φ_2 are given in Figs. 18.3 and 18.4.

All figures show that the fit is extremely good if one takes into consideration the fact that the sample size used is 150. Note that the same evaluations have been done for sample sizes equal to 500 and 1,000. Since we are primarily interested in testing procedures, the 95th percentiles of the empirical and asymptotic distributions are provided in Table 18.1 for all 6 competing tests (the $a = 0.01$ case in parentheses). Recall that the asymptotic percentiles follow from Theorems 1–3.

For checking now the accuracy of the Φ-family of tests and its asymptotic distribution, simulations using trinomial distributions are considered. For the null hypothesis, 10,000 simulations of size 150 are drawn from the trinomial distribution $\mathbf{p}_0 = (p_{10}, p_{20}, p_{30})' = (0.2, 0.6, 0.2)'$. Note that besides the examples presented here a number of other simulations have been performed (results not shown) with sample sizes equal to 50, 100, and 500 and for nonsymmetric null hypotheses (skewed either to the right or to the left) with similar conclusions.

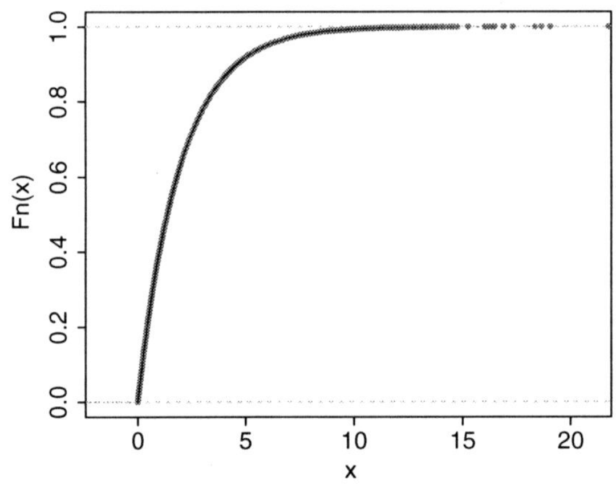

Figure 18.1. Empirical and asymptotic null distribution of Φ_1-test statistic ($a = 0.01$) for trinomial $M(150, \mathbf{p}_0)$

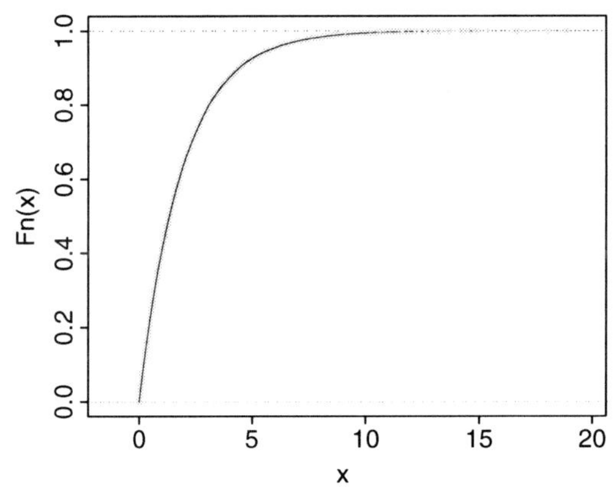

Figure 18.2. Empirical and asymptotic null distribution of Φ_1-test statistic ($a = 0.05$) for trinomial $M(150, \mathbf{p}_0)$

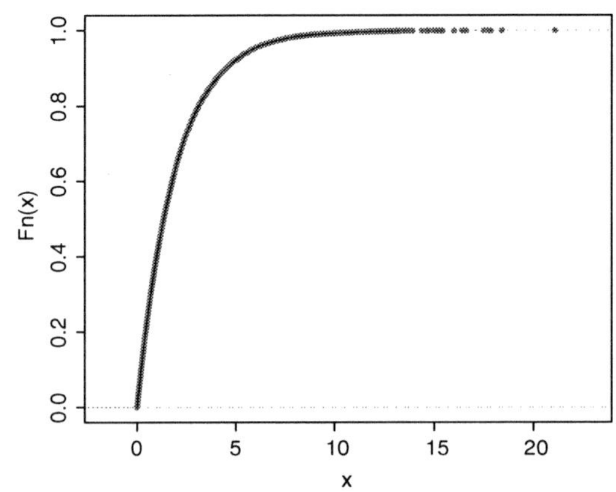

Figure 18.3. Empirical and asymptotic null distribution of Φ_2-test statistic ($a = 0.01$) for trinomial $M(150, \mathbf{p}_0)$

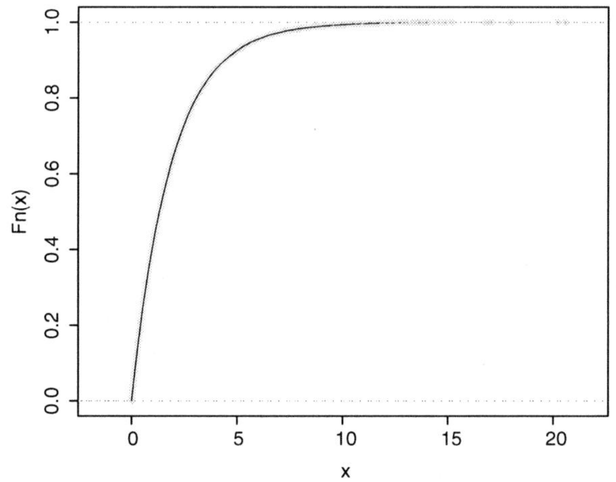

Figure 18.4. Empirical and asymptotic null distribution of Φ_2-test statistic ($a = 0.05$) for trinomial $M(150, \mathbf{p}_0)$

Table 18.1. 5% Upper critical values for tests of fit

Test	Empirical value	Asymptotic value
Mat	6.088768	5.991465
Pearson X^2	6.044444	5.991465
KL	5.961975	5.991465
CR	5.992962	5.991465
Φ_1	5.830634 (6.121232)	5.684291 (5.928373)
Φ_2	5.562362 (5.875057)	5.684291 (5.928373)

In Table 18.2, the powers based on the asymptotic distribution are presented. The results according to the empirical cdfs are presented in Table 18.3. Both tables provide the proportion of times the null hypothesis is rejected. The various alternatives considered are presented in the first column of the tables (p_3 is omitted). Observe though that the first row refers to the size of the test. The results from the power calculations reveal a number of conclusions which are stated below:

- The Φ_1-test performs better than the KL statistic irrespectively of the alternative hypothesis.
- The Φ_2-test for $a = 0.01$ is identical to the KL test. This is expected since as $a \to 0$, $\Phi_2(u) \to u \log u - u + 1$ which is the function associated with the KL test. More specifically observe that the powers of the resulting test increase as the index a approaches 0 and they reach their maximum value at $a = 0.01$. Notice that for both Φ_1 and Φ_2 tests, the larger the value of a the smaller the power of the test.
- The results based on the empirical distribution are very similar to the ones based on the asymptotic distribution for all cases examined.
- The Φ_1-test performs better than the other tests for all alternatives that are not far away from the null hypothesis. On the other hand it performs as good as all other tests for all alternatives that are far away from the null hypothesis.

Table 18.2. Powers (%) for Φ_1 and Φ_2 tests based on the asymptotic distribution ($\alpha = 0.05$)

H_a	Competing tests				Φ_1-test		Φ_2-test	
$H_O: p_{10} = 0.20, p_{20} = 0.60, p_{30} = 0.20,\ \chi^2_{2;0.05},\ n = 150$								
p_{1b} & p_{2b}	KL	X^2	Mat	CR	$a = 0.01$	0.05	$a = 0.01$	0.05
0.20, 0.60	5.12	5.67	5.78	5.44	5.75	5.58	5.12	4.97
0.21, 0.59	6.07	6.86	6.49	6.28	6.29	5.95	6.07	5.90
0.22, 0.60	9.44	10.00	10.13	9.69	10.09	9.68	9.44	8.96
0.25, 0.60	38.80	39.26	40.27	39.06	40.42	38.67	38.80	37.08
0.20, 0.70	89.15	88.39	91.10	88.53	91.87	91.82	89.15	88.91
0.10, 0.60	96.38	96.08	96.65	96.22	96.65	96.17	96.38	95.80
0.40, 0.36	99.99	100.00	99.99	100.00	99.99	99.99	100.00	100.00
0.45, 0.35	100.00	100.00	100.00	100.00	100.00	100.00	99.99	99.99
0.40, 0.30	100.00	100.00	100.00	100.00	100.00	100.00	100.00	100.00
0.55, 0.25	100.00	100.00	100.00	100.00	100.00	100.00	100.00	100.00

- The Matusita and the Φ_1 tests have a very similar behavior and in most cases are the most powerful tests among the ones examined. Both tests behave well for alternatives close to the null hypothesis and better than both the classical Pearson chi-squared test and the KL test. This observation indicates that the Φ_1-test, as well as the Matusita test, are able to distinguish between null and alternative hypotheses when they are very close.

Table 18.3. Powers (%) for Φ_1 and Φ_2 tests based on the empirical distribution ($\alpha = 0.05$)

H_a	Competing tests				Φ_1-test		Φ_2-test	
$H_O: p_{10} = 0.20, p_{20} = 0.60, p_{30} = 0.20,\ n = 150$								
p_{1b} & p_{2b}	KL	X^2	Mat	CR	$a = 0.01$	0.05	$a = 0.01$	0.05
0.20, 0.60	5.19	5.35	5.31	5.39	5.33	5.31	5.19	5.28
0.21, 0.59	6.14	6.32	6.07	6.22	5.67	5.70	6.14	6.14
0.22, 0.60	9.54	9.49	9.37	9.52	9.48	9.36	9.54	9.38
0.25, 0.60	38.84	38.24	38.38	38.59	38.40	38.20	38.84	38.40
0.20, 0.70	89.36	87.91	90.46	88.39	91.82	91.48	89.36	89.15
0.10, 0.60	96.38	95.80	96.20	96.06	96.11	96.07	96.38	96.16
0.40, 0.36	100.00	100.00	100.00	100.00	100.00	100.00	100.00	100.00
0.45, 0.35	99.99	100.00	99.99	100.00	99.99	99.99	99.99	100.00
0.40, 0.30	100.00	100.00	100.00	100.00	100.00	100.00	100.00	100.00
0.55, 0.25	100.00	100.00	100.00	100.00	100.00	100.00	100.00	100.00

The above results clearly show the appropriateness of the asymptotic distribution of the generalized test statistic for tests of fit presented in Theorem 1. It is important to point out that any type of data can be viewed as multinomial data by dividing the range of data into m categories. In that sense data related to biomedicine, engineering and reliability that usually come from continuous distributions can be transformed into multinomial data and tests of fit based on the above measures can be applied. Some of the most popular of such continuous distributions are the exponential, lognormal, Gamma, Inverse Gaussian, Weibull, Pareto, and Positive Stable distributions. For instance, the family of the two-parameter inverse Gaussian distribution (IG2) is one of the basic models for describing positively skewed data which arise in a variety of fields of applied research as cardiology, hydrology, demography, linguistics, employment service, etc. Such examples include the repair times of an airborne communication transceiver [CF77] and quality characteristics [SI03]. Recently, Huberman et al. [HPPL98] have argued and demonstrated the appropriateness of the inverse Gaussian family for studying the internet traffic and in particular the number of visited pages per user within an internet site. Most applications of IG2 are justified on the fact that the IG2 is the distribution of the first passage time in Brownian motion with positive drift. Furthermore, distributions like the Weibull, the Positive Stable and the Pareto are frequently encountered in survival modelling. The main problem of determining the appropriate distribution is extremely important for reducing the possibility of erroneous inference. In addition, the existence of censoring schemes in survival modelling makes the determination of the proper distribution an extremely challenging problem. Finally distributions like the exponential, the Gamma, the lognormal and others are very common in lifetime problems.

In Tables 18.4 and 18.5 we present the size of the Φ_1-test of fit for the Inverse Gaussian and the Gamma distributions. In particular, we compare, the size of the test for various null hypotheses. More specifically, for the Gamma distribution, we use a scale parameter equal to 1 and for the shape parameter we use the values 0.5, 1, 2, and 4. In regard to the Inverse Gaussian distribution, we examine a wide range of skewness values but report here the results only for small (skewness=1.414), medium (skewness=2) and large values (skewness=3 and 4) of skewness. The results are presented for both the empirical and the asymptotic distribution for comparative purposes. For the determination of the empirical percentiles, samples of size equal to 25000 have been used. For a better understanding, we provide also the size of a number of well known tests which are frequently used for testing the particular distributions. These tests include the R test [KCK10, KK10], the Anderson Darling test, the Mudholkar Z-test [MNC01] especially for the inverse Gaussian distribution and the minimum $\mathcal{X}^2$ test [KKC06] especially for the Gamma distribution. In all cases examined the range of values has been divided into $m = 3$ disjoint intervals such that the first and the last contain the upper and lower 20% of the distribution while the middle one the central 60% of the distribution.

The results clearly show that the empirical and asymptotic results for the Φ test are much closer than in all other tests. This indicates that the asymptotic distribution for the Φ test statistic given in Theorem 1 is quite satisfactory for continuous null distributions like the ones examined. Finally, we observe that the asymptotic distribution presents an underestimation for the KL test while in all other cases, the reversed is observed.

Table 18.4. Asymptotic (Asy) and empirical (Emp) size (%) for the Φ_1-test with $a = 0.01$ and 0.05 and competing tests for the inverse Gaussian distribution

Inverse Gaussian (shape=γ, scale=1)								
γ	4.5		2.25		1.00		0.56	
	Asy	Emp	Asy	Emp	Asy	Emp	Asy	Emp
Φ_1 ($a = 5\%$)	5.24	4.48	5.40	4.73	4.77	4.11	5.31	4.59
Φ_1 ($a = 1\%$)	5.51	5.30	5.60	5.48	5.00	4.92	5.57	5.30
CR test	5.19	4.97	5.26	5.03	4.78	4.46	5.71	5.41
Mat test	5.62	5.13	5.50	5.26	5.04	4.56	5.78	5.22
X^2 test	5.49	4.65	5.56	4.77	5.12	4.30	6.16	5.36
KL test	5.12	5.44	5.20	5.56	4.62	4.95	5.34	5.75
AD	4.90		5.00		4.70		4.90	
R test	5.50		5.40		5.10		5.60	
Z test	5.70		5.60		4.40		4.10	

Table 18.5. Asymptotic (Asy) and empirical (Emp) size (%) for the Φ_1-test with $a = 0.01$ and 0.05 and competing tests for the Gamma distribution

Gamma (shape=γ, scale=1)								
γ	4.0		2.0		1.0		0.5	
	Asy	Emp	Asy	Emp	Asy	Emp	Asy	Emp
Φ_1 ($a = 5\%$)	5.24	4.48	5.40	4.73	4.77	4.11	5.31	4.59
Φ_1 ($a = 1\%$)	5.51	5.30	5.60	5.48	5.00	4.92	5.57	5.30
CR test	5.19	4.97	5.26	5.03	4.78	4.46	5.71	5.41
Mat test	5.62	5.13	5.50	5.26	5.04	4.56	5.78	5.22
X^2 test	5.49	4.65	5.56	4.77	5.12	4.30	6.16	5.36
KL test	5.12	5.44	5.20	5.56	4.62	4.95	5.34	5.75
AD	5.20		5.30		4.70		5.30	
R test	5.00		6.00		3.50		5.60	
min $\mathcal{X}^2$ test	4.60		4.30		5.20		5.20	

18.4 Conclusions

The aim of this work is the investigation of generalized tests of fit for multinomial populations which are based on the Φ-divergence class of measures. In particular, we present various test statistics associated with the above testing problem and calculate the size and the power by simulating samples from trinomial distributions. For comparative purposes we are using both the asymptotic and the empirical distributions. We are mainly interested in investigating the behavior of the Φ-divergence class and establish its appropriateness.

The results show that the Φ_1-test as well as the Matusita test perform better than the Kullabck–Leibler test (KL) in most cases and also have the advantage of distinguishing between null and alternative hypothesis when they are very close. The empirical and asymptotic percentiles for all tests are extracted and their similarity is verified.

For checking the accuracy of the asymptotic distribution of the Φ_1 and Φ_2 tests the graphs of the associated empirical distributions are drawn together with the corresponding asymptotic ones. The graphs clearly show that the two distributions are identical in all cases which proves the appropriateness of the asymptotic distribution.

Finally, Inverse Gaussian and Gamma simulations are used with $m = 3$ disjoint intervals for establishing the appropriateness of the asymptotic distributions for continuous distributions often appearing in engineering systems and reliability theory.

References

[AN94] Aguirre, N., Nikulin, M.: Chi-squared goodness-of-fit test for the family of logistic distributions. Kybernetika **30**, 214–222 (1994)

[AS66] Ali, S. M., Silvey, S. D.: A general class of coefficients of divergence of one distribution from another. J. Roy. Statist. Soc. B **28**, 131–142 (1966)

[AR01] Arndt, C.: Information Measures. Springer, Berlin (2001)

[BN02] Bagdonavičius, V. Nikulin, M. S.: Goodness-of-fit tests for accelerated life models. In: Huber-Carol, C., Balakrishnan, N., Nikulin, M. S., and Mesbah, M. (eds) Goodness-of-fit tests and Model Validity. Birkhäuser, 281–297 (2002)

[BHHJ98] Basu, A., Harris, I. R., Hjort, N. L., Jones, M. C.: Robust and efficient estimation by minimising a density power divergence. Biometrika **85**, 549-559 (1998)

[CF77] Chhikara, R. S., Folks, J. L.: (1977). The inverse Gaussian distribution as a lifetime model. Technometrics **19**, 461–468 (1977)

[CR84] Cressie, N., Read, T. R. C.: Multinomial goodness-of-fit tests. JRSSB **5**, 440–454 (1984)

[CS63] Csiszar, I.: Eine Informationstheoretische Ungleichung und ihre Anwendung auf den Bewis der Ergodizitat on Markhoffschen Ketten. Publication of the Mathematical Institute of the Hungarian Academy of Sciences **8**, 84–108 (1963)

[GK78] Gokhale, D. V., Kullback, S.: The Information in Contingency Tables. Marcel Dekker, New York (1978)

[HV04] Huber-Carol, C., Vonta, F.: Frailty models for arbitrarily censored and truncated data. Lifetime Data Analysis **10 (4)**, 369–388 (2004)

[HPPL98] Huberman, B. A., Pirolli, P. L. T., Pitkow, J. E., Lukose, R. M.: Strong regularities in world wide web surfing. Science **280**, 95–97 (1998)

[KKC06] Kallioras, A.G., Koutrouvelis, I.A., Canavos, G.C.: Testing the fit of gamma distributions using the empirical moment generating function. Comm. Statist. Theory Methods **35**, 527–540 (2006)

[KCK10] Koutrouvelis, I. A., Canavos, G., Kallioras, A.: Cumulant plots for assessing the Gamma distribution, Commun. Statist. Theor. Meth. **39**, 626–641 (2010)

[KK10] Koutrouvelis, I.A., Karagrigoriou, A.: Cumulant plots and goodness-of-fit tests for the inverse Gaussian distribution (under revision) (2010)

[KL51] Kullback, S., Leibler, R.: On information and sufficiency. Annals of Math. Statist. **22**, 79–86 (1951)

[MAR06] Marsh, P.: Data driven likelihood ratio tests for goodness-to-fit with estimated parameters. Discussion Papers in Economics, Dept. of Economics and Related Studies, The University of York, **2006/20** (2006)

[MK10] Mattheou, K., Karagrigoriou, A.: A new family of divergence measures for tests of fit. Australian and New Zealand Journal of Statistics **52**(2), 187–200 (2010)

[MAT67] Matusita, K.: On the notion of affinity of several distributions and some of its applications. Ann. Inst. Statist. Math. **19**, 181–192 (1967)

[MPPP97] Menéndez, M. L., Pardo, J. A., Pardo, L., Pardo, M. C.: Asymptotic approximations for the distributions of the (h, ϕ)-divergence goodness-of-fit statistics: Applications to Rényi's statistic. Kybernetes **26**, 442–452 (1997)

[MNC01] Mudholkar, G. S., Natarajan, R., Chaubey, Y. P.: A goodness-of-fit test for the inverse Gaussian distribution using its independence characterization. Sankhya B **63**, 362–374 (2001)

[PA06] Pardo, L.: Statistical Inference Based on Divergence Measures, Chapman and Hall, London (2006)

[RC88] Read, T.R.C., Cressie, N.: Goodness-of-Fit Statistics for Discrete Multivariate Data. Springer, New York (1988)

[SI03] Sim, C.H.: Inverse Gaussian control charts for monitoring process variability. Commun. Statist. Simul. Comput. **32**, 223–239 (2003)

[VA89] Vajda, I.: Theory of Statistical Inference and Information. Kluwer, Dordrecht (1989)

[VA95] Vajda, I.: Information-theoretic methods in Statistics, in Research Report. Academy of Sciences of the Czech Republic. Institute of Information Theory and Automation. Prague (1995)

[ZH02] Zhang, J.: Powerful goddness-of-fit tests based on likelihood ratio. J. R. Stat. Soc. Ser. B **64 (2)**, 281–294 (2002)

[ZFP90] Zografos, K., Ferentinos, K., Papaioannou, T.: Φ-divergence statistics: Sampling properties, multinomial goodness of fit and divergence tests. Comm. in Statist. Theor. Meth. **19(5)**, 1785–1802 (1990)

19

Modeling and Scaling of Categorical Data

Henning Läuter* and Ayad M. Ramadan

Institute of Mathematics, University of Potsdam, Potsdam, Germany,
laeuter@uni-potsdam.de

Abstract: Estimation and testing of distributions in metric spaces are well known. R.A. Fisher, J. Neyman, W. Cochran, and M. Bartlett achieved essential results on the statistical analysis of categorical data. In the last 40 years, many other statisticians found important results in this field.

Often data sets contain categorical data, e.g., levels of factors or names. There does not exist any ordering or any distance between these categories. At each level there are measured some metric or categorical values. We introduce a new method of scaling based on statistical decisions. For this we define empirical probabilities for the original observations and find a class of distributions in a metric space where these empirical probabilities can be found as approximations for equivalently defined probabilities. With this method we identify probabilities connected with the categorical data and probabilities in metric spaces. Here, we get a mapping from the levels of factors or names into points of a metric space. This mapping yields the scale for the categorical data.

From the statistical point of view we use multivariate statistical methods, we calculate maximum likelihood estimations and compare different approaches for scaling.

Keywords and phrases: Multivariate scaling, Discrimination, Power of multivariate tests

19.1 Introduction

Estimation and testing for distributions of metric random variables are known since the end of the nineteenth century. R.A. Fisher and many other statisticians developed very efficient statistical methods for analyzing medical and biological data. These methods correspond to regression, multivariate analysis, and in general to data analysis. Many procedures, e.g. the procedures of the analysis of variance belong to the basic methods in applied statistics.

Essential contributions about statistics of categorical data were developed first by R.A. Fisher, J. Neyman, W. Cochran and M. Bartlett. One finds very different strong

V.V. Rykov et al. (eds.), *Mathematical and Statistical Models and Methods in Reliability: Applications to Medicine, Finance, and Quality Control*, Statistics for Industry and Technology, DOI 10.1007/978-0-8176-4971-5_19,

results for analyzing categorical data since the 1960s. Mostly data structures from social, biological, medical and technical areas are analyzed. In biomedical applications categories as sex, race, or social strata are considered, in technical problems one works with technical patterns or places. In social problems, one uses verbal assessments or marks, in political or philosophical context one finds arrangements as "liberal", "moderate" or "conservative".

In this paper, we introduce a method of scaling based on statistical decisions, especially classification methods are used. We will concentrate on methods and examples with categorical data. But it will be clear that the proposed procedures can be used as a pretreatment in other data structures for generating such transformed data which conform with assumptions in standard software.

Multidimensional scaling is considered by several authors. In most of the cases, they use similarities or dissimilarities and then they find scales for the categories [EvDu01]. Instead of such geometrical approaches we use here statistical decisions.

19.2 Basic Model

We consider the q-way classification model which is used mostly in the analysis of variance. At least the structure is interesting for us. The basic assumptions can be explained in the 2-way classification. We are given data in the following structure.

	Factor B				
Factor A	level 1	level 2	level 3	$\cdots$	level b
level 1	z_{11j} $j = 1, ..., m_{11}$	z_{12j} $j = 1, ..., m_{12}$	z_{13j} $j = 1, ..., m_{13}$	$\cdots$ $\cdots$	z_{1bj} $j = 1, ..., m_{1b}$
$\vdots$	$\vdots$	$\vdots$	$\vdots$	$\ddots$	$\vdots$
level a	z_{a1j} $j = 1, ..., m_{a1}$	z_{a2j} $j = 1, ..., m_{a2}$	z_{a3j} $j = 1, ..., m_{a3}$	$\cdots$ $\cdots$	z_{abj} $j = 1, ..., m_{ab}$

At level s of the factor A and level t of factor B there are m_{st} observations and the total sample size is

$$m = \sum_{s=1}^{a}\sum_{t=1}^{b} m_{st}.$$

Here, the categorical variables are the levels of the factors A and B. The $L := a \cdot b$ categories are described by

$$\{(s,t) : \quad s = 1, ..., a;\ t = 1, ..., b\}.$$

For finding a scale for the categories it is convenient to group the observations $\{z_{stj}\}$ in given classes. We assume that there are given k classes and h_{il} is the number of observations in the category $l = (s,t)$ falling in the class K_i. Then we have the contingency table

	Factor B								
Factor A	level 1		level 2		level 3		$\cdots$	level b	
level 1	$h_{1(1,1)}$	$...h_{k(1,1)}$	$h_{1(1,2)}$	$...h_{k(1,2)}$	$h_{1(1,3)}$	$...h_{k(1,3)}$	$\cdots$	$h_{1(1,b)}$	$...h_{k(1,b)}$
$\vdots$	$\vdots$		$\vdots$		$\vdots$		$\ddots$	$\vdots$	
level a	$h_{1(a,1)}$	$...h_{k(a,1)}$	$h_{1(a,2)}$	$...h_{k(a,2)}$	$h_{1(a,3)}$	$...h_{k(a,3)}$	$\cdots$	$h_{1(a,b)}$	$...h_{k(a,b)}$

This table of frequencies is our starting data set. The interpretation of this table is the following. At the level $l = (s,t)$ we have observations from different classes and from the

$$m_l = \sum_{i=1}^{k} h_{il} = \sum_{i=1}^{k} h_{i(s,t)}$$

observations the parts for the classes are given by

$$\frac{h_{1l}}{m_l}, ..., \frac{h_{kl}}{m_l}.$$

Such tables are obtained in a similar way if $q > 2$. The dimension of l depends on q in general.

19.2.1 Modeling of Categorical Data

We will find a model for data structures given in the last tables. For this we remember the discrimination of distributions or classes. There are given k distributions with densities $f_1, ..., f_k$ over a space $\mathbb{R}^p$ and for each point $y \in \mathbb{R}^p$ it is known that it can be a realization of one of the classes. Furthermore, let $\pi_1, ..., \pi_k$ be prior probabilities for the classes. Then it is known that a given realization y of the random variable Y with the density f with

$$f(y) = \sum_{j=1}^{k} \pi_j f_{\vartheta_j}(y)$$

corresponds to the class K_i with the probability

$$\mathsf{P}(Y \in K_i \mid Y = y)) = \frac{\pi_i f_{\vartheta_i}(y)}{\sum_{j=1}^{k} \pi_j f_{\vartheta_j}(y)} =: \widetilde{p}_i(y). \tag{19.1}$$

Consequently, under the assumption that a point y (or a point in a very near neighborhood of y) has the frequency $\tilde{m}$ in the data set then we expect $\tilde{m} \cdot \widetilde{p}_i(y)$ of these points corresponding to class K_i.

Modeling of Categories

We denote by $\mathsf{Mult}(\tilde{m}, p_1, ..., p_k)$ the multinomial distribution where p_i is the probability for the class K_i and we repeat the experiment $\tilde{m}$ times. Then we expect that under these $\tilde{m}$ observations approximately $\tilde{m} \cdot p_i$ correspond to K_i. As a consequence from the last two subsections we formulate the definition.

Definition 1. *The points* $\{s_l \in \mathbb{R}^p, l = 1, ..., L\}$ *are called scale points for the categories* $\{x_l, l = 1, ..., L\}$ *if*

$$(h_{1l}, ..., h_{kl})$$

are realizations of independent multinomial random variables W_l *with*

$$W_l \sim \mathsf{Mult}(h_{+l}, \widetilde{p}_1(s_l), ..., \widetilde{p}_k(s_l)),\ l = 1, ..., L$$

and

$$h_{+l} = \sum_{i=1}^{k} h_{il}.$$

This means that in the statistical model the same expected frequencies occur as in the table. In general, the densities depend on some parameters. Then, one has to estimate the distributional parameters and the scale parameters. This can be done using the likelihood principle. Here, we use another criterion. We will find such scales that the classes will be discriminated as well as possible. Voinov and Nikulin considered in [VoNi93] multivariate multinomial distributions for identically distributed W_l, here we use a more general model.

19.2.2 Determination of Observations

Scale points are to be constructed on the basis of the observations. The observations are those which are given by the categories and the frequencies. In our understanding, the categories are identified with points $x_1, ..., x_L \in \mathbb{R}^p$ and these points are to be determined in an optimal way. The observations express the correspondence to some classes, denoted by $\{y_{11}, ..., y_{kn_k}\}$. Explicitly, we have the observations

$$\{y_{11}, ..., y_{1n_1}\} = \{h_{11}\,\text{times}\,x_1, h_{12}\,\text{times}\,x_2, ..., h_{1L}\,\text{times}\,x_L\},$$

hence, we have $n_1 = h_{1+}$. Or we write

$$\begin{aligned} y_{1t} &= x_1, t = 1, ..., h_{11};\ y_{1t} = x_2, t = h_{11} + 1, ..., h_{11} + h_{12}; ...; \\ y_{1t} &= x_L, t = h_{1+} - h_{1L}, ..., h_{1+}. \end{aligned}$$

In an analogous way we have for the other classes $i = 2, ..., k$

$$\begin{aligned} y_{it} &= x_1, t = 1, ..., h_{i1};\ y_{it} = x_2, t = h_{i1} + 1, ..., h_{i1} + h_{i2}; ...; \\ y_{it} &= x_L, t = h_{i+} - h_{iL}, ..., h_{i+}. \end{aligned}$$

It holds $n_i = h_{i+}$. For statistical decisions one needs assumptions on the distributions.

19.2.3 Choice of Distributions

In general, one chooses $y_{ij} \in \mathbb{R}^q$ if one has data from a q-way classification model. This means $p = q$. But sometimes a lower-dimensional space is also possible especially if some factors have not a large influence on the results. The special case $p = 1$ is of interest if one likes to have an ordered scale for the categories.

Depending on the meaning of the observations we can choose the distributions. Quite often binomial, normal, or Poisson distributions are useful, but especially in reliability or survival analysis exponential or Weibull distributions are to be chosen.

Assuming that we are given k distributions $\mathsf{P}_{\vartheta_1}, ..., \mathsf{P}_{\vartheta_k}$ and for each distribution P_{ϑ_i} with a density f_{ϑ_i} we have a random sample $Y_{i1}, ..., Y_{in_i}$. All random variables should be independent. For testing

$$\mathcal{H} : \mathsf{P}_{\vartheta_1} = ... = \mathsf{P}_{\vartheta_k}$$

against $\mathcal{K}$, that not all distributions are the same, we use the likelihood ratio test. The joint density for $Y = (Y_{11}, ..., Y_{kn_k})$ is denoted by $f_{\vartheta_1,...,\vartheta_k}$. As usually, the LRT is given by

$$\varphi(y) = 1 \quad \text{if} \quad R_n(y) := \frac{\max_{\vartheta_1,...,\vartheta_k} f_{\vartheta_1,...,\vartheta_k}(y)}{\max_\vartheta f_{\vartheta,...,\vartheta}(y)} \geq c,$$

where c ensures the significance level.

Normal Distributions

We assume that

$$\begin{matrix} Y_{11} & \dots & Y_{1n_1} \\ \vdots & \ddots & \vdots \\ Y_{k1} & \dots & Y_{kn_k} \end{matrix}$$

are independent and normally distributed p-dimensional random variables, $Y_{ij} \sim \mathsf{N}_p(\mu_i, \Sigma)$. Then we consider the test problem

$$\mathcal{H}: \quad \mu_1 = ... = \mu_k \quad \text{against} \quad \mathcal{K}: \quad \text{not}\, \mathcal{H}. \tag{19.2}$$

We denote the sample mean for the ith distribution by $y_{i\cdot}$, $\quad i = 1, ..., k$, the total mean by

$$y_{\cdot\cdot} = \frac{1}{n} \sum_{i=1}^{k} \sum_{j=1}^{n_i} y_{ij} = \frac{1}{n} \sum_{j=1}^{k} n_j y_{j\cdot} \quad .$$

The unbiased estimator for the variance is

$$S = \frac{1}{n-k} \sum_{i=1}^{k} \sum_{j=1}^{n_i} (Y_{ij} - Y_{i\cdot})(Y_{ij} - Y_{i\cdot})^t.$$

Then

$$T_0^2(Y) = \frac{n-k-p+1}{(k-1)(n-k)p} \sum_{i=1}^{k} n_i (Y_{i\cdot} - Y_{\cdot\cdot})^t S^{-1} (Y_{i\cdot} - Y_{\cdot\cdot})$$

is approximately F-distributed. H. Ahrens and J. Läuter proposed in [AhLa81] the approximation $T_0^2(Y) \approx \mathsf{F}_{g_1,g_2}$ for

$$g_1 = \begin{cases} \frac{(k-1)(n-k-p)p}{n-(k-1)p-2} & \text{if} \quad n-(k-1)p-2 > 0 \\ \infty & \text{otherwise,} \end{cases}$$

$$g_2 = n-k-p+1.$$

Then an admissible test is given by

$$\varphi(y) = \begin{cases} 1 & \text{if } \; T_0^2(y) > \mathsf{F}_{g_1,g_2;\alpha} \\ 0 & \text{otherwise,} \end{cases}$$

for the α-fractile of the F_{g_1,g_2}-distribution.

Exponential Distributions

We choose $p = 1$. We consider independent exponentially distributed variables

$$\begin{matrix} Y_{11} & \dots & Y_{1n_1} \\ \vdots & \ddots & \vdots \\ Y_{k1} & \dots & Y_{kn_k}. \end{matrix}$$

With the densities

$$f_\mu(x) \; = \; \mu \exp(-\mu x) \, 1_{(0,\infty)}(x)$$

we assume that Y_{ij} has the density f_{μ_i}. The likeliood ratio test statistic has the form (up to a factor)

$$R(y) \; = \; \frac{y_{\cdot\cdot}^n}{y_{1\cdot}^{n_1} \cdot \dots \cdot y_{k\cdot}^{n_k}}.$$

If $R(y)$ is large enough, then we reject the hypothesis.

19.3 Criteria for Scaling

M.G. Kendall and A. Stuart [KeSt67] and later on H. Ahrens and J. Läuter in [AhLa81] introduced a method for scaling which bases on a test statistic. This will be generalized for higher dimensional q-way classification tables. This was considered by H. Läuter in [La07] too. At first we denote the levels of the q factors in an arbitrary way by real numbers. The factor i has ν_i levels. Then we put τ_{ij} for the level j of the factor i, all levels are described by

$$\tau \; = \; (\tau_{11}, ..., \tau_{1\nu_1}, ..., \tau_{q\nu_q})^t$$

and altogether we have $\nu \; = \; \sum_i \nu_i$ levels. In sect. 19.2.2, the categories were identified by $x_1, ..., x_L$ and we introduced the y_{ij}. For any x_l we find a $p \times \nu$ matrix C_l with $x_l = C_l \tau$. Every y_{ts} is one of the values $C_1\tau, ..., C_L\tau$. We use

$$h_{t\cdot} = \frac{1}{L}\sum_{l=1}^{L} h_{tl}, \quad h_{\cdot l} = \frac{1}{k}\sum_{t=1}^{k} h_{tl}, \quad h_{\cdot\cdot} = \frac{1}{kL}\sum_{t=1}^{k}\sum_{l=1}^{L} h_{tl},$$

$$h_{t\cdot} \cdot L = \sum_{l=1}^{L} h_{tl} = n_t, \qquad h_{\cdot\cdot} \cdot kL = n.$$

Then we calculate

$$y_{t\cdot} = \frac{1}{n_t}\sum_{s=1}^{n_t} y_{ts} = \frac{1}{n_t}\Big(h_{t1}C_1 + \cdots + h_{tL}C_L\Big)\tau, \quad y_{\cdot\cdot} = \frac{k}{n}\Big(h_{\cdot 1}C_1 + \cdots + h_{\cdot L}C_L\Big)\tau$$

$$y_{t\cdot} - y_{\cdot\cdot} = \Big(\big(\frac{h_{t1}}{n_t} - \frac{kh_{\cdot 1}}{n}\big)C_1 + \cdots + \big(\frac{h_{tL}}{n_t} - \frac{kh_{\cdot L}}{n}\big)C_L\Big)\,\tau =: D_t\,\tau.$$

These values are to be substituted in the test statistics. In the normal case we have T_0^2 as the test statistic. For calculating this statistic we use

$$H := \sum_{i=1}^{k} n_i\Big(y_{i\cdot} - y_{\cdot\cdot}\Big)\Big(y_{i\cdot} - y_{\cdot\cdot}\Big)^t = \sum_{i=1}^{k} n_i D_i\,\tau\,\tau^t D_i^t,$$

$$S := \frac{1}{n-k}\sum_{i=1}^{k}\sum_{s=1}^{n_i}\Big(y_{is} - y_{i\cdot}\Big)\Big(y_{is} - y_{i\cdot}\Big)^t = \frac{1}{n-k}\sum_{i=1}^{k}\sum_{l=1}^{L} h_{il}F_{il}\,\tau\,\tau^t F_{il}^t$$

for

$$F_{il} = C_l - \frac{1}{n_i}\Big(h_{i1}C_1 + \cdots + h_{iL}C_L\Big)$$

and

$$T_0^2 = \frac{n-k-p+1}{(k-1)(n-k)p}\sum_{i=1}^{k} n_i(y_{i\cdot} - y_{\cdot\cdot})^t S^{-1}(y_{i\cdot} - y_{\cdot\cdot})$$

$$= \frac{n-k-p+1}{(k-1)(n-k)p}\,tr\Big(HS^{-1}\Big),$$

$$tr\Big(HS^{-1}\Big) = \tau^t\Big[\sum_{i=1}^{k} n_i D_i^t S^{-1} D_i\Big]\tau$$

with

$$S = \frac{1}{n-k}\sum_{i=1}^{k}\sum_{l=1}^{L} m_{il}F_{il}\,\tau\,\tau^t F_{il}^t.$$

In the case of exponential distributions, we considered the LRT. There we had $p = 1$ and so we have $L = \nu$ and the statistic was

$$R(y) = \frac{y_{\cdot\cdot}^n}{y_{1\cdot}^{n_1}\cdot\ldots\cdot y_{k\cdot}^{n_k}}. \tag{19.3}$$

Substituting here the values with τ we obtain

$$R = \frac{\Big[\frac{k}{n}\Big(h_{\cdot 1}C_1 + \cdots + h_{\cdot L}C_L\Big)\tau\Big]^n}{\Big[\frac{1}{n_1}\Big(h_{11}C_1 + \cdots + h_{1L}C_L\Big)\tau\Big]^{n_1}\cdot\ldots\cdot\Big[\frac{1}{n_k}\Big(h_{k1}C_1 + \cdots + h_{kL}C_L\Big)\tau\Big]^{n_k}}. \tag{19.4}$$

In the case $q = 1$ it holds $C_l\tau = \tau_{1l} =: \tau_l$ and therefore we get

$$R = \frac{k^n n_1^{n_1} \cdot \ldots \cdot n_k^{n_k}}{n^n} \frac{\left(h_{.1}\tau_1 + \cdots + h_{.L}\tau_L\right)^n}{\left(h_{11}\tau_1 + \cdots + h_{1L}\tau_L\right)^{n_1} \cdot \ldots \cdot \left(h_{k1}\tau_1 + \cdots + h_{kL}\tau_L\right)^{n_k}}. \quad (19.5)$$

The aim is to find such a scale that the distributions or here classes can be discriminated as well as possible. Therefore, we have to determine such a vector τ^* that maximizes the corresponding test statistic. In the normal case the test bases on T_0^2, for the exponential distributions the likelihood ratio test statistic R was proposed.

Definition 2. *If LR denotes the test statistic where large values of LR lead to the rejection of the hypothesis then τ^* with*

$$LR(\tau^*) = \max_\tau LR(\tau) \quad (19.6)$$

is called a most separating scale.

19.4 Calculation of Most Separating Scales

In general, one has to use some optimization software for finding a maximal τ^*. We will consider in some detail the special case of normal distributions. In sect. 19.2.3 we considered the statistic T_0^2 as the statistic to be maximized. Up to a factor this coincides with

$$tr(HS^{-1}) = \tau^t \Big[\sum_{i=1}^k n_i D_i^t S^{-1} D_i\Big] \tau \quad (19.7)$$

with

$$S = \frac{1}{n-k} \sum_{i=1}^k \sum_{l=1}^L h_{il} F_{il}\, \tau\, \tau^t F_{il}^t.$$

19.4.1 One-Dimensional Normal Distributions

We consider $p = 1$. Then we have the $1 \times L$ matrices C_l, D_i, F_{il} and get

$$H = \tau^t A \tau, \quad S = \tau^t B \tau \quad (19.8)$$

for the matrices

$$A = \sum_{i=1}^k n_i D_i^t D_i, \quad B = \frac{1}{n-k} \sum_{i=1}^k \sum_{l=1}^L h_{il} F_{il}^t F_{il}. \quad (19.9)$$

Therefore, the $\tau^* = (\tau_1^*, \ldots, \tau_L^*)^t$ maximizing $tr(HS^{-1})$ is determined by the eigenvector to the maximal eigenvalue of

$$A\tau = \lambda B \tau. \quad (19.10)$$

Therefore, in the case $p = 1$ the optimal scale vector τ^* can be determined exactly. The level l corresponds to the number τ_l^*.

19.4.2 Higher-Dimensional Case

Now we consider q-way classification models and $p \le q$. Then we have the $p \times \nu$ matrices C_l, D_i, F_{il} and with $S_\tau := S$ we have

$$tr(HS^{-1}) = tr(HS_\tau^{-1}) = \tau^t \Big[\sum_{i=1}^{k} n_i D_i^t S_\tau^{-1} D_i \Big] \tau \tag{19.11}$$

for

$$S_\tau = \frac{1}{n-k} \sum_{i=1}^{k} \sum_{l=1}^{L} h_{il} F_{il} \tau \tau^t F_{il}^t. \tag{19.12}$$

Defining

$$\psi(\tau, a) := a^t \Big[\sum_{i=1}^{k} n_i D_i^t S_\tau^{-1} D_i \Big] a \tag{19.13}$$

and then τ^* fulfills

$$\psi(\tau^*, \tau^*) = \max_\tau \psi(\tau, \tau). \tag{19.14}$$

We see that ψ does not change if τ is substituted by $\mu\tau$ for any real μ.

Definition 3. *$\widetilde{\tau}$ is called a local extremum if*

$$\frac{\mathrm{d}}{\mathrm{d}\lambda} \psi\Big((1-\lambda)\widetilde{\tau} + \lambda v, (1-\lambda)\widetilde{\tau} + \lambda v \Big)|_{\lambda=0} \le 0 \quad \forall v \in \mathbb{R}^p.$$

We are interested in characterizing such a local extremum. This gives us the next theorem.

Theorem 1. *$\widetilde{\tau}$ is a local extremum if and only if $\alpha(\widetilde{\tau}) = 0$ with*

$$\alpha(\tau) := \sum_{i=1}^{k} n_i D_i^t S_\tau^{-1} D_i \tau - \frac{1}{n-k} \sum_{i=1}^{k} n_i \sum_{j=1}^{k} \sum_{l=1}^{L} h_{jl} F_{jl}^t S_\tau^{-1} D_i \tau \tau^t F_{jl}^t S_\tau^{-1} D_i \tau.$$

Proof. We put $\tau_\lambda = (1-\lambda)\widetilde{\tau} + \lambda v$ and obtain

$$\frac{\mathrm{d}}{\mathrm{d}\lambda} \tau_\lambda = v - \tau_\lambda, \quad \frac{\mathrm{d}}{\mathrm{d}\lambda} \tau_\lambda \tau_\lambda^t|_{\lambda=0} = (v - \widetilde{\tau})\widetilde{\tau}^t + \widetilde{\tau}(v - \widetilde{\tau})^t,$$

$$\frac{\mathrm{d}}{\mathrm{d}\lambda} S_{\tau_\lambda}^{-1} = -S_{\tau_\lambda}^{-1} \Big(\frac{\mathrm{d}}{\mathrm{d}\lambda} S_{\tau_\lambda}\Big) S_{\tau_\lambda}^{-1}$$

and consequently

$$\frac{\mathrm{d}}{\mathrm{d}\lambda} S_{\tau_\lambda}^{-1}|_{\lambda=0} = -\frac{1}{n-k} S_{\widetilde{\tau}}^{-1} \sum_{j=1}^{k} \sum_{l=1}^{L} h_{jl} F_{jl} (v \widetilde{\tau}^t + \widetilde{\tau} v^t - 2\widetilde{\tau}\widetilde{\tau}^t) F_{jl}^t S_{\widetilde{\tau}}^{-1}.$$

Now we calculate in a direct way

$$\frac{\mathrm{d}}{\mathrm{d}\lambda} \psi(\tau_\lambda, \tau_\lambda)|_{\lambda=0} = 2 v^t \alpha(\widetilde{\tau})$$

and so the theorem is proven.

This theorem gives us a proposal for the calculation of a local extremum.
Step 1: Choice of an initial point τ_0.
Step 2: Set $w := \frac{1}{|\alpha(\tau_0)|}\alpha(\tau_0)$ and $\widetilde{\tau}_\lambda = (1-\lambda)\tau_0 + \lambda w$ for euclidian norm $|\alpha(\tau_0)|$ of $\alpha(\tau_0)$.
Step 3: Determine such λ_1 that

$$\psi(\widetilde{\tau}_{\lambda_1}, \widetilde{\tau}_{\lambda_1}) = \max_\lambda \psi(\widetilde{\tau}_\lambda, \widetilde{\tau}_\lambda).$$

Step 4: Set $\tau_1 := \widetilde{\tau}_{\lambda_1}$ and calculate $\alpha(\tau_1)$. Now we set $w := \frac{1}{|\alpha(\tau_1)|}\alpha(\tau_1)$ and $\widetilde{\tau}_\lambda = (1-\lambda)\tau_1 + \lambda w$ and so on.
In this way we get a sequence of q-vectors $\tau_0, \tau_1, \tau_2, \ldots$ and have

$$\psi(\tau_0, \tau_0) \leq \psi(\tau_1, \tau_1) \leq \psi(\tau_2, \tau_2) \leq \ldots$$

In each step, one can check $\alpha(\tau_j)$ and decide to proceed in the sequential calculation or to break up. Under $\alpha(\tau_j) \approx 0$ one reaches the optimum.

Acknowledgment. The authors are very grateful to Prof. H. Liero and Dr. Martin Läuter for their helpful comments and suggestions.

References

[AhLa81] Ahrens, H. and Läuter, J.: Mehrdimensionale Varianzanalyse. Wiley, Akademie-Verlag, Berlin (1981)

[EvDu01] Everitt, B.S. and Dunn, G.: Applied Multivariate Data Analysis. Hodder Education, London (2001)

[KeSt67] Kendall, M.G. and Stuart, A.: The Advanced Theory of Statistics. Wiley, Griffin & Company, London (1967)

[La07] Läuter, H.: Modeling and scaling of categorical data. Preprint, University of Linz (2007)

[VoNi93] Voinov, V.G. and Nikulin, M.S.: Unbiased Estimators and Their Applications. Kluwer, Dordecht (1993)

20

Nonparametric Estimation and Testing the Effect of Covariates in Accelerated Life Time Models Under Censoring

Hannelore Liero

Institute of Mathematics, University of Potsdam, Potsdam, Germany,
liero@uni-potsdam.de

Abstract: We consider an accelerated life time model where the distribution of the basic life time is not specified and the function ψ describing the effect of the covariates has no parametric form. That is, the model is completely nonparametric. In this paper, a procedure for testing whether ψ has prespecified parametric form is proposed. The approach is based on a transformation to a regression model and the test statistic is a weighted L_2-distance between a nonparametric estimator for the regression function from the hypothetical regression. Since the problem is investigated for censored data the derivation of the test procedure requires the study of the asymptotic behavior of nonparametric regression estimators under censoring. A proposal for a Monte Carlo procedure for the realization of the test completes the considerations.

Keywords and phrases: Accelerated life time model, Beran estimator, Censoring, Goodness-of-fit tests, Nonparametric curve estimators

20.1 Introduction

We consider a life time model which describes the following situation: By some covariate X the time to failure may be accelerated or retarded relative to some baseline. The speeding up or slowing down is accomplished by some positive function ψ, and we may write

$$T = \frac{T_0}{\psi(X)},$$

where T_0 is the so-called baseline life time and T is the observable life time. We will assume that T is an absolute continuous random variable (r.v.) and that the covariate X does not depend on the time. For simplicity of presentation let X be one-dimensional.

For statistical application a suitable choice of the function ψ is important and the problem of testing ψ arises. A survey of test procedures for testing ψ under different model assumptions is given in the paper by Liero H. and Liero M. [LL08].

V.V. Rykov et al. (eds.), *Mathematical and Statistical Models and Methods in Reliability: Applications to Medicine, Finance, and Quality Control*, Statistics for Industry and Technology, DOI 10.1007/978-0-8176-4971-5_20,

The aim of the present paper is to propose a test procedures for testing whether the function ψ belongs to a pre-specified parametric class of functions

$$\mathcal{F} = \{\psi \mid \psi(\cdot) = \psi(\cdot;\beta), \quad \beta \in \mathbb{R}^d\}. \tag{20.1}$$

For data without censoring this problem was already considered by Liero in [Lie08]. In this paper we assume that the independent and identically distributed life times T_i are subject to random right censoring, i.e. the observations are

$$V_i = \min(T_i, C_i), \quad \Delta_i = 1(T_i \leq C_i) \quad \text{and} \quad X_i, \quad i = 1, \ldots, n,$$

where the C_i's are independent and identically distributed censoring times with distribution function G. Furthermore, we assume that the T_i's and the C_i's are conditionally independent given the X_i's.

The inference is based on the log transformation of the lifetime model to a regression model: The conditional expectation of $Y = \log T$ given the covariate X has the form

$$\mathsf{E}\,(Y|X = x) = \mu - \log \psi(x) \quad \text{with} \quad \mu = -\int \log z \, \mathrm{d}S_0(z) = \mathsf{E}\,(\log T_0),$$

where S_0 is the survival function of the baseline life time T_0, and we can translate the considered problem into a problem of testing the regression function in a nonparametric regression model

$$Y = \log T = m(X) + \varepsilon,$$

where $m(x) = \mu - \log \psi(x)$, and with $\varepsilon = \log(T_0) - \mathsf{E}(\log(T_0))$

$$\mathsf{E}\,(\varepsilon|X = x) = 0, \quad \text{and} \quad \mathsf{E}\,(\varepsilon^2|X = x) = \sigma^2$$

for some $\sigma^2 > 0$. For identifiability, we assume $\psi(0) = 1$.

Test problem (20.1) is translated into the following problem

$$H: \quad m \in \mathcal{M} \quad \text{versus} \quad K: \quad m \notin \mathcal{M}$$

where

$$\mathcal{M} = \{m \mid m(\cdot;\beta,\mu) = -\log \psi(\cdot;\beta) + \mu, \ \beta \in \mathbb{R}^d, \mu \in \mathbb{R}\},$$

that is we have to check whether the regression function has a parametric form or alternatively that this regression is nonparametric.

As test statistic a weighted L_2-distance between a parametric and the nonparametric regression estimator is proposed. To formulate the corresponding test procedure one has to investigate the properties of nonparametric estimators for regression functions under censoring. Therefore, in Sect. 20.2, the nonparametric estimation of the regression function under censoring is considered. In Sect. 20.3 asymptotic properties of the nonparametric regression estimator are presented; the main result is the asymptotic normality of the weighted L_2-distance of the estimator. This limit theorem is based on a so-called asymptotic (conditional) i.i.d. representation of the difference between the estimator and the regression function. The test procedure is given in Sect. 20.4.

20.2 Nonparametric Estimation of the Regression Function Under Censoring

We start with the a nonparametric estimator for the conditional distribution function of the transformed r.v. $Y = \log T$. Such an estimator was introduced by Beran [Ber81]. On one hand the Beran estimator can be regarded as an extension of the well-known Kaplan–Meier estimator proposed for models with censored data without covariates, on the other hand it is an extension of nonparametric estimators for conditional distributions functions studied for data sets without censoring. To define the Beran estimator it is useful to introduce the following functions and their estimators: The conditional distribution function of the r.v. $Z = \log V$ with $V = \min(T, C)$ is given by $H(z|x) = \mathsf{P}(Z \le z|X = x)$ and estimated by the kernel estimator

$$\hat{H}_n(z|x) = \sum_i W_{b_n i}(x, X_1, \ldots, X_n) 1(Z_i \le z), \tag{20.2}$$

where $W_{b_n i}$ are the kernel weights defined by

$$W_{b_n i}(x, X_1, \ldots, X_n) = \frac{\frac{1}{b_n} K\left(\frac{x - X_i}{b_n}\right)}{\frac{1}{b_n} \sum_{j=1}^n K\left(\frac{x - X_j}{b_n}\right)}.$$

Here $K : \mathbb{R} \to \mathbb{R}$ is a kernel function, and b_n is a sequence of bandwidths tending to zero as $n \to \infty$. The symbol "1" denotes the indicator function. The estimator of the conditional subdistribution function $H^U(z|x) = \mathsf{P}(Z \le z, \Delta = 1|X = x)$ is given by

$$\hat{H}_n^U(z|x) = \sum_i W_{b_n i}(x, X_1, \ldots, X_n) 1(Z_i \le z, \Delta_i = 1). \tag{20.3}$$

For the conditional cumulative hazard function Λ we have for $y \le \tau_x$

$$\Lambda(y|x) = \int_{-\infty}^{y} \frac{\mathrm{d}F(s|x)}{1 - F(s_-|x)} = \int_{-\infty}^{y} \frac{\mathrm{d}H^U(s|x)}{1 - H(s_-|x)}$$

where F denotes the conditional cdf of the transformed $Y = \log T$ and $H(s_-|x) = \lim_{t \uparrow s} H(t|x)$, $\tau_x = \inf\{y | H(y|x) = 1\}$ is the upper bound of the support of $H(\cdot|x)$. Replacing H^U and H by their estimators (20.2) and (20.3) leads to the weighted Nelson–Aalen type estimator for the conditional cumulative hazard function:

$$\hat{\Lambda}_n(y|x) = \int_{-\infty}^{y} \frac{\mathrm{d}\hat{H}_n^U(s|x)}{1 - \hat{H}_n(s_-|x)}.$$

Now from the well-known relation between the cumulative hazard function and the survival function we obtain as estimator for $S_Y(y|x) = 1 - F(y|x) = \mathsf{P}(Y > y|X = x)$

$$\hat{S}_{Yn}(y|x) = \prod_{t \le y} \left(1 - \Delta\hat{\Lambda}_n(t|x)\right) \tag{20.4}$$

where $\Delta\hat{\Lambda}_n(t|x) = \hat{\Lambda}_n(t|x) - \hat{\Lambda}_n(t_-|x)$ is the jump of $\hat{\Lambda}_n(\cdot|x)$ at t. An equivalent form of (20.4) is

$$\hat{F}_n(y|x) = 1 - \prod_{\substack{Z_i \le y \\ \Delta_i = 1}} \left\{ 1 - \frac{W_{b_n i}(x, \mathbb{X})}{\sum_j 1(Z_j \ge Z_i) W_{b_n j}(x, \mathbb{X})} \right\},$$

where $\mathbb{X} = (X_1, \ldots, X_n)$.

Note that for weights $W_{b_n i} = \frac{1}{n}$ the estimator $\hat{F}_n$ is the classical Kaplan–Meier estimator; for $\Delta_i = 1$ for all i the estimator $\hat{F}_n$ is the estimator of the conditional distribution function, and for $W_{b_n i} = \frac{1}{n}$ and $\Delta_i = 1$ for all i the estimator $\hat{F}_n$ is simply the empirical distribution function.

Several authors considered the asymptotic behavior of $\hat{F}_n$. Consistency of $\hat{F}_n(y|x)$ is proven for $y \le \tau_x$.

The regression function $m(x) = \mathsf{E}(Y|X = x)$ is defined by $\int y \mathrm{d}F(y|x)$. However, for the estimation of m and the investigation of the properties of the resulting estimator the following identities are useful:

$$\begin{aligned} m(x) &= \mathsf{E}(Y|X = x) \\ &= \int y \mathrm{d}F(y|x) && (20.5) \\ &= \mathsf{E}\left(\frac{1 - F(Z_-|X)}{1 - H(Z_-|X)} Z\Delta | X = x \right) && (20.6) \end{aligned}$$

and

$$m(x) = \int_0^1 F^{-1}(u|x)\,\mathrm{d}u \qquad (20.7)$$

where $F^{-1}(u|x) = \inf\{y|F(y|x) \ge u\}$.

To estimate m we replace $F(y|x)$ in (20.5) by the Beran estimator and obtain as nonparametric estimator for m

$$\hat{m}_n(x) = \int y\,\mathrm{d}\hat{F}_n(y|x).$$

One can show that for this estimator the empirical versions of (20.6) and (20.7) hold, i.e.,

$$\hat{m}_n(x) = \sum_{i=1}^n W_{b_n i}(x, \mathbb{X}) \frac{1 - \hat{F}_n(Z_{i-}|x)}{1 - \hat{H}_n(Z_{i-}|x)} Z_i \Delta_i$$

and

$$\hat{m}_n(x) = \int_0^1 \hat{F}_n^{-1}(u|x)\mathrm{d}u$$

where $\hat{F}_n^{-1}(u|x) = \inf\{y|\hat{F}_n(y|x) \ge u\}$. We see that as in the case without censoring the regression estimator is a weighted average; now, in the case with censoring an average of the uncensored observations. The weights depend on the kernel and on the ratio of the Kaplan–Meier estimator and the empirical df of the observations.

20.3 Properties of the Nonparametric Regression Estimator

Györfi et al. showed in [GKKW02] that an estimator of this type is L_2-consistent if the right endpoint of the support of F is smaller than that of G.[1] We follow another approach than those authors. We will use a conditional i.i.d. presentation of the difference between estimator and regression function; such a presentation is based on the corresponding result for the estimator $\hat{F}_n$ which is derived by Akritas and Du in [AD02]. Since this presentation holds only for $y \le y^*$, where $y^* < \sup_x \tau_x$ we will truncate the estimator (due to the right censoring):

Instead of $m(x) = \int_{-\infty}^{\infty} y \, \mathrm{d}F(y|x)$ we estimate the function

$$m^*(x) = \int_{-\infty}^{y^*} y \, \mathrm{d}F(y|x).$$

The function m^* is estimated by

$$\hat{m}_n^*(x) = \int_{-\infty}^{y^*} y \, \mathrm{d}\hat{F}_n(y|x).$$

Before we state the results let us formulate the assumptions

A1 The marginal density g of X is bounded on $\mathbb{R}$ and twice continuously differentiable in a neighborhood of a set $\mathfrak{I}$ and $g(x) \ge c > 0$ for some c and all $x \in \mathfrak{I}$.

A2 The kernel K is a symmetric density with compact support; furthermore, it is twice continuously differentiable.

A3 We will need typical smoothness conditions on the functions $H(\cdot|\cdot)$ and $H^U(\cdot|\cdot)$. We formulate here for a general (sub)distribution function L:
The derivatives

$$\ddot{L}(y|x) = \frac{\partial^2 L(y|x)}{\partial^2 x}, \quad L''(y|x) = \frac{\partial^2 L(y|x)}{\partial^2 y}, \quad \dot{L}'(y|x) = \frac{\partial^2 L(y|x)}{\partial y \partial x}$$

exist and are continuous for all y, and all x in a neighborhood of $\mathfrak{M}$.

Lemma 1. *Suppose that A1 and A2 hold and that A3 is satisfied by H and H^U. Then*

$$\hat{m}_n^*(x) - m^*(x) = \sum_{i=1}^{n} W_{b_n i}(x, \mathbb{X}) \, \eta(Z_i, \Delta_i|x) + R_n(x)$$

with

$$\eta(Z_i, \Delta_i|x) = y^*(1 - F(y^*|x))\xi(Z_i, \Delta_i, y^*|x) - \int_{-\infty}^{y^*} (1 - F(s|x))\xi(Z_i, \Delta_i, s|x)\mathrm{d}s,$$

where

$$\xi(Z_i, \Delta_i, s|x) = \frac{1(Z_i \le s, \Delta_i = 1)}{(1 - H(Z_i|x))} - \int_{-\infty}^{s} \frac{1(Z_i \ge w)\mathrm{d}H^U(w|x)}{(1 - H(w|x))^2},$$

[1] Instead of kernel weights considered here they used nearest neighbor weights.

and where

$$\sup_{x \in \mathfrak{I}} R_n(x) \;=\; O_{\mathsf{P}}\left((nb_n)^{-\frac{3}{4}}(\log n)^{\frac{3}{4}}\right) \qquad \text{as } n \to \infty.$$

Based on the presentation given in Lemma 1 we will prove the asymptotic normality of $\hat{m}_n(x)$ at an arbitrary fixed point x and a limit theorem for a weighted integrated squared error. Let us consider the conditional i.i.d. presentation as process and set

$$\mathcal{A}_n(x) \;=\; \sum_{i=1}^{n} W_{b_n i}(x, \mathbb{X})\, \eta(Z_i, \Delta_i | x).$$

In a first step, we will split $\mathcal{A}_n$ in a stochastic and in a systematic part:

$$\begin{aligned}\mathcal{A}_n(x) = & \sum_{i=1}^{n} W_{b_n i}(x, \mathbb{X})\; (\eta(Z_i, \Delta_i|x) - \mathsf{E}[\eta(Z_i, \Delta_i|x)|X_i]) \\ & + \sum_{i=1}^{n} W_{b_n i}(x, \mathbb{X}) \mathsf{E}[\eta(Z_i, \Delta_i|x)|X_i].\end{aligned}$$

Note that

$$W_{b_n i}(x, \mathbb{X}) \;=\; \frac{\frac{1}{n} K_{b_n}(x - X_i)}{\hat{g}_n(x)}$$

where

$$\hat{g}_n(x) \;=\; \frac{1}{n}\sum_{j=1}^{n} K_{b_n}(x - X_j)$$

is the estimator for the marginal density g of the covariate X.

The first part, the stochastic one, is approximated by

$$\mathcal{A}_{n1}(x) \;=\; \frac{1}{\mathsf{E}\hat{g}_n(x)}\,\frac{1}{n}\sum_{i=1}^{n} K_{b_n}(x - X_i)\; (\eta(Z_i, \Delta_i|x) - \mathsf{E}[\eta(Z_i, \Delta_i|x)|X_i])\,.$$

Using the well-known asymptotic properties of a nonparametric density estimator it is shown that the stochastic part of $\mathcal{A}_n$ and the statistic $\mathcal{A}_{n1}$ have the same asymptotic behavior. The stochastic behavior of $\mathcal{A}_{n1}$ is characterized by the covariance function

$$\mathcal{C}_n(x, y) \;=\; \mathsf{Cov}(\mathcal{A}_{n1}(x), \mathcal{A}_{n1}(y)).$$

Since this function plays a key role in proving limit theorems and deriving the corresponding standardizing terms an asymptotic expression for $\mathcal{C}_n(x, y)$ is presented in the following lemma:

Lemma 2. *Suppose that A1 and A2 hold, and H and H^U are Lipschitz continuous with respect to x. Set*

$$\beta_x(v) \;=\; \int_{-\infty}^{v} \frac{\mathrm{d}H^U(w|x)}{(1 - H(w|x))^2}$$

and

$$\gamma_{xy}(s,t) = \int_{-\infty}^{\infty} \beta_x(s \wedge z)\beta_y(t \wedge z)\mathrm{d}H(z|x)$$
$$- \int_{-\infty}^{s} \beta_y(t \wedge z)\mathrm{d}\Lambda(w|x) - \int_{-\infty}^{t} \frac{1 - H(z|x)}{1 - H(z|y)}\beta_x(s \wedge z)\mathrm{d}\Lambda(z|x)$$
$$+ \int_{-\infty}^{s\wedge t} \frac{\mathrm{d}H^U(z|x)}{(1 - H(z)|x)(1 - H(z|y))}.$$

Then

$$\begin{aligned}\mathcal{C}_n(x,y) &= \frac{1}{nb}(g(x))^{-1}(K * K)\left(\frac{y-x}{b}\right)\\ &\times\Big(y^{*2}(1 - F(y^*|x))(1 - F(y^*|y))\gamma_{xy}(y^*, y^*)\\ &-y^*(1 - F(y^*|x))\int_{-\infty}^{y^*}(1 - F(t|y))\gamma_{xy}(y^*, t)\mathrm{d}t\\ &-y^*(1 - F(y^*|y))\int_{-\infty}^{y^*}(1 - F(s|x))\gamma_{xy}(s, y^*)\mathrm{d}s\\ &+\int_{-\infty}^{y^*}\int_{-\infty}^{y^*}(1 - F(s|x))(1 - F(t|y))\gamma_{xy}(s,t)\mathrm{d}s\mathrm{d}t\Big) + o\left(n^{-1}\right),\end{aligned}$$

where $K * K$ denotes the convolution.

The approximating statistic $\mathcal{A}_{n1}(x)$ is a sum of i.i.d. r.v.'s. Applying the central limit theorem we obtain immediately the asymptotic normality at a fixed point x:

$$\frac{\mathcal{A}_{n1}(x)}{\mathcal{C}_n(x,x)} \xrightarrow{\mathcal{D}} \mathsf{N}(0,1).$$

After some transformations we obtain from Lemma 2 for $x = y$

$$\begin{aligned}\mathcal{C}_n(x,x) &= \frac{1}{nb_n}(g(x))^{-1}(K * K)(0)\\ &\times \int_{-\infty}^{y^*}(y^*(1 - F(y^*)) - A_x(s;y^*))^2\,\frac{\mathrm{d}H_x^U(s)}{(1 - H_x(s))^2} + o(n^{-1})\\ &= \frac{1}{nb_n}\kappa^2\rho^2(x) + o(n^{-1})\end{aligned}$$

with $A_x(s;y^*) = \int_s^{y^*}(1 - F(t|x))\mathrm{d}t$ and $\kappa^2 = (K * K)(0)$. Hence,

$$\sqrt{nb_n}\,\mathcal{A}_{n1}(x) \xrightarrow{\mathcal{D}} \mathsf{N}(0, \rho^2(x)\kappa^2).$$

Since $\hat{g}_n(x)$ is consistent,

$$\mathcal{A}_{n1}(x) = O_{\mathsf{P}}((nb_n)^{-1/2})$$

and

$$\hat{g}_n(x) - \mathsf{E}\hat{g}_n(x) = O_{\mathsf{P}}((nb_n)^{-1/2})$$

we obtain

$$\begin{aligned}(\mathcal{A}_n(x) - \sum_{i=1}^n W_{bi}(x,\mathbb{X})\mathsf{E}(\eta_x(Z_i,\Delta_i)|X_i)) \; - \; \mathcal{A}_{n1}(x) \\ = \mathcal{A}_{n1}(x)\frac{\mathsf{E}\hat{g}_n(x) - \hat{g}_n(x)}{\hat{g}_n(x)} \; = \; O_{\mathsf{P}}((nb_n)^{-1}).\end{aligned}$$

Hence,

$$\sqrt{nb_n}\left(\mathcal{A}_n(x) \; - \; \sum_{i=1}^n W_{bi}(x,\mathbb{X})\mathsf{E}(\eta_x(Z_i,\Delta_i)|X_i)\right) \xrightarrow{\mathcal{D}} \mathsf{N}(0,\rho^2(x)\kappa^2).$$

Now, to characterize the systematic part of the deviation define

$$B_1(s,x) \; = \; \int_{-\infty}^s \frac{\mathrm{d}\dot{H}^U(t|x)}{1-H(t|x)} + \int_{-\infty}^s \frac{\dot{H}(t|x)\mathrm{d}H^U(t|x)}{(1-H(t|x))^2}$$

and

$$B_2(s,x) \; = \; \int_{-\infty}^s \frac{\mathrm{d}\ddot{H}^U(t|x)}{1-H(t|x)} + \int_{-\infty}^s \frac{\ddot{H}(t|x)\mathrm{d}H^U(t|x)}{(1-H(t|x))^2}.$$

Using standard techniques for the investigation of a bias we obtain the following asymptotic expansion for the term $\sum_{i=1}^n W_{bi}(x,\mathbb{X})\mathsf{E}(\eta_x(Z_i,\Delta_i)|X_i)$:

$$\sum_{i=1}^n W_{bi}(x,\mathbb{X})\mathsf{E}(\eta_x(Z_i,\Delta_i)|X_i) \; = \; b_n^2\, B(x)\, \mu_2(K) \; + \; o_{\mathsf{P}}(b_n^2),$$

where

$$\begin{aligned}B(x) = \frac{g'(x)}{g(x)}\left(y^*(1-F(y^*))B_1(y^*,x) - \int_{-\infty}^{y^*}(1-F(s))B_1(s,x)\mathrm{d}s\right) \\ + \; \frac{1}{2}\left(y^*(1-F(y^*))B_2(y^*,x) - \int_{-\infty}^{y^*}(1-F(s))B_2(s,x)\mathrm{d}s\right)\end{aligned}$$

and $\mu_2(K) = \int u^2K(u)\mathrm{d}u$.

If $nb_n^5 \to 0$

$$\sqrt{nb_n}\sum_{i=1}^n W_{bi}(x,\mathbb{X})\mathsf{E}(\eta_x(Z_i,\Delta_i)|X_i) \; = \; o_{\mathsf{P}}\left((nb_n^5)^{1/2}\right) \; = \; o_{\mathsf{P}}(1),$$

in other words, the systematic part is asymptotically negligible.

If $nb_n^5 \to c > 0$ we have

$$\sqrt{nb_n}\sum_{i=1}^n W_{bi}(x,\mathbb{X})\mathsf{E}(\eta_x(Z_i,\Delta_i)|X_i) \to \sqrt{c}\,B(x)$$

and

$$\sqrt{nb_n}\,\mathcal{A}_n(x) \xrightarrow{\mathcal{D}} \mathsf{N}(\sqrt{c}B(x)\mu_2(K),\rho^2(x)\kappa^2).$$

By Lemma 1 we conclude from the asymptotic behavior of $\mathcal{A}_n(x)$ to that of the difference $\hat{m}_n^*(x) - m^*(x)$ and formulate the following theorem:

Theorem 1 (Asymptotic normality at a fixed point). *Under the assumptions given above and $b_n \to 0$ and $nb_n \to \infty$*

(i) If $nb_n^5 \to 0$ then

$$\sqrt{nb_n}\,(\hat{m}_n^*(x) - m^*(x)) \xrightarrow{\mathcal{D}} \mathsf{N}(0, \kappa^2\,\rho^2(x))$$

with

$$\rho^2(x) = (g(x))^{-1} \int_{-\infty}^{y^*} (y^*(1 - F(y^*|x)) - A_x(s; y^*))^2 \,\frac{\mathrm{d}H^U(s|x)}{(1 - H(s|x))^2}.$$

(ii) If $nb_n^5 \to c > 0$ then

$$\sqrt{nb_n}\,(\hat{m}_n^*(x) - m^*(x)) \xrightarrow{\mathcal{D}} \mathsf{N}(\sqrt{c}B(x)\mu_2(K), \rho^2(x)\kappa^2).$$

Remark 1. Consider the case without censoring. Using integration by parts we obtain for $y^* = \infty$, $H = H^U = F$

$$\int_{-\infty}^{\infty} A_x^2(s; \infty)\frac{\mathrm{d}F(s|x)}{(1 - F(s|x))^2} = \mathsf{Var}(Y|X = x) = \sigma^2.$$

Thus, in this case Theorem 1 coincides with the well-known limit theorem stating asymptotic normality of nonparametric kernel regression estimators.

The asymptotic normality at a fixed point characterizes the local behavior. For testing the formulated hypothesis it seems to be better to use a global deviation measure. So, let us consider the integrated squared difference, weighted by a known function a with $a(x) = 0$ for $x \notin \mathcal{I}$:

$$\mathrm{Q}_n = \int (\hat{m}_n^*(x) - m^*(x))^2 a(x)\mathrm{d}x.$$

Heuristically speaking this is an infinite sum of squares of asymptotically normally distributed r.v.'s. which are asymptotically independent as b_n tends to zero. Thus Q_n, properly standardized, converges in distribution to the standard normal distribution.

Applying the method proposed by P. Hall in [Hal84] for proving the asymptotic normality of the integrated squared error of kernel density estimators to the approximating $\mathcal{A}_n$ one can show the following limit theorem

Theorem 2 (Asymptotic normality of the ISE). *Under the assumptions formulated above and $nb_n \to \infty$ and $n^{\frac{2}{9}}b_n \to 0$*

$$\mathrm{Q}_n = \int (\hat{m}_n^*(x) - m^*(x))^2 a(x)\mathrm{d}x$$

$$nb_n^{1/2}\,(\mathrm{Q}_n - e_n) \xrightarrow{\mathcal{D}} \mathsf{N}(0, \nu^2)$$

with

$$e_n = e_n(g, H, H^U; K, a, b_n) = (nb_n)^{-1}\kappa^2 \int \rho^2(x)a(x)\mathrm{d}x$$

$$\nu^2 = \nu^2(g, H, H^U; K, a) = 2\,\kappa_1 \int \rho^4(x)\,a^2(x)\mathrm{d}x$$

with $\kappa_1 = \int (K * K)^2(x)\mathrm{d}x$.

20.4 Formulation of the Test Procedure

Let us apply the limit theorem for the L_2-type distance of the truncated estimator from the truncated regression function to formulate a test procedure for testing the hypothesis $m \in \mathcal{M}$.

The alternative is characterized by the nonparametric estimator $\hat{m}_n^*$. Suppose the null hypothesis is true. The hypothetical function $m^*(\cdot;\vartheta)$ is unknown. Firstly, one has to estimate the unknown parameter ϑ. There are several proposals in the literature to do this; we refer to Tsiatis [Tsi90], Ritov [Rit90] or Bagdonavičius and Nikulin [BN01]. The basic idea is to replace the unknown cumulative hazard function by an efficient estimator depending on ϑ and to estimate this unknown parameter then by the maximum likelihood method. The authors show that under suitable assumptions the resulting estimator is $\sqrt{n}$-consistent, i.e.

$$\sqrt{n}\,\left(\hat{\vartheta}_n - \vartheta\right) = O_P(1) \qquad \text{as } n \to \infty.$$

The next step is to determine $m^*(\cdot;\hat{\vartheta})$. With the estimator $\hat{\beta}$ the Breslow estimator for the cumulative hazard function of the unobservable r.v. T_0 is constructed as follows:

$$\hat{\Lambda}_{n0}(t;\hat{\beta}) = \int_0^t \frac{\mathrm{d}\hat{H}_{n0}^U(s;\hat{\beta})}{1 - \hat{H}_{n0}(s_-;\hat{\beta})},$$

where

$$\hat{H}_{n0}(t;\hat{\beta}) = \frac{1}{n}\sum_{i=1}^n 1(\hat{V}_{0i} \le t)$$

is the empirical distribution function of the estimated hypothetical baseline observations $\hat{V}_{0i} = V_i\psi(X_i,\hat{\beta})$, and

$$\hat{H}_{n0}^U(t;\hat{\beta}) = \frac{1}{n}\sum_{i=1}^n 1(\hat{V}_{0i} \le t, \Delta_i = 1)$$

is the corresponding estimator of the subdistribution of the uncensored observations. Then the baseline survival function is estimated by

$$\hat{S}_0(t;\hat{\beta}) = \prod_{s\le t}(1 - \Delta\hat{\Lambda}_0(s;\hat{\beta})),$$

and the hypothetical truncated regression function by

$$\begin{aligned}\tilde{m}^*(x;\hat{\vartheta}) &= -\int_{-\infty}^{y^*} y\mathrm{d}\hat{S}_0(\mathrm{e}^y\psi(x;\hat{\beta}))\\ &= -\int_0^{\mathrm{e}^{y^*}\psi(x;\hat{\beta})} \log\frac{z}{\psi(x;\hat{\beta})}\mathrm{d}\hat{S}_0(z)\\ &= -\int_0^{\mathrm{e}^{y^*}\psi(x;\hat{\beta})} \log z\mathrm{d}\hat{S}_0(z) + \log\psi(x;\hat{\beta})\left(\hat{S}_0(\mathrm{e}^{y^*}\psi(x;\hat{\beta})) - 1\right).\end{aligned}$$

We see, for $y^* \to \infty$ the function $\tilde{m}^*(x;\hat{\vartheta})$ converges to

$$-\int_0^{\infty} \log z\mathrm{d}\hat{S}_0(z) - \log\psi(x;\hat{\beta}) = \hat{\mu} - \log\psi(x;\hat{\beta}) = m(x;\hat{\vartheta}).$$

The test procedure has the following form: The hypothesis $m \in \mathcal{M}$, that is $\psi \in \mathcal{F}$ is rejected if the estimated L_2-distance

$$\hat{\mathrm{Q}}_n = \int(\hat{m}_n^*(x) - m^*(x;\hat{\vartheta}))^2 a(x)\mathrm{d}x$$

satisfies the inequality

$$\hat{\mathrm{Q}}_n \ge z_{1-\alpha}\frac{\hat{\nu}}{nb_n^{1/2}} + \hat{e}_n,$$

where $z_{1-\alpha}$ is the $(1-\alpha)$-quantile of the limiting distribution and the terms $\hat{e}_n = e_n(\hat{g}_n, \hat{H}_n, \hat{H}^U, K, a, b_n)$ and $\hat{\nu}^2 = \nu^2(\hat{g}_n, \hat{H}_n, \hat{H}^U, K, a)$ are the estimated standardizing terms.

20.4.1 A Proposal for a Monte Carlo Procedure

Finally, a Monte Carlo method for determining empirical p-values of the test procedure is proposed. The aim of this method is to generate data

$$(V_{ir}^*, X_{ir}^*, \Delta_{ir}^*), \qquad r = 1,\dots,R, \quad i = 1,\dots,n$$

according to the hypothetical model. Based on these data the test statistics

$$\hat{\mathrm{Q}}_{n1},\dots,\hat{\mathrm{Q}}_{nr},\dots,\hat{\mathrm{Q}}_{nR}$$

are computed and from their empirical distribution the p-value is determined.

The data can be constructed as follows:

1. Let $\hat{\beta}$ the estimator for β based on the original data. Construct the Breslow estimator $\hat{\Lambda}_0(\cdot;\hat{\beta})$ for the cumulative baseline hazard function. Then
$$\hat{S}_0(t;\hat{\beta}) = \prod_{s \le t}(1 - \Delta\hat{\Lambda}_0(s;\hat{\beta})).$$
2. Generate data T^*_{0ir} from the estimated survival function $\hat{S}_0(t;\hat{\beta})$ and set
$$T^*_{ir} = \frac{T^*_{0ir}}{\psi(X_i;\hat{\beta})}.$$
3. Estimate the distribution function of the C_i by the weighted Kaplan–Meier estimator $\hat{G}_n$ and generate censoring variables C^*_{ir} from the estimated survival function $\hat{G}_n$.
4. Finally, set
$$V^*_{ir} = \min(T^*_{ir}, C^*_{ir}), \quad \Delta^*_{ir} = 1(T^*_{ir} \le C^*_{ir}), \quad X^*_{ir} = X_i.$$

As of yet this MC procedure is only a proposal and further investigations should be pursued.

References

[AD02] M. G. Akritas and Y. Du. IID representation of the conditional Kaplan-Meier process for arbitrary distributions. *Mathematical Methods in Statistics*, 11:152–182, 2002.

[Ber81] R. Beran. Nonparametric regression with randomly censored survival data. Technical report, Univ. California, Berkeley, 1981.

[BN01] V. Bagdonavičius and M. Nikulin. *Accelerated Life Models.* Chapman and Hall, Boca Raton, London, New York, Washington, 2001.

[GKKW02] L. Györfi, M. Kohler, A. Krzyżak, and H. Walk. *A Distribution-Free Theory of Nonparametric Regression.* Springer Series in Statistics. Springer, Berlin, 2002.

[Hal84] P. Hall. Central limit theorem for integrated square error of multivariate nonparametric density estimators. *Journal of Multivariate Analysis*, 14:1–16, 1984.

[Lie08] H. Liero. Testing in nonparametric accelerated life time models. *Austrian Journal of Statistics*, 37:81–90, 2008.

[LL08] H. Liero and M. Liero. Testing the influence function in accelerated life time models. In F. Vonta, M. Nikulin, N. Limnios, and C. Huber, editors, *Statistical models and methods for biomedical and technical systems.* Boston, 2008.

[Rit90] Y. Ritov. Estimation in a linear regression model with censored data. *Annals of Statistics*, 18:303–328, 1990.

[Tsi90] A. A. Tsiatis. Estimating regression parameters using linear rank tests for censored data. *Annals of Statistics*, 18:354–372, 1990.

21

Nonparametric Estimation of Time Trend for Repairable Systems Data

Bo Henry Lindqvist

Department of Mathematical Sciences, Norwegian University of Science and Technology, Trondheim, Norway, bo@math.ntnu.no

Abstract: The trend-renewal-process (TRP) is defined to be a time-transformed renewal process, where the time transformation is given by a trend function $\lambda(\cdot)$ which is similar to the intensity of a nonhomogeneous Poisson process (NHPP). A nonparametric maximum likelihood estimator of the trend function of a TRP can be obtained in principle in a similar manner as for the NHPP using kernel smoothing. But for a TRP one must consider the simultaneous estimation of the renewal distribution, which is here assumed to belong to a parametric class such as the Weibull-distribution. A weighted kernel estimator for $\lambda(\cdot)$ is suggested and studied. Other approaches are also briefly discussed.

Keywords and phrases: Kernel estimator, Trend-renewal process

---

21.1 Introduction

Failures of a repairable system are usually modeled by a stochastic point process in time. The most common models are the renewal process (RP), the homogeneous Poisson process (HPP), and the nonhomogeneous Poisson process (NHPP) (see Ascher and Feingold [AF84]). As is well known, the RP model assumes what is called "perfect repair", indicating that after each failure, the system is renewed to its original condition. The NHPP model, on the other hand, assumes what is called "minimal repair". After each failure and following repair, the system is in the same state as it was just prior to that failure. This is often more plausible than the complete renewal assumption. Yet often the replaced part is not a minor part, or the repair may affect some other parts of the system, to the better or the worse. This could mean a small jump in the intensity, in either direction.

There is thus a need for models which allow the system to deteriorate (or improve) over time, yet still allow for the possibility that the system could have a drastic increase or decrease in its failure intensity just after a repair, because of damage done, or weak spots removed. Several models have been developed for this purpose.

A class of models of this kind is the trend-renewal process (TRP), which was defined and studied by Lindqvist, Elvebakk, and Heggland [LEH03], see also Lindqvist [Lin06].

V.V. Rykov et al. (eds.), *Mathematical and Statistical Models and Methods in Reliability: Applications to Medicine, Finance, and Quality Control*, Statistics for Industry and Technology, DOI 10.1007/978-0-8176-4971-5_21,

This model contains the RP and NHPP as special cases, and in a simple manner the TRP fills some of the gap between the two extreme repair models. While [LEH03] considered parametric estimation for TRPs, not much has been done on nonparametric estimation in the TRP model. An exception is here the paper Heggland and Lindqvist [HL07], where nonparametric estimation of time trend is done under the assumption that this trend is monotonic, leading to a problem of isotonic regression. The purpose of this chapter is to present an approach for nonparametric estimation without assuming monotonicity of the time trend, and using weighted kernel smoothing as the basic estimation technique. For completeness we also review the approach in [HL07].

21.2 Definitions and Preliminaries

Consider a repairable system, observed from time $t = 0$. Let $N(t)$ be the number of failures in $(0, t]$, let T_i be the time of the ith failure, where we define $T_0 = 0$, and let X_i be the time between failure number $i-1$ and failure number i, that is $X_i = T_i - T_{i-1}$. We assume that all repair times equal 0. This assumption is reasonable if the repair times are negligible compared to the times between failures, or if we let the time parameter be the operation time of the system. For a general treatment of repairable systems, see Ascher and Feingold [AF84] or Meeker and Escobar [ME98].

We next review the definitions of the RP and NHPP, and then we define the trend-renewal process which will be the main model used in this chapter and which can be seen as a generalization of the two first mentioned models.

21.2.1 Models for Repairable Systems

The Renewal Process, RP(F):

The process $N(t)$ is an RP(F) if $X_1, X_2, \ldots$ are independent and identically distributed with cumulative distribution function (cdf) F, where we assume $F(0) = 0$. If F is the exponential distribution with parameter λ, then RP(F)=HPP(λ), the homogeneous Poisson process with intensity λ.

The Nonhomogeneous Poisson Process, NHPP($\lambda(\cdot)$):

Let $\lambda(t),\ t \geq 0$ be a nonnegative function, called the intensity of the process. The cumulative intensity function is then $\Lambda(t) = \int_0^t \lambda(u)du$. The process $N(t)$ is an NHPP($\lambda(\cdot)$) if the time-transformed process $\Lambda(T_1), \Lambda(T_2), \ldots$ is an HPP(1).

The Trend-Renewal Process, TRP($F, \lambda(\cdot)$):

The idea behind the trend-renewal process is to generalize the property that was used above to define the NHPP($\lambda(\cdot)$), i.e. that $\Lambda(T_1), \Lambda(T_2), \ldots$ is an HPP(1).

The trend-renewal process (TRP) is defined simply by allowing the HPP(1) to be any renewal process RP(F) where F has expected value 1. Thus, in addition to

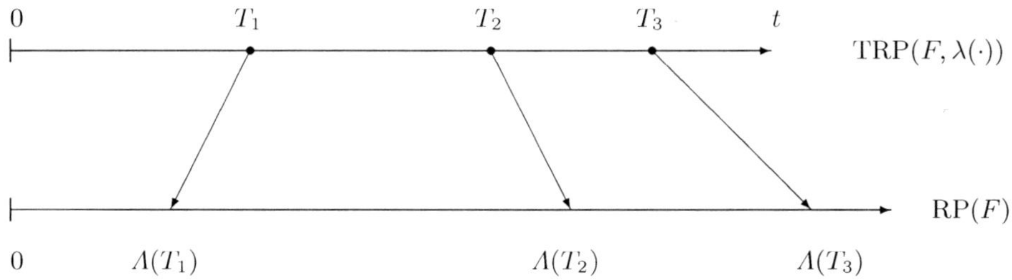

Figure 21.1. The defining property of the trend-renewal process

the intensity function $\lambda(t)$, for a TRP we need to specify a distribution function F of the inter-arrival times of this renewal process. Formally, we can define the process $\text{TRP}(F, \lambda(\cdot))$ as follows:

Let $\lambda(t)$ be a nonnegative function defined for $t \geq 0$, let $\Lambda(t) = \int_0^t \lambda(u)\mathrm{d}u$ and let F be a survival distribution with expected value 1. The process $T_1, T_2, \ldots$ is called $\text{TRP}(F, \lambda(\cdot))$ if the process $\Lambda(T_1), \Lambda(T_2), \ldots$ is $\text{RP}(F)$, that is if the $\Lambda(T_i) - \Lambda(T_{i-1})$; $i = 1, 2, \ldots$ are i.i.d. with distribution function F. The function $\lambda(\cdot)$ is called the trend function, while F is called the renewal distribution.

Figure 21.1 illustrates the definition. For the cited property of the NHPP, the points on the lower axis would correspond to an HPP with unit intensity, HPP(1). For the TRP, this process is instead taken to be any renewal process, RP(F), where F has expectation 1. This shows that the TRP includes the NHPP as a special case by letting F be the standard exponential distribution. Further, if G is a lifetime distribution with finite expectation μ, then with $\lambda(t) = 1/\mu$ and $F(t) = G(\mu t)$ we have $\text{RP}(G) = \text{TRP}(F, \lambda(\cdot))$. Thus, all RPs with inter-arrival times with finite expectation are TRPs.

21.2.2 The Likelihood Function for the TRP Model

The conditional intensity function of a point process (Andersen et al. [ABGK93]) is defined by

$$\gamma(t) = \lim_{\Delta t \to 0} \frac{\text{P}(\text{failure in } [t, t + \Delta t)|\mathcal{F}_{t-})}{\Delta t},$$

where $\mathcal{F}_{t-}$ is the history of the process $N(t)$ up to, but not including time t. The conditional intensity function will, in general, be stochastic.

For a $\text{TRP}(F, \lambda(\cdot))$ the conditional intensity function is given by

$$\gamma(t) = z(\Lambda(t) - \Lambda(T_{N(t-)}))\lambda(t), \tag{21.1}$$

where $z(t)$ is the hazard rate corresponding to F, i.e. $z(t) = f(t)/(1 - F(t))$ where the density function $f(\cdot)$ is assumed to exist.

Consider now a point process $N(t)$, observed from time $t = 0$ to time $t = \tau$, with corresponding failure times $T_1, T_2, \ldots, T_{N(\tau)}$ and conditional intensity function $\gamma(t)$. The likelihood function of the process is then given by (see [ABGK93])

$$L = \left\{ \prod_{i=1}^{N(\tau)} \gamma(T_i) \right\} \exp\left\{ - \int_0^\tau \gamma(u)\mathrm{d}u \right\}. \tag{21.2}$$

The likelihood function of a TRP is obtained by substituting (21.1) into (21.2), giving

$$L = \left\{ \prod_{i=1}^{N(\tau)} z(\Lambda(T_i) - \Lambda(T_{i-1}))\lambda(T_i) \right\} \exp\left\{ - \sum_{i=1}^{N(\tau)} \int_{T_{i-1}}^{T_i} z(\Lambda(u) - \Lambda(T_{i-1}))\lambda(u)\mathrm{d}u \right\}$$
$$\cdot \exp\left\{ - \int_{T_{N(\tau)}}^{\tau} z(\Lambda(u) - \Lambda(T_{N(\tau)}))\lambda(u)\mathrm{d}u \right\}.$$

By making the substitution $v = \Lambda(u) - \Lambda(T_{i-1})$ and taking the log we get the log likelihood function

$$l = \ln L = \sum_{i=1}^{N(\tau)} \{\ln(z(\Lambda(T_i) - \Lambda(T_{i-1}))) + \ln(\lambda(T_i)) - Z(\Lambda(T_i) - \Lambda(T_{i-1}))\}$$
$$- Z(\Lambda(\tau) - \Lambda(T_{N(\tau)})). \tag{21.3}$$

where $Z(t) = \int_0^t z(v)\mathrm{d}v$ is the cumulative hazard corresponding to F.

21.3 A Motivating Example

Meeker and Escobar [ME98, Table 16.4] display 71 times of unscheduled maintenance actions for the U.S.S. Halfbeak number 4 main propulsion diesel engine. Figure 21.2 shows the plot of cumulative failure number against time (in thousands of hours of operation). The plot can be interpreted as an estimate of the cumulative trend function $\Lambda(t)$ of a TRP, and hence the rate of increase of the plot at any time gives a rough estimate of the trend function $\lambda(t)$.

From Fig. 21.2 it is reasonable to conclude that the system is deteriorating, and an increasing $\lambda(\cdot)$ could be a fair assumption, even if we might suspect that $\lambda(t)$ could be decreasing at least for $t > 22$. Here and in the following we will make the usual convention of writing increasing to mean nondecreasing and decreasing to mean nonincreasing. By monotone we will then mean either increasing or decreasing.

A standard procedure for analysis of such data is to fit an NHPP model with, say, a power law intensity function $\lambda(t) = \alpha\beta t^{\beta-1}$, so $\Lambda(t) = \alpha t^{\beta}$. Figure 21.2 indicates, however, that such a model does not fit the data very well. In order to increase the flexibility of the model, Heggland and Lindqvist [HL07] fitted a parametric TRP with a power law trend function as given above, and with a renewal distribution F which is a Weibull-distribution with shape b. Again the fit was shown to be bad, but the interesting result was that b was estimated to 0.762 (with standard error 0.071), which clearly indicated that a power law NHPP is not appropriate. On the other hand, the Poisson property is reasonable to hold for a large system like this. This leads one to think that perhaps the rejection of the power law NHPP means rejection of the power law trend function, and not of the assumption of NHPP *per se*?

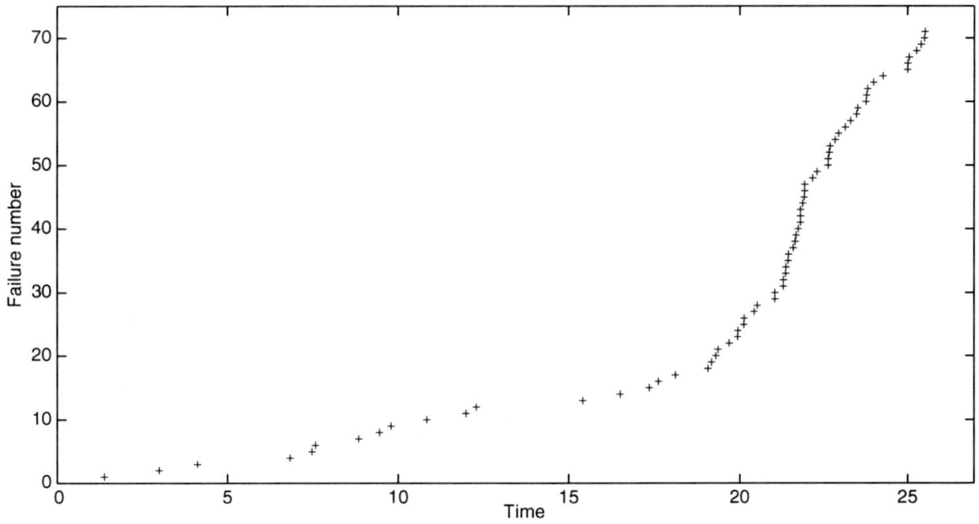

Figure 21.2. Failure number against time in thousands of hours of operation. U.S.S. Halfbeak data

The above motivates an approach where $\lambda(t)$ is allowed to be in principle any smooth positive function. We shall see that such an approach finds the NHPP property to be in fact very reasonable for these data. This will be made more precise in the following.

21.4 Nonparametric Estimation of a Monotone $\lambda(\cdot)$

Heggland and Lindqvist [HL07] considered the case where $\lambda(t)$ is monotone but otherwise is completely unspecified, and extended the approach by Bartozyński et al. [BBMT81] from the NHPP case to the TRP case.

More specifically, the problem considered by [HL07] was that of maximizing the log likelihood function l in (21.3) under the condition that $\lambda(\cdot)$ is nonnegative and monotone on $[0, \tau]$. As in Bartozyński et al. [BBMT81] it was first observed that the optimal $\lambda(t)$ must consist of step functions closed on the left, with no jumps except at failure time points.

Now, letting $n = N(\tau)$, $\lambda_i = \lambda(T_i)$, $X_i = T_i - T_{i-1}$; $i = 1, 2, \ldots, n$, and $X_{n+1} = \tau - T_n$, the problem of maximizing l is simplified to the problem of maximizing

$$l' = \sum_{i=1}^{n} \{\ln z(\lambda_{i-1}X_i) + \ln \lambda_i - Z(\lambda_{i-1}X_i)\} - Z(\lambda_n X_{n+1}), \tag{21.4}$$

subject to $0 \leq \lambda_0 \leq \lambda_1 \leq \ldots \leq \lambda_n$.

Heggland and Lindqvist [HL07] further assumed that $z(t)$ is given on parametric form $z(t; \theta)$. An iterative estimation technique was suggested, consisting in alternatively maximizing with respect to θ and the λ_i. It is readily seen that maximization of (21.4)

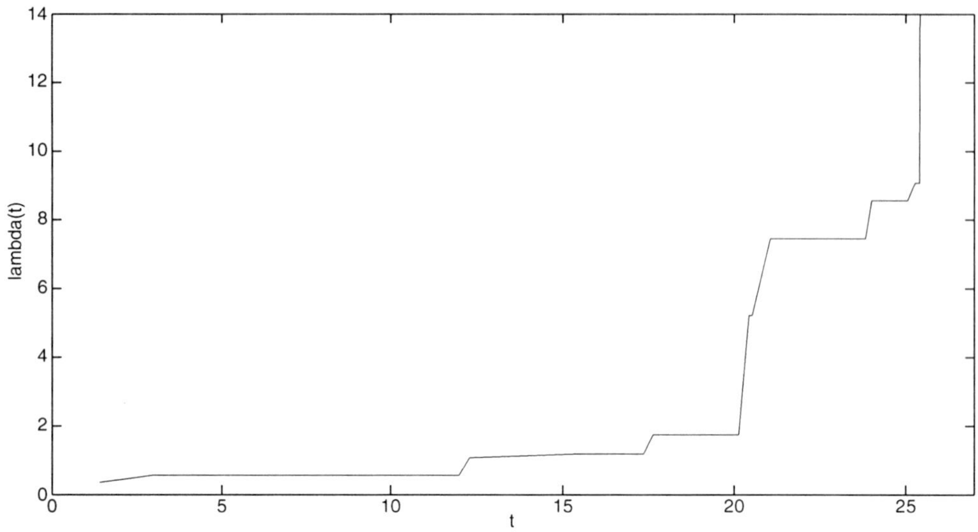

Figure 21.3. Nonparametric estimate of an increasing trend-function $\lambda(\cdot)$. U.S.S. Halfbeak data

with respect to θ for any given set of λ_i is the same as maximum likelihood estimation in the survival model corresponding to $z(\cdot;\theta)$ with completely observed "lifetimes" $\lambda_{i-1}X_i$ for $i = 1, 2, \ldots, n$, and a censored "lifetime" $\lambda_n X_{n+1}$ in the end. This simplifying property is due to the structure of the TRP model as a time-transformed renewal process.

Heggland and Lindqvist [HL07] considered in particular the case where F is a Weibull distribution with shape b. It was demonstrated that this leads to an isotonic regression problem, and an algorithm using the "minimum lower sets algorithm" of Robertson et al. [RWD88] was devised.

Example: For the U.S.S. Halfbeak data (see previous section), the b was estimated to 0.937, with a standard error computed by bootstrapping equal to 0.113. This estimate is much closer to 1 than the estimate 0.762 reported above for the fully parametric estimation. Since $b = 1$ corresponds to an NHPP, the nonparametric analysis therefore suggests that an underlying Poisson property is reasonable. The estimated increasing trend function $\lambda(\cdot)$ is shown in Fig. 21.3.

21.5 Kernel-Estimators for $\lambda(t)$ in the General Case

In the following, we assume that $\lambda(t)$ is completely nonparametric and not necessarily monotone. As in the previous section we assume, on the other hand, that the renewal distribution is given on parametric form with hazard rate $z(t;\theta)$ and expected value 1. For convenience we also assume that there is a parameter value $\theta^{(0)}$ such that $z(t;\theta^{(0)}) \equiv 1$, i.e. such that F is the standard exponential distribution.

The estimation in the next subsection will be done using an iterative scheme switching between maximizing the log likelihood (21.3) for estimation of θ, and estimating the trend function $\lambda(t)$ using in an ad hoc way the kernel smoothing method for counting process intensities due to Ramlau-Hansen [RH83].

21.5.1 The Naive Kernel Smoothing Algorithm

Let $K(t)$ be a positive density function and let h until further be a fixed window size. We consider the following algorithm for estimation of $\lambda(\cdot)$ and θ.

Step 1 Let $\theta = \theta^{(0)}$. Then $\{N(t)\}$ is an NHPP with intensity $\lambda(t)$, and it is well known that this can be estimated by a standard kernel estimator (Ramlau-Hansen [RH83]):

$$\lambda^{(1)}(t) = \frac{1}{h} \sum_{i=1}^{N(\tau)} K\left(\frac{t - T_i}{h}\right). \tag{21.5}$$

Let the cumulative time trend be $\Lambda^{(1)}(t) = \int_0^t \lambda^{(1)}(s)\mathrm{d}s$.

Step 2 Substitute $\lambda^{(1)}(\cdot)$ and $\Lambda^{(1)}(\cdot)$ in (21.3) and maximize the expression with respect to θ to find the maximum at $\theta = \theta^{(1)}$. (As already noted, this is the same as maximum likelihood estimation in the parametric model corresponding to the $z(\cdot;\theta)$.)

Step 3 Now use the estimated $\Lambda^{(1)}(\cdot)$ and $\theta^{(1)}$ as if they were known, to get from (21.1) an expression for the conditional intensity of $\{N(t)\}$ given by

$$z\left(\Lambda^{(1)}(t) - \Lambda^{(1)}(T_{N(t-)}); \theta^{(1)}\right) \lambda(t).$$

This is of the form of Aalen's multiplicative intensity model (see [ABGK93]), which assumes that the intensity of the counting process $\{N(t)\}$ is $Y(t)\lambda(t)$, with $\{Y(t)\}$ an observable stochastic process, adapted to the process $\{N(t)\}$ and predictable. In our case,

$$Y(t) = z\left(\Lambda^{(1)}(t) - \Lambda^{(1)}(T_{N(t-)}); \theta^{(1)}\right)$$

which clearly satisfies these criterions. From (3.12) in Ramlau-Hansen [RH83] we get from this a new kernel estimate for $\lambda(t)$ given by

$$\lambda^{(2)}(t) = \frac{1}{h} \sum_{i=1}^{N(\tau)} K\left(\frac{t - T_i}{h}\right) \cdot \frac{1}{z\left(\Lambda^{(1)}(T_i) - \Lambda^{(1)}(T_{i-1}); \theta^{(1)}\right)} \tag{21.6}$$

which is seen to be a weighted version of the kernel estimator $\lambda^{(1)}(t)$ in (21.5).

Step 4 Return to step 2 above with $\lambda^{(1)}(\cdot)$ and $\Lambda^{(1)}(\cdot)$ replaced by $\lambda^{(2)}(\cdot)$, $\Lambda^{(2)}(\cdot)$, to find $\theta^{(2)}$; then use these in step 3 and continue with steps 2 and 3 until convergence. Here, $\Lambda^{(k)}(\cdot)$ is defined in the obvious way by integrating $\lambda^{(k)}(\cdot)$.

The weights in (21.6) are thus of the form

$$\frac{1}{z\left(\Lambda(T_i) - \Lambda(T_{i-1}); \theta\right)}.$$

It is interesting to see that these have expected values 1. To see this, note first that the definition of the TRP implies that the $\Lambda(T_i) - \Lambda(T_{i-1})$ are i.i.d.$\sim F$ with expected value 1. Now it can be easily shown that for any lifetime $Y > 0$ with hazard rate $z(\cdot)$ we have $E(1/z(Y)) = E(Y)$ which proves the claim.

Example: A Weibull-distribution with shape parameter b and expected value 1 has hazard rate

$$z(t;b) = b[\Gamma(b^{-1}+1)]^b t^{b-1} \tag{21.7}$$

(so that $b = 1$ corresponds to an NHPP).

In this case, we have the (inverse) weights

$$z(\Lambda(T_i) - \Lambda(T_{i-1})) = b[\Gamma(b^{-1}+1)]^b (\Lambda(T_i) - \Lambda(T_{i-1}))^{b-1}.$$

Simulations using the Weibull-distribution has shown, however, that the given algorithm may not converge. The problem seems to be that the weights are too closely related to the values of the $K(\cdot)$ in corresponding terms of (21.6). Further, the estimation of $\lambda(\cdot)$ is done without regard to a possible maximization of the likelihood (21.3) (except in the estimation of θ). Still, the method suggests the use of a weighted kernel estimator, which will be studied in the next subsection.

21.5.2 Maximum Likelihood Weighted Kernel Estimation

Jones and Henderson [JH05, JH09] study kernel density estimation using variable weights and/or variable locations of the kernels, where the clue is to maximize the nonparametric likelihood with respect to these weights or locations. Since our estimator (21.6) in effect is a weighted kernel estimator, it might be worthwhile to search for better weights than the ones suggested by (21.6). We will do this by maximizing the log likelihood (21.3) simultaneously with respect to θ and the weights $w = (w_i; i = 1, 2, \ldots, N(\tau))$, using

$$\lambda(t;w) = \frac{1}{h}\sum_{i=1}^{N(\tau)} K\left(\frac{t-T_i}{h}\right) w_i. \tag{21.8}$$

We, thus, consider maximization of (21.3) with $z(\cdot) = z(\cdot;\theta)$ on parametric form and with $\lambda(\cdot) = \lambda(\cdot;w)$ given in the form (21.8). The maximization is with respect to $(\theta, w_1, \ldots, w_n)$, where $n = N(\tau)$. Note that, while in the density estimation considered by Jones and Henderson it is necessary to restrict the w_i to have sum 1, in our case such a restriction is not necessary. This follows from the definition of the TRP, where F is required to have expected value 1, and the scale parameter in question is instead included in the trend function $\lambda(\cdot)$.

We may maximize (21.3) in an iterative alternating manner, similar to what was done in the monotone case in Sect. 21.4. We then utilize the already noted fact that maximization with respect to θ for given $\lambda(\cdot)$ is the same as maximum likelihood estimation in the model $z(\cdot;\theta)$.

In order to simplify the algorithms we modify (21.3) by using the approximation

$$\Lambda(T_i;w) - \Lambda(T_{i-1}) \approx \lambda(T_i;w)(T_i - T_{i-1}) = \lambda(T_i;w)X_i.$$

The corresponding approximation of (21.3) is hence

$$\begin{aligned} l'' = &\sum_{i=1}^{n} \{\ln z(\lambda(T_i; w)X_i; \theta) + \ln \lambda(T_i; w) - Z(\lambda(T_i; w)X_i; \theta)\} \\ &- Z(\lambda(\tau; w)X_{n+1}; \theta), \end{aligned} \tag{21.9}$$

where $Z(t;\theta) = \int_0^t z(v;\theta)dv$ is the cumulative hazard corresponding to F.

21.5.3 Specializing to a Weibull Renewal Distribution

If the renewal distribution is Weibull with hazard rate (21.7), the log likelihood (21.9) can be written

$$\begin{aligned} l'' = &\, n \log b + nb \log \Gamma(b^{-1}+1) + \sum_{i=1}^{N(\tau)} \{b \log(X_i \lambda(T_i; w)) - \log X_i \\ &- [\Gamma(b^{-1}+1)\lambda(T_i; w)X_i]^b\} - [\Gamma(b^{-1}+1)\lambda(\tau; w)(\tau - T_{N(\tau)})]^b \end{aligned}$$

Differentiating with respect to w_ℓ we get

$$\begin{aligned} \partial l''/\partial w_\ell = &\, b \sum_{i=1}^{N(\tau)} \left\{ \frac{h^{-1}K((T_i - T_\ell)/h)}{\lambda(T_i, w)} \right. \\ &- \left. [\Gamma(b^{-1}+1)\lambda(T_i; w)X_i]^{b-1}\Gamma(b^{-1}+1)X_i h^{-1}K((T_i - T_\ell)/h) \right\} \\ &- b[\Gamma(b^{-1}+1)\lambda(\tau; w)X_{n+1}]^{b-1}\Gamma(b^{-1}+1)X_{n+1}h^{-1}K((\tau - T_\ell)/h) \end{aligned} \tag{21.10}$$

In the example below, we used these partial derivatives in a steepest ascent approach in order to maximize l'' with respect to the w_i for given value of b. The b was then, in alternating steps, estimated by maximum likelihood using the current values of $\lambda(T_i; w)$. The kernel used was the Epanechnikov kernel.

Example: Consider again the U.S.S. Halfbeak data. The results for three different choices of h are given in Table 21.1 and Fig. 21.4. As the table shows, there is a tendency for higher maximum likelihood for small values of h. This is reasonable as a small h means following too closely the obtained data (overfitting). This also indicates that in the approach of weighted kernel estimation based on maximizing a likelihood, it is not possible to include h among the parameters in the maximization. This issue is also discussed in [JH05]. It is also seen from the table that the value of b decreases as h increases. An intuitive reason for this could be that a too high degree of smoothing leads to "lifetimes" $\lambda(T_i)X_i$ which are (modulo a scale factor) closer and closer to the X_i themselves. When these are fitted to the Weibull, the shape parameter is estimated to 0.63. Thus, we would believe that increasing h further would lead to estimated values for b converging to 0.63.

The curves in Fig. 21.4 basically reflect the features of the cumulative plot given in Fig. 21.2 at least up to, say, $t = 22$. After that time, the curves drop due to the fact that no edge-correction has been implemented in the present analysis. For ordinary density estimation, such methods are considered for example by Cowling and Hall [CH96].

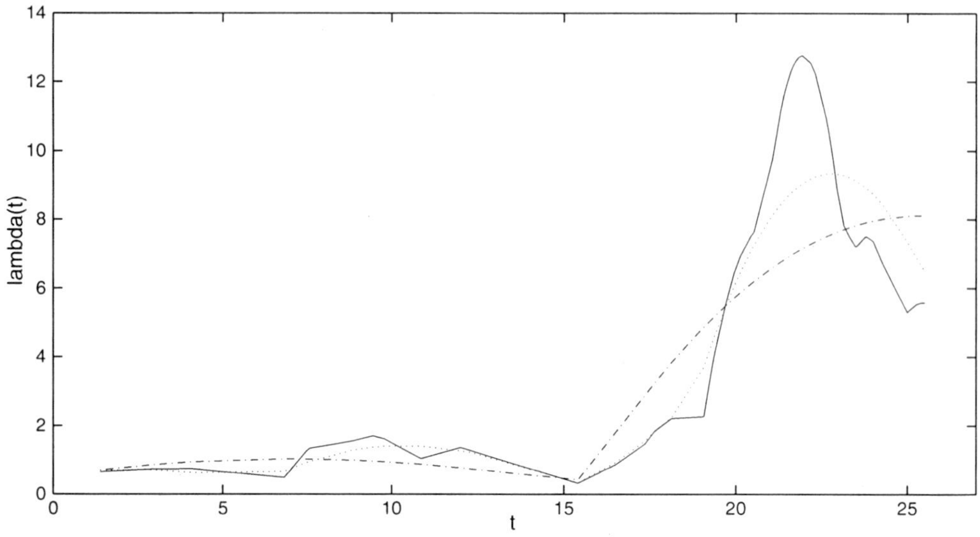

Figure 21.4. Weighted kernel estimation of U.S.S. Halfbeak data. Estimates of $\lambda(\cdot)$ for $h = 2$ (*solid*), $h = 5$ (*dotted*), $h = 10$ (*dot-dash*)

Table 21.1. Weighted kernel estimation of U.S.S. Halfbeak data. Estimates of the shape parameter b and the maximum value of the log likelihood ℓ'' in (21.10)

	$h = 2$	$h = 5$	$h = 10$
b	0.957	0.915	0.870
ℓ''	−116.91	−120.08	−123.38

It is noted in [JH05] that the weights obtained at the maximum of the log likelihood are non-zero only for fairly few observations. For the present example, the number of non-zero weights w_i are 20 for $h = 2$; 25 for $h = 5$; and just 9 for $h = 10$, while there are 71 observations. As also noted by Jones and Henderson, and as also found in our computations, the non-zero weights are usually clustered around similar values of the T_i, thus suggesting an even more parsimonious model.

21.6 Concluding Remarks

The aim of this chapter has been to develop useful nonparametric estimation techniques for trend analysis of repairable systems, particularly for systems modeled by TRPs.

The monotone NPMLE of $\lambda(\cdot)$ developed in [HL07] was briefly described. It is particularly simple when the renewal distribution F is a Weibull distribution function. The Weibull distribution was also used in the estimation of $\lambda(\cdot)$ in the general nonparametric case. This distribution was chosen because it is important and much used in reliability analyses, and because it gives a computationally tractable approach. In addition it implies a smooth connection with the exponential distribution which is the distribution inherent in an NHPP assumption. It would, though, be of interest to see how the suggested algorithms can be adapted to other distributions, such as the gamma or log-normal distributions.

According to [JH05] it may be even better, instead of using variable weights in kernel estimation, to change the location of the ith term of the kernel estimator to be an unknown value m_i instead of the observed point T_i. In this case, one may maximize the log likelihood (21.3) with respect to θ and the location points $m = (m_i; i = 1, 2, \ldots, N(\tau))$, using

$$\lambda(t; m) = \frac{1}{h} \sum_{i=1}^{N(\tau)} K\left(\frac{t - m_i}{h}\right).$$

Note then that each T_i has its own location m_i, while the weights are now all equal to 1. This approach is the main topic of [JH09], where it is noted the closeness to the kernel convolution sieve estimator proposed by Geman and Hwang [GH82].

As briefly mentioned, we have assumed that the kernel estimations are done with a fixed h, tacitly assuming that the actual value of h should be chosen by some additional procedure. This is in fact also the approach of Jones and Henderson, who advocate the use of the usual CV criterion. It is not quite clear, however, how to perform the cross validation for data from a TRP, where single failures can not easily be taken out without destroying the structure of the process. One possibility would be to choose an optimal value of h assuming an NHPP, in which case kernel estimation is essentially the same as density estimation. This value could be determined by cross-validation and could be kept throughout the iterative procedure.

It should be noted that Jones and Henderson [JH05] maximize a likelihood corresponding to assuming $z(t) \equiv 1$ in (21.3). Thus, the task of maximizing (21.3) for a general parametric $z(t; \theta)$ will obviously be a harder problem than theirs, although in principle they are similar.

We have in the present chapter not considered estimation of the standard error of the estimates, except for the estimates taken from [HL07]. Indeed, [HL07] suggest the use of bootstrapping in order to estimate bias, standard error and obtain confidence intervals, for both the parameter θ in the renewal distribution and the trend function $\lambda(\cdot)$. In fact, the TRP is particularly suited for bootstrapping. The obvious way to generate bootstrap samples from a TRP is to first simulate the estimated renewal process, and then transform the arrival times of this renewal process by the inverse mapping of the estimated cumulative trend function $\Lambda(\cdot)$. Both nonparametric and parametric bootstrapping are feasible in this way.

Throughout the chapter we have assumed that the failure data come from one single system. For the case of monotone $\lambda(\cdot)$, [HL07] suggested an approach based on the approximation of the superposition of several TRPs by an NHPP. This is of course exactly true only if the original processes are NHPPs. It is the purpose of future research to extend the weighted kernel methods of this chapter to cases with several systems.

References

[ABGK93] Andersen, P.K., Borgan, O., Gill, R.D., Keiding, N.: Statistical Models Based on Counting Processes. Springer, New York (1993)

[AF84] Ascher, H., Feingold, H.: Repairable Systems Reliability: Modeling, Inference, Misconceptions and their Causes. Marcel Dekker, New York (1984)

[BBMT81] Bartozyński, R., Brown, B.W., McBride, C.M., Thompson, J.R.: Some nonparametric techniques for estimating the intensity function of a cancer related nonstationary Poisson process. Ann. Stat., **9**(5), 1050–1060 (1981)

[CH96] Cowling, A., Hall, P.: On pseudodata methods for removing boundary effects in kernel density estimation. J. Roy. Stat. Soc. Ser. B, **58**, 551–563 (1996)

[GH82] Geman, S., Hwang, C.R.: Nonparametric maximum likelihood estimation by the method of sieves. Ann. Stat., **10**, 401–414 (1982)

[HL07] Heggland, K., Lindqvist, B.H.: A non-parametric monotone maximum likelihood estimator of time trend for repairable systems data. Reliab. Eng. Syst. Saf., **92**, 575–584 (2007)

[JH05] Jones, M.C., Henderson, D.A.: Maximum likelihood kernel density estimation. Technical Report 01/05, Department of Statistics, The Open University, UK (2005)

[JH09] Jones, M.C., Henderson, D.A.: Maximum likelihood kernel density estimation: On the potential of convolution sieves. Comput. Stat. Data Anal., **53**, 3726–3733 (2009)

[Lin06] Lindqvist, B.H.: On the statistical modelling and analysis of repairable systems. Stat. Sci., **21**, 532–551 (2006)

[LEH03] Lindqvist, B.H., Elvebakk, G., Heggland, K.: The trend-renewal process for statistical analysis of repairable systems. Technometrics, **45**, 31–44 (2003)

[ME98] Meeker, W.Q., Escobar, L.A.: Statistical Methods for Reliability Data. Wiley, New York (1998)

[RH83] Ramlau-Hansen, H.: Smoothing counting process intensities by means of kernel functions. Ann. Stat., **11**, 453–466 (1983)

[RWD88] Robertson, T., Wright, F.T., Dykstra, R.L.: Order Restricted Statistical Inference. Wiley, New York (1988)

22

Confidence Region for Distribution Function from Censored Data

Mikhail S. Nikulin[1]* and Valentin N. Solev[2]

[1] Université Victor Segalen Bordeaux 2, Bordeaux, France, mikhail.nikouline@u-bordeaux2.fr
[2] Institute of Mathematics, Russian Academy of Sciences, St. Petersburg, Russia, solev@pdmi.ras.ru

Abstract: The aim of this chapter is to consider the problem of construction of a confidence region for distribution function as we deal with censored data. This problem is very important for reliability theory and the survival analysis. We consider also a testing problem.

Keywords and phrases: Censored data, Confidence region, Dvoretsky–Kiefer–Wolfowitz inequality, Homogeneity problem, Kaplan–Meier estimator, Testing of hypothesis

22.1 Introduction

Let X be a random variable. We denote F the distribution function of X,

$$F(x) = \mathbf{P}\{X \le x\}, \quad x \in R,$$

and let Y be a random variable independent of X, and let G is the distribution function of Y.
We consider the model according to which instead of random variables X and Y, we observe the random vector

$$Z = (W, \delta), \ \text{ where } W \in R, \ \ \delta \in \{0,1\}. \tag{22.1}$$

Here,

$$W = X \wedge Y, \quad \delta = 1\!\!\mathrm{I}_{X \le Y}. \tag{22.2}$$

The problem is to estimate the distribution function F of the random variable X using the observations $Z_1, \ldots, Z_n$, where $Z_1, \ldots, Z_n$ are independent copies of the random vector Z.

The distribution of the vector Z has the density

$$p(w,t), \quad w \in R, \quad t \in \{0,1\},$$

V.V. Rykov et al. (eds.), *Mathematical and Statistical Models and Methods in Reliability: Applications to Medicine, Finance, and Quality Control*, Statistics for Industry and Technology, DOI 10.1007/978-0-8176-4971-5_22,

with respect to the measure

$$\mu(\mathrm{d}w \times \mathrm{d}t) = [t\mathrm{d}F(w) + (1-t)\mathrm{d}G(w)] \times \nu(\mathrm{d}t),$$

where $\nu(\mathrm{d}t)$ is the counting measure on the set $\{0, 1\}$,

$$p(w,t) = [1 - G(w)]^t \times [1 - F(w)]^{1-t}. \tag{22.3}$$

It is evident that if $t \in \{0, 1\}$ then

$$p(w,t) = t\,[1 - G(w)] + (1-t)\,[1 - F(w)]. \tag{22.4}$$

The distribution of the random variable W has the distribution function $H(w)$, where

$$1 - H(w) = (1 - F(w))(1 - G(w)). \tag{22.5}$$

Now we consider a statistical problem of estimation of the distribution function F. Let $X_1, \ldots, X_n$ are i.i.d.r.v., having a common distribution function F, and $Y_1, \ldots, Y_n$ are i.i.d.r.v., having a common distribution function G. We suppose that the samples

$$X_1, \ldots, X_n \quad \text{and} \quad Y_1, \ldots, Y_n$$

are independent.

Consider a sample $Z_1, \ldots, Z_n$, where $Z_i = (W_i, \delta_i)$,

$$W_i = X_i \wedge Y_i, \quad \delta_i = \mathbb{1}_{X_i \le Y_i}, \quad i = 1, 2, ..., n. \tag{22.6}$$

To estimate the distribution function F, Kaplan and Meier [KM58] proposed in 1958 their famous estimator $\widehat{F}_n$:

$$1 - \widehat{F}_n(x) = \prod_{\substack{i \in \{1, 2, \ldots, n\}, \\ W_i \le x}} \left(\frac{n - r_i}{n - r_i + 1} \right)^{\delta_i}, \tag{22.7}$$

where r_i - the rank of the statistic W_i in the vector of order statistics

$$(W_{(1)}, ..., W_{(i)}, ..., W_{(n)}), \quad W_{(1)} < ... < W_{(i)} < ... < W_{(n)},$$

associated with the sample $W_1, ...W_n$. The corresponding estimator $\widehat{G}_n$ for the unknown distribution function G is given by the next formula:

$$1 - \widehat{G}_n(x) = \prod_{\substack{i \in \{1, 2, \ldots, n\}, \\ W_i \le x}} \left(\frac{n - r_i}{n - r_i + 1} \right)^{1-\delta_i}. \tag{22.8}$$

Let H_n be the distribution function of the empirical distribution based on the sample $W_1, \ldots, W_n$. Then

$$1 - H_n(w) = \left(1 - \widehat{F}_n(w)\right)\left(1 - \widehat{G}_n(w)\right), \tag{22.9}$$

and

$$\widehat{F}_n(Z_j+0) - \widehat{F}_n(Z_j-0) = 0, \text{ if } \delta_j = 0; \; \widehat{G}_n(Z_j+0) - \widehat{G}_n(Z_j-0) = 0, \text{ if } \delta_j = 1.$$

We remind now the famous inequality of Dvoretsky–Kiefer–Wolfowitz [DKW56]. Let $X_1, \ldots, X_n$ be independent random variables having the common distribution function F, and let F_n is the empirical distribution function associated with the sample $X_1, \ldots, X_n$ defined by

$$F_n(x) = \frac{1}{n} \sum_{i=1}^{n} 1\!\!1_{(-\infty,x)}(X_i).$$

In this case, the next inequality holds:

$$\mathbb{P}\left\{\|F_n - F\|_\infty > \lambda\right\} \leq 2\exp\left\{-2n\lambda^2\right\}. \tag{22.10}$$

Here,

$$\|F_n - F\|_\infty = \sup_x |F_n(x) - F(x)|.$$

In the case of right censored data, Bitouzé et al. [BLM99] proved in 1999 that the next theorem holds.

Theorem 1. *Let $\widehat{F}_n$ be the Kaplan–Meier estimator of the distribution function F. There exists an absolute constant C such that $\forall \lambda > 0$ holds the next inequality:*

$$\mathbb{P}\left\{\|(1-G)(\widehat{F}_n - F)\|_\infty > \lambda\right\} \leq 2.5\exp\left\{-2n\lambda^2 + C\sqrt{n}\lambda\right\}. \tag{22.11}$$

It is evident that the next inequality holds:

$$\mathbb{P}\left\{\|(1-F)(\widehat{G}_n - G)\|_\infty > \lambda\right\} \leq 2.5\exp\left\{-2n\lambda^2 + C\sqrt{n}\lambda\right\}. \tag{22.12}$$

It is evident that there is one problem with the application of Theorem 1: we have the unknown $G(x)$. However, the inequality of Dvoretsky–Kiefer–Wolfowitz does not give the possibility to construct the confidence interval for the unknown function $F(x)$. Now we shall show how we may to change the situation.

Theorem 2. *Let $\widehat{F}_n$ be the Kaplan–Meier estimator for F, and $\widehat{G}_n$ be the Kaplan–Meier estimator for G. Then there exists an absolute constant C such that $\forall \lambda > 0$*

$$\mathbb{P}\left\{\|(1-\widehat{G}_n)(\widehat{F}_n - F)\|_\infty > \lambda\right\} \leq 5\exp\left\{-n\lambda^2/2 + C\sqrt{n}\lambda/2\right\}. \tag{22.13}$$

Now we reformulate the state of Theorem 2 as asymptotic upper bound for probabilities (see also [BLM99]).

Theorem 3. *Let $\widehat{F}_n$ be the Kaplan–Meier estimator for F, and $\widehat{G}_n$ be the Kaplan–Meier estimator for G. Then for all $\lambda > 0$*

$$\limsup_{n\to\infty} \mathbb{P}\left\{\sqrt{n}\,\|(1-\widehat{G}_n)(\widehat{F}_n - F)\|_\infty > \lambda\right\} \leq 4\exp\left\{-\lambda^2/2\right\}. \tag{22.14}$$

Suppose at first we have priori information according to which F belongs to the parametric family, i.e.,

$$F \in \mathscr{F} = \{F : F = F_\theta, \theta \in \Theta\}.$$

The confident interval

$$\Theta(\lambda) = \left\{\theta : \|(1 - \widehat{G}_n)(\widehat{F}_n - F_\theta)\|_\infty \le \lambda\right\}$$

gives the permission to construct the confidence set for the values of the unknown parameter θ with the confidence probability p,

$$p \ge 1 - 5\exp\{-n\lambda^2/2 + C\sqrt{n}\lambda/2\}.$$

If $\widehat{\theta}_n \in \Theta(\lambda)$ is an element of the confidence set, then the function $F_{\widehat{\theta}_n}$ is closed to F in the following sense:

$$\mathbb{P}\left\{\|(1 - \widehat{G}_n)(\widehat{F}_{\widehat{\theta}_n} - F)\|_\infty > \lambda\right\} \le 5\exp\{-n\lambda^2/2 + C\sqrt{n}\lambda/2\}.$$

Note that this method has a big advantage: we don't do any priori supposition about the distribution functions F and G, and hence our approach is very natural for statistical situations in estimation of parameters and testing hypotheses. It is evident that in such a case the width of the confidence zone around the estimator F_n depends on the observed values of the function $\widehat{G}_n$. If a statistician is not satisfied by this confidence zone, he needs to continue the observations!

22.1.1 Proofs

Proof of the Theorem 2. The value $(1 - \widehat{G}_n)(\widehat{F}_n - F)$ can be presented by the following way:

$$\begin{aligned} (1 - \widehat{G}_n)(F - \widehat{F}_n) &= (1 - \widehat{G}_n)((1 - \widehat{F}_n) - (1 - F)) \\ &= (1 - \widehat{G}_n)(1 - \widehat{F}_n) - (1 - \widehat{G}_n)(1 - F). \end{aligned} \tag{22.15}$$

Furthermore,

$$\begin{aligned} (1 - \widehat{H}_n) &= (1 - \widehat{F}_n)(1 - \widehat{G}_n), \\ (1 - H) &= (1 - F)(1 - G). \end{aligned}$$

It implies that

$$\begin{aligned} &(1 - \widehat{H}_n) - ((1 - \widehat{G}_n)(1 - F)) \\ &= \left[(1 - \widehat{H}_n) - (1 - H)\right] + \left[(1 - H) - (1 - \widehat{G}_n)(1 - F)\right] \\ &= \left[(1 - \widehat{H}_n) - (1 - H)\right] + \left[(1 - F)(\widehat{G}_n - G)\right]. \end{aligned}$$

Hence,

$$(1 - \widehat{G}_n)(F - \widehat{F}_n) = \left[(1 - \widehat{H}_n) - (1 - H)\right] + \left[(\widehat{G}_n - G)(1 - F)\right], \tag{22.16}$$

and we obtain for the uniform norm of the function

$$(1-\widehat{G}_n(x))(\widehat{F}_n(x)-F(x))$$

the estimator

$$\|(1-\widehat{G}_n)(\widehat{F}_n-F)\|_\infty \le \|H-\widehat{H}_n\|_\infty + \|(1-F)(\widehat{G}_n-G)\|_\infty.$$

From the inequality of Dvoretsky–Kiefer–Wolfowitz (1956) it follows that

$$\mathbb{P}\left\{\|(H-\widehat{H}_n)\|_\infty > \lambda/2\right\} \le 2\exp\{-n\lambda^2/2\}.$$

From the Theorem 2 it follows that

$$\mathbb{P}\left\{\|(1-F)(\widehat{G}_n-G)\|_\infty > \lambda/2\right\} \le 2.5\exp\{-n\lambda^2/2 + C\sqrt{n}\lambda/2\}.$$

Since

$$\begin{aligned}&\left\{\|H-\widehat{H}_n\|_\infty + \|(1-F)(\widehat{G}_n-G)\|_\infty > \lambda)\right\}\\ &\quad\subset \left\{\|(H-\widehat{H}_n)\|_\infty > \lambda/2\right\}\bigcup\left\{\|(1-F)(\widehat{G}_n-G)\|_\infty > \lambda/2\right\},\end{aligned}$$

we obtain

$$\begin{aligned}\mathbb{P}\left\{\|(H-\widehat{H}_n)\|_\infty + \|(1-F)(\widehat{G}_n-G)\|_\infty > \lambda\right\} &\le \mathbb{P}\left\{\|(1-\widehat{H}_n)-(1-H)\|_\infty > \frac{\lambda}{2}\right\}\\ &\quad+\mathbb{P}\left\{\|(1-F)(\widehat{G}_n-G)\|_\infty > \frac{\lambda}{2}\right\}.\end{aligned} \tag{22.17}$$

From the last inequality it follows that

$$\mathbb{P}\left\{\sqrt{n}\|(1-\widehat{G}_n)(\widehat{F}_n-F)\|_\infty > \lambda\right\} \le 2.5\exp\{-\lambda^2/2 + C\lambda/2\} + 2\exp\{-\lambda^2/2\}, \tag{22.18}$$

which we present by the following way:

$$\mathbb{P}\left\{\|(1-\widehat{G}_n)(\widehat{F}_n-F)\|_\infty > x\right\} \le 5\exp\{-nx^2/2 + C\sqrt{n}x/2\}, \tag{22.19}$$

from where it follows that the Theorem 2 is proved. ■

Proof of the Theorem 3. It is proved in [BLM99] that for all $\lambda > 0$

$$\limsup_{n\to\infty} \mathbb{P}\left\{\sqrt{n}\,\|(1-G)(\widehat{F}_n-F)\|_\infty > \lambda\right\} \le 2\exp\{-2\lambda^2\}. \tag{22.20}$$

Using the same arguments as above we deduce from (22.20) the inequality (22.14). ■

22.2 Testing of the Homogeneity Hypothesis

As before let suppose that $X_1, \ldots, X_n$ be i.i.d.r.v., having a common distribution function F, and $Y_1, \ldots, Y_n$ be i.i.d.r.v., having a common distribution function G. We suppose that the samples

$$X_1, \ldots, X_n \quad \text{and} \quad Y_1, \ldots, Y_n$$

are independent.

Consider a sample $Z_1, \ldots, Z_n$, where $Z_i = (W_i, \delta_i)$,

$$W_i = X_i \wedge Y_i, \quad \delta_i = \mathbb{1}_{X_i \le Y_i}, \quad i = 1, 2, ..., n. \tag{22.21}$$

To estimate unknown distribution function F and G we may, as before, to construct the Kaplan–Meier estimators $\widehat{F}_n, \widehat{G}_n$ based on the sample $Z_1, \ldots, Z_n$, given by formulas (6) and (7) respectively.

Also let suppose that $X_1^*, \ldots, X_m^*$ be i.i.d.r.v., having a common distribution function F^*, and $Y_1^*, \ldots, Y_m^*$ be i.i.d.r.v., having a common distribution function G^*. We suppose that samples

$$X_1^*, \ldots, X_m^* \quad \text{and} \quad Y_1^*, \ldots, Y_m^*$$

are independent. We assume also that

$$X_1, \ldots, X_n, Y_1, \ldots, Y_n \quad \text{and} \quad X_1^*, \ldots, X_m^*, Y_1^*, \ldots, Y_m^*$$

are independent.

Suppose the sample $Z_1^*, \ldots, Z_m^*$, $Z_i^* = (W_i^*, \delta_i^*)$, is constructed by usual way on observations $X_1^*, \ldots, X_m^*$ and $Y_1^*, \ldots, Y_n^*$. Denote by $\widehat{F_m^*}, \widehat{G_m^*}$ Kaplan–Meier estimators of unknown distribution functions F^*, G^*, that constructed on the sample $Z_1^*, \ldots, Z_m^*$:

$$1 - \widehat{F_m^*}(x) = \prod_{\substack{i \in \{1, 2, \ldots, m\}, \\ W_i^* \le x}} \left(\frac{m - r_i}{m - r_i + 1} \right)^{\delta_i^*}, \tag{22.22}$$

$$1 - \widehat{G_m^*}(x) = \prod_{\substack{i \in \{1, 2, \ldots, m\}, \\ W_i^* \le x}} \left(\frac{m - r_i}{m - r_i + 1} \right)^{1-\delta_i^*}, \tag{22.23}$$

where r_i - the rank of the statistics W_i^*.

For simplicity, further we shall assume that $n = m$ and denote

$$\left(1 - \tilde{G}(t)\right) = (1 - G(t)) \wedge (1 - G^*(t)), \tag{22.24}$$

$$\left(1 - \tilde{G}_n(t)\right) = \left(1 - \widehat{G}_n(t)\right) \wedge \left(1 - \widehat{G_n^*}(t)\right). \tag{22.25}$$

To test the hypothesis $H_0 : F = F^*$ we consider the statistics

$$T_n = \sup_t \left(1 - \tilde{G}_n(t)\right) \left|\widehat{F}_n(t) - \widehat{F_n^*}(t)\right|. \tag{22.26}$$

We note that we want to test H_0 without any suggestions on G and G^*. From the Theorem 2 it is easily to deduce following results concern the asymptotical upper bound.

Theorem 4. *Under hypothesis $H_0 : F = F^*$, there exists an absolute constant C such that for all $\lambda > 0$*

$$\mathbb{P}\{T_n > \lambda\} \leq 10 \exp\{-n\lambda^2/2 + C\sqrt{n}\lambda/2\}. \tag{22.27}$$

Theorem 5. *Under alternative*

$$H_1 : \sup_t \left(1 - \tilde{G}(t)\right) |F(t) - F^*(t)| \geq r > 0,$$

there exists an absolute constant C such that for all $0 < \lambda < r$

$$\mathbb{P}\{T_n < \lambda\} \leq 10 \exp\{-n(r-\lambda)^2/8 + C\sqrt{n}\,(r-\lambda)/4\}. \tag{22.28}$$

From the Theorem 3 it is easily to deduce following results

Theorem 6. *Under hypothesis $H_0 : F = F^*$, for all $\lambda > 0$*

$$\limsup_{n\to\infty} \mathbb{P}\left\{\sqrt{n}\,T_n > \lambda\right\} \leq 8 \exp\{-\lambda^2/2\}. \tag{22.29}$$

Theorem 7. *Under alternative*

$$H_1 : \sup_t \left(1 - \tilde{G}(t)\right) |F(t) - F^*(t)| \geq \frac{1}{\sqrt{n}} r > 0,$$

for all $0 < \lambda < r$

$$\limsup_{n\to\infty} \mathbb{P}\left\{\sqrt{n}\,T_n < \lambda\right\} \leq 8 \exp\{-(r-\lambda)^2/8\}. \tag{22.30}$$

Acknowledgments. The investigation was supported in part by RFBR grants 08-01-00692, NSH – 4472.2010.1, RFBR–DFG 09-01-91331 and Federal targeted program of the Ministry of Education and Science of the Russian Federation “Scientific and scientific-educational personnel of innovation Russia”.

References

[BLM99] D. Bitouzé, B. Laurent, and P. Massart: A Dvoretsky–Kiefer–Wolfowitz type inequality for the Kaplan–Meier estimator, Ann. Inst. Henry Poincaré, **35**(6), 735–763 (1999)

[DKW56] A. Dvoretsky, J.C. Kiefer, and J. Wolfowitz: Asymptotic minimax character of the sample distribution function and of the classical multinomial estimator, Ann. Math. Statist., **33**, 642–669 (1956)

[KM58] E.L. Kaplan and P. Meier: Nonparametric estimation from incomplete observations, J. Am. Stat. Assoc., **53**, 457–481 (1958)

23

Empirical Estimate with Uniformly Minimal d-Risk for Bernoulli Trials Success Probability

E. D. Sherman and I. N. Volodin*

Kazan State University, Kremlevskaya St., 18, Kazan, Russia, Igor.Volodin@ksu.ru

Abstract: In the framework of empirical d-posterior approach, two methods of a construction of estimates with uniformly minimal d-risk for a Bernoulli trials success probability is considered when the prior distribution is completely unknown. The first one is based on a modification of the historical data. The second is based on the invariance property of an estimate with uniformly minimal d-risk for a scalar parameter of a discrete exponential family.

Keywords and phrases: Empirical d-posterior approach, Estimates with uniformly minimal d-risk, Convergence of empirical d-risk, Success probability for Bernoulli trials

23.1 Introduction

The classical probability model for a guarantee control by a quality attribute is connected with a random selection of products from a finite population with subsequent classification into two classes as good and defective. Usually, the population size is significantly bigger than the number of selected items n. Assume that an organization of the population is fulfilled in correspondence with a particular technological process. For example, assume that the population contains a metallic products from the same melting and the controlled attribute is connected with the metal characteristics. Then, the control procedure is fulfilled in framework of Bernoulli trials with a constant probability θ of "success" for a product testing (the probability of the present of a defect for a randomly chosen product from the population).

For L.N. Bolshev's d-posterior approach, the notion of guarantee estimation is based on the definition of d-risk of the estimation as a mean value of losses among the experiments that result in an acceptance of the same decision. From the point of view of a consumer of the controlled production, this values correspond to the proportion of defective production among all that was accepted as good after the quality control procedure. A solution of the problem of the minimization of such loses is possible only in the framework of Bayesian model of production. Therefore, estimating probability θ we assume that its value, that changes from an experiment to experiment, is a realization of a random variable ϑ with some prior distribution.

V.V. Rykov et al. (eds.), *Mathematical and Statistical Models and Methods in Reliability: Applications to Medicine, Finance, and Quality Control*, Statistics for Industry and Technology, DOI 10.1007/978-0-8176-4971-5_23,

Usually, there is no information to specify this prior distribution. Hence, for the d-risk minimization problem, an approach similar to empirical Bayesian approach by Robbins [Rob55] is explored. First an estimation of θ is calculated under the known prior distribution and after its empirical analog is constructed under which the minimization procedure for d-risk is substituted by its minimization by the historical data.

Estimates with the minimal d-risk (EUMD) were considered in papers by Volodin, Novikov, and Simushkin (see [Sim83, SV83, VN93, VS87]) as a Bayesian alternative to unbiased estimates with uniformly minimal risk (UEUMR) [VN83].

Possible variants of empirical approach to d-guarantee procedures with unknown prior distribution suggested in article [SV83] are used in the current article for the problem of EUMD construction. Such estimates were considered for the first time in article [SV83]. The existence, unbiasedness and other properties of EUMD were studied in article [VN93] it is proved that maximum likelihood estimates have asymptotically minimal d-risk. Finally, in article [She08] an empirical version of EUMD (EEUMD) was obtained for discrete exponential families.

For the case of known prior distribution, an EUMD construction is based on the minimization of the posterior risk according to the values of the random sample (compare with Bayesian approach where minimization is obtained over all possible values of the estimate). As it is mentioned in article [SV83], such minimization is possible only if there exists a sufficient statistics of the same dimension as the parameter, otherwise a reduction of sample values to a statistic is required (for example, to a Bayesian estimate). A similar problem arises also in the case of unknown prior distribution for a construction of empirical versions of EUMD (the same as in the empirical Bayesian approach of Robbins).

In this article, the binomial distribution with probability function

$$f(x \mid \theta, n) = C_n^x \, \theta^x \, (1-\theta)^{n-x}, \; x \in \{0, \dots, n\}, \; \theta \in (0, 1), n \in \mathbb{N} \tag{23.1}$$

is considered. For the case of quadratic loss function, two methods of empirical variant of EUMD are constructed for the success probability θ (parameter n is assumed to be known) according to the historical data $X^{[m]}$ of size m. The first method is based on the posterior risk estimate minimization, which is constructed by a modification of then historical data. The second method based on the invariance property of EUMD for a scalar parameter of one dimensional discrete exponential family. For the case of prior Beta distribution, the accuracy properties of the two constructed empirical estimates, UEUMR, and empirical Bayesian estimate are compared by the method of stochastic modeling.

23.2 An Estimate with Uniformly Minimal d-Risk

The Binomial distribution family (23.1) has a sufficient statistics with the same distribution type, hence without loss of generality we assume the sample size equals one. Let $g(\theta)$, $\theta \in \mathrm{R}_+$ be density functions of the prior distribution G of parameter ϑ. For the problem of θ estimation under the quadratic loss function $L(\theta, d)$, the d-posterior risk (or simply d-risk) $\Re(d \mid \delta)$ of the decision function (estimate) $\delta(x)$, is defined as the conditional mathematical expectation $L(\vartheta, d)$ with respect to the σ-algebra generated by the statistic $\delta(x)$. Obviously, $\Re(d \mid \delta)$ could be represented as the conditional mathematical expectation of the posterior risk

$$R\,(d \mid x) = \frac{\int\limits_{\Theta} L(\theta,\, d) f(x \mid \theta) g(\theta) \mathrm{d}\theta}{f(x)}\,,$$

where $f(x)$ is the marginal density function of the random variable x.

It is known (see [SV83]) that an estimate that takes some given value d_0 only at the sample space points that provide the minimum of posterior risk $R\,(d_0 \mid x)$ (denote the set of such points as $X(d_0)$) possesses the minimal d-risk at the point $d = d_0$). This means that for an EUMD $\delta^*(x)$ construction, it is required for each fixed $d \in \mathrm{R}_+$ to find the corresponding set $X(d)$ and after to make the decision $\delta^*(x) = d$ only in the case when the observation value $x \in X(d)$. Note that the d-risk of the estimate constructed in such way is $\Re(d \mid \delta^*) = \min_x\, R\,(d \mid x)$. Therefore, the problem of EUMD construction is reduced to a search of sets $X(d)$.

It may be the case that there exists a subset $D \in \mathrm{R}_+$ for which

$$Z = \bigcap_{d \in D} X(d) \neq \emptyset,$$

then for $x \in Z$ an arbitrary decision $d \in D$ should be taken in accordance with an arbitrary randomized rule. An existence of such subsets D is common for discrete distributions, moreover for each $d \in D$, the sets $X(d)$ coincide and consist from one point.

The projection of the graph of the posterior risk $R\,(d \mid x)$ function (for $n = 5$ and quadratic loss function) on the plane (d, R) is presented as Fig. 23.1 above.

The shape of the graph suggests that there is a finite number of intervals D_i for which $X_i(d) = i,\ i = 0, \ldots, n$

At the bottom of Fig. 23.1 the function $x(d),\ d \in (0,\, 1)$ is presented, that provides the points of achieving the minimum by $x \in \mathfrak{X} = \{0, \ldots, n\}$ of the posterior risk $R\,(d \mid x)$ for each fixed $d \in (0,\, 1)$.

Therefore, for each value of the current observation x_0 it is allowed to take an arbitrary decision d for which $x(d) = x_0$ and the problem of choosing a concrete decision arises. For the considered example it seems to be reasonable to chose for a estimate the value $d_0 = \min\{d :\ x(d) = x_0\}$ because the posterior risk function $R\,(d \mid x_0)$ achieves the minimum at the point $d = d_0$ (see Fig. 23.1).

If another deterministic way of the decision is used, then we obtain a different rule with the uniformly minimal d-risk, which has a support on a different subset of the decision space. Since these two rules have different supports, it is not possible to compare them from the point of view of d-risks comparison.

Because of that, the randomized rule seems to be universal. According to it, a decision from the interval $\{d :\ x(d) = x_0\}$ is chosen randomly with some (for example, unform) distribution with the support on the whole interval.

For the empirical estimate with uniformly minimal d-risk (EEUMD) construction in the case of completely unknown prior distribution, an estimate $\hat{x}(d)$ instead of the function $x(d)$ should be used. In the above suggested method, we substitute the minimization of the posterior risk on the minimization of its estimate $\hat{R}(d \mid x)$, which is constructed by the historical data $X^{[m]} = (x_1, \ldots, x_m, x)$. In connection with this, the

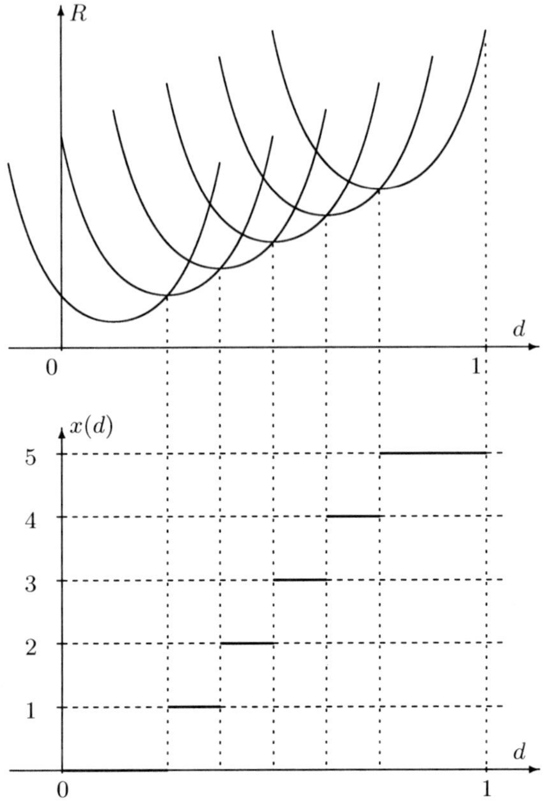

Figure 23.1. Functions $R\,(d\,|\,x)$ (*on the top*) and $x\,(d)$ (*at the bottom*)

question raises of the convergence of the minimum of $\hat{R}(d\,|\,x)$ over all possible outcomes of the experiment to the correspondent minimum of the posterior risk (which coincides with the minimal value of the d-risk). Such convergence takes place (see [VDV98], p.45, Theorem 5.7), if for each fixed $d \in \mathrm{R}_+$

$$\lim_{m\to\infty} \mathbf{P}\left(\sup_{x\in\mathfrak{X}} |\,\hat{R}(d\,|\,x) - R(d\,|\,x)\,| > \varepsilon\right) = 0, \tag{23.2}$$

Where $\mathbf{P}$ is the probability that corresponds to the marginal distribution of the date set.

23.3 Empirical Estimate Based on the Modification of the Historical Data

In this article, the problem of EEUMD construction for the Binomial distribution (23.1) parameter is considered under the quadratic loss function and completely unknown prior distribution.

It is not difficult to show that the posterior risk

$$R(d\,|\,x) = \frac{\int_0^1 (\theta-d)^2 C_n^x\,\theta^x\,(1-\theta)^{n-x} g(\theta)\mathrm{d}\theta}{\int_0^1 C_n^x\,\theta^x\,(1-\theta)^{n-x} g(\theta)\mathrm{d}\theta}$$

can be represented in the form

$$\frac{(x_{n-2}+1)(x_{n-2}+2)}{(n-1)n}\cdot\frac{f_n(x_{n-2}+2)}{f_{n-2}(x_{n-2})} - 2d\frac{(x_{n-2}+1)}{(n-1)}\cdot\frac{f_{n-1}(x_{n-2}+1)}{f_{n-2}(x_{n-2})} + d^2, \quad (23.3)$$

where

$$f_k(y) = \int_0^1 C_k^y\,\theta^y\,(1-\theta)^{k-y} g(\theta)\mathrm{d}\theta$$

is unconditional probability of y successes in k Bernoulli trials, x_{n-2} is the number of successes in the first $(n-2)$ out of n trials. The representation (23.3) of the posterior risk we denote as $R(d\,|\,x_{n-2})$.

It is possible to obtain an estimate of the posterior risk by substituting into (23.3) instead of unknown prior densities $f_k(y)$ their frequential estimates based on the historical data. For this goal, the data set $X^{[m]}$ which contains the result of $(m-1)$-st observation of the random variable x (which corresponds to the number of successes for n Bernoulli trials) and also the result of the current observation, is necessary to represent in the form

$$X^{[m]} = \begin{pmatrix} y_{1,\,n-2} & y_{1,\,n-1} & y_{1,\,n} \\ \vdots & \vdots & \vdots \\ y_{m-1,\,n-2} & y_{m-1,\,n-1} & y_{m-1,\,n} \\ x_{n-2} & x_{n-1} & x_n \end{pmatrix},$$

where $y_{i,\,k}$ is the number of successes for k trials for i-th historical data element, and x_k is the number of successes for k trials of the current experiment, $i = 1, \ldots, m-1;\quad k = n-2,\ n-1,\ n$. The frequency estimate of the marginal density $f_{n+i-3}(y)$ $(i = 1, 2, 3)$ is $\hat{f}_{n+i-3}(y) = \nu_i/m$, where ν_i is the number of elements of the i-th column of the data set $Y^{[m]}$ which are equal to y and the posterior risk estimate is

$$\hat{R}(d\,|\,x_{n-2}) = \frac{(x_{n-2}+1)(x_{n-2}+2)}{(n-1)n}\cdot\frac{\hat{f}_n(x_{n-2}+2)}{\hat{f}_{n-2}(x_{n-2})} -$$

$$2d\frac{(x_{n-2}+1)}{(n-1)}\cdot\frac{\hat{f}_{n-1}(x_{n-2}+1)}{\hat{f}_{n-2}(x_{n-2})} + d^2 \quad (23.4)$$

The following result is true.

Theorem 1. *The minimal value of the posterior risk estimate (23.4) over all* $x \in \{0, \ldots, n-2\}$ *converges in probability* $\mathbf{P}$ *to the minimal value of the d-risk.*

Proof. For the proof of the theorem it is sufficient to show that condition (23.2) is satisfied. It is simple to show that for estimate (23.4) the following asymptotic representation (as $(\hat{f}_{n-2}(x_{n-2}) - f_{n-2}(x_{n-2}))/f_{n-2}(x_{n-2}) \to 0$) is true

$$\hat{R}(d\,|\,x_{n-2}) = \frac{(x_{n-2}+1)(x_{n-2}+2)}{n(n-1)f_{n-2}(x_{n-2})}\left(\hat{f}_n(x_{n-2}+2) - f_n(x_{n-2}+2)\right) -$$

$$\frac{(x_{n-2}+1)(x_{n-2}+2)}{n(n-1)f_{n-2}^2(x_{n-2})}\left(\hat{f}_{n-2}(x_{n-2}) - f_{n-2}(x_{n-2})\right) -$$

$$2d\,\frac{(x_{n-2}+1)}{(n-1)f_{n-2}(x_{n-2})}\left(\hat{f}_{n-1}(x_{n-1}+1) - f_{n-1}(x_{n-1}+1)\right) +$$

$$2d\,\frac{(x_{n-2}+1)f_{n-1}(x_{n-2}+1))}{(n-1)f_{n-2}^2(x_{n-2})}\left(\hat{f}_{n-2}(x_{n-2}) - f_{n-2}(x_{n-2})\right) +$$

$$o\left(\frac{\hat{f}_{n-2}(x_{n-2}) - f_{n-2}(x_{n-2})}{f_{n-2}(x_{n-2})}\right) + R(d\,|\,x_{n-2}).$$

The probability in (23.2) is bounded above by

$$\mathbf{P}\left(\sup_x |\,\hat{R}(d\,|\,x_{n-2}) - R(d\,|\,x_{n-2})\,| > \varepsilon\right) \leqslant$$

$$\leqslant \mathbf{P}\left(\sup_x \frac{(x_{n-2}+1)(x_{n-2}+2)}{n(n-1)f_{n-2}(x_{n-2})}\left|\hat{f}_n(x_{n-2}+2) - f_n(x_{n-2}+2)\right| > \frac{\varepsilon}{5}\right) +$$

$$\mathbf{P}\left(\sup_x \frac{(x_{n-2}+1)(x_{n-2}+2)}{n(n-1)f_{n-2}^2(x_{n-2})}\left|\hat{f}_{n-2}(x_{n-2}) - f_{n-2}(x_{n-2})\right| > \frac{\varepsilon}{5}\right) +$$

$$\mathbf{P}\left(\sup_x 2d\,\frac{(x_{n-2}+1)}{(n-1)f_{n-2}(x_{n-2})}\left|\hat{f}_{n-1}(x_{n-1}+1) - f_{n-1}(x_{n-1}+1)\right| > \frac{\varepsilon}{5}\right) +$$

$$\mathbf{P}\left(\sup_x 2d\,\frac{(x_{n-2}+1)f_{n-1}(x_{n-2}+1))}{(n-1)f_{n-2}^2(x_{n-2})}\left|\hat{f}_{n-2}(x_{n-2}) - f_{n-2}(x_{n-2})\right| > \frac{\varepsilon}{5}\right) +$$

$$\mathbf{P}\left(\sup_x o\left(\frac{\hat{f}_{n-2}(x_{n-2}) - f_{n-2}(x_{n-2})}{f_{n-2}(x_{n-2})}\right) > \frac{\varepsilon}{5}\right). \tag{23.5}$$

Show that the first term in (23.5) (denote it $\mathbf{P_1}$) converges to zero as $m \to \infty$. For other terms the same procedure is applied. The following inequalities are true.

$$\mathbf{P_1} \leqslant \mathbf{P}\left(\max_{x_{n-2}} |\hat{f}_n(x_{n-2}+2) - f_n(x_{n-2}+2)\,| > \frac{\varepsilon\, f_{n-2}(x_{n-2})n(n-1)}{5\,(x_{n-2}+1)(x_{n-2}+2)}\right) \leqslant$$

$$\leqslant \mathbf{P}\left(\max_{x_{n-2}} |\hat{f}_n(x_{n-2}+2) - f_n(x_{n-2}+2)\,| > \varepsilon\, C\right),$$

where the last inequality is true because the maximum is taken over the finite set of values $x_{n-2} \in \{0, \ldots, n-2\}$ for which the expression $f_{n-2}(x_{n-2})\, n\,(n-1)/(5\,(x_{n-2}+1)(x_{n-2}+2))$ is bounded below by a positive constant C. Applying to the expression obtained the known estimate for the frequency estimate $\hat{f}(x)$ of the density $f(x)$

$$\mathbf{P}\left(\,|\hat{f}(x) - f(x)\,| > \gamma\right) < \exp\left(-2\,m\,\gamma^2\right),$$

We obtain the following upper bound for $\mathbf{P_1}$:

$$\mathbf{P_1} \leqslant \exp\left(-2\,m\,\varepsilon^2 C^2\right).$$

Hence, $\mathbf{P_1} \to 0$ as $m \to \infty$.

This means that condition (23.2) is satisfied and hence the statement of the theorem is true.

In the framework of the method suggested in Sect. 1, the algorithm of finding EUMD is reduced to a construction of a function $x(d)$, for which the posterior risk $R(d\,|\,x)$ is minimized for each fixed $d \in (0,\,1)$ over $x \in \{0, \ldots, n\}$.

Hence, for the empirical variant of EUMD it is necessary to find an estimate $\hat{x}(d)$ of the function $x(d)$ substituting the posterior risk minimization on its estimate $\hat{R}(d\,|\,x_{n-2})$ minimization over all $x \in \{0, \ldots, n-2\}$. Obviously, the function $\hat{x}(d)$, $d \in (0,\,1)$ represents the points of achieving the minimum by x of the posterior risk estimate $\hat{R}(d\,|\,x_{n-2})$ for each fixed $d \in (0,\,1)$. For an estimate of the parameter θ for the current result of the observation $x \in \{0, \ldots, n-2\}$ the value

$$\widehat{\theta}_1 = \arg \min_{\{d:\,\hat{x}(d)=x\}} \hat{R}(d\,|\,x_{n-2})$$

should be taken.

In order to find $\hat{x}(d)$, for each $x_i \in \{0, \ldots, n\}$ we correspond the interval $D_i = \{d:\, \hat{R}(d\,|\,x_i) \leqslant \hat{R}(d\,|\,x_{n-2})\ \ \forall x \in \{0, \ldots, n-2\}\}$ for which $x_m(d) = x_i$. It is possible that for some values of $x_0 \in \{0, \ldots, n-2\}$ the correspondent interval $D_0 = \emptyset$ and the function $\hat{x}(d) \neq x_0$ for all values $d \in (0,\,1)$. For such results of the current observation we use empirical Bayesian estimate for the estimate $\widehat{\theta}_1$ of the parameter θ.

23.4 Empirical Estimate Based on the Invariance of EUMD for the Scalar Parameter of a Discrete Exponential Family

In [She08], a method of EEUMD construction under the quadratic loss function for the scalar parameter of a discrete exponential family with the density function

$$f(x\,|\,\lambda) = h(x)\,\lambda^x\,b(\lambda),\ \ x \in \mathfrak{X} \subseteq \{0,\,1,\,2,\ldots\},\ \ \lambda \in \Lambda \subseteq \mathrm{R}_+,\ \ h(x) > 0\ \forall x \in \mathfrak{X}, \quad (23.6)$$

was developed. It is based on the minimization of the posterior risk

$$R_\lambda(d_\lambda \,|\, x) = \frac{\int_\Lambda (\lambda - d)^2 f(x \,|\, \lambda) g_\lambda(\lambda) \mathrm{d}\lambda}{\int_\Lambda f(x \,|\, \lambda) g_\lambda(\lambda) \mathrm{d}\lambda}$$

(where $g_\lambda(\lambda)$ is the prior density of the parameter λ).

Density function (23.1) can be written in the form (23.6) with substitution $\lambda = \theta/(1-\theta)$ and if it is possible to state that the EUMD has the invariance property under such type of transformations, then the problem of EEUMD construction for parameter θ is possible to reduce to the problem of EEUMD construction for the parameter λ.

The next theorem provides the invariance condition for the EUMD.

Theorem 2. *Let the transformation $a : \Lambda \to \Theta$ be a bijection, moreover $(a(\lambda)-a(l))^2 = m(\lambda, l)(\lambda-l)^2 \ \forall\, \lambda, l \in \Lambda$, where function $m(\lambda, l)$ is such that $\forall\, \lambda, l \in \Lambda \ \int_\Lambda m(\lambda, l)\mathrm{d}\lambda < \infty$. If d_λ is the EUMD of the parameter λ of a discrete exponential family (23.6), then $d_\theta = a(\lambda)$ is the EUMD of the parameter θ.*

Proof. Let x_* be the result of current observation of random variable x with the density function (23.6). The following inequality is true for the posterior risk d_λ of the EUMD $R_\lambda(d_\lambda \,|\, x_*) \leqslant R_\lambda(d_\lambda \,|\, x) \ \forall\, x \in \mathfrak{X}$. For the proof of the statement it is sufficient to show that the similar inequality $R_\theta(d_\theta \,|\, x_*) \leqslant R_\theta(d_\theta \,|\, x)$ is true for $\forall\, x \in \mathfrak{X}$ for the posterior risk of the estimate $d_\theta = a(\lambda)$, which has the form

$$R_\theta(a(\lambda) \,|\, x) = \frac{\int_\Theta (a(d_\lambda) - \theta)^2 f(x \,|\, \theta) g_\theta(\theta) \mathrm{d}\theta}{\int_\Theta f(x \,|\, \theta) g_\theta(\theta) \mathrm{d}\theta}, \tag{23.7}$$

where $g_\theta(\theta)$ is the prior density of θ.

Change the variable in (23.7) as $\theta = a(\lambda)$ and by the assumptions of the proposition, transform the posterior risk

$$R_\theta(a(\lambda) \,|\, x) = \frac{\int_\Lambda (a(d_\lambda) - a(\lambda))^2 f(x \,|\, a(\lambda)) g_\lambda(\lambda) \mathrm{d}\lambda}{\int_\Lambda f(x \,|\, a(\lambda)) g_\lambda(\lambda) \mathrm{d}\lambda} =$$

$$\frac{\int_\Lambda m(\lambda, d_\lambda)(d_\lambda - \lambda)^2 f(x \,|\, a(\lambda)) g_\lambda(\lambda) \mathrm{d}\lambda}{\int_\Lambda f(x \,|\, a(\lambda)) g_\lambda(\lambda) \mathrm{d}\lambda}.$$

Applying the mean value theorem, we obtain

$$R_\theta(a(\lambda) \,|\, x) = \frac{C \int_\Lambda (d_\lambda - \lambda)^2 f(x \,|\, a(\lambda)) g_\lambda(\lambda) \mathrm{d}\lambda}{\int_\Lambda f(x \,|\, a(\lambda)) g_\lambda(\lambda) \mathrm{d}\lambda} = C\, R_\lambda(d_\lambda \,|\, x),$$

where C is a positive constant. The theorem is proved.

In our case, the function $a(\lambda) = \lambda/(1+\lambda)$ obviously satisfies the assumptions of Theorem 2; hence, EEUMD of the parameter θ is

$$\widehat{\theta}_2 = \hat{d}_\lambda/(1 + \hat{d}_\lambda),$$

where $\hat{d}_\lambda$ is the EEUMD of the parameter λ. In the case when EEUMD $\hat{d}_\lambda$ does not exist (see [She08] for conditions of its existence), as the estimate $\widehat{\theta}_2$ of the parameter θ we use empirical Bayesian estimate, the same as in Sect. 23.3.

23.5 Accuracy Properties Investigation for the Estimates Based on the Statistical Modeling

Under the assumption that the probability θ has the prior beta-distribution $\mathrm{B}(\alpha, \beta)$, for some values of parameters α and β and historical data of various size m, by the method of statistical modeling, the estimates $\widehat{\theta}_1$ and $\widehat{\theta}_2$ were calculated for $n = 10$. The values $\alpha = 10$ and $\beta = 1$ are of special interest because the most of the mass of beta-distribution is concentrated near 0 and for the quality control the true values of θ are usually small.

For each fixed historical data the values of the current observation were generated, that is, 100 values of probably θ and correspondent observed values of the random variable X. Based on them, for each current observation the values of estimates $\widehat{\theta}_1$ and $\widehat{\theta}_2$ were calculated. In Table 23.1, the arithmetic averages for $\theta, \widehat{\theta}_1$, and $\widehat{\theta}_2$ are presented.

The obtained modeling results allow us to make the following conclusions.

1. Sufficiently accurate values of both estimates for all considered sizes of the data set are obtained in the case when prior beta-distribution is concentrated around one ($\alpha = 5$, $\beta = 1$), while for $\alpha = 1$, $\beta = 10$ (the case interesting for quality control) with small sample sizes, the estimate appear to be underestimating. Hence if there is an information about possible small values of probability θ, then it is better to estimate $(1 - \theta)$.

2. In the case of symmetric prior distribution ($\alpha = 2$, $\beta = 2$) the estimate $\widehat{\theta}_2$ possesses a significant bias and hence is not recommended for applications.

3. It was noted during the modeling process that EEUMD $\widehat{\theta}_1$ does not exist more frequently than EEUMD $\widehat{\theta}_2$ (remind that in the case when EEUMD does not exist, we use empirical Bayesian estimate for the probability θ). For example, for small sizes of the historical data ($m = 250, 500$) EEUMD $\widehat{\theta}_1$ does not exist in approximately 60% of cases, while the EEUMD $\widehat{\theta}_2$ does not exist for 40%. When the size of the historical data is large ($m = 10\,000, 100\,000$) EEUMD $\widehat{\theta}_1$ does not exist in approximately 30% of cases and EEUMD $\widehat{\theta}_2$ in approximately 10%.

Table 23.1. Arithmetic averages for $\theta, \widehat{\theta}_1$ and $\widehat{\theta}_2$

m	$\alpha = 1$, $\beta = 10$			$\alpha = 5$, $\beta = 1$			$\alpha = 2$, $\beta = 2$		
	θ	$\widehat{\theta}_1$	$\widehat{\theta}_2$	θ	$\widehat{\theta}_1$	$\widehat{\theta}_2$	θ	$\widehat{\theta}_1$	$\widehat{\theta}_2$
250	0,082	0,034	0,037	0,821	0,823	0,834	0,491	0,482	0,565
500	0,102	0,060	0,040	0,841	0,890	0,862	0,473	0,450	0,515
1,000	0,091	0,057	0,061	0,820	0,882	0,847	0,489	0,503	0,542
10,000	0,105	0,064	0,066	0,830	0,829	0,845	0,496	0,497	0,559
100,000	0,089	0,081	0,081	0,814	0,819	0,833	0,506	0,505	0,578

References

[Rob55] Robbins H. An empirical Bayes approach to statistic. Proc. Third Berkeley Symp. Math. Statist. Probab. Univ. of Calif. Press, **1**, 157–164 (1955)

[Sim83] Simushkin, S.V. Empirical *d*-posterior approach to the problem of the statistical guarantee inference. Sov. Math.(Iz. VUZ), **27**, no.11, 68–78 (1983)

[SV83] Simushkin, S.V.; Volodin, I.N. Statistical inference with a minimal *d*-risk. Lecture Note in Math., **1021**, 107–114 (1983)

[VDV98] van der Vaart A.W. Asymptotic Statistics. Cambridge University Press, Cambridge (1998)

[VN83] Voinov, V.G.; Nikulin M.S. Unbiased Estimators and Their Applications. **1**, Univariate case, Kluwer, Dordrecht (1983)

[VN93] Volodin, I.N.; Novikov, An.A. Statistical estimates with asymptotically minimal *d*-risk. Theory Probab. Appl., **38**, no.1, 118–128 (1993)

[VS87] Volodin, I.N.; Simushkin, S.V. Unbiasedness and the Bayes property. Soviet Math.(Iz VUZ), **31**, no.1, 1–8 (1987)

[She08] Sherman, E.D. An empirical estimator with the uniform minimal *d*-risk for parameter of one-dimensional discrete exponential families Sov. Math.(Iz VUZ) (submitted 2008)

24

Estimation of Archival Lifetime Distribution for Writable Optical Disks from Accelerated Testings

Watalu Yamamoto, Chiharu Kumazaki and Kazuyuki Suzuki*

Department of Informatics, University of Electro-Communications, Tokyo, Japan,
suzuki@se.uec.ac.jp

Abstract: ISO/IEC 10995:2008 specifies a procedure for estimating the distribution of archival lifetimes of optical disks from accelerated lifetime test results. In this article we derive the maximum likelihood estimator of the lifetime distributions based on an accelerated lifetime model, an Eyring model, and lognormal distribution, from a reliability engineering perspective. We also propose a procedure to analyze a data set from an accelerated lifetime test.

Keywords and phrases: Accelerated lifetime testing, Eyring model, Lognormal distribution

24.1 Introduction

ISO/IEC 10995:2008 [1] specifies a procedure for estimating the distribution of archival lifetimes of writable optical disks on the basis of accelerated lifetime test results. The standard implements the procedure using a sort of bootstrap technique and also a nonparametric estimator of density functions. However, it is not usually recommended to use those techniques in small sample cases. We estimate the lifetime distribution from a reliability engineering perspective and derive a slightly different and simple method of estimation for writable optical disks.

24.2 Models

As specified in Annex B of ISO/IEC 10995:2008 [1], we assume a set of models to investigate the problem of estimation of archival lifetime distributions. We denote the random variable for the archival lifetime of an optical disk by Y and its observations as $y_1, \ldots, y_n$.

V.V. Rykov et al. (eds.), *Mathematical and Statistical Models and Methods in Reliability: Applications to Medicine, Finance, and Quality Control*, Statistics for Industry and Technology,
DOI 10.1007/978-0-8176-4971-5_24,

24.2.1 Lognormal Distribution

We assume that Y is distributed with lognormal distribution $\mathrm{LN}\left(\mu,\sigma^2\right)$, where μ and σ^2 are the expected value of $\log Y$ and the variance, respectively. The density function of this distribution is defined as

$$f_{\mathrm{LN}}\left(y;\mu,\sigma^2\right)=\frac{1}{\sqrt{2\pi\sigma^2}}\exp\left[-\frac{\left(\log y-\mu\right)^2}{2\sigma^2}\right]. \tag{24.1}$$

The expected value of Y is $\exp\left(\mu+\sigma^2/2\right)$, and the variance is $\left(\exp\left(\sigma^2\right)-1\right)\exp\left(2\mu+\sigma^2\right)$. The logarithm of Y is distributed with normal distribution $\mathrm{N}\left(\mu,\sigma^2\right)$. Although the expected value of Y is not $\exp\left(\mu\right)$, the percentiles of the lognormal distribution coincide with the exponentially transformed percentiles of the normal distribution.

24.2.2 Accelerated Lifetime Model

Accelerated lifetime tests are widely used when the product to be tested is highly reliable under regular usage conditions and it is not realistic to conduct lifetime tests under such usage conditions. The product lifetime is shortened by increasing the use rate, the aging-rate, or the levels of stresses under which the test units operated. The effects of changes in the test conditions on the lifetime or lifetime distributions is modeled either theoretically or empirically.

ISO/IEC 10995:2008 [1] specifies for optical disk testing temperature T (K) and relative humidity $H(\%)$ and assumes an Eyring model for determining the relationship between the lifetime under regular usage conditions and that under acceleration. Accelerated lifetime models assume that the method of acceleration affects the lifetime distribution through acceleration factor $\alpha\left(x\right)$, where x is a vector of variables each of which affects the acceleration of the lifetime.

Meeker and Escobar [2] gives a concise yet sufficient overview of accelerated lifetime models and accelerated lifetime tests. Bagdonavičious and Nikulin [3] described more advanced topics of accelerated lifetime models. If we denote the cumulative distribution function of the lifetime distribution under the target condition as $F_0\left(y_0;\theta\right)$, the lifetime distribution under acceleration condition x is given by

$$F_x\left(y_x;\theta\right)=F_0\left(y_0\alpha\left(x\right);\theta\right). \tag{24.2}$$

The lifetime of the product is $\alpha\left(x\right)$ times shorter under test condition x than under the regular usage condition. The functional form of $\alpha\left(x\right)$ depends on the mechanism of acceleration used in the test. Under the assumption of lognormality, acceleration factor $\alpha\left(x\right)$ affects only μ. For we have $\log Y_x=\log Y_0+\log\alpha\left(x\right)$. If we denote the distribution under the regular usage conditions as $\mathrm{LN}\left(\mu,\sigma^2\right)$, the distribution under test condition T and H is given as $\mathrm{LN}\left(\mu_0-\log\alpha\left(T,H\right),\sigma^2\right)$ for the problem considered here.

24.2.3 Eyring Model

The Eyring model assumed in ISO/IEC 10995:2008 [1] is a generalized Eyring relationship between the accelerated lifetime and the lifetime under regular usage conditions:

$$Y_{T,H} = Y_0 \exp\left\{-\beta_1\left(\frac{1}{T_0}-\frac{1}{T}\right)-\beta_2\left(H_0-H\right)\right\}, \tag{24.3}$$

where β_1 $(= -E_a/k)$ is an unknown constant representing the activation energy divided by Boltzmann's constant, and β_2 is another unknown constant. The terms in the exponents containing T_0 or H_0 are also constants. We can thus reparameterize this relationship as

$$Y_{T,H} = Y_0 \exp\left\{\kappa_0 + \frac{\beta_1}{T} + \beta_2 H\right\} \tag{24.4}$$

by defining $\kappa_0 = -\beta_1/T_0 - \beta_2 H_0$.

24.3 Estimation of Parameters and Percentiles

By combining the three models described in the previous section, we obtain a model for the archival lifetime distribution for optical disks:

$$Y_{T,H} \sim \text{LN}\left(\mu_0 + \kappa_0 + \beta_1/T + \beta_2 H, \sigma^2\right).$$

It is not necessary to estimate μ and κ_0 separately; we simply need to estimate $\beta_0 = \mu_0 + \kappa_0$. Therefore we have an accelerated lifetime model for our problem:

$$Y_{T,H} \sim \text{LN}\left(\beta_0 + \beta_1/T + \beta_2 H, \sigma^2\right). \tag{24.5}$$

The resulting model is illustrated in Fig. 24.1.

ISO/IEC 10995:2008 [1] specifies four pairs of temperature T and relative humidity H with which the accelerated lifetime tests are to be conducted, and also specifies the minimum sample size for each condition, as shown in Table 24.1. We assume that the

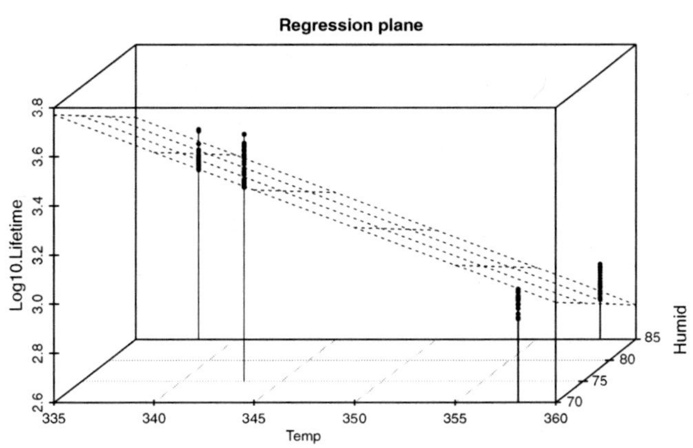

Figure 24.1. Lognormal plots of example data

Table 24.1. The test conditions specified in ISO/IEC 10995:2008 [1]

Group	Temperature (C)	Relative humidity	Minimum sample size
A	85	0.85	20
B	85	0.70	20
C	65	0.85	20
D	70	0.75	30

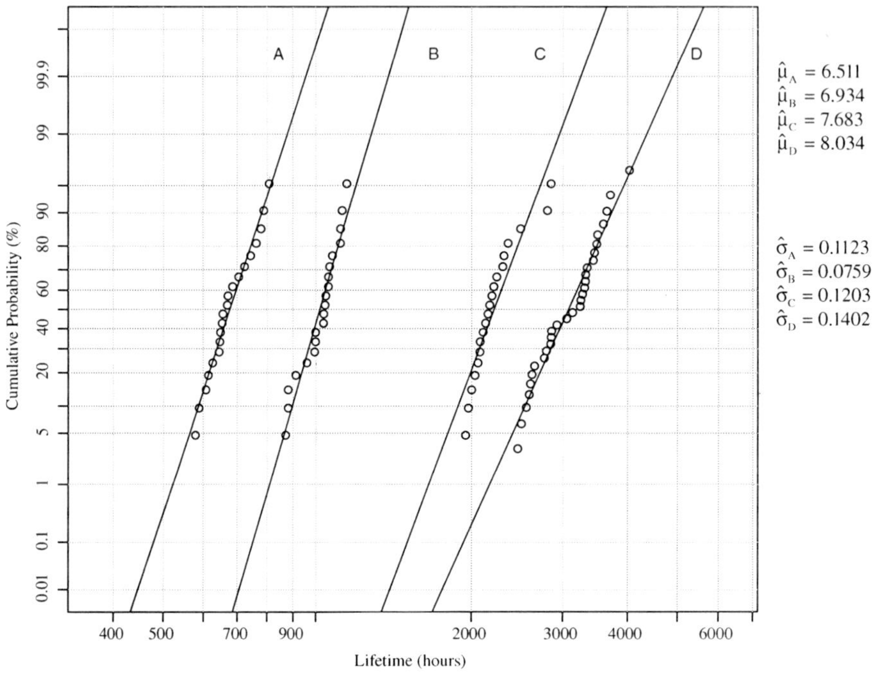

Figure 24.2. Lognormal probability plots for the example data set from ISO/IEC 10995:2008 [1]

lifetimes are observed or well predicted for all products and that there is no censoring, as is assumed in the standard.

Let y_{ik} denote the observed lifetime of the i-th product for the k-th set of test conditions, n be the sample size for group k, and T and H be the temperature and relative humidity for group k. The example data set included in ISO/IEC 10995:2008 [1] is analyzed, and the lognormal probability plots as in Fig. 24.2.

Since it is rather straightforward, we omit the mathematical details and describe the criterion and the resulting estimators of β_0, β_1, β_2, and σ^2. The maximum likelihood principle is an important principle for statistical inference. The simultaneous distribution function as a function of unknown parameters is called the likelihood function. The principle states that all information about the unknown parameters are contained in that the likelihood function and the estimator is obtained as the maximizer of the likelihood function and its logarithm. See Meeker and Escobar [2] or Lawless [4] for further details. The maximum likelihood estimators of β_0, β_1, and β_2 are the minimizers of the sum of squares:

$$\sum_{k=1}^{4}\sum_{i=1}^{n_k}\left\{y_{ki}-\left(\beta_0+\frac{\beta_1}{T_k}+\beta_2 H_k\right)\right\}^2. \tag{24.6}$$

The minimizers, $\hat{\beta}_0$, $\hat{\beta}_1$, $\hat{\beta}_2$, coincide with the least square estimators and are easily obtained by any software, such as spreadsheet software, that can fit linear functions to data using least square methods. The maximum likelihood estimator of σ^2 is

$$\hat{\sigma}^2=\frac{1}{\sum_{k=1}^{4}n_k}\sum_{k=1}^{4}\sum_{i=1}^{n_k}\left\{y_{ki}-\left(\hat{\beta}_0+\frac{\hat{\beta}_1}{T_k}+\hat{\beta}_2 H_k\right)\right\}^2, \tag{24.7}$$

which is the sum of the squared residuals divided by the total number of observations. It has been proven that the set $\hat{\beta}_0$, $\hat{\beta}_1$, $\hat{\beta}_2$, and $\hat{\sigma}^2$ maximizes the likelihood function.

The p percentile of the lifetime distribution, so called B_p life, is widely used in reliability engineering as a measure of reliability. A point estimator of $\log B_p$ is

$$\log \hat{B}_p(T,H)=\hat{\beta}_0+\frac{\hat{\beta}_1}{T}+\hat{\beta}_2 H+z_{p/100}\hat{\sigma}, \tag{24.8}$$

where $z_{p/100}$ is the lower p percentile of the standard normal distribution, $\mathrm{N}(0,1)$. Further, it is easy to derive the lower confidence limit of this percentile. If we denote the covariance matrix of $\hat{\beta}_0$, $\hat{\beta}_1$, $\hat{\beta}_2$ and $\hat{\sigma}^2$ as V and the row vector of $(1, 1/T, H, z_{p/100})$ as $\mathbf{u}'$, standard statistical calculations lead to the variance formula of $\log B_p(T,H)$ as $\mathbf{u}'V\mathbf{u}$. The estimate of V is obtained using the Fisher information formula. Furthermore, we can obtain the $100-a$ percent lower confidence limit of $\log B_p$ as $\log \hat{B}_p(T,H) - z_{1-a/100}\sqrt{\mathbf{u}'V\mathbf{u}}$ by using normal approximation. That for $B_p(T,H)$ is obtained by exponential transformation.

24.4 An Example

As mentioned, we analyze the data set included in Annex B of ISO/IEC 10995:2008 [1].

First, we plot the observations for each group on lognormal probability paper to determine whether the lognormality assumption is valid (Fig. 24.2). Then we compare the plots as is done in Fig. 24.2. We temporarily conclude that each group of plots is fairly linear and that the difference among slopes are due to sampling variations. Finally we fit the models described in the previous sections. The resulting estimates are listed as Table 24.2 and fitted lines are added to Fig. 24.2.

The point estimates and 95% lower confidence limits of B_5 and B_{50} under the regular usage condition are given in Table 24.3. The point estimates obtained by the procedure, stated in Annex B of ISO/IEC 10995:2008 [1], are 12.64 and 31.06, respectively.

Table 24.2. Maximum likelihood estimates

$\hat{\beta}_0$	$\hat{\beta}_1$	$\hat{\beta}_2$	$\hat{\sigma}$
−13.934	8561.7	−0.041601	0.16778

Table 24.3. Point estimates and lower confidence limits of percentile lifes under the regular usage condition

Estimate	B_5	B_{50}
Point estimate	28.34	37.50
95% Lower confidence limit	21.62	28.53

24.5 Concluding Remarks

We derived the maximum likelihood estimator of the archival lifetime distributions by using an accelerated lifetime model, an Eyring model, and lognormal distribution and using the data obtained from accelerated lifetime tests for writable optical disks. It is fairly easy and simple to implement the proposed estimation procedure in any spreadsheet software. With this method, we can use the following steps to analyze a data set from an accelerated lifetime test.

1. Plot the data set for each accelerated condition on lognormal probability paper and check the distributional assumption.
2. Fit the accelerated lifetime model (24.5) to the entire data set so as to minimize (24.6) and the variance estimate (24.7).
3. Draw the fitted line on the probability paper used in Step 1 and check the assumption of accelerated lifetime model (24.6).
4. Estimate B_5 and B_{50} and their confidence limits.

The final step is intended to use B_5 and B_{50} as characteristic quantities of the estimated lifetime distributions to compare the reliability of various items.

The possible directions of further investigation of this problem include (1) developing methods for model assessment and selection, (2) simultaneous modeling of μ and σ^2, (3) developing statistically optimal designs for accelerated lifetime testing, and (4) extending our estimation procedure to other families of lifetime distributions, e.g., Weibull distribution. It is also interesting to develop the optimal designs of accelerated lifetime testings for optica disks, as is developed in Escobar and Meeker [5].

Furthermore, ISO/IEC 10995:2008 [1] describes a scheme for accelerated degradation testing of optical media. Though we assume here that we can use the results of a model fitting as observed lifetimes as we would those from an accelerated lifetime testing, it would be interesting to investigate this further using accelerated degradation models.

References

[1] ISO/IEC: Information Technology - Digitally recorded media for information interchange and storage - Test method for the estimation of the archival lifetime of optical media, ISO/IEC (2008)

[2] Meeker, W. Q. and Escobar, L. A.: Statistical Methods for Reliability Data, Wiley, New York (1998)

[3] Bagdonavičious, V. and Nikulin, M.: Accelerated Life Models: Modeling and statistical analysis, Chapman & Hall/CRC, Boca Raton (2002)

[4] Lawless, J. F.: Statistical Models and Methods for Lifetime Data, 2nd Ed., Wiley, New York (2003)

[5] Escobar, L. A. and Meeker, W. Q. : Planning accelerated life tests with two or more experimental factors. Technometrics, **37**, 411–427 (1995)

Part III

Applications

25

Ages in Reliability and Bio Systems, Interpretations, Control, and Applications

Boyan Dimitrov

Kettering University, Flint, Michigan

Abstract: In this article, we define age and compare ages for objects from two populations by making use of their lifetime distributions. The main purposes of this talk are to propose various definitions, based on relationships between important different but equivalent probability components, which allow tractable use in engineering and humane practice. These definitions admit different meaning important in numerous applications. Respectively, there are various areas of use. Most of the statements and Examples are straightforward consequences of the Definitions. For instance, there are several ways for evaluation of the true age of an object on the background of its population. An approach to the time scaling in the age evaluation based on accelerated testing is offered, using the concept of accumulated total stress. Examples with biological age control similar to reliability age-correcting factor illustrate the theory. We generalize this approach to two-dimensional and multi-dimensional distributions that also may represent life, and sketch areas where this knowledge can be applied.

Keywords and phrases: Ages and aging, Equivalent ages, Determination of individual age, Multi-dimensional ages and comparison, Process ages comparison, Reliability applications

25.1 Introduction

Aging begins at the birth. The common approach to measure the age is by measuring the time of existence (survival) since the birth/creation of the items. However, age is a heterogeneous and individual property which is better tolerated by some individuals than others, as noted by Campbell [CC06]. It is perhaps better expressed in biological individuals than among technical items. However, we claim that technical items in engineering have individual ages and these can be assessed. There exists an internal clock (IC) which indicates is what the real age of each item. To be able to read the counts in the IC for a particular individual means that we understand the age. The most popular method of measuring the age of objects is according to the time of its

V.V. Rykov et al. (eds.), *Mathematical and Statistical Models and Methods in Reliability: Applications to Medicine, Finance, and Quality Control*, Statistics for Industry and Technology, DOI 10.1007/978-0-8176-4971-5_25,

creation and existence (otherwise, this is the elapsed life since the birth). However, individuals have their internal (biological) clock, the IC, which shows their individual age, also counted since the birth, but running with a speed which may be different from the astronomical time.

Definition 1. *The age of an individual is the count on its Internal Clock.*

The only problem is to clearly define the ways of measuring the age, i.e. the ways to read the counts on an IC. The approaches of probability theory, statistics, and reliability offer some mathematical tools to use for the purpose. This is the main goal of our discussion here. In the probability textbooks we see problems like this: The times of first failure of brand new printers are approximately normally distributed with a mean of 1,500 h, and a standard deviation of 200 h. What should be the guarantee time for these printers if the manufacturer wants only 5% to fail within the guarantee period? Well, but what if these were used autos, or refrigerators, or repaired planes, or houses? What is common between these items in similar questions? The common is that we are talking about comparable ages. Our review of various references, articles, books, monographs and encyclopedia did not find any satisfactory answer of the question, *what is age*? We could not find sources specifying such question and addressing its possible answer for the area of potential unified application. However, the resent book of Finkelstein [MF08] contains some ideas which are close to the presented here, but not the same. This encouraged us to write this material here, and pretend that it contains some fresh ideas.

25.2 Main Definitions and Equivalent Representations

Age of a particular individual is impossible to understand without its comparison to (or, say, projection on) some background of other individuals. To get to the idea of particular individual's age one should understand the concept of ages of populations, and the approaches to age measuring. Then, in order to find the ways for evaluation of individual ages one needs tools for comparing the ages. Thus, we start with comparison of ages for populations, and with definition of equivalent ages. It is fair to notice that actually, in age comparison, we do not compare the ages of two particular individuals, but the respective ages of two average individuals from two populations to which these individuals belong. Thus, we imagine that there are many copies similar to these two which we do compare, and the comparison is made based on the images about the respective populations. We consider biological individuals, or technical items, etc., anything that experiences aging. With X, Y, ... we will denote life times of objects, and by $F_X(x)$, $F_Y(y)$, ... their cumulative distribution functions (c.d.f.).

Definition 2. *We say, the age a of the individual with life time X is equivalent to the age b of an individual with life time Y if the probabilities to survive the respective ages are equal, i.e., if it is fulfilled*

$$P\{X > a\} = P\{Y > b\}. \tag{25.1}$$

In the sense of Definition 2, every age a of one of the two individuals (with continuous c.d.f. of its life) has equivalent comparable age b to any other individual. Ages a and b are equivalent when the probabilities to survive these respective values a and b are equal for the individuals of the two populations.

Example 1. Let $X \sim Exp(1/1000)$ be the life time of Y be the life time of a microfilm with a uniform distribution over the interval $[3, 5]$ in years (we write $Y \sim U[3, 5]$, and know that the respective average life time is 4 years). Then we have age $a = 800\,\text{h}$ of a bulb equivalent to what the age b of a microfilm? Let solve the (1) for this particular case. We have

$$P\{X > 800\} = e^{-\frac{800}{1000}} = P\{Y > b\} = 1 - \frac{b-3}{5-3},$$

or $b = 5 - (2)e^{-.8} = 4.10$ years is the equivalent age of the microfilm, compare to the age of 800 h of the considered electrical bulb. Inversely, the age of $b = 4$ years of a microfilm is compared to the age $a = (1,000)\ln\,2 = 693.12\,\text{h}$ of one of the considered electric bulbs.

Notice, that our approach requires the knowledge of the probability distributions of the life for populations of items whose ages we like to compare.
Next, we focus on some equivalent representations of Definition 2 for equivalent ages. Directly from Definition 2 by making use of some probability identities, we see the truth of the following:

Definition 3. *The ages a, and b of two individuals with life times X and Y are equivalent if and only if*

$$F_X(a) = F_Y(b) \tag{25.2}$$

i.e. when it is true that the probabilities not to survive age a for the first individual, and not to survive age b for the second individual, are equal. In other words, the equivalent ages are equal quantiles of the respective life time distributions.

Definition 3 is equivalent to Definition 2. The interpretation there is that two ages a and b are equivalent when the probabilities to survive these values are equal. In other words, the equivalent ages are equal quantiles of the respective life time distributions.

Remark 1. There is an easy graphical presentation of the equivalent ages from two populations with life time distributions $F_X(x)$ and $F_Y(x)$, which immediately follows from Definition 3. Introduce the *inverse function* of the cumulative probability distribution by the equation

$$x_p = F^{-1}(p) = \inf\{x; \quad F(x) \geq p\} \text{ for } p \in (0,\ 1).$$

Then *the curve of equivalent ages* of the two populations is the curve **C** defined by the parametric equation

$$\mathbf{C} := \left\{(x_p,\ y_p);\quad x_p = F_X^{-1}(p);\ y_p = F_Y^{-1}(p),\quad p \in (0,\ 1)\right\}. \tag{25.3}$$

Coordinates of each point on the curve **C** are the pair of equivalent ages in the two populations. We call such curve a *nomogram.* Such presentation would make an easy, graphical nomogram for comparing the ages between any two populations with life times X and Y.

Example 2. Now we look at another popular class of life distributions called *location* and *scale equivalent family of probability distributions.* Two random variables X, and Y belong to the same scale and location parameter family of probability distributions when there exists a principal (standard) distribution function $F(t)$ such that it is fulfilled

$$F_X(t) = F\left(\frac{t-\mu_1}{\sigma_1}\right) \text{ and } F_Y(t) = F\left(\frac{t-\mu_2}{\sigma_2}\right) \text{ for all } t \geq 0. \qquad (25.4)$$

Here, μ_i and σ_i, $i = 1, 2$ are some positive, non-zero constants (called *location* parameter, μ_i, and *scale* parameter, σ_i, $i = 1, 2$).

Items whose life times X and Y have scale and location equivalent probability distributions with parameters μ_i and σ_i respectively, $i = 1, 2$, will have equivalent ages a, and b if and only if these ages are related by the equations

$$a = \mu_1 + \sigma_1 \frac{b-\mu_2}{\sigma_2}, \text{ and } b = \mu_2 + \sigma_2 \frac{a-\mu_1}{\sigma_1}, \text{ or simply, by } \frac{a-\mu_1}{\sigma_1} = \frac{b-\mu_2}{\sigma_2} \qquad (25.5)$$

Note for instance, that all exponential distributions are scale equivalent. Despite the lack of memory property which may urge us to believe that such objects are tending to have no age, they do have ages, and we can compare ages in these cases too. The equivalent ages for two populations $X \sim \exp(\lambda)$ and $Y \sim \exp(\mu)$ are related by the equalities $\lambda a = \mu b$.

As another note we remark, that we may have linear dependence between equivalent ages. The precise legal transfer of one age in terms of the other needs appropriate shifts and scale changes. In the last equation of (25.5) we notice (in some analogy with the standardization in the case of the normal distribution), that the ratios in both sides of (25.5) can be called z-scores. Thus, we find out that two life times from one and the same scale and location equivalent family of probability distributions would have equivalent ages a and b only when these numbers have the same z-scores in their populations. Relationships (25.5) say that the smaller the variance is within a population, the younger its individuals are compare to individuals of same average but larger variance in another population.

Example 3. There is a popular dog – human life comparison, the belief in proportional age relationship (like 1 dog's year of age is equivalent to 7 years of life of a human). If we assume normally distributed life times, is not true anymore. Let the probability distributions of the humans life and the dog's life have values $\mu_1 = 65$ years, and $\mu_2 = 12$ years (by an analogy with the normal distribution, these values of the location parameters are chosen to be approximately equal to the average ages for the two populations). The respective scale parameters σ_1, and σ_2 (usually understood as the standard deviations in the two normally distributed populations) satisfy conditions (25.5), and can be appropriately calculated. Assume that they have values $\sigma_1 = 8$, and $\sigma_2 = 3$. Then the age of a dog at $b = 12$ years is, according to (6), equivalent to $a = 65$ years age of man. But, the age of a 5 years old dog is equivalent to $a = 46,3$ years old man. Reversely, an $a = 80$ years old man has a "dog's comparable age" of $b = 17.5$ years. A 1 year old dog has an equivalent of $a = 35.3$ years age of a human. The table bellow gives a numeric comparison of the ages between the dog' and human populations for quantiles multiple to 0.10. It is based on simulated 150 observations of the human population (assumed normal with mean 65 years and standard deviation

8 years) and 95 observations for dog' population (assumed normal with mean 12 years and standard deviation 3 years). The same numbers are also theoretically obtained as in Example 3. What is not shown in this table, and is a fact easy to understand, that more drastic differences appear in the zones of low (0.01, 0.02, ... 0.08) and high (0.91, ...,0.99) probability levels of the equivalent ages

Probability not to survive an age, p	0.1	0.2	0.3	0.4	0.5	0.6	0.7	0.8	0.9
Theoretical Human Age $N(65, 8)$	54.75	58.27	60.80	62.97	65.00	67.03	69.20	71.73	75.25
Theoretical Human Age $N(65, 8)$	54.75	58.27	60.80	62.97	65.00	67.03	69.20	71.73	75.25
Theoretical Dog' Age $N(12, 3)$	8.16	9.48	10.43	11.24	12.00	12.76	13.57	14.52	14.52
Empirical Human Age	56.02	59.12	60.96	64.33	65.64	67.28	68.80	70.83	73.58
Empirical Dog' Age	7.84	9.26	10.36	11.17	12.07	13.09	14.22	14.76	16.26

Comparing the numbers in proportional, and the scale and location rules of comparison of ages, we see significant differences in the results obtained by two approaches. The true model of age comparison is an important factor, and we must be aware of it. We should revise some simple rules in comparing ages, based straightforward on the average life expectancy only. There is a simple explanation: The IC for dog population runs much faster from the start of their life than the IC for the humans, and rates of these runs are not uniform, as the calendar rate (for convenience of comparison) is.

25.3 Empirical Distribution Functions and Empirically Equivalent Ages

It is clear from the definitions of equivalent ages that one may use samples from each population, and the empirically estimated quantiles to compare ages of items from different populations. The accuracy in determination of the empirical quantiles will determine the precision in the evaluation of the equivalent ages from both populations. We have no intention to enter in more details here, considering the case as intuitively clear and practically usable. The obtained estimations are based on consistent non-parametric sample estimators of the p-quantiles of probability distributions. The equivalent ages are equal empirical quantiles of the respective life time distributions.

This approach is convenient to compare equivalent ages in two groups of similar objects, e.g. same purpose products from different providers, ages among ethnic populations based on data collected within a census, or for comparing ages from different countries, etc. The empirical approach is expected to be useful to compare the market value of products based on the duration of their free of failures use. It can be used for purposes of warranty, life support, quality evaluation, and similar risk areas of application.

25.4 Equivalent Ages Have Equivalent/Equal Accumulated Stress: Comparing Ages of Processes

There are other convenient descriptions of the life times, which offer additional interpretations suitable in life comparison and broaden its possible area of applications. They are using the terms of the failure rate function and related to it hazard function associated with any life time distribution. The idea comes from reliability theory and from the survival analysis.

Let the initial lifetime, X, of an individual be a continuous r.v. with c.d.f. $F(x)$ and p.d.f. $f(x)$. Then its *hazard function* is defined by the equation

$$\Lambda(x) = -\ln[1 - F(x)], \quad x \geq 0. \tag{25.6}$$

Its *failure rate function* is respectively defined by

$$\lambda(x) = \frac{\mathrm{d}}{\mathrm{d}x}\Lambda(x) = \frac{f(x)}{1 - F(x)}, \quad \text{for all } x \geq 0, \text{ where } 1 - F(x) \neq 0. \tag{25.7}$$

The relationship between $\Lambda(x)$ and $\lambda(x)$ is remarkable:

$$\Lambda(x) = \int_0^x \lambda(t)\,\mathrm{d}t. \tag{25.8}$$

It allows to introduce in our terminology time dependent processes, which seem more suitable for the age equivalence purposes and age comparison establishment. Living is a process. In demography and survival analysis the function $\lambda(x)$ is called *mortality rate* among the individuals at calendar age x. Here we propose to call it *risk function* (risk to fail, risk to die, risk of something to happen at age x since the process has started). Another suitable terminology seems to be *stress function.* Respectively, equation (14) offers an intuitively clear, natural name for the function $\Lambda(x)$. We call $\Lambda(x)$ *accumulated stress* (or *accumulated risk*) during the life up to age x. This is what we believe is the count one may read at the IC. Now, the following statement holds.

Theorem 1. *The ages a, and b of two individuals with life times X and Y are equivalent if and only if the accumulated stress up to age a of the first individual, and the accumulated stress up to age b of the second individual are equal, i.e. when it is fulfilled*

$$\Lambda_X(a) = \int_0^a \lambda_X(t)\,\mathrm{d}t = \int_0^b \lambda_Y(t)\,\mathrm{d}t = \Lambda_Y(b). \tag{25.9}$$

In other words, *equivalent ages have equal accumulated stress within each of the individuals* (one may read it as having the meaning that *equivalent ages have equivalent readings on their respective Internal Clocks*).

Proof. From Definition 3 we know that two ages a and b are equivalent when equation (2) is fulfilled. Then, it is also true that $1 - F_X(a) = 1 - F_Y(b)$. Therefore, it is true $-\ln[1 - F_X(a)] = -\ln[1 - F_Y(b)]$. By making use of relationships (25.7) and (25.8) we conclude that (25.9) holds.

Definition 4. *The ages a, and b of two individuals with life time variables X and Y are equivalent if and only if they have equal accumulated stress up to the moments a and b respectively*

$$\Lambda_X(a) = \Lambda_Y(b). \tag{25.10}$$

i.e. when it is true that the incurred stresses up to age a, and to age b respectively, are equal. In other words, the equivalent ages have equal quantities of the respective accumulated stresses.

Definition 4 is equivalent to Definitions 2 and 3. However, we now gain the opportunity to find a way for determination of the true age of an individual within the population where this individual belongs.

Remark 2. To understand the concept of incurred stress, also what we think can be called accumulated risk, expressed by (25.10). One needs to admit, that the way to measure it is unique, and does not depend on the nature of the population to which it is related. The measuring unit for the incurred stress is one and the same for the entire world of random variables representing life times, and should correspond to one of the presentations (25.6) or (25.8) used in reliability theory. Respectively, the rate of accumulation the stress per unit time may get its measuring unit from the relations in (25.7).

25.4.1 About Assessment of Individual Life: Age of Bio-Systems

Assume, that the accumulated stress function $\Lambda(t)$ is analytically known, and also that it can be particularly evaluated, i.e. that it is possible to measure the accumulated stress of a particular individual when he is at calendar age t. Equation (25.10) allows to evaluate the real age of this individual on the background of his own population, no matter what its calendar age t is at the moment.

Corollary 1. *If the accumulated stress up to some calendar time T of a particular individual within his population equals A, then its actual particular age is T_{A}, obtained as solution of the equation*

$$\Lambda(t) = A, \text{ i.e. } T_{\text{A}} = \Lambda^{-1}(A), \tag{25.11}$$

where $\Lambda^{-1}(t)$ is the inverse function of the function $\Lambda(t)$.

We may call $\Lambda(t)$ also as *function of the wasted resource* on behalf of this individual within the "calendar" time interval $[0, t]$. As more resources an individual wastes, as more age he/she gets. The picture reminds the one in counting the accumulated work-load of technical equipment (as it is for instance, in airplane motors, nuclear reactors, heating facilities, mileage for cars) when attempting to find out the true value/age of the items. However, there is no clear picture in how to transform the load amount into a calendar age, and our approach here can be a helpful tool. An open question raises: How would be the age prolonged with adding extra resources to an individual.

Remark 3. The Nobel Laureates in Medicine for 2006 are A. Z. Fire and C. C. Mello "for their discovery of RNA interference – gene silencing by double-stranded RNA".

Professor G. Hansson [HG06], Chairman of the Nobel Committee for Medicine, in his presentation speech says: "In their brilliant paper from 1998, Andrew Fire and Craig

Mello [FXKDM98] demonstrated that double-stranded RNA activates an enzymatic mechanism that leads to gene silencing, with the genetic code in the RNA molecule determining which gene to silence. Today, we call this mechanism RNA interference.

Continued research has shown that our cells use RNA interference to regulate thousands of genes. Through RNA interference, the pattern of gene expression is fine-tuned in such a way that each cell uses precisely those genes that are needed for building its proteins. Today we also know that RNA interference helps to protect us against viruses and jumping genes."

The Nobel Laureates have discovered a fundamental mechanism for controlling the flow of genetic information. Our genome operates by sending instructions conveyed by messenger RNA (mRNA). RNA interference occurs in plants, animals, and humans. It is of great importance for the regulation of gene expression, participates in defense against viral infections, and keeps jumping genes under control. RNA interference is already being widely used in basic science as a method to study the function of genes and it may lead to novel therapies in the future, especially in slowing the aging.

Since 1988, some gigantic steps towards understanding the mechanism of aging have been made. My search through the available literature gives to me some optimistic hopes. The following excerpt may be in support: "Many aging-related diseases are linked to shortened telomeres. Organs deteriorate as more and more of their cells die off or enter cellular senescence. Telomeres are the physical ends of linear eukaryotic chromosomes. They are specialized nucleo-protein complexes that have important functions, primarily in the protection, replication, and stabilization of the chromosome ends. In most organisms studied, telomeres contain lengthy stretches of tandemly repeated simple DNA sequences composed of a G- rich strand and a C-rich strand (called terminal repeats). These terminal repeats are highly conserved; in fact all vertebrates appear to have the same simple sequence repeat in telomeres: (TTAGGG)n.

Telomerase is an enzyme that adds specific DNA sequence repeats ("TTAGGG" in all vertebrates) to the $3'$ ("three prime") end of DNA strands in the telomere regions, which are found at the ends of eukaryotic chromosome. The telomeres contain condensed DNA material, giving stability to the chromosomes. The enzyme is a reverse transcriptase that carries its own RNA molecule, which is used as a template when it elongates telomeres, which are shortened after each replication cycle." Telomerase was discovered by Greider and Blackburn in 1985 [GB85].

Advocates of human life extension promote the idea of lengthening the telomeres in certain cells through temporary activation of telomerase (by drugs), or possibly permanently by gene therapy. They reason that this would extend human life. So far these ideas have not been proven in humans. In 2006, Geron corporation's web site (http://www.hoovers.com/generic-drugs/--HICID_1488--/free-ind-factsheet.xhtml) indicated that it had at least two candidate drugs able to activate telomerase.

There are several techniques currently employed to assess average telomere length in eukaryotic cells. These methods, however, require significant amounts of genomic DNA (2–20 micrograms) and labor which renders its use limited in large epidemiological studies.

Laboratory studies show, that human individuals come to life at birth with a telomere of length between 9,000 and 18,000. Due to stress and deceases, the cells reproduction etc., the length of the telomere shortens at each of these actions by several letters in their *"TTAGGG"* sequence. By the end of the natural life span the length of telomere becomes short to the size of 400–700 letters.

The statements in this remark indicate, that the telomeres are these hidden sources of counts on our *IC* since there is a way to measure telomeres length. A combined statistical work is needed to find out the exact form of the function $\Lambda(t)$ which corresponds to the true life time of a human' subpopulation. The *Gompertz–Makenham law of mortality* discussed below offers a useful opportunity since it contains just a few of constants to be estimated, and is proven to work for all live organisms (bio systems). The Gompertz–Makenham mortality law has been confirmed for people and for other mammals, flies, mollusks with specific values of its parameters.

Bio-systems are obviously subject of wearing and aging. Bio-systems naturally have a proven life-span (something like a maximal value of life beyond which no copy of the bio-system can pass). Life span for people is, for instance, 120 years. Life spans have also most of the functional components of the bio-systems. Life span for people's brain is 250 years. For technical devices the exponential, the Weibull, the Gamma, and even the Normal distributions frequently fit for modeling the life times. For the bio-systems, despite of their complexity, according to Koltover [VK04] there exists this "universal kinetics of the growth of mortality with the age", expressed by the Gompertz–Makenham law of mortality

$$\lambda(t) = \beta + \alpha \mathrm{e}^{\gamma t}. \tag{25.12}$$

Here, the parameters β, α, and $\gamma > 0$ are independent on time. From Koltover [VK04] we understand that for people parameter $\beta \neq 0$ if the age is less than 35 years, and $\beta = 0$ if the age is greater than 35. Values of the other parameters, according to that same source are given as $\alpha \approx 42.8271 \pm 8.85$ years, and $\gamma \approx .094 \pm .0014$ years^{-1}

We use (25.12) and immediately find the form of the function of the accumulated stress the Gompertz–Makenham law of mortality

$$\Lambda(t) = \beta t + \frac{\alpha}{\gamma}(1 - \mathrm{e}^{\gamma t}) \quad t \geq 0. \tag{25.13}$$

In order to find the values of the parameters β, α, and γ and to relate them to the length of telomeres and the age distribution, it seems sufficient to use the normalizing conditions for the probability functions i.e. solve the equations $\Lambda(120) = 400$, $\Lambda(0) = 18{,}000$, and $\exp(-\Lambda(120)) - \exp(-\Lambda(0)) = 1$. Similar approach seem to be appropriate and for any other live population, where the same mortality law is valid, and just the parameters and measured numbers differ in values. After the constants β, α, and γ are established, a measurement of the respective telomere length, L, of an individual and the solution of the equation

$$\beta T + \frac{\alpha}{\gamma}(1 - \mathrm{e}^{\gamma T}) = L \tag{25.14}$$

would give the assessed particular age T_L of the individual.

The last equation here is a transcendental equation and des not have an explicit solution. However, its solution is unique. The same will be and in the most common case. We mark here its general solution.

Let $\Lambda(t)$ be the population accumulated stress function, and let $\Lambda_i(T)$ be the accumulated stress in an individual i. Then its individual age, compare to the average individual age in (on the background of) the population, is

$$\hat{T}_i = \Lambda^{-1}(\Lambda_i(T)). \tag{25.15}$$

Notice, that (25.15) allows to forecast the ages of items (individuals) which are operating in different work environment, or in different workload conditions. This same approach should be applicable and in accelerated life testing in reliability.

25.4.2 Process Ages Comparison

In the theory of the Non-stationary Poisson processes (NPP), the function $\Lambda(t)$ discussed in the previous section is called *Leading function.* It constantly increases with t from 0 to ∞ and characterizes the entire development of the respective process onto any time interval $[0,t]$, as well as what may happen on any other interval $[t,t+s]$. Every function $\Lambda(t)$ with such properties defines a NPP. A NPP describes flows of events which occur one at a time while the process exists. Age of such process can be called the time variable T_{A} which satisfies the equation $T_{\mathrm{A}} = \Lambda^{-1}(A)$ where A is certain amount of accumulated "work" achieved by the leading function, and $\Lambda^{-1}(t)$ is the inverse function of $\Lambda(t)$. The less known fact is that there is a random variable, say X, defined by a probability distribution $F(t)$, related to $\Lambda(t)$ by the equation $F_X(t) = 1 - \mathrm{e}^{-\Lambda(t)}$, and this is the distribution of the time X from the start of the process up to the occurrence of its first event. We call this variable associated with the considered NPP.

Let two processes have leading functions $\Lambda_i(t)$, $i = 1,2$. The above considerations give us the reasonable opportunity to compare the ages of the two processes. Actually, these are the equivalent ages of the underlying random variables X_i associated with each of the two processes. We have the following Definition true:

Definition 5. *The ages T_1 and T_2 of two processes are equivalent whenever the equality between the values of their leading functions*

$$\Lambda_1(T_1) = \Lambda_2(T_2). \tag{25.16}$$

at these times holds.

Definition 5 is equivalent to Definitions 2, 3, and 4. Definition 5 requires the knowledge of the growth of the leading functions for the processes whose ages we are willing to compare.

Based on Definition 5 one gets the opportunity to do lots of useful practical applications. For instance, the *time scaling* between laboratory (possibly, overloaded, accelerated, intensive) testing and real world (in the field) transfer of results. If the field conditions allow to determine how the resource function $\Lambda_2(T)$ will be wasted (used), then the true *field age* up to the death of an item which dies in accelerated (laboratory) conditions at age T_1 and has had used resources $\Lambda_1(T_1)$ will be predicted by an equation similar to (25.15), i.e. the predicted death age T_2 will be $T_2 = \Lambda_2^{-1}(\Lambda_1(T_1))$.

Example 4. The Gompertz–Makenham life-time distribution with an age-affecting factor.

Drug use activities may improve the performance of the individuals and give them a "new life". The specifics of the drugs, its regular intake, amount of labor, recovery time, or money invested in the health care may have significant impact on the health improvement, which directly affects the longevity of life. If assume that health improvement shortens the current age of such individuals by certain percentage δ, we call it *an age- reducing factor*. Drug abuses may affect the future performance of the individuals, and make them look older. We introduce *age-accelerating factor* δ, which means making the overall life of such individuals shorter by certain percentage δ. The model and the theorem below explain how it happens and the reflection of the changes on the individual mortality rate and on the integrated risk function (Dimitrov et al. [DGCK04, DCK04]).

Let X_i denote (calendar) time intervals between successive epochs of drugs intake that affect the individual. Assume that δ_i denote the lack of perfection the live system may get as result of the i-th action. The values

$$T_0 = 0, \quad T_i = T_{i-1} + \delta_i X_i, \quad i = 1, 2, \ldots \tag{25.17}$$

are understood as virtual ages of the individual right after the i-th action. When $\delta_i = 1$, then no improvement or deterioration of the virtual age of the individual occurs at the i-th epoch of action. When $\delta_i < 1$ (or if $\delta_i > 1$), then an improvement (or a deterioration) of the virtual age of the individual occurs at that epoch of the i-th action. The model described here is also known in Reliability as Kijima [MK89] model II.

Dimitrov et al. [DCK04] consider this model with the assumption that $\delta_i = \delta \neq 0$, and call this δ an age-correcting factor. It has been noticed that when the individual is at calendar age t its virtual age measured on the calendar age scale is $x_\delta(t) = \delta t$. Based on some calculations, the following relations are found: The virtual hazard rate (accumulated risk) $\Lambda^*(t)$ and the original hazard rate function (the "normal run" of accumulation) $\Lambda(t)$, as well as their rates of change, are related by the equalities

$$\Lambda^*(t) = \frac{1}{\delta}\Lambda(\delta t) \text{ and } \lambda^*(t) = \lambda(\delta t) \quad t \geq 0, \quad \delta \neq 0. \tag{25.18}$$

Here, the function $\Lambda^*(t)$ also represents the accumulated risk up to calendar time t of an individual maintaining his life as in the Kijima model (25.17), and the function $\lambda^*(t)$ represents the risk (mortality) at the same calendar age t.

When considering the Gompertz–Makenham life-time distribution Dimitrov et al. [DHKS07] show that for individuals of a population who maintain their life with same age-correcting factor δ, their accumulated stress is represented by the function

$$\Lambda^*(t) = \beta t + \frac{\alpha}{\delta\gamma}(1 - e^{\gamma\delta t}) \quad t \geq 0. \tag{25.19}$$

By making use actually, of for equivalent ages of processes, The authors apply Definition 6 with the shown in the previous section values of the parameters α, β, γ and obtain a nomogram (a graph) of related ages. It shows how equivalent ages between human populations are (could be) transferred at any calendar age t. The table bellow is an excerpt of these results. The equivalent ages depend on the values of the correcting factor δ and on the levels p of survival probability.

p δ	1.5	1.25	1.1	**1**	0.9	0.6	0.3
0.9	28	30	31	**33**	34	37	39
0.7	42	45	50	**58**	61	80	112
0.5	48	55	61	**66**	72	98	158
0.25	53	62	69	**75**	82	114	197
0.1	57	66	75	**81**	89	125	220

We selected the equivalent ages for several probability levels. In bold black digits are shown the ages for the normal population where $\delta = 1$. In the same line are shown the ages in the other populations, at which (we intuitively understand) an average individual would look (has the age) equivalent to the shown for the normal population. One may see some unreal numbers which we need to comment. For instance, at level $p = 0.25$ (the age of 75 years, which will be survived only by a quarter of the individuals in the normal population) same properties would be in possession by the individuals at 82 years age, if they reduce their ages regularly by a factor $\delta = 0.9$ (an improvement by 10% compare to the normal). In the same line, we see that one quarter would survive 114 year old people in the population who got 40% improvements regularly, i.e. have $\delta = 0.6$. Finally, we see that as a 75 year old normal individual would have feel an individual at age 197 if it was possible to reduce the age regularly by 70%. Numbers in the last column are somewhat unreal, because they represent a mystic dream for such high level of age reduction. If one chooses a calendar age, it should be as shown for the normal population. His/her real age is the one read in the column of his/her population.

Notice that the telomeres length L discussed in the previous section is a random process, which shows the existing (initial minus wasted) resources within an individual. Measuring the telomeres length L_i, $i = 1, 2$ and using the equation (25.14) one is able to compare the ages of individuals from two populations, and get an information on how to evaluate the age correction factor δ when such problem is of interest.

On the other hand, biological systems can be treated similarly as the devices constructed to perform some preset functions considered in reliability. They perform their functions in the presence of a great number of random factors, which may disturb the normal operations. In terms of reliability, keeping most of the biological objects under control is maintenance. The malfunction of the bio-system is equivalent to the failure in a technical device. The applied treatment in the bio-system corresponds to respective repair in reliability.

Bio-systems can be considered as a specifically organized devices constructed to perform some preset functions, according to their genetic programs. These functions are performed in the presence of a great number of random factors (environmental conditions). Biological discoveries materialize the vision of imagination demonstrated in mathematical models. There are connections between the two fields which have to be appropriately used. There are number of interesting problems associated with the discussed above mechanisms of control of process age and its prolongation. Here we describe one of these.

Problem 1. How frequently to use the aid tool (e.g. telomerase in bio cells) in order to prolong the total cells life (e.g. keep the telomere above the critical level). You understand that if you miss the moment, the cell will be reproduced, and the telomere will be shortened. If you use the aid tool more than once before the reproduction instant, there will be waste of material. Set up a problem either to maximize the probability for staying above the critical level, or to minimize the limited aid tool quantity within a given calendar time interval, or to minimize related expected cost functional, and similar goal function.

If the expenses are at least particularly covered by some life, or medical insurance, then just the calculation of the expected associated costs would be of significant importance.

If $\Lambda(t)$ is the amount of wasted resources during the life of an individual, what measures how adding extra resources prolongs the total life?

This section was induced by the number of presentations and articles presented at the First French-Russian Conference on Longevity, Aging and Degradation Models in Reliability, Public Health, Medicine and Biology (LAD'2004 as it is well described by Auperin and Nikulin in [AN04]). It intended to combine the latest biological and medical results with the probability and statistical methods.

25.5 Multidimensional Life Time Equivalence

There are objects (technical items, bio individual) where the age gets a two, three, and possibly more dimensional form. The life time almost any system can be measured in more than one time scale. For instance, for the automobiles people use two "parallel scales", the calendar age, and the total mileage. For airplanes there could be three "time" scales, the calendar age, the amount of fly time (in the air) and the number of takeoffs and landings. At the time of death every one of these "age variables", say X, and Y, have certain values. The death itself is an event which is just indirectly related to the meaning of X, and Y, i.e. these variables may have no "time meaning". The probability of the pair (X, Y) not to survive certain values (x, y), namely $P(X \leq x, Y \leq y)$ is called *joint probability distribution* and is denoted by $F_{X,Y}(x, y)$. Its value, roughly speaking, represents the proportion of individuals from the population which will have $X \leq x$ and $Y \leq y$ when die. The probability to survive both values (x, y), we call *Survival Function*, and denote by $F_{X,Y}(x, y)$. Its value, roughly speaking, represents the proportion of individuals from the population which will have $X > x$ and $Y > y$ when die. In other words the variables X, and Y are just *collateral to the age.* Ways to transfer these values into calendar measurement of the age are still expected. However, the measurements on variables X, and Y are giving important information for use in the same way as calendar time can be used in the risk assessment, and utilization. The following of this section uses the ideas described in Dimitrov [BD07], and in Dimitrov et al. [DGS07].

Now we introduce two definitions of equivalent ages, and claim that despite of the differences, each may have its reason to be used.

Definition 6. *(Optimistic): We say, two individuals with two dimensional dependent age variables* (X_1, Y_1) *and* (X_2, Y_2) *have equivalent ages* $(T_1^{(1)}, T_2^{(1)}) = (T_1^{(2)}, T_2^{(2)})$ *if it is fulfilled*

$$F_{(X_1,Y_1)}(T_1^{(1)}, T_2^{(1)}) = F_{(X_2,Y_2)}(T_1^{(2)}, T_2^{(2)}). \tag{25.20}$$

Notice, that now we have two sets of *equivalent ages*, not just a unique pair $(T_1^{(1)}, T_2^{(1)}) = (T_1^{(2)}, T_2^{(2)})$. Even in the frame of just one population with a two dimensional age (X, Y), for any number p within the interval $(0, 1)$ all the points on the level p curve

$$\mathbf{C_p} : \ = \{(x_p, y_p);\ \ F_{X,Y}(x_p, y_p) = p\} \tag{25.21}$$

are points whose coordinates trace all *individuals of the same (we say equivalent) ages*. Only individuals who have $(X = x, Y = y)$ for which the value of the probability $F_{X,Y}(x, y) = q$ is less than p can be called *younger*, i.e. we can say that then $(x, y) < (x_p, y_p)$. The set $\mathbf{C_p}$ given by (25.21) is the *curve of equivalent ages* of the individuals *at level* p. Individuals whose measurements in X and Y are located on the lower level curves are younger than individuals with measurements on higher level curves. Analogously, in the three (and higher) dimensional ages we will have *level surfaces of equivalent ages*, and an order between ages can be given according to the level of the respective surface, on which these ages belong.

Example 5. Fair Pricing of the Used Cars

Assume that X is the calendar age of a car, and Y is its mileage. When buying a car people usually are interested of these two car characteristics. Assume, other car' features are equivalent (color, impression), and the joint distribution of (X, Y) is known. Then all the used cars of same brand with individual values of age x and mileage y laying on the curve of equivalent ages should have same price. More on this example can be seen in [DGS09]

Example 6. Warranty costs for used items – an estimation.

Assume, we are in situation as in above example, and the age of the car is $X = x_0$, and the mileage $Y = y_0$.. The conventional "extended warranty" is offered for an addition a ti the one component, and addition b to the second whichever comes first.. In other words, if the measured pair (X, Y) fails in the rectangle $R = [x_0, x_0 + a] \times [y_0, y_0 + b]$, some of the repair expenses will be covered by the warranter.

By noticing that the variables X and Y can only increase in time, and using the curves of equivalent ages one can see that the expected warranty costs will be the same from any initial point on the curve $\mathbf{C_0} : = \{(x, y);\ F_{X,Y}(x, y) = F_{X,Y}(x_0, y_0)\}$ to any end point on the curve $\mathbf{C_w} : = \{(x_w, y_w);\ F_{X,Y}(x_w, y_w) = F_{X,Y}(x_0 + a, y_0 + b)\}$. Then in the advertisement of the extended warranty can be used any pair of numbers $(a = x_w - x_0, b = y_w - y_0)$. The ads may use the most attractive pair without hurting the expected warranty costs. Detailed numeric examples are given in Dimitrov et al. [DGS09] We do believe, that similar situation can be build on an example with life insurance, or in guarantees related to yearnings from investments, or from other portfolio.

Remark 4. In two, and higher dimension the value of the joint distributions decrease when adding an additional dimension. In other words, adding an additional time (age) characteristic (component) to an already existing set of time characteristics (age components) will make this object to look "younger" compare to what it would be with less components in consideration. This is due to the fact that

$$F_{(X_1,\ldots,X_{n-1})}(T_1, \ldots, T_{n-1}) \geq F_{(X_1,\ldots,X_{n-1},X_n)}(T_1, \ldots, T_{n-1}, T_n),$$

whose interpretation e.g. is "chances to die before the expiration of certain level of n age components $(T_1, \ldots, T_{n-1}, T_n)$ are less than the chances to die before the expiration of same level of $n-1$ age components (no matter which exactly are chosen) $(T_1, \ldots, T_{n-1})$". Therefore, it looks that it is more likely to survive a highest dimension age than an age of lower dimension. This sounds as a paradox, but fact is proven. By adding an additional component into consideration in your life complex you low the level on the surface of equivalent ages.

And this is our reason to call this definition of age equivalence (and model for age comparison) as *optimistic definition.* It decreases the level of portion of dead individuals in the population with the increase of the number of age components we care for.

Definition 7. *(Pessimistic): We say, two individuals with two age components* (X_1, Y_1) *and* (X_2, Y_2) *have equivalent ages* $(T_1^{(1)}, T_2^{(1)}) = (T_1^{(2)}, T_2^{(2)})$ *if the values of their survival functions at these points fulfill the requirement*

$$S_{(X_1,Y_1)}(T_1^{(1)}, T_2^{(1)}) = S_{(X_2,Y_2)}(T_1^{(2)}, T_2^{(2)}). \tag{25.22}$$

Notice, that as in the optimistic case, we have sets of *equivalent ages*, not just unique pairs $(T_1^{(1)}, T_2^{(1)}) = (T_1^{(2)}, T_2^{(2)})$. Even in the frame of one population with two age components (X, Y), for any number p within the interval (0, 1) all the points on the curve $\mathbf{G_p}$:

$$\mathbf{G_p} := \{(x, y);\ \ S_{(X,Y)}(x, y) = p\} \quad p \in (0, 1), \tag{25.23}$$

are points whose coordinates trace the *level curve of equivalent ages* at survival level p. Only individuals whose values $(X = x, Y = y)$ satisfy the inequality $S_{(X,Y)}(x, y) = q > p$ can be called *younger* than any individual whose measurements are on the curve $\mathbf{G_p}$. Then we can say that $(x, y) < (x_p, y_p) \in \mathbf{G_p}$. Individual whose life components measurements are on higher level curve are younger than individual whose life components measurements are on lower level curves. The curve $\mathbf{C_p}$, $0 < p < 1$ given by equation (25.23) is the set of equivalent ages for the individuals at the survival level p. And this curve is quite different from the curve given by (25.21). More precisely it is always fulfilled

$$S_{(X,Y)}(x, y) + F_{(X,Y)}(x, y) \leq 1.$$

Therefore, the proportion, say p of individuals in the population not surviving level $(x_p, y_p) \in \mathbf{C_p}$ is always less then proportion of individuals surviving the level $(x_{1-p}, y_{1-p}) \in \mathbf{G_{1-p}}$. It says that the curve of equivalent ages $\mathbf{C_p}$ is located below the curve of equivalent ages $\mathbf{G_{1-p}}$ $p \in (0, 1)$. Some graphs can be shown to illustrate better visually these possibilities.

Analogously to the case of previous definition, in the three (and higher) dimensional ages we will have level surfaces of equivalent ages, and an opportunity to put order between ages of individuals by making use of values of the respective survival functions at the measured values of age components. Age rating is given according to the level of the respective surface of equivalent ages, $\mathbf{G_p}$, to which these ages belong, when using the survival function, and Definition 7.

Example 7. The two examples discussed for the case of Definition 6 can be exploited here too. Including the survival function instead of the c.d.f. in consideration just changes the vision angle, possibly the goals and tools of analysis. We do not focus here on any particular problem. More detailed examples can be found in Dimitrov et al. [DGS09].

Remark 5. In two, and higher dimension the values of the joint survival function also decrease when adding an additional dimension. But now inclusion of an additional age characteristic to an already chosen set of age components will give this object less chances to survive. In other words, this object will "look older with more components considered in his age than if less components are taken into consideration" (even when compared to itself). This is due to the fact that

$$S_{(X_1,\ldots,X_{n-1})}(T_1,\ldots,T_{n-1}) \geq S_{(X_1,\ldots,X_{n-1},X_n)}(T_1,\ldots,T_{n-1},T_n),$$

An interpretation of the above facts could be "probability to survive any fixed levels of a set of n age components $(T_1,\ldots,T_{n-1},T_n)$ is less than the probability to survive the same fixed levels of a subset of $n-1$ time variables $(T_1,\ldots,T_{n-1})$". Therefore, it is less likely to survive the highest dimension age than ages of lower dimension. Practically it comes to an advise: "The more age characteristics you take into account, the more age to yourself you will add. Your age will not change only if you are "perfect" in regard to the newly added characteristic, i.e. when $P\{X_n > T_n\} = 1$."

This is our reason to call Definition 7 (of age equivalence and model for age comparison) a *pessimistic definition.* However, we do expect this definition to be used more frequently in determining ages of various dimensions, because it seems to us a little bit more realistic (by properties) than Definition 6. More about multidimensional life in Dimitrov et al. [DGS07].

Problem 2. In two, and higher dimension joint distributions the introduction of the concepts of failure rates and integral hazard rates has various approaches (see e.g. Galambos and Kotz [GK78]). Here, for the purposes of age studies we can not propose any specific approach, which could be of equivalent value to the meaning and interpretations offered in the one dimensional case, or to the one of the multidimensional cases in this section. This is a hard problem, and any progress in it could be a significant contribution.

Another question of general issue that makes the bridge between studies on ages and related risks is the question of modeling dependence between age component. The *Copula approach in the study of multi dimensional dependence* should have some specific when related to ages and aging. The reason is justified by the importance of Gompertz–Makenham distribution in the 1-dimensional case. Copulas with use of this distribution are not known to me.

Remark 6. The web site http://message.realage.com/ offers a newsletter periodically e-mailed to RealAge members. There are several options, e.g. to take a RealAge test and to receive health information. They say: "The RealAge test is a science-based health assessment that calculates your biological age (or RealAge) and includes an Age Reduction. The page where you take the test is http://www.realage.com/ralong/qa/HI.aspx. It contains more than 60 questions (one is what is your current age), many with multiple answers, and at the and a number is given as your real age. I went in contact with them and was told that the algorithm of their calculations is their company's secret. However, part of the questions are about frequency of taking medications, like aspirin, drinking habits (in quantity), exercises, etc., and almost immediately change your prognoses of the age. This is what reminds the Kijima models, but it could be another story, possibly multi-dimensional.

25.6 Conclusions

Age of technical items, as well as the age of live individuals may not correspond to their calendar age. To answer lots of questions related to determination of the true age, and also what is aging and how to keep aging under control we need to define age.

In the present article, we give several equivalent definitions of age based on the ways to compare ages. Then we find corresponding equivalent forms of the definition of age, and claim that each form offers an opportunity to see different aspects of age and aging.

Some interpretations and recent medical discoveries trace the way for new research in this area. It also offers numerous opportunities to get into cooperation in studying ages by means of mathematics, probability and statistics, reliability, engineering, medicine and biology.

Multidimensional ages are briefly discussed. Some additional open problems related to the ideas present here are sketched.

References

[AN04] Auperin C. and Nikulin M., Unified Bio-Reliability Approach in Statistical Modeling of Aging, Longevity, Degradation and Mortality with Dynamic Environment, in *Proceedings of LAD'2004,* Volume **2,** Edits Antonov et al., Saint Petersburg, pp. 346-361, (2004)

[CC06] Campbell C.A., Two Americas: Country's healthiest live 3 decades longer than those most ailing, *The Flint Journal,* Nov. 26, p. C1–C2, (2006)

[DGCK04] Dimitrov B., Green D., Chukova S. and Khalil Z., Age Affecting Repairs and Warranty Costs, in *Proceedings of LAD'2004,* Volume **2,** Edits Antonov et al., Saint Petersburg, pp. 101–114, (2004a)

[DCK04] Dimitrov B., Chukova S. and Khalil Z., Warranty Costs: An Age-Dependent Failure/Repair Model, *Naval Research Logistic Quarterly,* **51**, pp. 959–976, (2004b)

[BD07] Dimitrov B., Lifetime Distributions and Equivalent Ages. Applications in Insurance, Risk, and More, *Proceedings of the Third Brazilian Conference on Statistical Modelling in Insurance and Finance*, Maresias, Brazil, pp. 21–40, (2007)

[DHKS07] Dimitrov B, Hayrapetyan G., Khalil Z. and Stanchev P., Aging and Longevity Control of Biological Systems via Drugs – a Reliability Model. e-journal: Reliability: Theory & Applications No 2, Vol.2, June 2007, pp. 10–20, (2007)

[DGS09] Dimitrov B., Green D. and Stanchev P., Warranty and Fair Pricing for Used Items in Two-Dimensional Life Time, VI International Conf. MMR'2009, Proceedings, Moscow, pp. 262–267, (2009)

[DGS07] Dimitrov B., Green D. Jr. and Stanchev P., Multidimensional Life Time Equivalence, *Proceedings of the International Workshop "Distributed Computer and Communication Networks: Theory and Applications*, Moscow, Russia Sept. 10–12, **1**, pp. 26–33, (2007)

[MF08] Finkelstein M. (2008) *Failure Rate Modelling for Reliability and Risk,* Springer series in reliability engineering, London, (2008)

[FXKDM98] Fire A., Xu S.Q., Montgomery M.K., Kostas S.A., Driver S.E., Mello C.C., Potent and specific genetic interference by double-stranded RNA in Caenorhabditis elegans. *Nature,* **391,** pp. 806–811 (1998)

[GK78] Galambos J. and Kotz S., *Characterizations of Probability Distributions.* Lecture Notes in Mathematics 675, Springer, Berlin, (1978)

[GB85] Greider C.W. and Blackburn E.H., Identification of a specific telomere terminal transferase – activity in tetrahymena extracts, *Cell,* **43(2),** pp. 405–413, (1985)

[HG06] Hansson G., Chairman of the Nobel Committee for Medicine, Presentation-speech on December 10,
http://nobelprize.org/nobel_prizes/medicine/laureates/2006/
presentation-speech.html) (2006)

[MK89] Kijima M., Some Results for Repairable Systems with General Repair, *J. Appl. Probab.,* **26,** pp. 89–102, (1989)

[VK04] Koltover V. K., Reliability of Biological Systems: Terminology and Methodology, in *Proceedings of LAD'2004,* Volume **2,** Edits Antonov et al., Saint Petersburg, pp. 98–113, (2004)

26

Shocks in Mixed Populations

Maxim Finkelstein[1*] and Ji Hwan Cha[2]

[1] Department of Mathematical Statistics, University of the Free State, South Africa/ Max Planck Institute for Demographic Research, Bloemfontein, South Africa, FinkelM@ufs.ac.za

[2] Department of Statistics, Ewha Womans University, Seoul, Korea, jhcha@ewha.ac.kr

Abstract: We consider shocks as a method of burn-in in discrete and continuous heterogeneous populations. Burn-in is a widely used engineering method of elimination of defective items before they are shipped to customers or put into field operation. In conventional burn-in procedures, components or systems are subject to a period of simulated operation prior to actual usage and those which failed during this period are scrapped and discarded. In this paper, we assume that the 'weak' items are more susceptible to elimination via shocks and therefore this method can be considered as burn-in. Optimal severity levels of these shocks that minimize the defined expected costs are investigated.

Keywords and phrases: Burn-in, Heterogeneous populations, Optimal severity, Shocks

26.1 Introduction

In this chapter, we consider shocks (i.e., "instantaneous" stresses of "high" level) as a *method of burn-in in heterogeneous populations*. Burn-in is a method of elimination of initial failures in field usage. To burn-in a component or a system usually means to subject it to a period of simulated operation prior to actual usage. Due to the high failure rate in the early stages of component's life, burn-in has been widely accepted as a method of screening out early failures [JP82]; [KK83]; [BS97].

As burn-in is usually expensive, one of the major tasks is to decide for how long this procedure should be performed. The corresponding optimal problem has been considered in numerous publications. See, for example, [BMS93], [BSS02], [NM82], [CS91], [M94a], [M96], [M97], [C00], [C01] and [C03].

In order to shorten the duration of this procedure, burn-in is often performed in an accelerated environment. In this case, obviously, the larger values of stress should correspond to the shorter duration of burn-in. By letting the stress to increase, we can end up (as some limit) with very short (negligible) durations, in other words, *shocks*.

V.V. Rykov et al. (eds.), *Mathematical and Statistical Models and Methods in Reliability: Applications to Medicine, Finance, and Quality Control*, Statistics for Industry and Technology, DOI 10.1007/978-0-8176-4971-5_26,

We discuss two settings: the case of two ordered subpopulations and the continuous "pattern" of a heterogeneous population composition. We present here only some of our findings in this direction. The extended version with numerous additional results and generalizations will be published elsewhere.

26.2 Two Ordered Subpopulations

Assume that a population is a mixture of two ordered subpopulations – the strong subpopulation and the weak subpopulation. Let the lifetime of a component from the strong subpopulation be denoted by X_S and its Cdf, pdf and the failure rate function be $F_1(t)$, $f_1(t)$ and $\lambda_1(t)$, respectively. Similarly, the lifetime, the Cdf, pdf and the failure rate function of a weak component are denoted by X_W, $F_2(t)$, $f_2(t)$ and $\lambda_2(t)$, respectively. Let the lifetimes in these subpopulations be ordered in the sense of the failure rate ordering [SS06]:

$$\lambda_1(t) \leqslant \lambda_2(t), \text{ for all } t \geqslant 0,$$

and the mixing proportion (distribution) $\pi(z)$ is defined as $\pi(z_1) = \pi$, $\pi(z_2) = 1 - \pi$ where z_1 and z_2, $z_1 < z_2$ are variables that represent the strong and the weak subpopulations, respectively, and $0 \leqslant \pi \leqslant 1$. Then the mixture failure rate is [F08]:

$$\lambda_m(t) = \frac{\pi f_1(t) + (1-\pi) f_2(t)}{\pi \overline{F}_1(t) + (1-\pi)\overline{F}_2(t)} = \pi(z_1|t)\lambda_1(t) + \pi(z_2|t)\lambda_2(t),$$

where the time-dependent probabilities are

$$\pi(z_1|t) = \frac{\pi \overline{F}_1(t)}{\pi \overline{F}_1(t) + (1-\pi)\overline{F}_2(t)}, \quad \pi(z_2|t) = 1 - \pi(z_1|t) = \frac{(1-\pi)\overline{F}_2(t)}{\pi \overline{F}_1(t) + (1-\pi)\overline{F}_2(t)}.$$

and $\overline{F}_i(t) = 1 - F_i(t)$, $i = 1, 2$.

Assume that a shock with complementary probabilities either "kills" an item (i.e., a failure occurs), or "leaves it unchanged". It is also reasonable to assume that *items with larger failure rates have larger probabilities of failures.*

Let $\pi_s(z)$ denote the mixing distribution *after a shock* ($\pi_s(z_1) = \pi_s$, $\pi_s(z_2) = 1 - \pi_s$) and let $\lambda_{ms}(t)$ be the corresponding mixture (observed) failure rate. Denote the probabilities of failures caused by each shock for two subpopulations as:

$$p(z) = \begin{cases} p_1, z = z_1, \\ p_2, z = z_2. \end{cases} \tag{26.1}$$

In accordance with our assumption, $p_1 \leqslant p_2$. It is easy to show [F08] that

$$\lambda_{ms}(t) = \frac{\pi_s f_1(t) + (1-\pi_s) f_2(t)}{\pi_s \overline{F}_1(t) + (1-\pi_s)\overline{F}_2(t)} = \pi_s(z_1|t)\lambda_1(t) + \pi_s(z_2|t)\lambda_2(t), \tag{26.2}$$

where

$$\pi_s(z_1|t) = \frac{\pi_s \overline{F}_1(t)}{\pi_s \overline{F}_1(t) + (1-\pi_s)\overline{F}_2(t)}, \quad \pi_s(z_2|t) = 1 - \pi_s(z_1|t) = \frac{(1-\pi_s)\overline{F}_2(t)}{\pi_s \overline{F}_1(t) + (1-\pi_s)\overline{F}_2(t)}.$$

The following simple result justifies the fact that a shock can be considered as a burn-in, as reliability characteristics of items after a shock are better than before.

Theorem 1. *Let* $p_1 \leqslant p_2$. *If* $\lambda_1(t) \leqslant \lambda_2$, *for all* $t \geqslant 0$, *then* $\lambda_{ms}(t) \leqslant \lambda_m(t)$, $\forall t \in [0, \infty)$.

Proof. Observe that $\lambda_m(t)$ and $\lambda_{ms}(t)$ are the weighted averages of $\lambda_1(t)$ and $\lambda_2(t)$. Then it is sufficient to show that $\pi_s(z_1|t) \geqslant \pi(z_1|t)$, which follows from the fact that

$$\pi(z_1|t) = \frac{\pi \overline{F}_1(t)}{\pi \overline{F}_1(t) + (1-\pi)\overline{F}_2(t)} = \frac{\overline{F}_1(t)}{\overline{F}_1(t) + (1/\pi - 1)\overline{F}_2(t)} \tag{26.3}$$

is increasing in π and that $\pi_s \geqslant \pi$.

The optimal burn-in time is the main characteristic of interest in conventional burn-in procedures. In our model, the "severity" of a shock will somehow correspond to this parameter. We will suggest now an approach for determining an optimal magnitude of a shock that maximizes the "quality" of our population after burn-in.

Denote the magnitude of a shock by $s \in [0, \infty]$. Assume that the "strength" of the component in a strong subpopulation is a continuous random variable, which is denoted by U, i.e., if $s > U$, then the failure occurs. Let the Cdf, the survival function and the failure rate function of U are denoted by $G(s)$, $\overline{G}(s)$, and $r(s)$, respectively. Similarly, let the strength of the component in a weak subpopulation be denoted by U_w. In accordance with our assumption, let $U \geqslant_{st} U_w$, which is equivalent to

$$G_w(s) = G(\rho(s)), \text{ for all } s \geqslant 0, \tag{26.4}$$

where $G_w(s)$ is the Cdf of U_w, $\rho(s)$ is an increasing function, $\rho(s) \geqslant s$ for all $s \geqslant 0$, and $\rho(0) = 0$. Thus from (26.1) and (26.4): $p(z_1, s) = p_1 = G(s)$, $p(z_2, s) = p_2 = G(\rho(s))$, and $p_1 \leqslant p_2$ holds for all $s \in [0, \infty)$.

Denote the corresponding mixture failure rate by $\lambda_{ms}(t; s)$ and consider now its uniform minimization for all fixed $t \geqslant 0$, with respect to $s \in [0, \infty]$:

$$s^* = \operatorname{arginf}_{s \in [0,\infty]} \lambda_{ms}(t; s), \text{ for all fixed } t \geqslant 0.$$

It is clear that this optimal s^* will maximize also the important reliability characteristics such as the *expected lifetime* of an item that has survived a shock and the probability of success (*survival probability*) for a mission time τ.

Denote by $R(s) \equiv \int_0^s r(u)du$ the cumulative failure rate that corresponds to the Cdf $G(s)$. Then the following result defines the optimal severity s^*.

Theorem 2. *Let* $\lambda_1(t) \leqslant \lambda_2(t)$, *for all* $t \geqslant 0$. *Then the optimal* s^* *is the value which maximizes* $R(\rho(s)) - R(s)$. *In particular,*
(i) If $r(s)$ *is increasing and* $\rho'(s) > 1$, *then* $s^* = \infty$.
(ii) If $\frac{\rho'(s)r(\rho(s))}{r(s)} > 1$, *for* $s < s_0$, *and* $\frac{\rho'(s)r(\rho(s))}{r(s)} < 1$, *for* $s > s_0$, *then* $s^* = s_0$.

Proof. As $\lambda_{ms}(t; s)$ is the weighted average of $\lambda_1(t)$ and $\lambda_2(t)$ (see (26.2)) and, similar to (26.3), $\pi_s(z_1|t)$ is increasing in π_s, the minimum of $\lambda_{ms}(t; s)$ for $\forall t \geqslant 0$ is obtained by maximizing

$$\pi_s = \frac{(1 - G(s))\pi}{(1 - G(s))\pi + (1 - G(\rho(s)))(1 - \pi)}, \tag{26.5}$$

which is equivalent to minimization of $\exp\{-[R(\rho(s)) - R(s)]\}$ (maximization of $R(\rho(s)) - R(s)$).

(i) Denote $\phi(s) \equiv R(\rho(s)) - R(s)$. Then $\phi'(s) \equiv \rho'(s)r(\rho(s)) - r(s)$. As $\rho'(s) > 1$ and $r(x)$ is increasing,

$$\phi'(s) = \rho'(s)r(\rho(s)) - r(s) > r(\rho(s)) - r(s) \geqslant 0,$$

where assumption $\rho(s) \geqslant s$ is used. Thus, in this case, $s^* = \infty$.
(ii) Now assume that $\frac{\rho'(s)r(\rho(s))}{r(s)} > 1$, for $s < s_0$, and $\frac{\rho'(s)r(\rho(s))}{r(s)} < 1$, for $s > s_0$. Then $\phi'(s) > 0$, for $s < s_0$, and $\phi'(s) < 0$, for $s > s_0$, which implies $s^* = s_0$.

Remark 1. In practice, obviously, there exists a maximum level of stress $s_a < \infty$ that can be applied to items without destroying the whole population or without the non-negligible damage in the survived items.

Consider now a model of determining the optimal severity minimizing the expected cost function, which takes into account burn-in and field operation. An item is chosen at random from our heterogeneous population and is exposed to a shock. If it survives, it is considered to be ready for usage, otherwise the failed item is discarded and the new one is chosen from the population, etc. This procedure is repeated until the first survived item is obtained.

Let c_{sr} be the shop replacement cost and c_s be the cost for conducting a single shock. Let $c_1(s)$, as a function of s, be the expected cost for eventually obtaining a component which has survived a shock. It is easy to show that

$$c_1(s) = \frac{c_s + c_{sr}P}{1 - P},$$

where $P = G(s)\pi + G(\rho(s))(1 - \pi)$ is the probability that an item from the mixture population does not survive the shock.

Let the cost c_m be incurred by the event $\{T_s \leqslant \tau\}$ (failure of a mission with duration τ) and the gain g_m result from the event $\{T_s > \tau\}$ (success of a mission). Then the expected cost during field operation, $c_2(s)$, is

$$c_2(s) = -g_m\left(\pi_s\overline{F}_1(\tau) + (1 - \pi_s)\overline{F}_2(\tau)\right) + c_m\left(\pi_s F_1(\tau) + (1 - \pi_s)F_2(\tau)\right),$$

where π_s is defined by (26.5). Denote $c(s) = c_1(s) + c_2(s)$

Let s^* be the optimal severity level that satisfies

$$s^* = \operatorname{arginf}_{s\in[0,\infty]}c(s).$$

The following theorem defines properties of optimal s^*:

Theorem 3. *Let $\overline{F}_1(t) \geqslant \overline{F}_2(t)$, for all $t \geqslant 0$. If $R(\rho(s)) - R(s)$ strictly decreases for $s > s_0$, then $s^* \leqslant s_0$. In particular,*

(i) If $\rho'(s) > 1$ and $r(x)$ is increasing, then $s^* < \infty$.
(ii) If $\frac{\rho'(s)r(\rho(s))}{r(s)} < 1$, for $s > s_0$, then $s^* \leqslant s_0$.

Proof. Note that $c_1(s)$ strictly increases from $c_1(0) = c_s$ to $c_1(\infty) = \infty$. Also observe that $c_2(s) = -(g_m + c_m)\overline{F}_{ms}(\tau; s) + c_m$, where $\overline{F}_{ms}(t; s)$ (the survival function of a component after a shock) is the weighted average of $\overline{F}_1(t)$ and $\overline{F}_2(t)$ with the corresponding weights π_s and $1 - \pi_s$, respectively. If $R(\rho(s)) - R(s)$ strictly decreases for $s > s_0$, then, by similar arguments as those described in the proof of Theorem 2, $c_2(s)$ strictly increases for $s > s_0$. This imply that $c(s)$ strictly increases for $s > s_0$ and thus we can conclude that $s^* \leqslant s_0$. The specific cases are easily obtained based on the proof of Theorem 2.

26.3 Continuous Mixtures

Consider a general "continuous" mixing model for a heterogeneous population, i.e.,

$$F_m(t) = \int_0^\infty F(t, z)\pi(z)\mathrm{d}z,$$

$$f_m(t) = \int_0^\infty f(t, z)\pi(z)\mathrm{d}z,$$

where $F(t, z)$, $f(t, z)$ are the Cdf and the pdf of subpopulations indexed by the frailty parameter Z and $\pi(z)$ is the pdf of Z with support in $[0, \infty)$. Then the mixture (the *observed* or the *population*) failure rate $\lambda_m(t)$, similar to the discrete case (26.2) is

$$\lambda_m(t) = \frac{\int_0^\infty f(t, z)\pi(z)\mathrm{d}z}{\int_0^\infty \overline{F}(t, z)\pi(z)\mathrm{d}z} = \int_0^\infty \lambda(t, z)\pi(z|t)\mathrm{d}z, \tag{26.6}$$

where

$$\pi(z|t) \equiv \pi(z)\frac{\overline{F}(t, z)}{\int_0^\infty \overline{F}(t, z)\pi(z)\mathrm{d}z}. \tag{26.7}$$

Assume that an instantaneous shock with complementary probabilities either "kills]' an item, or "leaves it unchanged']. It is natural to suppose (as in the discrete case) that *the more frail (e.g., with larger failure rate) items are, the more susceptible they are to be destroyed by a shock.*

- *Burn-in procedure by means of a shock*: An item is chosen at random from a heterogeneous population and is exposed to a shock. If it survives, it is considered to be ready for usage, otherwise the failed item is discarded and the new one is chosen from the population, etc.

This setting can be formalized in the following way: Let $\pi_s(z)$ denote the pdf of the frailty Z_s (with support in $[0, \infty)$) after a shock and let $\lambda_{ms}(t)$ be the corresponding mixture (observed) failure rate. In accordance with (26.6):

$$\lambda_{ms}(t) = \int_0^\infty \lambda(t, z)\pi_s(z|t)\mathrm{d}z,$$

where, similar to (26.7), $\pi_s(z|t)$ is defined by the right-hand side of (26.7) with $\pi(z)$ substituted by $\pi_s(z)$.

Let each item fails with probability $p(z)$ and is survived (as good as new) with probability $q(z) = 1 - p(z)$. Assume, similar to the discrete case, that $p(z)$ is an increasing function of z $(0 \leq p(z) \leq 1)$: larger values of frailty correspond to larger values of the failure rate. Therefore, population densities before and after a shock are obviously related as

$$\pi_s(z) = \frac{q(z)\pi(z)}{\int_0^\infty q(z)\pi(z)\mathrm{d}z}, \tag{26.8}$$

which means that population frailties before and after a shock are ordered in the sense of the likelihood ratio: $Z \geq_{LR} Z_s$ (see, e.g., [SS06]) where $g(z)$ is a decreasing function and therefore $\pi_s(z)/\pi(z)$ is decreasing.

Theorem 4. *Let relationship (26.8), defining the mixing density after a shock, where $q(z)$ is a decreasing function, hold.*
Assume that

$$\lambda(t, z_1) \leq \lambda(t, z_2), \quad z_1 < z_2, \forall z_1, z_2 \in [0, \infty], t \geq 0.$$

Then

$$\lambda_{ms}(t) \leq \lambda_m(t); \ \forall t \geq 0. \tag{26.9}$$

where $\lambda_{ms}(t)$ is a population (mixture) failure rate after a shock.

Proof. It can be shown ([F08], page 164) that:

$$\begin{aligned} &sign[\lambda_{ms}(t) - \lambda_m(t)] \\ &= sign \int\int_{u>s} \bar{F}(t,u)\bar{F}(t,s)(\lambda(t,u) - \lambda(t,s))(\pi_s(u)\pi(s) - \pi_s(s)\pi(u))duds, \end{aligned} \tag{26.10}$$

which is negative due to (26.8) and assumptions of this theorem.

Remark 2. Inequality (26.9) is a "natural" ordering in the family of failure rates $\lambda(t, z), z \in [0, \infty)$, and trivially holds, e.g., for the specific multiplicative model:

$$\lambda(t, z) = z\lambda(t). \tag{26.11}$$

To consider the corresponding optimization, we must define the costs and probabilities of interest. A simple and convenient model for $p(z)$ is the step function:

$$p(z) = \begin{cases} 0, & 0 \leq z \leq z_b, \\ 1, & z > z_b. \end{cases} \tag{26.12}$$

It means that all weak items with $z > z_b$ will be eliminated and only strong items will remain in the population. In accordance with (26.12), the probability of not surviving the shock in this case is

$$P_{z_b} \equiv \bar{\Pi}(z_b) = \int_{z_b}^\infty \pi(z)\mathrm{d}z,$$

where $\Pi(z)$ is the Cdf that corresponds to the pdf $\pi(z)$. Obviously, for a general form of $p(z)$, this probability is defined by the following mixture

$$P = \int_0^\infty p(z)\pi(z)\mathrm{d}z. \tag{26.13}$$

It is clear that parameter z_b in the specific model (26.12) can be considered as a parameter of *severity*: the larger values of z_b correspond to the smaller severity. For a more general setting, define the functions $p(z)$ and $q(z)$ as functions of a frailty variable z and the severity parameter $s \in [0, \infty)$, $p(z, s)$ and $q(z, s)$. Assume that $q(z, s)$ is decreasing in z for each fixed s and is decreasing in s for each z. The assumption that $q(z, s)$ is decreasing in s for each fixed z is also quite natural and implies that items characterized by the same value of frailty have larger failure probabilities under larger severity levels.

Denote the corresponding failure rate and survival functions by $\lambda_{ms}(t; s)$ and $\overline{F}_{ms}(t; s)$, respectively. Similar to (26.8) and (26.7):

$$\pi_s(z, s) = \frac{q(z, s)\pi(z)}{\int_0^\infty q(u, s)\pi(u)\mathrm{d}u}, \pi_s(z, s|t) \equiv \pi_s(z, s)\frac{\bar{F}(t, z)}{\int_0^\infty \bar{F}(t, u)\pi_s(u, s)\mathrm{d}u}.$$

To compare two severity levels, we need the following definition.

Definition 1. *The severity (stress) level s is said to be dominated if there exists another level s′ such that*

$$\lambda_{ms}(t; s) \geq \lambda_{ms}(t; s'), for\ all\ t \geq 0.$$

Otherwise, the severity (stress) level s is called non-dominated.

Theorem 5. *Assume that $q(z, s)$ is decreasing in z for each fixed s and is decreasing in s for each z. Consider two stress levels s and s'. Let*

$$q(u, s')q(v, s) - q(v, s')q(u, s) \leq 0,\ for\ all\ u > v, \tag{26.14}$$

which means that $q(z, s')/q(z, s)$ is decreasing in z. If

$$\lambda(t, z_1) \leq \lambda(t, z_2),\ z_1 < z_2, \forall z_1, z_2 \in [0, \infty], t \geq 0,$$

then the severity level s is dominated.

Proof. Similar to (26.10):

$$\begin{aligned} sign[\lambda_{ms}(t; s') - \lambda_{ms}(t; s)] &= sign \int\int_{u>v} \bar{F}(t, u)\bar{F}(t, v)(\lambda(t, u) \\ &\quad -\lambda(t, v))(\pi_s(u, s')\pi_s(v, s) - \pi_s(v, s')\pi_s(u, s))\mathrm{d}u\mathrm{d}v. \end{aligned}$$

Thus, if (26.14) holds, then

$$\pi_s(u, s')\pi_s(v, s) - \pi_s(v, s')\pi_s(u, s) \leq 0,$$

which implies the result.

Remark 3. Intuitively, it can be believed that a higher level of severity results in a better "quality of a population" but it is not always true. A similar observation holds for the conventional burn-in in homogeneous populations when the larger time of burn-in does not necessarily improve the "quality of a population".

Similar to the discrete case, the optimal severity of a shock that minimizes the average cost incurred during the burn in and field usage will be now considered.

As previously, a new component randomly selected from the heterogeneous population is burned-in by means of a shock. If the first one did not survive then we take another one from infinite heterogeneous population and burn-in again. This procedure is repeated until we obtain the first component which survives burn-in. Then this component is put into field operation. The expected cost of burn-in until obtaining the first item that has survived shocks, $c_1(s)$, is given by (26.6). This function increases when P increases in $[0, 1)$. Note that P (see (26.13) is now a function of the stress level s , that is, $P(s)$. Let, as in the discrete case, c_m and g_m are the cost and the (gain) that correspond to the failure (success) of the mission. Obviously, as in the discrete case, the expected cost during field operation is:

$$c_2(s) = -(g_m + c_m)\overline{F}_{ms}(\tau; s) + c_m.$$

Therefore, the total expected cost function (as a function of the severity level s) for the burn-in and the field operation phases is $c(s) = c_1(s)+c_2(s)$. The values c_{sr}, c_s, g_m, c_m are assumed to be known. Thus, similar to the discrete case:

$$s^* = \operatorname{arginf}_{s\in[0,\infty]} c(s). \tag{26.15}$$

The following result immediately follows from Theorem 5:

Theorem 6. *Suppose that*

$$\overline{F}(t, z_1) \geq \overline{F}(t, z_2), \quad z_1 < z_2, \forall z_1, z_2 \in [0, \infty], t \geq 0.$$

If, for any $s_2 > s_1$, $q(u, s_2)q(v, s_1) - q(v, s_2)q(u, s_1) \leq 0$, for all $u > v$, i.e., $q(z, s_2)/q(z, s_1)$ decreases in z for all $s_2 > s_1$, then there exists the finite optimal level $s^* < \infty$ for the optimization problem (26.15).

Example 1. Consider the multiplicative model (26.11) with constant baseline failure rate: $\lambda(t, z) = z\lambda$. Note that many electronic components have a constant failure rate which is varying from component to component due to production instability, etc.

Assume (for simplicity) that Z is also exponentially distributed: $\Pr(Z \leq z) = 1 - \exp\{-\alpha z\}$. It is well known that the mixture failure rate in this case is

$$\lambda_m(t) = \frac{\int_0^\infty z\lambda \exp\{-z\lambda t\}\pi(z)\mathrm{d}z}{\int_0^\infty \exp\{-z\lambda t\}\pi(z)\mathrm{d}z} = \frac{\lambda}{\lambda t + \alpha}.$$

Let the impact of a shock be defined by the specific $p(z)$ in (26.12). Then

$$\begin{aligned}\lambda_{ms}(t, z_b) &= \frac{\int_0^{z_b} z\lambda \exp\{-z\lambda t\}\pi(z)dz}{\int_0^{z_b} \exp\{-z\lambda t\}\pi(z)dz} \\ &= \frac{\lambda}{\lambda t+\alpha}\left(1 - \frac{z_b(\lambda t+\alpha)}{\exp\{z_b(\lambda t+\alpha)\}-1}\right).\end{aligned}$$

The corresponding total expected cost function is $c(z_b) = c_1(z_b) + c_2(z_b)$, where

$$c_1(z_b) = \frac{c_{sr}\exp\{-\alpha\, z_b\} + c_s}{1 - \exp\{-\alpha\, z_b\}},$$

$$c_2(z_b) = -(g_m + c_m)\exp\left\{-\int_0^{\tau} \frac{\lambda}{\lambda u + \alpha}\left(1 - \frac{z_b(\lambda u + \alpha)}{\exp\{z_b(\lambda u + \alpha)\} - 1}\right) du\right\} + c_m.$$

A simple analysis of the shape of $c(z_b)$ (as a function of z_b) shows that there exists a finite optimal stress level $s^* < \infty$ (positive optimal $z_b^* > 0$).

26.4 Concluding Remarks

Conventional burn-in procedures are usually performed during specified intervals of time for items with decreasing or bathtub failure rates in order to eliminate early failures or (and) to improve reliability characteristics. This can be done in a "normal" or accelerated environment or with the help of high environmental stresses applied for short (or "instantaneous") duration that are often called "shocks". The latter method of burn-in was not considered previously in the literature on burn-in modeling.

In this paper, we assume that population of items is heterogeneous (discrete and continuous) and different subpopulations have different resistances to shocks. Items with larger failure rates are assumed to have larger probabilities of failures when exposed to shocks and also larger values of stress levels result in larger probabilities of failures. Our modeling is based on these 'natural assumptions'. Based on this reasoning the optimal severity levels that minimize relevant cost functions are studied.

References

[BS97] Block, H.W., Savits, T.H.: Burn-in. Statistical Science, **12**, 1–19 (1997)

[BMS93] Block, H.W., Mi, J., Savits, T.H.: Burn-in and mixed populations. Journal of Applied Probability, **30**, 692–702 (1993)

[BSS02] Block, H.W., Savits, T.H., Singh, H.: Criterion for burn-in that balances mean residual life and residual variance. Operations Research, **50** 290–296 (2002)

[C00] Cha, J.H.: On a better burn-in procedure. Journal of Applied Probability, **37** 1099–1103 (2000)

[C01] Cha, J.H.: Burn-in procedures for a generalized model. Journal of Applied Probability, **38** 542–553 (2001)

[C03] Cha, J.H.: A further extension of the generalized burn-in model. Journal of Applied Probability, **40** 264–270 (2003)

[C06] Cha, J.H.: A stochastic model for burn-in procedures in accelerated environment. Naval Research Logistics, **53** 226–234 (2006)

[CS91] Clarotti, C.A., Spizzichino, F.: Bayes burn-in decision procedures. Probability in the Engineering and Informational Sciences, **4** 437–445 (1991)

[F08] Finkelstein, M.S.: Failure Rate Modelling for Reliability and Risk. Springer, London (2008)

[JP82] Jensen, F., Petersen, N.E.: Burn-in. Wiley, New York (1982)

[KK83] Kuo, W., Kuo, Y.: Facing the headaches of early failures: A state-of-the-art review of burn-in decisions. Proceedings of IEEE., **71** 1257–1266 (1983)

[M91] Mi, J.: Optimal burn-in. Ph.D. Thesis, Univ. Pittsburgh, Pittsburgh (1991)

[M94a] Mi, J.: Burn-in and maintenance policies. Advances in Applied Probability, **26** 207–221 (1994)

[M94b] Mi, J.: Maximization of a survival probability and its application. Journal of Applied Probability, **31** 1026–1033 (1994)

[M96] Mi, J.: Minimizing some cost functions related to both burn-in and field use. Operations Research, **44** 497–500 (1996)

[M97] Mi, J.: Warranty policies and burn-in. Naval Research Logistics, **44** 199–209 (1997)

[NM82] Nguyen, D.G., Murthy, D.N.P.: Optimal burn-in time to minimize cost for products sold under warranty. IIE Transactions, **14** 167–174 (1982)

[SS06] Shaked, M., Shanthikumar, J.G.: Stochastic Orders. Springer, New York (2006)

27

Bayesian Estimation of Degradation Model Defined by a Wiener Process

Fabrice Guérin[1*], Mihaela Barreau[1], Amel Demri[1], Sylvain Cloupet[1], Julien Hersant[1] and Ridha Hambli[2]

[1] LASQUO, University of Angers, Angers, France, fabrice.guerin@univ-angers.fr
[2] Polytech Orléans, Orléans, France

Abstract: The constantly increasing market request of high quality vehicles ask the automotive manufacturers to perform lifetime testing in order to verify the reliability levels of new products. In this paper, we deal with two difficulties in reliability assessment for mechanical parts. On one hand, there is the small number of parts available for testing. On the other hand, there is the problem of wear. In the automotive applications, mechanical components subjected to relative motion of parts have to be designed against wear. In this paper, the Bayesian estimation of Wiener process parameters (usually used to define the degradation process) is studied to improve the estimation accuracy in incorporating the available knowledge on the product. In particular, the finite element results and expert knowledge are considered as "a priori". For wear prediction by FEM, a model based on Archard law was developed for the brake disc wear.

Keywords and phrases: Archard model, Bayesian estimation, Brake, FEM, Reliability, Wear, Wiener process

27.1 Introduction

Reliability assessment is becoming an integral part of the design process of complex systems in order to highlight potential risk areas so they can be dealt with at the design stage of the project. Indeed, the early control of system specifications allows diminishing operating (either financial or safety) risks. Since systems must be more and more reliable and offer longer guarantees, it is necessary to check the compliance of their performances as early as possible.

One can analyze two failure types:

- Material failures, often appearing all of a sudden.
- Soft failures, meaning a performance drift in time, until unacceptable levels.

Testing prototypes allows evaluating the reliability of a system before it is mass-produced. This process requires long testing times and huge numbers of prototypes since

V.V. Rykov et al. (eds.), *Mathematical and Statistical Models and Methods in Reliability: Applications to Medicine, Finance, and Quality Control*, Statistics for Industry and Technology, DOI 10.1007/978-0-8176-4971-5_27,

the latter are more and more reliable, therefore extremely diminishing the probability of failure.

One alternative solution is the study of a performance drift in time, in order to characterize failure probability. This is done by testing a number of systems and by measuring the evolution of their performance in time, $z(t)$. The systems are considered as failed when their performance has reached the critical value denoted z_0.

The constantly increasing market request for high quality products compels the manufacturers to verify, before starting mass production, if new components or parts attain a field reliability target. To this end, reliability testing is used to estimate the lifetime distribution (Meeker et al [MEE98]). Common problems in lifetime distribution estimation by testing are the total time required to test and the available number of examples for testing to demonstrate reliability to a customer's satisfaction. This paper proposes to use Bayes estimation [SAN91] in incorporating the prior expert opinions and finite elements results.

27.2 Reliability Testing

Degradation tests for reliability estimation consist in measuring the evolution of the degradation during the testing of a sample of products or systems. We thus obtain a degradation path, $z(t)$, for each tested system and a network of degradation paths for the entire sample (see Fig. 27.1).

The system is considered as failed when its degradation reaches a critical value, denoted z_0. Reaching this critical value allows obtaining pseudo-failure times, denoted $t_i$which are then used to asses reliability function.

Degradation processes are paths of some stochastic process with independent increments. Wiener process [COU05], [NIK10] characterizes average monotonic degradations. In this paper, we consider the case of a Wiener process with linear leaning μ and variance σ^2, with following hypothesis:

- $W(0) = 0$.
- Increment law $W(t + h) - W(t)$ is normal distribution $N(\mu h, \sigma^2 h)$.
- If W_0 is a standard Wiener process, i.e., $\mu = 0$ and $\sigma = 1$, then $W(t) = \mu t + \sigma W_0(t)$ is a Wiener process of linear leaning μ and variance σ^2.

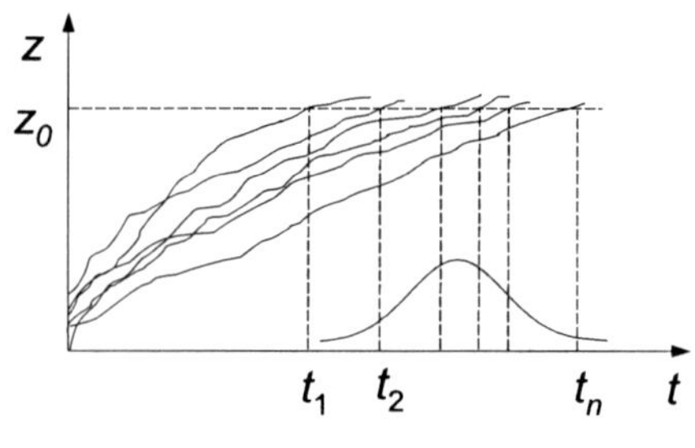

Figure 27.1. Degradation paths example

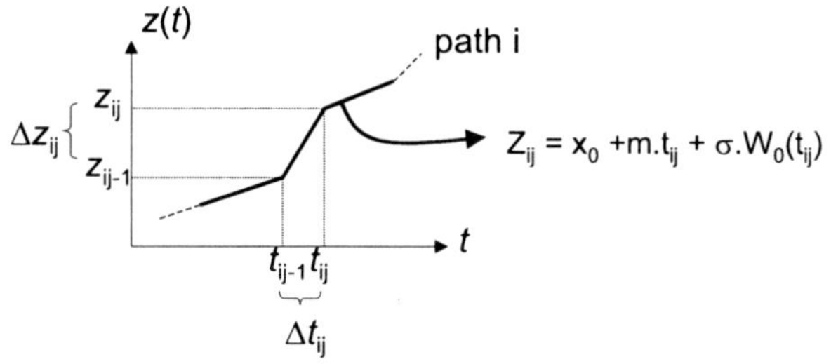

Figure 27.2. Example of data

The distribution of pseudo instants of failure, T, is an inverse normal distribution $\mathrm{IG}(z_0/\mu,\ z_0^2/\sigma^2)$, of density given by:

$$f(T, z_0, \mu, \sigma) = \frac{z_0}{\sqrt{2\pi\sigma}} T^{-\frac{3}{2}} \mathrm{e}^{\left(-\frac{(z_0-\mu t)^2}{2\sigma^2 T}\right)} \tag{27.1}$$

The estimation of μ and σ is obtained by maximum likelihood, using the observed increments; the degradation increments are denoted Δz_{ij} (for path i (m paths) and time j (q_i measures on path), as shown Fig. 27.2).

Since degradation increment Δz_{ij} is characterized by a normal distribution (of mean $\mu \Delta t_{ij}$ and variance $\Delta t_{ij}.(1/\theta^2)$ with $\sigma^2 = 1/\theta^2$), the likelihood is:

$$g(\Delta z|\mu, \theta) = \prod_{i=1}^{m} \prod_{j=1}^{q_i} \frac{\theta^{1/2}}{\sqrt{2\pi \Delta t_{ij}}} \mathrm{e}^{-\left(\frac{\theta(\Delta z_{ij}-\mu\Delta t_{ij})^2}{2\Delta t_{ij}}\right)} \tag{27.2}$$

In test, the periodicity of degradation measurements is often constant ($\Delta \mathrm{t}_{ij} = \Delta \mathrm{t}$). Considering this assumption, the likelihood function can be written

$$g(x|\mu, \theta) = \prod_{i=1}^{m} \prod_{j=1}^{q_i} \frac{\theta^{1/2}}{\sqrt{2\pi}} \mathrm{e}^{-\left(\frac{\theta(x_{ij}-\mu\Delta t_{ij})^2}{2\Delta t_{ij}}\right)} \tag{27.3}$$

with $x_{ij} = \Delta z_{ij}/\Delta t$.

The function $g(x|\mu, \theta$ is characteristic of normal likelihood function. Usually, a probability distribution is defined by its parameters which are often unknown constants. Based on a random sample, one can use the maximum likelihood method to estimate and obtain confidence intervals for the parameters and the reliability function.

From a random sample of n observations ($n = \sum_{i=1}^{m} q_i$) the sample mean $\bar{x}$ and sample standard deviation s are computed [AYY97]:

$$n = \sum\nolimits_{i=1}^{m} q_i \tag{27.4}$$

$$\bar{x} = \frac{\sum\limits_{i=1}^{m} \sum\limits_{j=1}^{q_i} x_{ij}}{n} \tag{27.5}$$

and

$$s = \sqrt{\frac{\sum_{i=1}^{n} \sum_{j=1}^{q_i} (x_{ij} - \bar{x})^2}{n-1}}. \tag{27.6}$$

27.3 Bayesian Estimation of Wiener Process Parameters

Recently, a rising interest in the Bayesian approach to reliability and life parameter estimation has emerged. To statisticians and reliability engineers this approach is appealing since it provides a method of using their past experiences and/or prior convictions in describing the studied parameter x stochastically. On some situations the parameter is not known, but can be treated as a random variable with a known prior probability density. Under this scenario, one can combine information from the random sample and prior probability distributions to obtain $(l - \gamma)$ Bayesian confidence intervals for the parameters. The objective of this section is to obtain the Bayesian estimators for the parameters μ and θ of the normal distribution [CHE97], [AHM95].

27.3.1 Bayesian Principle

The probability density function $f(\mu, \theta|x)$ of the posterior *pdf* of μ and θ obtained from the sample of observations $x = \{x_{ij}\}$ and the *pdf* $f(\mu)$ and $f'(\theta)$ of the prior distribution of μ and θ is given by

$$f(\mu, \theta/x) = \frac{g(x/\mu, \theta) \cdot f(\mu) \cdot f'(\theta)}{\int\limits_{D(\mu)} \int\limits_{D(\theta)} g(x/\mu, \theta) \cdot f(\mu) \cdot f'(\theta) \cdot \mathrm{d}\mu \mathrm{d}\theta} \tag{27.7}$$

where

- μ and θ: parameters to estimate
- $x = \{x_{ij}\}$: observed data
- $f(\mu)$ and $f'(\theta)$: prior probability density functions (available knowledge from the experts)
- $g(x|\mu, \theta)$: likelihood function
- $f(\mu, \theta|x)$: posterior density function
- $D(\mu)$ and $D(\theta)$: set of nature states

Now, two cases are studied to define the posterior *pdf*:

- No knowledge on μ and θ
- Available knowledge on μ and θ

Case 1: No knowledge on μ and θ. When there is no information about the mean μ and the inverse variance θ, the uniform uninformative density (Fig. 27.3) is used to define the prior *pdf* $f(\mu)$ and $f'(\theta)$. The selection of this uniform probability density is based on the fact that this *pdf* has maximum entropy among all *pdf* that are non zero in a given range.

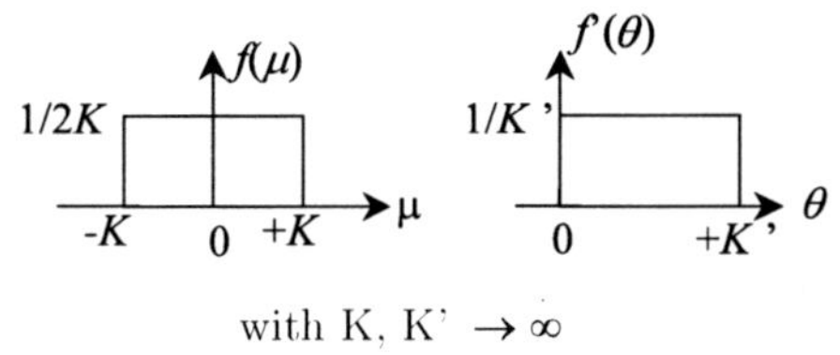

Figure 27.3. Uniform uninformative pdf

The likelihood function is given for a sample of size n by

$$g(x|\mu,\theta)=\prod_{i=1}^{m}\prod_{j=1}^{q_i}\frac{\theta^{1/2}}{\sqrt{2\pi}}\mathrm{e}^{-\left(\frac{\theta(x_{ij}-\mu\Delta t_{ij})^2}{2\Delta t_{ij}}\right)} \tag{27.8}$$

Thus, the posterior *pdf* is written as

$$f(\mu,\theta/x)=\frac{g\left(x/\mu,\theta\right)f(\mu)f(\theta)}{\int\limits_{-K}^{+K}\int\limits_{0}^{+K'}g\left(x/\mu,\theta\right)f(\mu)f(\theta)\mathrm{d}\mu\mathrm{d}\theta} \tag{27.9}$$

with K,K' $\rightarrow\infty$

Following Congdon, Ahmad et al. and Chen et al. [CON01], [AHM95] and [CHE97], the posterior *pdf* is given by

$$f(\mu,\theta|x)=\frac{b^a}{\Gamma(a)}\theta^{a-1}\mathrm{e}^{-b\theta}\sqrt{\frac{n\theta}{2\pi}}\mathrm{e}^{-\left(\frac{n\theta}{2}(\mu-c)\right)}=G(a,b)N\left(c,\frac{1}{\theta(2a-1)}\right) \tag{27.10}$$

where $a=\frac{n+1}{2}$ $b=\frac{1}{2}\sum\limits_{i=1}^{m}\sum\limits_{j=1}^{q_i}(x_{ij}-\overline{x})^2$ and $c=\overline{x}$ which is a combination of the gamma and normal distributions.

Case 2: Available knowledge about μ and θ. Following Congdon, Ahmad et al. and Chen et al. [CON01], [AHM95] and [CHE97], we propose to choose the prior *pdf* defined by the relationship (27.10). Then the posterior *pdf* is written

$$f(\mu,\theta|x)=\frac{b'^{a'}}{\Gamma(a')}\theta^{a'-1}\mathrm{e}^{-b'\theta}\sqrt{\frac{n\theta}{2\pi}}\mathrm{e}^{-\left(\frac{n\theta}{2}(\mu-c')\right)}=G(a',b')N\left(c',\frac{1}{\theta(2a'-1)}\right) \tag{27.11}$$

where $a'=\frac{n}{2}+a$, $c'=M=\frac{n\overline{x}+(2a-1)c}{n+2a-1}$ $b'=\frac{1}{2}\sum\limits_{i=1}^{m}\sum\limits_{j=1}^{q_i}(x_{ij}-\overline{x})^2+b+\frac{n(2a-1)(\overline{x}-c)^2}{2(n+2a-1)}$ which is a combination of the gamma and normal distributions.

Note that the form of relationship (27.11) is identical to (27.10). The prior *pdf* defined by (27.10) is the natural conjugate.

27.3.2 Bayesian Estimators

Once the posterior distribution is defined, the estimators of the parameters μ and θ can be obtained by using the marginal distributions associated to μ and θ.

The marginal distribution of θ is written as

$$f(\theta/x) = \int_{-\infty}^{+\infty} f(\mu,\theta/x)\mathrm{d}\mu f(\theta/x)$$

$$f(\theta/x) = G(a',b')\int_{-\infty}^{+\infty} N\left(c',\frac{1}{\theta(2a'-1)}\right)\mathrm{d}\mu f(\theta/x)$$

$$f(\theta|x) = \frac{b'^{a'}\theta^{a'-1}\mathrm{e}^{-b'\theta}}{\Gamma(a')} = G(a',b'). \tag{27.12}$$

The point estimate for θ is defined by the mode of $f(\theta/x)$ and the point estimate of variance are given by

$$\hat{\theta} = \frac{a'-1}{b'} \tag{27.13}$$

and

$$\hat{s}^2 = \frac{b'}{a'-1}. \tag{27.14}$$

The two-sided confidence interval (defined by $\theta_{\min}$ and $\theta_{\max}$) is evaluated such that:

$$\begin{cases} \frac{\gamma}{2} = \int_0^{\theta_{\min}} f(\theta/x)\mathrm{d}\theta \Rightarrow \theta_{\min} \\ 1-\frac{\gamma}{2} = \int_0^{\theta_{\max}} f(\theta/x)\mathrm{d}\theta \Rightarrow \theta_{\max} \end{cases} \tag{27.15}$$

with (1-γ) the given confidence level.

The marginal distribution of μ is written

$$f(\mu/x) = \int_0^{+\infty} f(\mu,\theta)\mathrm{d}\theta = \frac{b^a\sqrt{2a-1}}{\alpha^{a+\frac{1}{2}}\sqrt{2\pi}}\frac{\Gamma\left(a+\frac{1}{2}\right)}{\Gamma(a)} \tag{27.16}$$

with

$$\alpha = b + \frac{1}{2}(2a-1)(\mu-c)^2 \tag{27.17}$$

The point estimate for μ is defined by the mode of *f(μ/x)*

$$\hat{\mu} = c' \tag{27.18}$$

The two-sided confidence interval (defined by $\mu_{\min}$ and $\mu_{\max}$) is evaluated such that

$$\begin{cases} \frac{\gamma}{2} = \int_0^{\mu_{\min}} f(\mu/x)\mathrm{d}\mu \Rightarrow \mu_{\min} \\ 1-\frac{\gamma}{2} = \int_0^{\mu_{\max}} f(\mu/x)\mathrm{d}\mu \Rightarrow \mu_{\max} \end{cases} \tag{27.19}$$

with $(1-\gamma)$ the given confidence level. In the case when no knowledge is available, the point estimates of mean and variance are defined by known relationships

$$\hat{\mu} = c = \overline{x} \tag{27.20}$$

and

$$\hat{s}^2 = \frac{b}{a-1} = \frac{\sum_{i=1}^{n}\sum_{j=1}^{q_i}(x_{ij}-\overline{x})^2}{n-1} \tag{27.21}$$

with a, b and c defined by (27.11).

In the case in which prior knowledge is available, the point estimates of mean and variance are defined by:

$$\hat{\mu} = c' = \frac{n\bar{x}+(2a-1)c}{n+2a-1} \tag{27.22}$$

and

$$\hat{s}^2 = \frac{b'}{a'-1} = \frac{\frac{1}{2}\sum_{i=1}^{n}\sum_{j=1}^{q_i}(x_{ij}-\overline{x})^2 + b + \frac{n(2a-1)(\overline{x}-c)^2}{2(n+2a-1)}}{\frac{n}{2}+a-1} \tag{27.23}$$

where a, b and c are the parameters of the prior *pdf* and a', b' and c' the parameters of the posterior *pdf.*

27.3.3 Determination of the Prior Distribution from Available Information

The standard deviation (s) interval and prior mean μ are provided by an expert [COO91] or the results of a previous analysis. The prior knowledge is given by a believed estimation of mean and a range believed to contain the inverse of the variance

$$[\mu] \text{ and } \left[\theta_{\min} = 1/s^2_{\max}, \theta_{\max} = 1/s^2_{\min}\right].$$

The prior *pdf* $f(\mu,\theta/x)$ is defined by the relationship (27.10) with the unknown parameters a, b and c.

Evaluation of a and b by Moments Method

The marginal distribution of θ is written

$$f(\theta/x) = \frac{b^a\theta^{a-1}\mathrm{e}^{-b\theta}}{\Gamma(a)} = G(a,b) \tag{27.24}$$

The interval $[\theta_{\min}, \theta_{\max}]$ defines an uniform distribution. The mean and variance for this distribution are

$$E(\theta) = \frac{(\theta_{\min}+\theta_{\max})}{2} \tag{27.25}$$

and

$$V(\theta) = \frac{(\theta_{\max}-\theta_{\min})^2}{12} \tag{27.26}$$

The mean and variance for a gamma distribution $G(a,b)$ are given by

$$E'(\theta) = \frac{a}{b} \tag{27.27}$$

and

$$V'(\theta) = \frac{a}{b^2} \tag{27.28}$$

By evaluating of means and variances ($E(\theta)$ =$E'(\theta)$ and $V(\theta)$ =$V'(\theta)$), the values of parameters a and b are deducted to be

$$a = \frac{3\left(\theta_{\min} + \theta_{\max}\right)^2}{\left(\theta_{\max} - \theta_{\min}\right)^2} \tag{27.29}$$

and

$$b = \frac{6\left(\theta_{\min} + \theta_{\max}\right)}{\left(\theta_{\max} - \theta_{\min}\right)^2} \tag{27.30}$$

Evaluation of c. The marginal distribution of μ is written

$$f\left(\mu/x\right) = \frac{b^a\sqrt{2a-1}}{\alpha^{a+\frac{1}{2}}\sqrt{2\pi}}\frac{\Gamma\left(a+\frac{1}{2}\right)}{\Gamma(a)} \tag{27.31}$$

with $\alpha = b + \frac{1}{2}(2a-1)(\mu - c)^2$.

This marginal distribution is symmetric around the c value (a and b are also defined). Then the c value is given by

$$c = \mu \tag{27.32}$$

In the following section, we propose to define the μ value by FEM modeling.

27.4 Estimation of Prior Mean of Wiener Process by FEM Method: Application to Brake Disc Wear

The aim of this section is to propose a method for the estimation of the prior mean of Wiener process by FEM. The method is illustrated by an application on the brake disc wear.

Brakes are common structural components (Fig. 27.4) that often require analysis [MEE98]. Finite element analyses can provide information needed to determine the brake performance and behavior. This information includes a contact pressure distribution, friction and temperature distribution.

The relative motion of the pad-disc surfaces may result in a loss of tool material through adhesive wear [ARC53], [CHO96] and [CAR94]. This wear processes are initiated by the interfacial adhesive junctions formed in contact zone. As a normal load is applied, local pressure at the contact area become extremely high. Therefore, the surfaces adhere together. The sliding between the surfaces leads to the generation of wear particles. The presence of hard particles accelerates wear by abrasion.

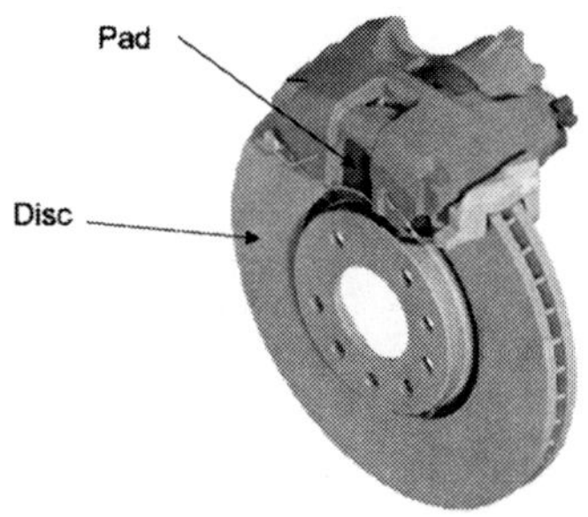

Figure 27.4. Disc brake system

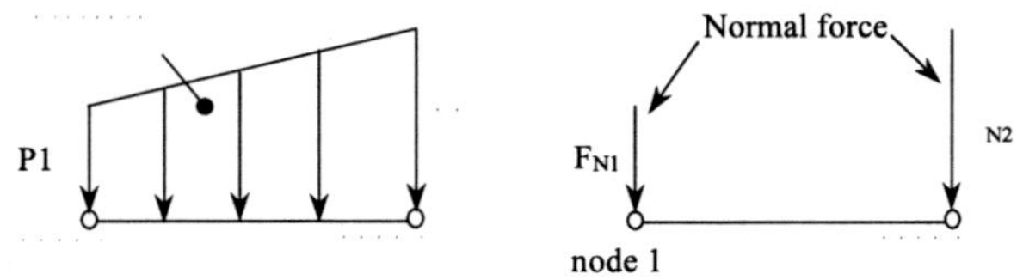

Figure 27.5. Normal pressure and normal force acting on contact elements

27.4.1 Wear Modeling

The wear resulting from adhesive wear process has a phenomenological description by the equation [ARC53]

$$W_{ad} = \frac{V}{s} = k\frac{F_N}{3H} \tag{27.33}$$

W_{ad} is the worn volume per unit sliding distance, V is the volume of the material removed by wear from surface, k is a material constant that expresses the probability of generating a wear particle (dimensionless), s is the sliding distance, H is the hardness of the sheet and F_N is the normal load applied on the tool.

If the parameters of the wear models are assumed to be constant in time, the above wear models can be obtained as:

$$V = \gamma_w F_N s \tag{27.34}$$

where γ_w denotes a wear coefficient.

Equation (27.33) employs hardness as the only material property. Typical values of the wear coefficient k for a combination of materials are given in [ARC53], [ABA62].

From a numerical point of view, at each node "i" of the contact elements of the tool mesh, the above equation can be written as:

$$V_i = (\gamma_w)_i (F_N)_i s_i \tag{27.35}$$

With use of FEM, it is easier to use normal contact pressure instead of normal force. At each node "i", the normal force $(F_N)_i$, can be obtained from the normal stress as (Fig. 27.5):

$$(F_N)_i = \int_{\Omega} (P)_i \, d\Omega_i \tag{27.36}$$

Ω_i is the area of each contact element "i" and P the normal contact pressure acting on the disc.

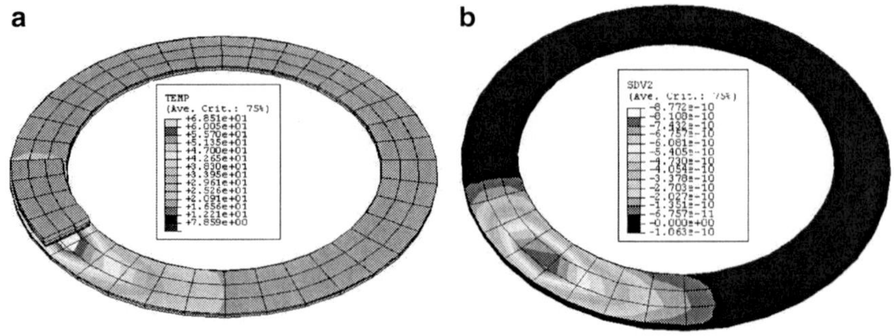

Figure 27.6. Temperature and wear contours of the disc brake

Fully coupled thermal-stress analysis is needed when the stress analysis is dependent on the temperature distribution and the temperature distribution depends on the stress solution. When contact conditions exist in some problems where the heat conducted between surfaces may depend strongly on the separation of the surfaces or the pressure transmitted across the surfaces, it is necessary to take into account such coupled procedure. For such cases the thermal and mechanical solutions must be obtained simultaneously rather than sequentially. Coupled temperature-displacement elements are provided for this purpose. More details can be found in Jensen [JEN98] about the analysis procedure used in the work.

27.4.2 Wear Analysis of a Brake Disc

The problem consists in simulating of a 360° rotation of disk brake with an inner radius of 90 mm, an outer radius of 135 mm and disk thickness of 10 mm. The pad has a 10 mm thickness.

The 3D model including brick elements can be seen Fig. 27.6. Frictional contact between the pads and the disc is modeled by contact pairs between surfaces defined on the element faces in the contact region. The pad is a resin-bonded composite friction material and the disk from steel. Although material occurs in the pad in time and because the thermal degradation, one can assume the pad has the characteristics of the unused material. The pad in the model is fixed and is pressed against the disc with a specified pressure. The second step consists in applying a prescribed rotation of the disc.

27.5 Results

Figure 27.6.a shows the temperature distribution within the disc generated by the friction. It can be observed that the temperature is at maximum level in the contacting surfaces.

Figure 27.6.b shows the wear contour for the disc-pad brake model expressed in (27.33). It can be observed that the wear profile is not uniform on the disk surface.

The wear is higher in the center of the disc as a consequence of the non-uniform pressure stress generated by the pad contact. It can be observed that the maximum wear value is 8.772×10^{-10} mm per revolution.

27.5.1 Bayesian Analysis for Brake Disc Wear

The aim of this study is to compare the classic method with Bayesian method (introduced in Sect. 27.3). The brake disc life is assumed to be described by a Wiener Process. The pseudo failure times are defined by an Inverse Normal distribution. For this purpose, a Monte Carlo simulation is used to generate random wear paths. The Wiener process and testing conditions are defined in Table 27.1.

The prior distribution is defined with the following information:

- The mean lifetime is estimated by the FEM Modeling (see Sect. 27.4): $\mu = 8.772 \times 10^{-10}$ mm per revolution.
- An expert provides an interval on the variance: $s^2 \in [3\times10^{-10}, 3.7\times10^{-10}]$ then $\theta \in [2.7\times10^9, 3.33\times10^9]$.

From these values and relationships (27.29) and (27.30), the parameters a, b and c are deducted

- $a = 274.8367$
- $b = 8.2041\times10^{-11}$
- $c = 8.772\times10^{-10}$

A series of tests is performed. In each test, the wear depth of an example brake is measured. Figure 27.7 presents the results of wear random simulation.

This measured values are used to update the *pdf* of normal distribution (Wiener process) using the equation in Sect. 27.3. The Wiener parameters are evaluated by classical and Bayesian estimations for μ and σ^2 (see Table 27.2).

With these estimations, the reliability functions (see Fig. 27.8) are computed, considering the theoretical, Bayesian, and classical approaches (defined by (27.1).

We observe that the result obtained by Bayesian estimation is closer to the theoretical distribution than classical estimation (Bayesian estimation is merged with theoretical function).

The choice of prior distribution is very important. The advantage of the proposed method is that it enables one to take into account the available knowledge and to reduce the number of tests.

Table 27.1. Simulation parameters

Physical parameters	Wear depth $=8.772\times10^{-10}$ mm per revolution; $\sigma^2 = 3.34 \times 10^{-10}$
Wiener process	$\mu = 8.772 \times 10^{-10}$ mm per revolution; $\sigma^2 = 3.34 \times 10^{-10}$
Test conditions	End time $= 1\times10^{88}$ revolutions; Number of paths = 10; Number of points for each path = 1; Critical wear $z_0 = 2$ mm

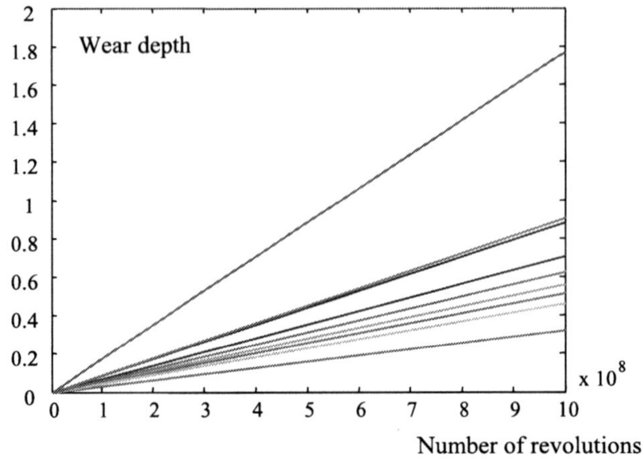

Figure 27.7. Results of wear random simulation

Table 27.2. Estimation results

	Bayesian method	Classical method
Point estimator	$\hat{\mu} = 8.7682 \times 10^{-10}$ $\hat{\sigma} = 2.4882 \times 10^{-10}$	$\hat{\mu} = 8.3094 \times 10^{-10}$ $\hat{\sigma} = 2.6169 \times 10^{-10}$

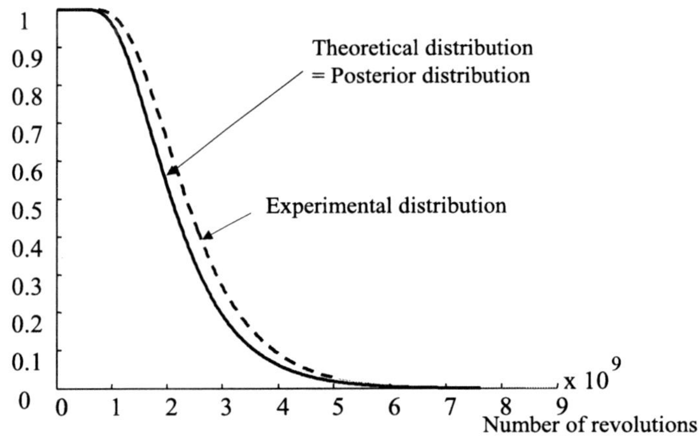

Figure 27.8. Reliability functions

27.6 Conclusion

In this paper, the Bayesian estimation of Wiener process parameters (usually used to define the wear) has been studied, in order to improve the estimation accuracy by incorporating the available knowledge on the product. In particular, the finite elements results and expert opinions have been considered prior knowledge. For life time prediction by FEM, a model based on Archard law was developed for brake disc wear.

The analysis of results obtained by numerical simulations demonstrates the efficiency of the Bayesian approach compared to the classical approach.

References

[AYY97] B.M. Ayyub and R.H. McCuen, *Probability, Statistics & Reliability for engineers*, CRC, New York, (1997).

[CHE97] K.W. Chen and A.S. Papadopoulos, Shortest bayes credibility intervals for the lognormal failure model, Journal of Microelectronic Reliability, Vol. 37, No. 12, pp. 1859–1863, (1997).

[MEE98] William Q. Meeker and Luis A. Escobar, Statistical Methods for Reliability Data, Wiley, (1998).

[SAN91] P. Sander and R. Badoux, Bayesian Methods in Reliability, Kluwer, (1991).

[COO91] R.M. Cooke, Experts in uncertainty : Expert opinion and subjective probability in science, New York, Oxford University Press, (1991).

[CON01] P. Congdon, Bayesian statistical modelling, Wiley, (2001).

[SHA07] A. Shah and B.S. Dhillon, Reliability and Availability Analysis of Three-state Device Redundant Systems with Human Errors and Common-cause Failures. International Journal of Performability Engineering, **3**, 411–418 (2007).

[SRI06] S.K. Srinivasan and R. Subramanian, Reliability analysis of a three unit warm standby redundant system with repair Annals of Operations Research, **143**, 227–235 (2006).

[AHM95] K.E. Ahmad and Z.F Jaheen, Approximate bayes estimators applied to the inverse gaussian lifetime model, Computers Math. Applic, Vol. 29, No. 12, pp. 39–47, (1995).

[COU05] V. Couallier, Degradation modeling and semi parametric estimation with non independent noises. Qualita, Bordeaux, 2005.

[ABA62] ABAQUS/Standard - HKS, theory manual, version 6.2

[ARC53] J.F. Archard, Contact and rubbing of flat surfaces, Journal of Applied Physics, **24**, 981–988, (1953).

[CHO96] C.M. Choy and R. Balendra, Experimental analysis of parameters influencing sheared-edge profiles, p: 101,110, The 5th International Conference on Sheet Metal, University of Twente, Netherland, (1996).

[CAR94] W.T. Carter, A model for friction in metal forming, Journal of Engineering Materials and Technology, **113**, 8–13, (1994).

[JEN98] M.R. Jensen, F.F. Damborg, K.B. Nielsen, and J. Danckert, Applying the finite element method for determination of tool wear in conventional deep-drawing, Journal of Materials Processing Technology, **83**, 98–105, (1998).

[NIK10] M.S. Nikulin, N. Limnios, N. Balakrishnan, W. Kahle, and C. Huber-Carol, Advances in Degradation Modeling : Applications to Reliability, Survival Analysis, and Finance, Series: Statistics for Industry and Technology, Ed Birkhäuser, (2010)

28

Benefits of Threshold Regression: A Case-Study Comparison with Cox Proportional Hazards Regression

Mei-Ling Ting Lee[1*], G. A. Whitmore[2], and Bernard Rosner[3]

[1] University of Maryland, College Park, USA, MLTLEE@UMD.EDU
[2] McGill University, Montreal, Canada
[3] Harvard Medical School, Boston, MA, USA,

Abstract: Cox proportional hazards (PH) regression is a well-known model for analyzing survival data and its strengths are widely recognized. *Threshold regression* (TR) is a relatively new methodology but one that is receiving greater attention and being used successfully by researchers in different fields, including biopharmaceutical statistics. In threshold regression, event times are modeled by a stochastic process reaching a boundary threshold. The TR model does not require the proportional hazards assumption. It also can provide more insights into data than the Cox model, even where the PH assumption holds. Thus, threshold regression deserves consideration by investigators and their analysts as a serious alternative to Cox regression. In this article, we demonstrate the benefits of the TR model using a large cohort data set drawn from the Nurses' Health Study (NHS). The TR results for the NHS data set show the anticipated link between lung cancer and smoking for women. The TR model allows this link to be understood with substantial insight and clarity and with a refined attribution of disease progression to particular influences. We compare TR results with those obtained from Cox proportional hazards regression. The adequacies of the TR and Cox models in fitting the data set are examined using a new analytical approach. We also present *Stata* programs to compare the models.

Keywords and phrases: Endpoint, First hitting time, Lifetime, Lung cancer, Maximum likelihood, Smoking, Stochastic process, Survival analysis, Threshold regression, Time-to-event, Wiener diffusion process

28.1 Introduction

Cox proportional hazards (PH) regression is a well-known model for analyzing survival data and its strengths are widely recognized. *Threshold regression* (TR) is a relatively new methodology but one that is receiving greater attention and being used successfully by researchers in different fields, including biopharmaceutical statistics. In threshold regression, event times are modeled by a stochastic process reaching a boundary

V.V. Rykov et al. (eds.), *Mathematical and Statistical Models and Methods in Reliability: Applications to Medicine, Finance, and Quality Control*, Statistics for Industry and Technology, DOI 10.1007/978-0-8176-4971-5_28,

threshold. The TR model does not require the proportional hazards assumption. It also can provide more insights into data than the Cox model, even where the PH assumption holds. Thus, threshold regression deserves consideration by investigators and their analysts as a serious alternative to Cox regression. Lee, Chang, Whitmore [LCW08] used a threshold regression mixture model for assessing treatment efficacy in a multiple myeloma clinical trial. They did not compare the benefits of TR with those of Cox PH regression.

In this article, we compare the TR and Cox models and demonstrate the benefits of the TR model using a large cohort data set drawn from the Nurses' Health Study (NHS). The TR results for the NHS data set show the anticipated link between lung cancer and smoking for women. The TR model allows this link to be understood with substantial insight and clarity and with a refined attribution of disease progression to particular influences. Specifically, we compare TR results with those obtained from Cox proportional hazards regression. The adequacies of the TR and Cox models in fitting the data set are examined using a new analytical approach. We also present *Stata* programs to compare the models.

28.2 First-Hitting Time (FHT) and Threshold Regression (TR) Model

Threshold regression (TR) refers to a statistical model for time-to-event data in which the time to the event is defined as the first hitting time of an absorbing boundary by an underlying stochastic process. In our application of the TR model, the health status of each subject with respect to lung cancer follows a latent Wiener diffusion process $\{X(t)\}$ where t denotes time measured from the baseline of an observation interval. The initial health status of the subject at baseline is $X(0) = x_0 > 0$, which is a parameter to be estimated. The mean rate of change of health status over the interval is denoted by μ. Lung cancer occurs when the health status process first decreases to the zero-level, which is taken as an absorbing boundary or *threshold* for the process. The time of this first encounter, denoted by S, is called the *first hitting time* (FHT). If $\mu > 0$ then the lung cancer endpoint is not assured because the process would tend to drift away from the threshold. Other causes of death are competing with lung cancer and, hence, death from another cause will produce a right censored observation. In threshold regression, statistical techniques are used to estimate the effects of covariates on the parameters of the FHT model. See Aalen and Gjessing [AG01, AG04, ABG08] and Lee and Whitmore [LW06] for a review of FHT models and threshold regression.

28.2.1 The Nurses Health Study

We consider a large cohort data set drawn from the Nurses' Health Study (NHS) to compare the benefits of the TR model to the Cox model. The Nurses' Health Study (NHS) was established in 1976 when a cohort of 121,700 female registered nurses, aged 30–55 years, returned a mailed questionnaire reporting on disease history, personal characteristics and behaviors, and then updated the information by completing follow-up questionnaires on a biannual basis. The study was designed to allow prospective

Table 28.1. Summary statistics for the failure variable and covariates used in the fitted threshold regression model. Covariate *pkyrs_sq* = $(pkyrs0 - 28)^2$. All variables have 115,768 readings

Variable		Statistics			
Name	Units	Mean	Std. Dev.	Min.	Max.
Response					
fail	indicator (0,1)	0.0104174		0	1
Covariates					
age0	years	52.343	7.217	39	72
pkyrs0	pack years	13.072	18.456	0	122
pkyrs_sq	pack years squared	563.471	447.294	0	8836
dpkyrs	pack years	0.138	0.346	0	4

examination of the influences of lifestyle on the occurrence of disease, especially heart disease and cancers. Every two years in follow-up questionnaires they have updated and extended these data. In this article, data from the NHS for the period 1986–2000 are considered. The data set consists of observation sequences for 115,768 women which represent 1,577,382 person-years at risk. The endpoint of interest here is a diagnosis of primary lung cancer as confirmed from medical records or death certificates. This endpoint was experienced by 1206 of the women by the year 2000. For more details about this study analyzed by the Cox PH model, see Bain, Feskanich et al. [BFSTHRC04]. We examine the link between lung cancer and smoking using the TR model and discuss the benefits of the TR model.

Our model assumes that the logarithm of initial health status $\ln(x_0)$ is a linear regression function of two covariates, namely, cumulative smoking at baseline *pkyrs0* (in pack years) and baseline age *age0* (in years). These covariates are selected for the initial health status on the assumption that initial health can only depend on conditions that prevail at baseline. The parameter μ describes the mean rate of change in health status with time. We assume that μ is a linear regression of the same covariates, *pkyrs0* and *age0*. In addition to these two covariates, we include a covariate that is an affine quadratic term for cumulative smoking, denoted by *pkyrs_sq*. The affine adjustment involves subtracting a constant from *pkyrs0* before squaring it. This adjustment reduces collinearity between the linear and quadratic terms. We have chosen the constant to be 28 pack years because this value reduces the correlation between *pkyrs0* and *pkyrs_sq* to almost zero. The quadratic term assesses the presence, if any, of a curvature effect in parameter μ for cumulative smoking. A further covariate, denoted by *dpkyrs*, is also included in the regression function for μ. This covariate represents the average annual rate of additional smoking by the subject between baseline and the endpoint (in pack years). The expectation is that this covariate will capture the influence on the rate of change in health status of continued smoking. Table 28.1 shows summary statistics for the failure indicator variable and the covariates.

28.2.2 Regression Link Functions and Sample Log-Likelihood Function

Because the health status process is latent here, it can be given an arbitrary measurement unit. Thus, in general, one parameter may be fixed and we choose to set the variance parameter σ^2 to unity. We link parameters μ and x_0 to baseline covariates that are represented by row vector $\boldsymbol{z} = (1, z_1, \ldots, z_k)$. The leading 1 in $\boldsymbol{z}$ allows for a constant term in the regression relationship. An identity link function of form

$$\mu = \boldsymbol{z\beta} = \beta_0 + \beta_1 z_1 + \ldots + \beta_k z_k$$

is our choice for the mean parameter μ. A logarithmic link function

$$\ln(x_0) = \boldsymbol{z\gamma} = \gamma_0 + \gamma_1 z_1 + \ldots + \gamma_k z_k$$

is our choice for the initial health parameter x_0. Here $\boldsymbol{\beta} = (\beta_0, \beta_1, \ldots, \beta_k)'$ and $\boldsymbol{\gamma} = (\gamma_0, \gamma_1, \ldots, \gamma_k)'$, where β_0 and γ_0 are regression constants.

We now set up the sample log-likelihood function for censored inverse Gaussian regression. We assume that censoring is uninformative. Also, we assume that any nurse who contracts primary lung cancer during the study period does so at the end of the last reporting interval. We denote μ and x_0 for the ith subject by $\mu^{(i)}$ and $x_0^{(i)}$. We let $t^{(i)}$ denote the survival time of the ith subject for whom *fail* equals 1 or the right censoring time of the ith subject for whom *fail* equals 0. Hence, a subject i for whom *fail* $= 1$ contributes probability density $f(t^{(i)}|\mu^{(i)}, x_0^{(i)})$ to the sample likelihood function, for $i = 1, \ldots, n_1$, where $n_1 = 1,206$ here. For subject i for whom *fail* $= 0$, the survival probability $\overline{F}(t^{(i)}|\mu^{(i)}, x_0^{(i)}) = 1 - F(t^{(i)}|\mu^{(i)}, x_0^{(i)})$ is the contribution to the sample likelihood function, for $i = n_1 + 1, \ldots, n_1 + n_0$. The sum $n = n_1 + n_0$, which equals 115,768 here, is the total number of subjects. Note that the variance parameter has been set to 1 and, hence, is suppressed in the preceding notation. The sample log-likelihood function to be maximized therefore has the form:

$$\ln L(\boldsymbol{\beta}, \boldsymbol{\gamma}) = \sum_{i=1}^{n_1} \ln f(t^{(i)}|\mu^{(i)}, x_0^{(i)}) + \sum_{i=n_1+1}^{n_1+n_0} \ln \overline{F}(t^{(i)}|\mu^{(i)}, x_0^{(i)}). \tag{28.1}$$

Numerical gradient methods can be used to find maximum likelihood estimates for $\boldsymbol{\beta}$ and $\boldsymbol{\gamma}$ and estimates of their asymptotic standard errors. We have used a numerical optimization routine in *Stata* for this purpose.

28.3 Threshold Regression (TR) Investigations of Lung Cancer

Basic Regression Output

Table 28.2 shows output for our chosen regression model. A parsimonious model has been chosen to avoid overfitting and to simplify the interpretation of effects. For parameter $\ln(x_0)$, representing the logarithm of the initial health level, it is seen that covariate *pkyrs0* is significant, with a P-value of 0.000. Its regression coefficient is negative, signifying that baseline health status (with respect to lung cancer) tends to be lower for subjects with a larger amount of cumulative smoking at baseline. In other words, heavier smokers tend to be closer to a diagnosis of primary lung cancer. The covariate *age0* has a negative regression coefficient but it is not significant with the conventional 0.05 rule (a P-value of 0.066). For the mean parameter μ, the regression

Table 28.2. Threshold regression output for a model in which parameter $\ln(x_0)$ depends on baseline age *age0* and baseline cumulative smoking *pkyrs0*. Parameter μ depends on the same covariates as well as an affine quadratic term for cumulative smoking *pkyrs_sq* and a covariate *dpkyrs* which represents the average annual smoking rate of the subject between baseline and the endpoint

Parameter	Variable	Estimate	Std. Error	P-value
$\ln(x_0)$				
	age0	−.0033470	.0018225	0.066
	pkyrs0	−.0030204	.0004752	0.000
	constant	1.792918	.1008379	0.000
μ				
	age0	−.0106827	.0012021	0.000
	pkyrs0	−.0036346	.0003566	0.000
	pkyrs_sq	.0000508	.0000058	0.000
	dpkyrs	−.1457822	.0086916	0.000
	constant	1.146989	.0674920	0.000

coefficients of all covariates are significant with P-values of 0.000. The coefficients of *age0*, *pkyrs0*, and *dpkyrs* are all negative, indicating the adverse effects on lung cancer health status of baseline age, baseline cumulative smoking, and continued smoking after baseline. The regression coefficient of *pkyrs_sq* for cumulative smoking is positive. The combined linear and curvature effects for cumulative smoking suggest that heavier smoking is increasingly harmful to health but that the rate of increase moderates slightly with the amount of smoking.

Estimated Risks of Developing Primary Lung Cancer

The fitted threshold regression model provides estimates for the absolute risk or probability of developing primary lung cancer for women of different baseline ages and smoking habits. Women in the cohort have been followed for at most sixteen years so the fitted model can only provide reliable estimates within a forward time horizon of about 15 years. Figure 28.1 shows the absolute risk of developing primary lung cancer within the next five years (panel a) and the next ten years (panel b) at different baseline ages *age0* for three smoking profiles: (1) a nonsmoker, (2) a smoker who has smoked one pack each day since age 18 and who will continue to smoke at the same rate, and (3) a smoker who has smoked two packs each day since age 18 and who will continue to smoke at the same rate. The figure shows the small risk of developing primary lung cancer for nonsmokers and the much greater risks for smokers, with the risk escalating with heavier smoking and advancing years. For the smoker of two packs per day, for example, the risk for the next decade of life rises to about 20% for women who are over 60 years old at baseline. The probabilities in Fig. 28.1 take no account of competing risks of death and therefore will be larger than recorded mortality rates for lung cancer. It is clear that a potential death from lung cancer will not be observed and recorded for women who happen to die of other causes before the specified time horizon.

Another comparison of risks is offered by Fig. 28.2. The figure shows estimated lung cancer survival functions over a 20-year horizon for a 45-year old nurse for four smoking profiles: (1) a nonsmoker, (2) a smoker who has smoked one pack each day since age 20 and who will continue to smoke at the same rate, (3) a smoker who has smoked

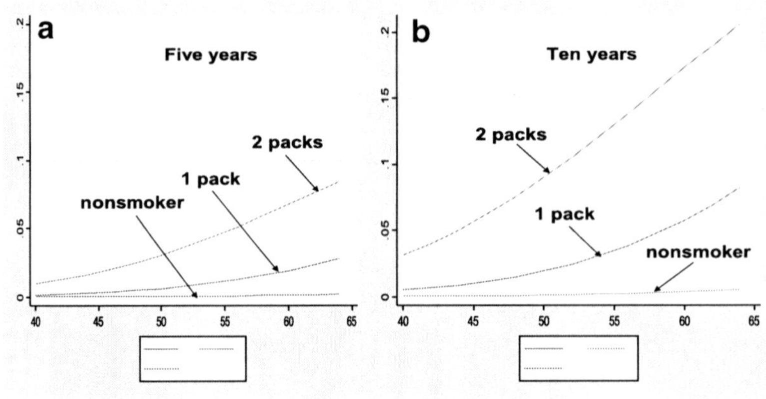

Figure 28.1. The absolute risk (probability) of developing primary lung cancer within the next five years (panel a) and the next ten years (panel b) at different baseline ages *age0* for three smoking profiles: (1) a nonsmoker, (2) a smoker who has smoked one pack each day since age 18 and who will continue to smoke at the same rate, and (3) a smoker who has smoked two packs each day since age 18 and who will continue to smoke at the same rate. The probabilities take no account of competing risks of death

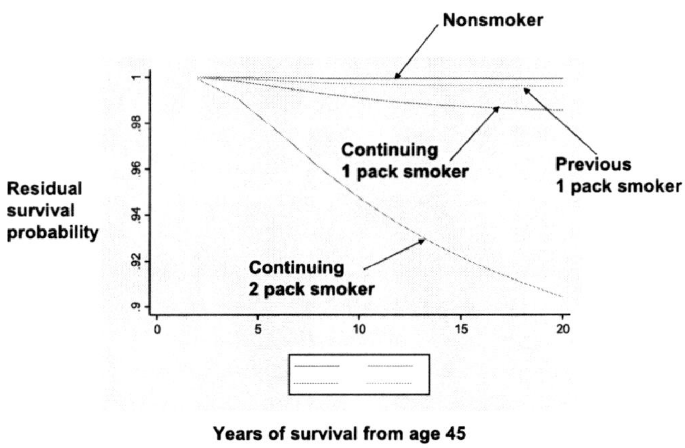

Figure 28.2. Estimated lung cancer survival functions over a 20-year horizon for a 45-year old nurse for four smoking profiles: (1) a nonsmoker, (2) a smoker who has smoked one pack each day since age 20 and who will continue to smoke at the same rate, (3) a smoker who has smoked two packs each day since age 20 and who will continue to smoke at the same rate, and (4) a smoker who has smoked one pack each day since age 20 but quits smoking at age 45. The survival curves take no account of competing risks of death

two packs each day since age 20 and who will continue to smoke at the same rate, and (4) a smoker who has smoked one pack each day since age 20 but quits smoking at age 45. The survival curves take no account of competing risks of death. As expected, survival prospects are worst for the continuing smoker who has a two-pack per day habit. Her probability of developing primary lung cancer reaches close to 10% at the 20-year horizon when she would be 65 years of age.

A final comparison of risks is offered by hazard functions. Figure 28.3 presents plots of hazard functions for three smoking profiles: (1) a nonsmoker, (2) a smoker who has

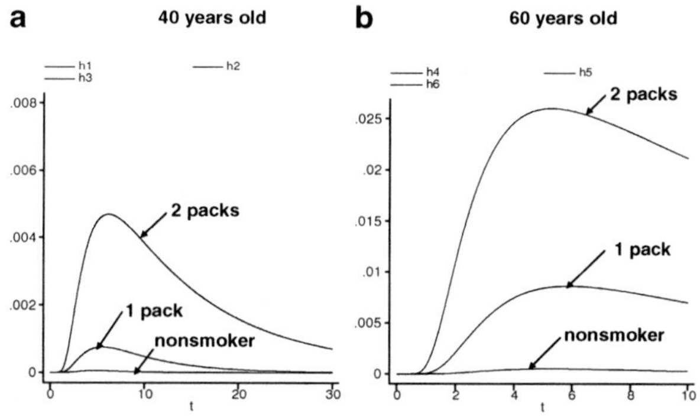

Figure 28.3. Plots of hazard functions for three smoking profiles: (1) a nonsmoker, (2) a smoker who has smoked one pack each day since age 18 and who will continue to smoke at the same rate, and (3) a smoker who has smoked two packs each day since age 18 and who will continue to smoke at the same rate. The functions in panel (**a**) span the 30-year period from a baseline age of 40 until age 70. Those in panel (**b**) span the 10-year period from a baseline age of 60 until age 70

smoked one pack each day since age 18 and who will continue to smoke at the same rate, (3) a smoker who has smoked two packs each day since age 18 and who will continue to smoke at the same rate. Panels (a) and (b) show these functions for nurses who are 40 and 60 years old at baseline, respectively. Observe that scales of the graphs are not comparable. The time horizon in each panel is 70 years of age and, thus, the hazard window in panel (a) is 30 years while that in panel (b) is 10 years. Also, the hazard levels in panel (b) are much larger than in panel (a) because the risk increases sharply with age.

A comparison of the ratios of the hazard functions for different smoking profiles over the age ranges presented in Fig. 28.3 shows that the hazard functions are far from proportional. This observation is relevant for our comparison of TR and Cox proportional hazard regression that we address in a later section.

28.4 Comparisons of Results Obtained from the Cox PH Model

In this article, we compare TR results with those from the Cox proportional hazards regression model for the case of fixed covariates. The Cox model is the conventional one for this kind of application in time-to-event analysis – see, for example, Kalbfleisch and Prentice [KP80] and Cox and Oakes [CO84]. Comparisons of the TR model with the Cox regression for longitudinal data with time-dependent covariates will be discussed in a subsequent article.

$$h(t|\boldsymbol{\zeta z}) = h_0(t)\exp(\boldsymbol{\zeta z}) \tag{28.2}$$

Here, $h(t|\boldsymbol{\zeta z})$ is the hazard function of a subject with covariate vector $\boldsymbol{z}$, $h_0(t) = h(t|0)$ is an arbitrary baseline hazard function, and $\boldsymbol{\zeta}$ is a vector of regression coefficients.

We have fitted the Cox model (28.2) to the data using the same covariates as for the TR model, namely, baseline age *age0*, baseline cumulative smoking *pkyrs0*, an affine quadratic term for cumulative smoking *pkyrs_sq*, and the average annual smoking rate of the subject between baseline and the endpoint *dpkyrs*. The regression

Table 28.3. Cox proportional hazards regression results for the study using covariates: baseline age *age0*, baseline cumulative smoking *pkyrs0*, an affine quadratic term for cumulative smoking *pkyrs_sq*, and the average annual smoking rate of the subject between baseline and the endpoint *dpkyrs*

Variable	Estimate	Std. Error	P-value
age0	.0967993	.005142	0.000
pkyrs0	.0434696	.0019047	0.000
pkyrs_sq	−.0005211	.000041	0.000
dpkyrs	1.0036	.0564842	0.000

results appear in Table 28.3. The signs of the regression coefficients for the covariates in Table 28.2 are the reverse of those in Table 28.3, which confirms that the effects for the covariates are in agreement with respect to the direction of effect. The signs of the regression coefficients in Table 28.3 show increasing hazard with increasing *age0*, *pkyrs0*, and continued smoking *dpkyrs*. The quadratic effect for cumulative smoking moderates the linear effect. Direct comparisons of the actual magnitudes of the coefficients are not meaningful, however, because they represent effects on parameters in completely different models.

28.4.1 Checking Model Fit

As a check on the TR model fitted in Table 28.2, we have examined the difference between actual and fitted lung cancer outcomes at different baseline ages *age0*. To compute the differences, we consider all subjects with a given baseline age a, where a ranges over 40–65 years. If $Y_a^{(i)}$ is an indicator variable for development of primary lung cancer for subject i of that age and $P_a^{(i)}$ is the true survival probability for the observation interval then the difference $Y_a^{(i)} - (1 - P_a^{(i)})$ has expected value 0 and variance $P_a^{(i)}(1 - P_a^{(i)})$. Our model is presumed to estimate the survival probability $P_a^{(i)}$ without bias. To check this claim, we have summed the differences $Y_a^{(i)} - (1 - \hat{P}_a^{(i)})$ at each distinct year of baseline age a, where $\hat{P}_a^{(i)}$ denotes the estimate of $P_a^{(i)}$. The approximate variance of this sum for each age is calculated as the sum of the individual variances based on the assumption that the sum components are independent. Independence is a reasonable approximation as the estimation errors for the parameters impart little dependence to the $\hat{P}_a^{(i)}$. The ratio of the sum of differences to its standard deviation at each age should be (approximately) a random standard normal number if the chosen TR model is correct. These ratios are plotted in panel (a) of Fig. 28.4, with the points connected by straight lines to assist visual interpretation. The plot shows a zigsaw pattern that appears random and unbiased with no outliers. For a comparison, panel (b) of Fig. 28.4 shows the standardized excess of actual lung cancer cases over predicted lung cancer cases for the fitted Cox regression model in Table 28.3. The zigsaw patterns are almost indistinguishable, showing that both models provide equally good fits to actual lung cancer outcomes as a function of baseline age.

The check on fit was repeated with covariate *age0* replaced by the covariate *pkyrs0*. Figure 28.5 shows the resulting plots. Again we see similar fits, except for the region where *pkyrs0* exceeds about 100 pack years. In this upper region, the Cox model is biased as shown by the standardized excess being quite large for several points. Both graphs show one major outlier that happens to occur at the point where *pkyrs0* is 106.

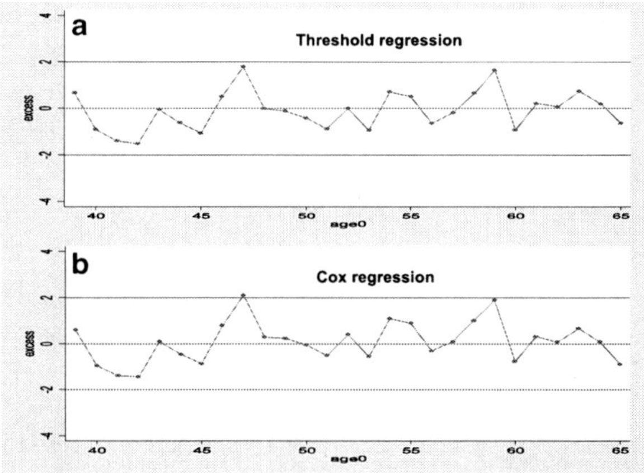

Figure 28.4. Ratios representing the standardized excess of actual lung cancer cases over predicted lung cancer cases for each baseline year *age0* from 40 to 65 for (**a**) the fitted TR regression model in Table 28.2 and (**b**) the fitted Cox regression model in Table 28.3

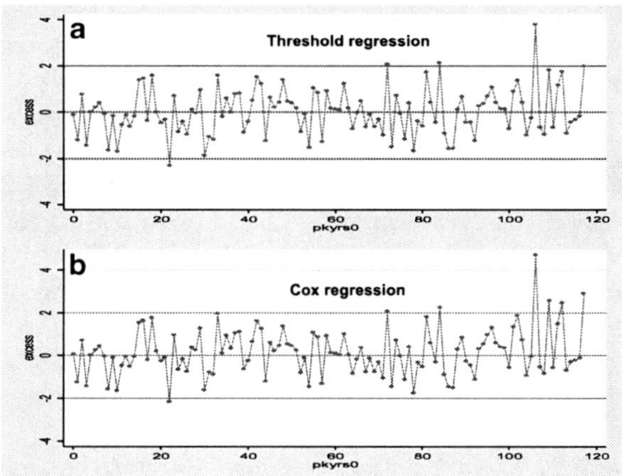

Figure 28.5. Ratios representing the standardized excess of actual lung cancer cases over predicted lung cancer cases for each baseline cumulative pack years of smoking *pkyrs0* for (**a**) the fitted TR regression model in Table 28.2 and (**b**) the fitted Cox regression model in Table 28.3

A global test of the proportional hazards assumption was also performed. The result is a chi-square statistic of 1.59 for df $= 4$, giving a P-value of 0.811. The finding suggests the assumption is quite adequate for this application.

28.4.2 Benefits of Threshold Regression Over Cox PH Regression

As noted already, the Cox proportional hazards (PH) regression model has been the model of choice for many studies involving time-to-event and survival data. We have seen that in spite of their different mathematical structures, threshold regression and Cox regression give qualitatively similar findings and similar fits to the data in this study. We quickly add that this similarity is not assured in other settings. Yet, in this setting, the clear similarity leads to the obvious question of why an alternative to Cox regression should be considered. We certainly recognize the strengths of Cox regression and, where its assumptions are valid, it should be used. On the other hand, there are benefits to threshold regression that should be considered by investigators and their analysts.

1. It can be shown that variants of the first hitting time model can be constructed that do have the PH property. Thus, adopting a threshold regression framework may enrich the interpretation of a Cox regression application. For example, setting Cox regression within a first hitting time context, if that is appropriate, can give a meaningful interpretation to the baseline hazard function.
2. Where a first hitting time model is appropriate and its survival functions do not have the PH property then threshold regression finds immediate application and the Cox model is disqualified. The inverse Gaussian survival distribution that is implicit in our TR regression application here does not possess the PH property. Yet, the Cox model and TR model do not differ sufficiently over the range of data to be statistically distinct. Women in this cohort have been monitored for only 16 years at most, which is not a long survival window.
3. The TR model is actually more parsimonious than the Cox regression model. The TR model is fully parametric and, in this application, has two set of coefficients for covariate effects, namely, those associated with the $\ln(x_0)$ and μ parameters. In contrast, Cox regression is a semi-parametric procedure because the baseline hazard function is arbitrary. The Cox model is rich in the parameters that define the baseline hazard function $h_0(t)$ and numerous degrees of freedom are absorbed in estimating that function, although this fact is not explicit in estimation routines based on the partial likelihood approach. Thus, the Cox model (28.2) involves estimating the regression coefficient vector $\boldsymbol{\zeta}$ as well as the baseline hazard function $h_0(t)$. Although the latter is often viewed as being of secondary interest, it deserves more attention than it receives. Criticism is sometimes leveled at investigators who use Cox regression without examining or attempting to understand the nature of its unspecified baseline hazard function.
4. As noted already, TR formulations force investigators to consider the actual causal mechanism of survival. Is a first hitting time involved? If so, what is the parent process? What is the nature of the absorbing boundary? What is the appropriate regression structure for each parameter? Which covariates affect initial health status $\ln(x_0)$ and which influence the mean parameter μ that determines the course of disease progression after baseline. What are the relative magnitudes and directions of these influences? Threshold regression answers these important questions that are aimed at the scientific foundation of the analysis. In contrast, Cox regression provides only a regression structure for the log-hazard ratio without forcing investigators to dig deeper.
5. Most standard software packages contain a handy Cox regression routine. Some investigators may worry that the TR method will require a major investment in programming. We can allay this worry with two remarks. First, TR requires little programming. For example, in *Stata* software, the programs used for this study involve only the following core program lines for TR and Cox regression.

1. Threshold regression for the inverse Gaussian model
```
ml model lf tr_mle
(lnx0: age0 pkyrs0)
(m: age0 pkyrs0 pkyrs_sq dpkyrs)
```

2. Cox regression with fixed covariates
```
stset t, failure(f)
stcox age0 pkyrs0 pkyrs_sq dpkyrs
```

The Cox subroutine is built into *Stata* while TR uses a subroutine (called *tr_mle* here) to compute the sample log-likelihood function in (28.1). The latter is not complicated in *Stata* as it only involves the normal density and distribution functions given earlier. Second, even this amount of programming is not required for TR as a new *Stata* command called *threg* has been developed for doing the kind of TR analysis described here. The command *threg* may be found at the website *http://sph.umd.edu/epib/faculty/mltlee/test/trprograms.html*.
The TR command estimates regression coefficients of a TR model based on the first-hitting-time of a Wiener diffusion process. The command uses a maximum likelihood estimation routine in *Stata* for calculating estimates of regression coefficients, asymptotic standard errors and *P*-values. Optionally, the routine can provide estimates of hazard ratios at selected time points for specified scenarios and plots of estimated hazard functions, survival functions and probability density functions of first-hitting-times.

Acknowledgement. This research is supported in part by National Institutes of Health grant HL40619 (Rosner), OH008649 (Lee), and by a research grant from the Natural Sciences and Engineering Research Council of Canada (Whitmore).

References

[BFSTHRC04] Bain C, Feskanich D, Speizer FE, Thun M, Hertzmark E, Rosner BA, Colditz GA (2004). Lung cancer rates in men and women with comparable histories of smoking, *Journal of the National Cancer Institute*, **96 (11)**, June 2.

[AG01] Aalen OO, Gjessing HK (2001). Understanding the shape of the hazard rate: a process point of view, *Statistical Science*, **16**, 1–22.

[AG04] Aalen OO, Gjessing HK (2004). Survival models based on the Ornstein-Uhlenbeck process, *Lifetime Data Analysis*, **10**, 407–423.

[ABG08] Aalen OO, Borgan A, Gjessing HK (2008). *Survival and Event History Analysis: A Process Point of View*. New York: Springer.

[LW06] Lee M-LT, Whitmore GA (2006). Threshold regression for survival analysis: Modeling event times by a stochastic process reaching a boundary, *Statistical Science*, **21**, 501–513.

[LCW08] Lee M-LT, Chang M, Whitmore GA (2008). A threshold regression mixture model for assessing treatment efficacy in a multiple myeloma clinical trial, *Journal of Biopharmaceutical Statistics*, **18**, 1136–1149.

[LW10] Lee M-LT, Whitmore GA (2010). Proportional hazards and threshold regression: their theoretical and practical connections. *Lifetime Data Analysis*, **16**, 196–214.

[LWR10] Lee M-LT, Whitmore GA, Rosner B (2010). Threshold regression for survival data with time-varying covariates. *Statistics in Medicine*, **29**, 896–905.

[CF89] Chhikara RS, Folks JL (1989). *The Inverse Gaussian Distribution: Theory, Methodology and Applications*, New York: Marcel Dekker.

[KP80] Kalbfleisch JD, Prentice RL (1980). *The Statistical Analysis of Failure Time Data*, Wiley.

[CO84] Cox DR, Oakes D (1984). *Analysis of Survival Data*, Chapman and Hall.

29

Optimal Stopping and Reselling of European Options

Robin Lundgren[1*] and Dmitrii S. Silvestrov[2]

[1] Mälardalen University, Västerås, Sweden, robin.lundgren@mdh.se
[2] Stockholm University, Stockholm, Sweden, silvestrov@math.su.se

Abstract: We consider the problem of optimal reselling of European options. A bivariate exponential diffusion process is used to describe the reselling model. In this way, the reselling problem is imbedded to the model of finding optimal reward for American type option based on this process. Convergence results are formulated for optimal reward functionals of American type options for perturbed multi-variate Markov processes. An approximation bivariate tree model is constructed and convergence of optimal expected reward for this tree model to the optimal expected reward for the corresponding reselling model is proved.

Keywords and phrases: American option, Binomial–trinomial approximation, Convergence, European option, Optimal stopping, Reselling problem

29.1 Introduction

European options can only be exercised at maturity; however, there exists the possibility for the holder to sell the option on the second hand market. The question then arises at which moment of time is it optimal for the holder to sell the option, this is the reselling problem.

We use the classical geometric Brownian motion to model the price process and an exponential mean reverse Ornstein–Uhlenbeck process correlated with the price process to describe stochastic dynamics for implied volatility. We also assume that a market price for option is given by the Black–Scholes formula where implied volatility is used instead its initial value.

The problem of optimal reselling of European option is treated as the problem of finding optimal expected reward for American type option for this bivariate exponential diffusion process with asset price and implied volatility components.

The reselling model considered in this paper has been introduced in the recent paper Lundgren et al. [LSK08]. In present paper, we essentially improve results of this paper and give the complete solution of an approximation problem that is to build up an effective approximation algorithms for evaluation of optimal reward functionals for the reselling model.

V.V. Rykov et al. (eds.), *Mathematical and Statistical Models and Methods in Reliability: Applications to Medicine, Finance, and Quality Control*, Statistics for Industry and Technology, DOI 10.1007/978-0-8176-4971-5_29,

The paper contains five sections. In Sect. 29.2, we give exact formulation of reselling problem. In Sect. 29.3, we present general convergence results for reward functional of American type options for perturbed multivariate exponential Markov price processes. We refer to the recent paper Lundgren and Silvestrov [LS09], where one can find the corresponding proofs. In Sect. 29.4, we construct the bivariate binomial–trinomial model approximating the bivariate diffusion process used to describe the reselling model and prove convergence of the optimal expected rewards for this tree model and the corresponding bivariate diffusion process. In Sect. 29.5, we give numerical examples for the behavior of the reselling reward given by the algorithm presented in Sect. 29.4.

In conclusion, we also would like to mention works related to the subject of the present paper. For recent general results about optimal stopping we refer to the papers by Dayanik and Karatzas [DK03], Henderson and Hobson [HH08], Ekström et al. [ELTW09]. Optimal stopping problems for American type options have been studies in the works by Jacka [Jac91], Kim [Kim90], Peskir and Shiryaev [PS06], in Zhang and Lim [ZL06] for models with stochastic volatility, which concept were introduced in Stein, E. and Stein, J. [StSt91], Heston [Hes93], Schöbel and Zhu [SZ99], in Gau et al. [GHS00] for American barrier options, in Lundgren [Lun07] for generalized American knock out option, in Shepp and Shiryaev [SS93] for Russian options, and in Xia and Zhou [XZ07] for related stock loans models. A simpler reselling model was considered in Kukush et al. [KMS06], where an usual geometric Brownian motion was used as a model for implied volatility. Convergence of option rewards have been studied in works Amin and Khanna [AK94], Coquet and Toldo [CT07], Dupuis and Wang [DW05], Jönsson [Jon01], [Jon05], Jönsson et al. [JKS04], [JKS05], Kukush and Silvestrov [KS00], [KS01], [KS04], Prigent [Pri03], Silvestrov et al. [SGM01], [SGS99], [SJS08a], [SJS08b].

29.2 Formulation of the Reselling Problem

We consider the geometric Brownian motion as a price process given by the stochastic differential equation

$$\mathrm{d}\ln S(t) = \mu \mathrm{d}t + \sigma \mathrm{d}W_1(t),\ t \geqslant 0, \tag{29.1}$$

where $\mu \in \mathbb{R}, \sigma > 0$; $W_1(t)$ is a standard Brownian motion, and the initial state $S(0) = s_0 > 0$ is a constant.

It is also assumed that the continuously compounded interest model with a riskless interest rate $r > 0$ is used.

In this case, the price (at moment t and under condition that $S(t) = S$) for a European option, with the strike price $K > 0$ and maturity $T > 0$, is given by the Black–Scholes formula,

$$C(t, S, \sigma) = SN(d_t) - K\mathrm{e}^{-r(T-t)}N(d_t - \sigma\sqrt{T-t}), \tag{29.2}$$

where $d_t = \frac{\ln(S/K)+r(T-t)}{\sigma\sqrt{T-t}} + \frac{\sigma\sqrt{T-t}}{2}$, $N(x) = \frac{1}{\sqrt{2\pi}}\int_{-\infty}^{x} \mathrm{e}^{-y^2/2}\mathrm{d}y$.

The theoretical volatility σ in (29.2) is usually some kind of historical volatility. The theoretical price of the European option at moment t is given by formula (29.2), where S should be replaced the price value $S(t)$. However, it is well known that the market price of European option at moment t deviates from the theoretical price. In fact, the market price randomly oscillates around the theoretical price. In this case, the implied volatility $\sigma(t)$ corresponding to the market price is used. It, respectively, randomly oscillates around σ. A reverting character of these oscillations makes it natural to use a model given by a mean reverting Ornstein–Uhlenbeck process for modeling of stochastic dynamics of implied volatility,

$$d(\ln \sigma(t) - \ln \sigma) = -\alpha(\ln \sigma(t) - \ln \sigma)dt + \nu dW_2(t), \; t \geqslant 0, \tag{29.3}$$

where $\alpha, \nu > 0$, $W_2(t)$ is also a standard Brownian motion, and the boundary condition is $\sigma(0) = \sigma$. It is also naturally to assume that the process $\mathbf{W}(t) = (W_1(t), W_2(t))$ is the bivariate Brownian motion with correlated components, i.e., $\mathsf{E}W_1(t)W_2(t) = \rho t, \; t \geqslant 0$, where $\rho \in [-1, 1]$. Note that the process $(S(t), \sigma(t))$ is a diffusion process.

The model for stochastic volatility described above possesses necessary reverting properties and, at the same time, let one get the explicit solution for the corresponding system of stochastic differential equations suitable for effective constructing of approximation tree models.

The use of the market price $C(t, S(t), \sigma(t))$ is an approach used in practice. In this case, $C(t, S, \sigma)$ may be interpreted as some kind of utility reward function commonly recognized and used by market agents for evaluation of market option prices in the extended, as above, Black–Scholes model.

The use of market option prices actualizes the problem of reselling for European options. In this case, it is assumed that an owner of an option can resell the option at some stopping time from the class $\mathcal{M}_T$ which includes all stopping times $0 \leqslant \tau \leqslant T$ that are Markov moments with respect to the filtration $\mathcal{F}_t = \sigma((S(s), \sigma(s)), s \leqslant t)$ generated by the vector process $(S(t), \sigma(t))$. It is worth to note that the process $\sigma(t)$ is indirectly observable as an implied volatility corresponding to the observable market price of an option. The object of our studies is the reward functional,

$$\Phi(\mathcal{M}_T) = \sup_{\tau \in \mathcal{M}_T} \mathsf{E}\mathrm{e}^{-r\tau} C(\tau, S(\tau), \sigma(\tau)). \tag{29.4}$$

Thus, the problem of reselling the European option is imbedded in the problem of optimal execution of American type option with the pay-off function $\mathrm{e}^{-rt}C(t, S, \sigma)$ for the two-dimensional process $(S(t), \sigma(t))$.

It should also be noted that the problem is considered not before an option is bought but under the assumption that the option is already bought. In the former case, the risk-neutral considerations are actual for evaluation the fair price of the option. In the latter case, the owner is only interested in finding the optimal expected reward for reselling the option. In the this case, his estimation of the actual trend of the price process should be involved is the analysis. This analysis is realized in the paper.

Our approach is based on the approximation of process $(S(t), \sigma(t))$ by a properly fitted bivariate binomial–trinomial model. This approach requires to solve three problems. First, appropriate results concerning convergence for reward functional of American type options should be developed for multivariate Markov price processes. Second, the bivariate binomial–trinomial model satisfying the corresponding recombination conditions and a polynomial (quadratic for bivariate trees) rate of growth of the number of nodes as a function of the number of steps should be constructed. Third, the conditions of convergence for the reward functionals mentioned above should be verified.

29.3 Convergence of Rewards for Multivariate Markov Price Processes

To show the convergence of the bivariate binomial–trinomial tree we apply the results regarding convergence of optimal rewards for American type options for multivariate Markov price processes and multivariate price processes with independent increments given in the recent paper Lundgren and Silvestrov [LS09]. We summarize in this section the results we need and refer to this paper, where one can find the corresponding proofs.

For every $\varepsilon \geqslant 0$, let $\mathbf{Y}^{(\varepsilon)}(t) = (Y_1^{(\varepsilon)}(t), \ldots, Y_k^{(\varepsilon)}(t))$, $t \geqslant 0$ be a càdlàg Markov process with the phase space $\mathbb{R}^k$ and transition probabilities $P^{(\varepsilon)}(t, \mathbf{y}, t+s, A)$ and a constant initial state $\mathbf{Y}^{(\varepsilon)}(0) = \mathbf{y}_0 \in \mathbb{R}^k$. We interpret $\mathbf{Y}^{(\varepsilon)}(t)$ as a vector log-price process. Now, we define a vector price process $\mathbf{S}^{(\varepsilon)}(t) = (S_1^{(\varepsilon)}(t), \ldots, S_k^{(\varepsilon)}(t))$, $t \geqslant 0$ with the phase space $\mathbb{R}_+^k = \mathbb{R}_+ \times \cdots \times \mathbb{R}_+$, where $\mathbb{R}_+ = (0, \infty)$. Let us use the notation $\mathrm{e}^{\mathbf{y}} = (\mathrm{e}^{y_1}, \ldots, \mathrm{e}^{y_k}), \mathbf{y} = (y_1, \ldots, y_k) \in \mathbb{R}^k$. The price process $\mathbf{S}^{(\varepsilon)}(t)$ and the log-price process $\mathbf{Y}^{(\varepsilon)}(t)$ are connected by the relation, $\mathbf{S}^{(\varepsilon)}(t) = \mathrm{e}^{\mathbf{Y}^{(\varepsilon)}(t)}$, $t \geqslant 0$. Due to the one-to-one mapping and continuity property of exponential function, $\mathbf{S}^{(\varepsilon)}(t)$ is also a càdlàg Markov process.

Let $g(t, \mathbf{s})$, $(t, \mathbf{s}) \in \mathbb{R}_+ \times \mathbb{R}_+^k$ be a pay-off function. We assume that $g(t, \mathbf{s})$ is a real valued Borel measurable function. Note that we do not assume pay-off functions to be non-negative. The first condition assumes the absolute continuity of pay-off functions and imposes power type upper bounds on their partial derivatives:

$\mathbf{A_1}$: (**a**) Function $g(t, \mathbf{s})$ is absolutely continuous in t with respect to the Lebesgue measure on $[0, T]$ for every fixed $\mathbf{s} \in \mathbb{R}_+^k$ and in $\mathbf{s}$ with respect to the Lebesgue measure on $\mathbb{R}_+^k$ for every fixed $t \in [0, T]$; (**b**) For every $\mathbf{s} \in \mathbb{R}_+^k$, the partial derivative $\left|\frac{\partial g(t,\mathbf{s})}{\partial t}\right| \leqslant R_1 + R_2 \sum_{j=1}^k s_j^{\gamma_0}$ for almost all $t \in [0, T]$ with respect to the Lebesgue measure on $[0, T]$, where $0 \leqslant R_1, R_2 < \infty$ and $\gamma_0 \geqslant 0$; (**c**) For every $t \in [0, T]$, the partial derivative $\left|\frac{\partial g(t,\mathbf{s})}{\partial s_m}\right| \leqslant R_3 + R_4 \sum_{j=1}^k s_j^{\gamma_m}$ for almost all $\mathbf{s} \in \mathbb{R}_+^k$ with respect to the Lebesgue measure on $\mathbb{R}_+^k$, where $0 \leqslant R_3, R_4 < \infty$ and $\gamma_1, \ldots, \gamma_k \geqslant 0$, $m = 1, \ldots, k$. (**d**) For every $t \in [0, T]$, the function $g(t, \mathbf{0}) = \overline{\lim}_{\mathbf{s} \to \mathbf{0}} g(t, \mathbf{s}) \leqslant R_5$, where $0 \leqslant R_5 < \infty$.

It is useful to note that condition $\mathbf{A_1}$ implies that the function $g(t, \mathbf{s})$ is continuous in $(t, \mathbf{s}) \in [0, T] \times \mathbb{R}_+^k$.

Let $\mathcal{F}_t^{(\varepsilon)} = \sigma(\mathbf{Y}^{(\varepsilon)}(s), s \leqslant t)$ be the natural filtration of σ-fields, associated with the vector log-price process $\mathbf{Y}^{(\varepsilon)}(t), t \geqslant 0$. It is useful to note that this filtration coincides with the natural filtration generated by the price process $\mathbf{S}^{(\varepsilon)}(t), t \geqslant 0$.

We consider Markov moments $\tau^{(\varepsilon)}$ with respect to the filtration $\mathcal{F}_t^{(\varepsilon)}, t \geqslant 0$. It means that $\tau^{(\varepsilon)}$ is a random variable which takes values in $[0, \infty]$ and with the property $\{\omega : \tau^{(\varepsilon)}(\omega) \leqslant t\} \in \mathcal{F}_t^{(\varepsilon)}, t \geq 0$. Let $\mathcal{M}_{\max,T}^{(\varepsilon)}$ be the class of all Markov moments $\tau^{(\varepsilon)} \leqslant T$, where $T > 0$, and consider a class of Markov moments $\mathcal{M}_T^{(\varepsilon)} \subseteq \mathcal{M}_{\max,T}^{(\varepsilon)}$. The main object of our studies is the reward functional, that is, the maximal expected pay-off over different classes of Markov moments, $\mathcal{M}_T^{(\varepsilon)}$,

$$\Phi(\mathcal{M}_T^{(\varepsilon)}) = \sup_{\tau^{(\varepsilon)} \in \mathcal{M}_T^{(\varepsilon)}} \mathsf{E} g(\tau^{(\varepsilon)}, \mathbf{S}^{(\varepsilon)}(\tau^{(\varepsilon)})). \tag{29.5}$$

We use notations $\mathsf{E}_{\mathbf{y},t}$ and $\mathsf{P}_{\mathbf{y},t}$ for expectation and probability calculated under condition that $\mathbf{Y}^{(\varepsilon)}(t) = \mathbf{y}$. For $\beta, c, T > 0, i = 1, \ldots, k$, define the exponential moment modulus of compactness for the càdlàg process $Y_i^{(\varepsilon)}(t), t \geqslant 0$,

$$\Delta_\beta(Y_i^{(\varepsilon)}(\cdot), c, T) = \sup_{0 \leqslant t \leqslant t+u \leqslant t+c \leqslant T} \sup_{\mathbf{y} \in \mathbb{R}^k} \mathsf{E}_{\mathbf{y},t}(\mathrm{e}^{\beta |Y_i^{(\varepsilon)}(t+u) - Y_i^{(\varepsilon)}(t)|} - 1).$$

We use the following conditions for exponential moment modulus of compactness for log-price processes:

C$_1$: $\lim_{c \to 0} \overline{\lim}_{\varepsilon \to 0} \sum_{i=1}^k \Delta_\beta(Y_i^{(\varepsilon)}(\cdot), c, T) = 0$ for some $\beta > \gamma = \max(\gamma_0, \gamma_1 + 1, \ldots, \gamma_k + 1)$, where γ_0 and $\gamma_1, \ldots, \gamma_k$ are the parameters introduced in condition **A$_1$**,

Condition **C$_1$** implies that for any constant $\mathrm{e}^{-\beta} < L_0 < 1$ one can choose $c = c(L_0) > 0$ and then $\varepsilon_0 = \varepsilon_0(c)$ such that $\frac{\Delta_\beta(Y_i^{(\varepsilon)}(\cdot), c, T) + 1}{\mathrm{e}^\beta} \leqslant L_0$ for $\varepsilon \leqslant \varepsilon_0$, and $i = 1, \ldots, k$.

The following lemma gives asymptotically uniform upper bounds for moments of maximum of price processes, with respect to perturbation parameter and guarantee that the reward functionals $\Phi(\mathcal{M}_T^{(\varepsilon)})$ take finite values for all ε small enough.

Lemma 1. *Let conditions* **A$_1$** *and* **C$_1$**, *hold. Then there exists a constant $L_1 < \infty$ such that for every $\varepsilon \leqslant \varepsilon_0$,*

$$\sup_{\tau^{(\varepsilon)} \in \mathcal{M}_{max,T}^{(\varepsilon)}} \mathsf{E}|g(\tau^{(\varepsilon)}, \mathbf{S}^{(\varepsilon)}(\tau^{(\varepsilon)}))| \leqslant \mathsf{E} \sup_{0 \leqslant u \leqslant T} |g(u, \mathbf{S}^{(\varepsilon)}(u))|^{\frac{\beta}{\gamma}} \leqslant L_1. \tag{29.6}$$

Let us now assume the following condition of weak convergence (denoted by the symbol $\Rightarrow$) for the transition probabilities:

B$_1$: There exist measurable sets $\mathbb{Y}_t \subseteq \mathbb{R}^k, t \in [0, T]$ such that: **(a)** $P^{(\varepsilon)}(t, \mathbf{y}^{(\varepsilon)}, t+u, \cdot) \Rightarrow P^{(0)}(t, \mathbf{y}, t+u, \cdot)$ as $\varepsilon \to 0$, for any $\mathbf{y}^{(\varepsilon)} \to \mathbf{y} \in \mathbb{Y}_t$ as $\varepsilon \to 0$ and $0 \leqslant t < t+u \leqslant T$; **(b)** $P^{(0)}(t, \mathbf{y}, t+u, \mathbb{Y}_{t+u}) = 1$ for every $\mathbf{y} \in \mathbb{Y}_t$ and $0 \leqslant t < t+u \leqslant T$; **(c)** $\mathbf{Y}^{(\varepsilon)}(0) = \mathbf{y}_0 \in \mathbb{Y}_0$.

The following theorem presents our main convergence result. It gives conditions of convergence for reward functionals $\Phi(\mathcal{M}_{\max,T}^{(\varepsilon)})$.

Theorem 1. *Let conditions* **A$_1$**, **B$_1$**, *and* **C$_1$** *hold. Then*

$$\Phi(\mathcal{M}_{\max,T}^{(\varepsilon)}) \to \Phi(\mathcal{M}_{\max,T}^{(0)}) \text{ as } \varepsilon \to 0. \tag{29.7}$$

Let $\Pi = \{0 = t_0 < t_1 < \ldots t_N = T\}$ be a partition on the interval $[0, T]$ and $d(\Pi) = \max_{1 \leqslant i \leqslant N}(t_i - t_{i-1})$.

We consider the class $\mathcal{M}_{\Pi,T}^{(\varepsilon)}$ of all Markov moments from $\mathcal{M}_{\max,T}^{(\varepsilon)}$, which only take the values $t_0, t_1, \ldots t_N$, and such that the event $\{\omega : \tau^{(\varepsilon)}(\omega) = t_j\} \in \sigma(\mathbf{Y}^{(\varepsilon)}(t_0), \ldots, \mathbf{Y}^{(\varepsilon)}(t_j))$ for $j = 0, \ldots N$. By definition, $\mathcal{M}_{\Pi,T}^{(\varepsilon)} \subseteq \mathcal{M}_{\max,T}^{(\varepsilon)}$. This relation implies that, under conditions of Lemma 1, $-\infty < \Phi(\mathcal{M}_{\Pi,T}^{(\varepsilon)}) \leqslant \Phi(\mathcal{M}_{\max,T}^{(\varepsilon)}) < \infty$.

The reward functionals $\Phi(\mathcal{M}^{(\varepsilon)}_{\max,T})$, and $\Phi(\mathcal{M}^{(\varepsilon)}_{\Pi,T})$ correspond to American type option in continuous time, and American type option in discrete time, respectively. In the first case, the underlying price process is a continuous time Markov type price process, while in the second case, the corresponding price process is a discrete time Markov type process. The random variables $\mathbf{Y}^{(\varepsilon)}(t_0), \mathbf{Y}^{(\varepsilon)}(t_1), \dots, \mathbf{Y}^{(\varepsilon)}(t_N)$ are connected in a discrete time inhomogeneous Markov chain with the phase space $\mathbb{R}^k$, transition probabilities $P^{(\varepsilon)}(t_n, \mathbf{y}, t_{n+1}, A)$, and the initial state $\mathbf{Y}^{(\varepsilon)}(t_0) = \mathbf{y}_0$. Note that we have slightly modified the standard definition of a discrete time Markov chain by counting moments $t_0, \dots, t_N$ as the moments of jumps for the Markov chain $\mathbf{Y}^{(\varepsilon)}(t_n)$, instead of the moments $0, \dots, N$. This is done in order to synchronize the discrete and continuous time models. Thus, the optimization problem (29.5) for the class $\mathcal{M}^{(\varepsilon)}_{\Pi,T}$ is really a problem of optimal expected reward for American type options in discrete time.

The following theorem gives a skeleton approximation for reward functionals $\Phi(\mathcal{M}^{(\varepsilon)}_{\max,T})$ which is asymptotically uniform with respect to perturbation parameter.

Theorem 2. *Let conditions* $\mathbf{A}_1$, *and* $\mathbf{C}_1$ *hold. Let also* $\varepsilon \leqslant \varepsilon_0$ *and* $d(\Pi) \leqslant c$ *where* ε_0 *and* c *are defined above. Then there exist constants* $L_2, L_3 < \infty$ *such that the following skeleton approximation inequality holds, for every* $\varepsilon \leqslant \varepsilon_0$,

$$\Phi(\mathcal{M}^{(\varepsilon)}_{\max,T}) - \Phi(\mathcal{M}^{(\varepsilon)}_{\Pi,T}) \leqslant L_2 d(\Pi) + L_3 \sum_{i=1}^{k} \Delta_\beta(Y_i^{(\varepsilon)}(\cdot), d(\Pi), T)^{\frac{\beta-\gamma}{\beta}}.$$

The following theorem gives conditions of convergence for reward functionals $\Phi(\mathcal{M}^{(\varepsilon)}_{\Pi,T})$.

Theorem 3. *Let conditions* $\mathbf{A}_1$, $\mathbf{B}_1$, *and* $\mathbf{C}_1$ *hold. Then, the following asymptotic relation holds for any partition* $\Pi = \{0 = t_0 < t_1 \cdots < t_N = T\}$ *on the interval* $[0, T]$ *such that* $d(\Pi) \leqslant c$, *where* c *is defined above,*

$$\Phi(\mathcal{M}^{(\varepsilon)}_{\Pi,T}) \to \Phi(\mathcal{M}^{(0)}_{\Pi,T}) \text{ as } \varepsilon \to 0. \tag{29.8}$$

Let us now formulate some useful sufficient conditions for an important condition of moment compactness $\mathbf{C}_1$.

Let us introduce the modulus of J-compactness, for $h, c > 0, i = 1, \dots, k$,

$$\Delta(Y_i^{(\varepsilon)}(\cdot), h, c, T) = \sup_{0 \leqslant t \leqslant t+u \leqslant t+c \leqslant T} \sup_{\mathbf{y} \in \mathbb{R}^k} \mathsf{P}_{\mathbf{y},t}\{|Y_i^{(\varepsilon)}(t+u) - Y_i^{(\varepsilon)}(t)| \geqslant h\}.$$

The following condition of J-compactness plays the key role in functional limit theorems for Markov type càdlàg processes:

$\mathbf{D}_1$: $\lim_{c \to 0} \overline{\lim}_{\varepsilon \to 0} \Delta(Y_i^{(\varepsilon)}(\cdot), h, c, T) = 0,\ h > 0,\ i = 1, \dots, k.$

Introduce also the quantity, which represents the maximum of moment generating functions for increments of the log-price processes $Y_i^{(\varepsilon)}(t), t \geqslant 0,\ i = 1, \dots, k$,

$$\Xi_\beta(Y_i^{(\varepsilon)}(\cdot), T) = \sup_{0 \leqslant t \leqslant t+u \leqslant T} \sup_{\mathbf{y} \in \mathbb{R}^k} \mathsf{E}_{\mathbf{y},t} e^{\beta(Y_i^{(\varepsilon)}(t+u) - Y_i^{(\varepsilon)}(t))},\ \beta \in \mathbb{R}.$$

The following condition formulated in terms of these moment generating functions can be effectively verified in many cases:

$\mathbf{C_2}$: $\overline{\lim}_{\varepsilon\to 0}\, \Xi_{\pm\beta'}(\mathbf{Y}_i^{(\varepsilon)}(\cdot), T) < \infty, i = 1, \ldots, k$, for some $\beta' > \gamma$, where γ is the parameter introduced in condition $\mathbf{A_1}$.

Lemma 2. *Conditions* $\mathbf{D_1}$ *and* $\mathbf{C_2}$ *imply that condition* $\mathbf{C_1}$ *holds for any* $\beta \in (\gamma, \beta')$.

In order to illustrate the results given in Theorems 1–3, let us consider the model where the log-price process $\mathbf{Y}^{(\varepsilon)}(t), t \geqslant 0$ is a càdlàg processes with independent increments. The process $\mathbf{Y}^{(\varepsilon)}(t)$ is a càdlàg Markov process with transition probabilities which are connected with the distributions of increments for this process $P^{(\varepsilon)}(t, t+u, A)$ by the following relation,

$$\begin{aligned} P^{(\varepsilon)}(t, \mathbf{y}, t+u, A) &= P^{(\varepsilon)}(t, t+u, A - \mathbf{y}) \\ &= \mathsf{P}\{\mathbf{y} + \mathbf{Y}^{(\varepsilon)}(t+u) - \mathbf{Y}^{(\varepsilon)}(t) \in A\}. \end{aligned} \tag{29.9}$$

Let us assume the following standard condition of weak convergence for distributions of increments for log-price processes:

$\mathbf{B_2}$: $P^{(\varepsilon)}(t, t+u, \cdot) \Rightarrow P^{(0)}(t, t+u, \cdot)$ as $\varepsilon \to 0$, $0 \leqslant t \leqslant t+u \leqslant T$.

Representation (29.9) implies that condition $\mathbf{B_1}$ holds with the sets $\mathbb{Y}_t = \mathbb{R}^k, t \in [0, T]$, i.e., distributions of increments for the processes $Y_i^{(\varepsilon)}(t)$ locally uniformly weakly converge if and only if condition $\mathbf{B_2}$ holds. Thus, in the case of processes with independent increments, the condition $\mathbf{B_1}$ with the sets $\mathbb{Y}_t = \mathbb{R}^k$ pointed above is, in fact, equivalent to the standard condition of weak convergence for such processes. In this case the J-compactness modulus $\Delta(Y_i^{(\varepsilon)}(\cdot), h, c, T)$ takes the following form:

$$\Delta'(Y_i^{(\varepsilon)}(\cdot), h, c, T) = \sup_{0 \leqslant t \leqslant t+u \leqslant t+c \leqslant T} \mathsf{P}\{|Y_i^{(\varepsilon)}(t+u) - Y_i^{(\varepsilon)}(t)| \geqslant h\}.$$

Thus, condition $\mathbf{C_2}$ is reduced to the standard J-compactness condition for the log-price processes:

$\mathbf{D_2}$: $\overline{\lim}_{c\to 0} \lim_{\varepsilon\to 0} \Delta'(Y_i^{(\varepsilon)}(\cdot), h, c, T) = 0,\ h > 0, i = 1, \ldots, k$.

Note that conditions $\mathbf{B_2}$ and $\mathbf{D_2}$ are necessary and sufficient for J-convergence of processes $\mathbf{Y}^{(\varepsilon)}(t),\ t \in [0, T]$ to process $\mathbf{Y}^{(0)}(t), t \in [0, T]$ as $\varepsilon \to 0$ and stochastic continuity of the limit process.

Also, the quantities $\Xi_\beta(Y_i^{(\varepsilon)}(\cdot), T), i = 1, \ldots, k$ take a simplified form,

$$\Xi'_\beta(Y_i^{(\varepsilon)}(\cdot), T) = \sup_{0 \leqslant t \leqslant t+u \leqslant T} \mathsf{E} e^{\beta(Y_i^{(\varepsilon)}(t+u) - Y_i^{(\varepsilon)}(t))},\ \beta \in \mathbb{R}.$$

Therefore, condition $\mathbf{C_2}$ takes the following form:

$\mathbf{C_3}$: $\overline{\lim}_{\varepsilon\to 0}\, \Xi'_{\pm\beta'}(Y_i^{(\varepsilon)}(\cdot), T) < \infty, i = 1, \ldots, k$, for some $\beta' > \gamma$, where γ is the parameter in condition $\mathbf{A_1}$.

According to Lemma 2, conditions $\mathbf{C_3}$ and $\mathbf{D_2}$ imply that condition $\mathbf{C_1}$ holds for any $\beta \in (\gamma, \beta')$.

The following theorem summarize the remarks above.

Theorem 4. *Let conditions* $\mathbf{A_1}$, $\mathbf{B_2}$, $\mathbf{D_2}$, *and* $\mathbf{C_3}$ *hold for the exponential price processes with independent increments* $\mathbf{S}^{(\varepsilon)}(t)$. *Then*

$$\Phi(\mathcal{M}^{(\varepsilon)}_{max,T}) \to \Phi(\mathcal{M}^{(0)}_{max,T}) \text{ as } \varepsilon \to 0. \tag{29.10}$$

It is worth to note that conditions $\mathbf{A_1}$, $\mathbf{B_2}$, $\mathbf{D_2}$, and $\mathbf{C_3}$ imply that conditions of Theorems 2 and 3 hold for the exponential price processes with independent increments $\mathbf{S}^{(\varepsilon)}(t)$ and, therefore, the skeleton approximation inequality given in Theorem 2 as well as the convergence relation given in Theorem 3 also take place.

29.4 Binomial–Trinomial Approximations for Reselling Model

Let us continue consideration of reselling model introduced in Sect. 29.2. In this model, there exists the unique solution to the system of stochastic differential equations (29.1) and (29.3). It is a diffusion process given by the following explicit formulas,

$$S(t) = S(0)\mathrm{e}^{\mu t + \sigma W_1(t)}, \ \sigma(t) = \sigma \mathrm{e}^{\nu \mathrm{e}^{-\alpha t} \int_0^t \mathrm{e}^{\alpha s} \mathrm{d}W_2(s)}, \ t \geqslant 0, \tag{29.11}$$

where $\mathbf{W}(t) = (W_1(t), W_2(t)), t \geqslant 0$ is the bivariate Brownian motion defined in (29.1) and (29.3).

Therefore, our object is the reward functional $\Phi(\mathcal{M}_T)$ for American type option with the pay-off function $\mathrm{e}^{-rt}C(t, S, \sigma)$ for this bivariate diffusion process $(S(t), \sigma(t))$.

The problem can be however reduced to the simpler case of a bivariate process with independent increments using suitable transformations for the price processes and the payoff functions. Let us consider processes,

$$S_1^{(0)}(t) = \mathrm{e}^{\sigma W_1(t)}, \ S_2^{(0)}(t) = \mathrm{e}^{\nu \mathrm{e}^{-\alpha T} \int_0^t \mathrm{e}^{\alpha s} \mathrm{d}W_2(s)}, \ t \geqslant 0. \tag{29.12}$$

By the definition, $S(t) = s_0 \mathrm{e}^{\mu t} S_1^{(0)}(t), t \geqslant 0$ and $\sigma(t) = \sigma (S_2^{(0)}(t))^{\mathrm{e}^{\alpha(T-t)}}, t \geqslant 0$, i.e., the process $(S(t), \sigma(t))$ is a non-random one-to-one continuous transformation of the process $(S_1^{(0)}(t), S_2^{(0)}(t))$ given by the above formulas. The vector process $\mathbf{S}^{(0)}(t) = (S_1^{(0)}(t), S_2^{(0)}(t)), t \geqslant 0$ is a bivariate continuous non-homogeneous exponential Gaussian process with independent increments. In some sense, this process is simpler than the process $(S(t), \sigma(t))$. It is more suitable for construction of the corresponding tree approximations.

The filtration $\mathcal{F}_t = \sigma((S(s), \sigma(s)), s \leqslant t)$ generated by the vector process $(S(t), \sigma(t))$ coincides with the filtration $\mathcal{F}_t = \sigma((S_1^{(0)}(s), S_2^{(0)}(s)), \ s \leqslant t), t \geqslant 0$ generated by the bivariate process $\mathbf{S}^{(0)}(t)$. Thus, the class $\mathcal{M}_T$, which includes all stopping times $0 \leqslant \tau \leqslant T$ that are Markov moments with respect to the filtration $\mathcal{F}_t, t \geqslant 0$, does not depend on which bivariate process is taken as a generator of this filtration, i.e., $\mathcal{M}_T = \mathcal{M}^{(0)}_{\max,T}$. Let us now define a pay-off function,

$$g(t, \mathbf{s}) = \mathrm{e}^{-rt} C(t, s_0 \mathrm{e}^{\mu t} s_1, \sigma s_2^{\mathrm{e}^{\alpha(T-t)}}). \tag{29.13}$$

Note that its derivatives have not more than polynomial rates of growth. More precisely, condition $\mathbf{A_1}$ holds for this function with some constants $R_i, i = 1, \ldots, 5$ and the parameters $\gamma_0 = 2 + \mathrm{e}^{2\alpha T}, \gamma_1 = 0$, and $\gamma_2 = \mathrm{e}^{2\alpha T}$, and, therefore, $\gamma = 2 + \mathrm{e}^{2\alpha T}$.

It follows from the remarks above that the reward functional,

$$\Phi(\mathcal{M}_T) = \sup_{\tau \in \mathcal{M}_T} \mathsf{E}\mathrm{e}^{-r\tau} C(\tau, S(\tau), \sigma(\tau)) = \sup_{\tau \in \mathcal{M}^{(0)}_{\max,T}} \mathsf{E} g(\tau, \mathbf{S}^{(0)}(\tau)). \tag{29.14}$$

Therefore, the reward functional $\Phi(\mathcal{M}_T) = \Phi(\mathcal{M}^{(0)}_{\max,T})$ is the optimal expected reward for American type option with the payoff function $g(t, \mathbf{s})$ for this bivariate exponential process with independent increments $\mathbf{S}^{(0)}(t)$.

Let us now consider the corresponding bivariate log-price process $\mathbf{Y}^{(0)}(t) = (Y_1^{(0)}(t), Y_2^{(0)}(t)), t \geqslant 0$ with the components

$$Y_1^{(0)}(t) = \sigma W_1(t), \ Y_2^{(0)}(t) = \nu \mathrm{e}^{-\alpha T} \int_0^t \mathrm{e}^{\alpha s} \mathrm{d}W_2(s), \ t \geqslant 0. \tag{29.15}$$

We approximate the process $\mathbf{Y}^{(0)}(t), t \geqslant 0$ by a bivariate binomial–trinomial sum-process $\mathbf{Y}^{(\varepsilon)}(t) = (Y_1^{(\varepsilon)}(t), Y_2^{(\varepsilon)}(t)), t \geqslant 0$ with components

$$Y_i^{(\varepsilon)}(t) = \sum_{1 \leqslant n \leqslant [t/\varepsilon]} Y_{n,i}^{(\varepsilon)}, \ t \geqslant 0, \ i = 1, 2. \tag{29.16}$$

Here, $\mathbf{Y}_n^{(\varepsilon)} = (Y_{n,1}^{(\varepsilon)}, Y_{n,2}^{(\varepsilon)})$, $n = 1, 2, \ldots$ are, for every $\varepsilon > 0$, independent random vectors which have the following structure,

$$(Y_{n,1}^{(\varepsilon)}, Y_{n,2}^{(\varepsilon)}) = \begin{cases} (+u_{n,1}^{(\varepsilon)}, +u_{n,2}^{(\varepsilon)}) & \qquad p_{n,++}^{(\varepsilon)} \\ (+u_{n,1}^{(\varepsilon)}, 0) & \qquad p_{n,+\cdot}^{(\varepsilon)} \\ (+u_{n,1}^{(\varepsilon)}, -u_{n,2}^{(\varepsilon)}) & \qquad p_{n,+-}^{(\varepsilon)} \\ & \text{with probability} \\ (-u_{n,1}^{(\varepsilon)}, +u_{n,2}^{(\varepsilon)}) & \qquad p_{n,-+}^{(\varepsilon)} \\ (-u_{n,1}^{(\varepsilon)}, 0) & \qquad p_{n,-\cdot}^{(\varepsilon)} \\ (-u_{n,1}^{(\varepsilon)}, -u_{n,2}^{(\varepsilon)}) & \qquad p_{n,--}^{(\varepsilon)} \end{cases} \tag{29.17}$$

Respectively, the process $\mathbf{S}^{(0)}(t), t \geqslant 0$ is approximated by a bivariate exponential binomial–trinomial process $\mathbf{S}^{(\varepsilon)}(t) = \mathrm{e}^{\mathbf{Y}^{(\varepsilon)}(t)}, t \geqslant 0$.

Let assume for simplicity that $\varepsilon = T/N$. We shall try to fit the bivariate binomial–trinomial sum-process $\mathbf{Y}^{(\varepsilon)}(t)$ to the bivariate process $\mathbf{Y}^{(0)}(t)$ by fitting expectations, variances, and covariance between for increments $\mathbf{Y}^{(\varepsilon)}(n\varepsilon) - \mathbf{Y}^{(\varepsilon)}((n-1)\varepsilon) = \mathbf{Y}_n^{(\varepsilon)}$ and $\mathbf{Y}^{(0)}(n\varepsilon) - \mathbf{Y}^{(0)}((n-1)\varepsilon)$, for every $n = 1, \ldots, N$. The corresponding quantities are given by the following formulas,

$$\mathsf{E}\,\sigma(W_1(n\varepsilon) - W_1((n-1)\varepsilon)) = 0, \ \mathsf{E}\,\nu \mathrm{e}^{-\alpha T} \int_{(n-1)\varepsilon}^{n\varepsilon} \mathrm{e}^{\alpha s} \mathrm{d}W_2(s) = 0. \tag{29.18}$$

$$\begin{aligned} \sigma^2 \varepsilon &= \mathsf{Var}\,(\sigma(W_1(n\varepsilon) - W_1((n-1)\varepsilon))), \\ \sigma_{n,\varepsilon}^2 &= \mathsf{Var}\,\nu \mathrm{e}^{-\alpha T} \int_{(n-1)\varepsilon}^{n\varepsilon} \mathrm{e}^{\alpha s} \mathrm{d}W_2(s) \\ &= \nu^2 \mathrm{e}^{-2\alpha T} \int_{(n-1)\varepsilon}^{n\varepsilon} \mathrm{e}^{2\alpha s} \mathrm{d}s = \nu^2 \mathrm{e}^{-2\alpha T} \mathrm{e}^{2\alpha n \varepsilon} \frac{1 - \mathrm{e}^{-2\alpha\varepsilon}}{2\alpha}, \end{aligned} \tag{29.19}$$

$$\begin{aligned}\varrho_{n,\varepsilon} &= \mathsf{E}\,\sigma(W_1(n\varepsilon) - W_1((n-1)\varepsilon)) \cdot \nu \mathrm{e}^{-\alpha T} \int_{(n-1)\varepsilon}^{n\varepsilon} \mathrm{e}^{\alpha s} \mathrm{d}W_2(s) \\ &= \rho\sigma\nu \mathrm{e}^{-\alpha T} \int_{(n-1)\varepsilon}^{n\varepsilon} \mathrm{e}^{\alpha s} \mathrm{d}s = \sigma\rho\nu \mathrm{e}^{-\alpha T} \mathrm{e}^{\alpha n\varepsilon} \frac{1 - \mathrm{e}^{-\alpha\varepsilon}}{\alpha}.\end{aligned} \tag{29.20}$$

The following system of $6N$ equations with $8N$ unknowns should be solved,

$$\begin{cases} \mathsf{E}[Y_{n,1}^{(\varepsilon)}] & = u_{n,1}^{(\varepsilon)}(2(p_{n,++}^{(\varepsilon)} + p_{n,+-}^{(\varepsilon)} + p_{n,+\cdot}^{(\varepsilon)}) - 1) & = 0, \\ \mathsf{Var}[Y_{n,1}^{(\varepsilon)}] & = (u_{n,1}^{(\varepsilon)})^2 & = \sigma^2\varepsilon, \\ \mathsf{E}[Y_{n,2}^{(\varepsilon)}] & = u_{n,2}^{(\varepsilon)}(p_{n,++}^{(\varepsilon)} + p_{n,-+}^{(\varepsilon)} - p_{n,+-}^{(\varepsilon)} - p_{n,--}^{(\varepsilon)}) & = 0, \\ \mathsf{Var}[Y_{n,2}^{(\varepsilon)}] & = (u_{n,2}^{(\varepsilon)})^2(p_{n,++}^{(\varepsilon)} + p_{n,-+}^{(\varepsilon)} + p_{n,+-}^{(\varepsilon)} + p_{n,--}^{(\varepsilon)}) & = \sigma_{n,\varepsilon}^2, \\ \mathsf{E}Y_{n,1}^{(\varepsilon)}Y_{n,2}^{(\varepsilon)} & = u_{n,1}^{(\varepsilon)}u_{n,2}^{(\varepsilon)}(p_{n,++}^{(\varepsilon)} + p_{n,--}^{(\varepsilon)} - p_{n,-+}^{(\varepsilon)} - p_{+-}^{(\varepsilon)}) & = \varrho_{n,\varepsilon}, \\ & \quad p_{n,++}^{(\varepsilon)} + p_{n,-+}^{(\varepsilon)} + p_{n,+-}^{(\varepsilon)} + p_{n,--}^{(\varepsilon)} + p_{n,+\cdot}^{(\varepsilon)} + p_{n,-\cdot}^{(\varepsilon)} & = 1, \\ n = 1, \dots, N. \end{cases} \tag{29.21}$$

It follows from the relations above that the only possible choice for $u_{n,1}^{(\varepsilon)} = \sigma\sqrt{\varepsilon}$. We also try to search a solution for $u_{n,2}^{(\varepsilon)} = u_n\sqrt{\varepsilon}$, where $u_n > 0, n = 1, \dots, N$ are parameters under our control, due to the fact that the number of unknowns in the system (29.21) exceeds the number of equations. It is also natural to take into account the symmetric property of the process $\mathbf{Y}^{(0)}(t)$ and to search for unknown probabilities satisfying the additional conditions, $p_{n,++}^{(\varepsilon)} = p_{n,--}^{(\varepsilon)}$, $p_{n,+-}^{(\varepsilon)} = p_{n,-+}^{(\varepsilon)}$, and $p_{n,+\cdot}^{(\varepsilon)} = p_{n,-\cdot}^{(\varepsilon)}$, for $n = 1, \dots, N$. In this case, it can be checked that the system above has the following solution,

$$\begin{cases} u_{n,1}^{(\varepsilon)} & = \sigma\sqrt{\varepsilon}, \\ u_{n,2}^{(\varepsilon)} & = u_n\sqrt{\varepsilon}, \\ p_{n,++}^{(\varepsilon)} = p_{n,--}^{(\varepsilon)} & = \frac{1}{4}\left(\frac{\sigma_{n,\varepsilon}^2}{u_n^2\varepsilon} + \frac{\varrho_{n,\varepsilon}}{\sigma u_n \varepsilon}\right), \\ p_{n,+-}^{(\varepsilon)} = p_{n,-+}^{(\varepsilon)} & = \frac{1}{4}\left(\frac{\sigma_{n,\varepsilon}^2}{u_n^2\varepsilon} - \frac{\varrho_{n,\varepsilon}}{\sigma u_n \varepsilon}\right), \\ p_{n,+\cdot}^{(\varepsilon)} = p_{n,-\cdot}^{(\varepsilon)} & = \frac{1}{2} - \frac{\sigma_{n,\varepsilon}^2}{2u_n^2\varepsilon}, \\ n = 1, \dots, N. \end{cases} \tag{29.22}$$

It is also necessary to find conditions under which the solutions $p_{n,++}^{(\varepsilon)}$, $p_{n,+-}^{(\varepsilon)}$, $p_{n,+\cdot}^{(\varepsilon)}$ of the system (29.22) are probabilities, i.e., $0 \leqslant p_{n,++}^{(\varepsilon)}, p_{n,+-}^{(\varepsilon)}, p_{n,+\cdot}^{(\varepsilon)} \leqslant 1$, $n = 1, \dots, N$. This holds if and only if the following system of inequalities holds,

$$\begin{cases} \frac{\sigma_{n,\varepsilon}^2}{u_n^2\varepsilon} & = \frac{\nu^2\mathrm{e}^{-2\alpha T}}{u_n^2\varepsilon}\int_{(n-1)\varepsilon}^{n\varepsilon} \mathrm{e}^{2\alpha s}\mathrm{d}s \\ & = \frac{\nu^2\mathrm{e}^{-2\alpha T}}{u_n^2}\mathrm{e}^{2\alpha(n-1)\varepsilon}\frac{\mathrm{e}^{2\alpha\varepsilon}-1}{2\alpha\varepsilon} \leqslant 1, \\ \frac{\sigma_{n,\varepsilon}^2}{u_n^2\varepsilon} \pm \frac{\varrho_{n,\varepsilon}}{\sigma u_n\varepsilon} & = \frac{\nu^2\mathrm{e}^{-2\alpha T}}{u_n^2\varepsilon}\int_{(n-1)\varepsilon}^{n\varepsilon} \mathrm{e}^{2\alpha s}\mathrm{d}s \pm \frac{\rho\nu\mathrm{e}^{-\alpha T}}{u_n\varepsilon}\int_{(n-1)\varepsilon}^{n\varepsilon}\mathrm{e}^{\alpha s}\mathrm{d}s \\ & = \frac{\nu\mathrm{e}^{-\alpha T}}{u_n}\frac{\mathrm{e}^{\alpha(n-1)\varepsilon}(\mathrm{e}^{\alpha\varepsilon}-1)}{\alpha\varepsilon}\left(\frac{\nu\mathrm{e}^{-\alpha T}}{u_n}\mathrm{e}^{\alpha(n-1)\varepsilon}\frac{\mathrm{e}^{\alpha\varepsilon}+1}{2} \pm \rho\right) \geqslant 0, \\ n = 1, \dots, N. \end{cases} \tag{29.23}$$

Taking the inequality $\frac{e^{2\alpha\varepsilon}-1}{2\alpha\varepsilon} = \frac{e^{\alpha\varepsilon}+1}{2}\frac{e^{\alpha\varepsilon}-1}{\alpha\varepsilon} < \left(\frac{e^{\alpha\varepsilon}+1}{2}\right)^2$ into account, we can conclude that the system of inequalities (29.23) holds if the following system of two-sided inequalities holds,

$$\begin{cases} \nu e^{-\alpha T} e^{\alpha n\varepsilon} \frac{1+e^{-\alpha\varepsilon}}{2} \leqslant u_n \leqslant \nu|\rho|^{-1} e^{-\alpha T} e^{\alpha n\varepsilon} \frac{1+e^{-\alpha\varepsilon}}{2}, \\ n = 1, \ldots, N. \end{cases} \tag{29.24}$$

Thus, a solution of the system (29.22) satisfying the system of inequalities (29.23) exists for any value $-1 \leqslant \rho \leqslant 1$.

It follows, that we consider a model with a M-bounded solution which satisfies for some constant $M \geqslant 1$ the following inequalities,

$$\nu \leqslant u_n \leqslant \nu M, \ n = 1, \ldots, N. \tag{29.25}$$

If $|\rho| > 0$, one can take $M = |\rho|^{-1}$. If $|\rho| = 0$, any number $M \geqslant 1$ can be taken.

The defining relation (29.17) implies that for any $\delta > 0$ if ε is small enough, namely, if $(\sigma \wedge M)\sqrt{\varepsilon} \leqslant \delta$, then

$$\sum_{n \leqslant [T/\varepsilon]} (\mathsf{P}\{|Y_{n,1}^{(\varepsilon)}| > \delta\} + \mathsf{P}\{|Y_{n,2}^{(\varepsilon)}| > \delta\}) = 0. \tag{29.26}$$

Also, by the definition of processes $\mathbf{Y}^{(\varepsilon)}(t)$, for $\varepsilon \geqslant 0$ and $0 \leqslant t \leqslant T$,

$$\mathsf{E}Y_1^{(\varepsilon)}(t) = 0, \ \mathsf{E}Y_2^{(\varepsilon)}(t) = 0. \tag{29.27}$$

and, for every $0 \leqslant t \leqslant T$,

$$\begin{aligned} \mathsf{Var}Y_1^{(\varepsilon)}(t) &= [t/\varepsilon]\sigma\varepsilon \to \mathsf{Var}Y_1^{(0)}(t) = t\sigma \text{ as } 0 < \varepsilon \to 0, \\ \mathsf{Var}Y_2^{(\varepsilon)}(t) &= \nu^2 e^{-2\alpha T} \int_0^{[t/\varepsilon]\varepsilon} e^{2\alpha s} ds = \nu^2 e^{-2\alpha T} \frac{e^{2\alpha[t/\varepsilon]\varepsilon} - 1}{2\alpha} \\ &\to \mathsf{Var}Y_2^{(0)}(t) = \nu^2 e^{-2\alpha T} \frac{e^{2\alpha t} - 1}{2\alpha} \text{ as } 0 < \varepsilon \to 0, \\ \mathsf{E}Y_1^{(\varepsilon)}(t)Y_2^{(\varepsilon)}(t) &= \rho\sigma\nu e^{-\alpha T} \int_0^{[t/\varepsilon]\varepsilon} e^{\alpha s} ds = \rho\sigma\nu e^{-\alpha T} \frac{e^{\alpha[t/\varepsilon]\varepsilon} - 1}{\alpha} \\ &\to \mathsf{E}Y_1^{(0)}(t)Y_2^{(0)}(t) = \rho\sigma\nu e^{-\alpha T} \frac{e^{\alpha t} - 1}{\alpha} \text{ as } 0 < \varepsilon \to 0. \end{aligned} \tag{29.28}$$

Since the functions given on the left hand side in (29.28) are monotone, and the corresponding limit functions are continuous, this convergence is also uniform in interval $[0, T]$ and, therefore, conditions of Ascoli–Arzelá theorem, in particular, condition of compactness in uniform topology holds as $\varepsilon \to 0$.

The above remarks imply that conditions of convergence theorems for vector sum-processes with independent increments, given, for example, in Skorokhod [Sko64], hold for processes $\mathbf{Y}^{(\varepsilon)}(t), t \in [0, T]$ with parameters given in (29.22), (29.24), and (29.25). Thus, the processes $\mathbf{Y}^{(\varepsilon)}(t), t \in [0, T]$ weakly and, moreover, J-converge to the process $\mathbf{Y}^{(0)}(t), t \in [0, T]$ as $\varepsilon \to 0$.

Therefore, conditions $\mathbf{B_2}$ and $\mathbf{D_2}$ hold for step-sum processes with independent increments $\mathbf{Y}^{(\varepsilon)}(t)$.

The moment generating functions $\mathsf{E}\exp\{\beta'(Y_i^{(\varepsilon)}(t+s) - Y_i^{(\varepsilon)}(t))\}$, exists for any $\beta' \in \mathbb{R}$, for every $0 \leqslant t \leqslant t+s \leqslant T$ and $i = 1, 2$ and can be given in the explicit form. Namely,

$$\mathsf{E}\exp\{\beta'(Y_1^{(\varepsilon)}(t+s) - Y_1^{(\varepsilon)}(t))\} \\ = \begin{cases} \left(e^{\beta'\sigma\sqrt{\varepsilon}}\frac{1}{2} + e^{-\beta'\sigma\sqrt{\varepsilon}}\frac{1}{2}\right)^{[(t+s)/\varepsilon]-[t/\varepsilon]} & \text{if } \varepsilon > 0, \\ e^{\frac{\beta'^2\sigma^2 s}{2}} & \text{if } \varepsilon = 0, \end{cases} \tag{29.29}$$

where it is taken into account that $p_{n,++}^{(\varepsilon)} + p_{n,+-}^{(\varepsilon)} + p_{n,+\cdot}^{(\varepsilon)} = p_{n,-+}^{(\varepsilon)} + p_{n,--}^{(\varepsilon)} + p_{n,-\cdot}^{(\varepsilon)} = \frac{1}{2}$, for $n = 1, \ldots, N$. Also,

$$\mathsf{E}\exp\{\beta'(Y_2^{(\varepsilon)}(t+s) - Y_2^{(\varepsilon)}(t))\} \\ = \begin{cases} \prod_{n=[t/\varepsilon]+1}^{[(t+s)/\varepsilon]}(e^{\beta' u_n\sqrt{\varepsilon}}p_{n,+}^{(\varepsilon)} + e^{-\beta' u_n\sqrt{\varepsilon}}p_{n,-}^{(\varepsilon)} + p_{n,\cdot}^{(\varepsilon)}) & \text{if } \varepsilon > 0, \\ e^{\frac{1}{2}\beta'^2\nu^2 e^{-2\alpha T}\int_t^{t+s} e^{2\alpha v}dv} & \text{if } \varepsilon = 0. \end{cases} \tag{29.30}$$

where $p_{n,+}^{(\varepsilon)} = p_{n,++}^{(\varepsilon)} + p_{n,-+}^{(\varepsilon)} = \frac{\sigma_{n,\varepsilon}^2}{2u_n^2\varepsilon}$, $p_{n,-}^{(\varepsilon)} = p_{n,+-}^{(\varepsilon)} + p_{n,--}^{(\varepsilon)} = \frac{\sigma_{n,\varepsilon}^2}{2u_n^2\varepsilon}$, and $p_{n,\cdot}^{(\varepsilon)} = p_{n,+\cdot}^{(\varepsilon)} + p_{n,-\cdot}^{(\varepsilon)} = 1 - \frac{\sigma_{n,\varepsilon}^2}{u_n^2\varepsilon}$, for $n = 1, \ldots, N$.

Using formulas (29.29) and (29.30) it is possible to check that condition $\mathbf{C_3}$ holds for processes $\mathbf{Y}^{(\varepsilon)}(t)$ for any $\beta' > \gamma$. In this case,

$$\begin{aligned} \varXi_{\pm\beta'}(Y_1^{(\varepsilon)}(\cdot), T) &= \sup_{0\leqslant t\leqslant t+u\leqslant T} \mathsf{E}e^{\pm\beta'(Y_1^{(\varepsilon)}(t+u) - Y_1^{(\varepsilon)}(t))} \\ &\leqslant \left(e^{\beta'\sigma\sqrt{\varepsilon}}\frac{1}{2} + e^{-\beta'\sigma\sqrt{\varepsilon}}\frac{1}{2}\right)^{T/\varepsilon} \to e^{\frac{\beta'^2\sigma^2 T}{2}} < \infty \text{ as } \varepsilon \to 0. \end{aligned} \tag{29.31}$$

Also, using that $\frac{\sigma_{n,\varepsilon}^2}{2u_n^2\varepsilon} \leqslant \frac{1}{2}$ and $\nu < u_n \leqslant M\nu$, for $n = 1, \ldots, N$, we get,

$$\begin{aligned} \varXi_{\pm\beta'}(Y_1^{(\varepsilon)}(\cdot), T) &= \sup_{0\leqslant t\leqslant t+u\leqslant T} \mathsf{E}e^{\pm\beta'(Y_1^{(\varepsilon)}(t+u) - Y_1^{(\varepsilon)}(t))} \\ &\leqslant \prod_{n=1}^{[T/\varepsilon]}\left(1 + \frac{\sigma_{n,\varepsilon}^2}{2u_n^2\varepsilon}\left(e^{\beta' u_n\sqrt{\varepsilon}} + e^{-\beta' u_n\sqrt{\varepsilon}} - 2\right)\right) \\ &\leqslant (1 + \frac{1}{2}(e^{\beta' M\nu\sqrt{\varepsilon}} + e^{-\beta' M\nu\sqrt{\varepsilon}} - 2))^{T/\varepsilon} \to e^{\frac{\beta'^2 M^2\nu^2 T}{2}} < \infty \text{ as } \varepsilon \to 0. \end{aligned} \tag{29.32}$$

Relations (29.31) and (29.32) imply condition $\mathbf{C_3}$ to hold.

Summarizing the remarks above, one can conclude that the conditions and, therefore, the statement of Theorem 4 holds for the corresponding bivariate exponential price processes with independent increments $\mathbf{S}^{(\varepsilon)}(t), t \geqslant 0$, i.e.,

$$\Phi(\mathcal{M}_{\max,T}^{(\varepsilon)}) \to \Phi(\mathcal{M}_{\max,T}^{(0)}) = \Phi(\mathcal{M}_T) \text{ as } \varepsilon \to 0. \tag{29.33}$$

We are going now to construct the corresponding bivariate binomial–trinomial tree approximation model. As above, let assume that $\varepsilon = T/N$ and consider the partition $\Pi_\varepsilon = \langle t_0 = 0 < t_1 = \varepsilon < \cdots < t_{N-1} = (N-1)\varepsilon < t_N = T\rangle$ on the interval $[0, T]$.

Let us try to find condition, which would make it possible to choose the jump values $u_n = u, n = 1, 2, \dots, N$ independent of n. This would automatically provide a very important recombining condition to hold for the corresponding bivariate binomial–trinomial tree. In this case, the number of nodes (as a function of the number of tree steps) in the tree would have not more than quadratic rate of growth.

The system of inequalities (29.24) takes in this case the form $\nu e^{-\alpha T} e^{\alpha n\varepsilon \frac{1+e^{-\alpha\varepsilon}}{2}} \leqslant u \leqslant \nu|\rho|^{-1} e^{-\alpha T} e^{\alpha n\varepsilon \frac{1+e^{-\alpha\varepsilon}}{2}}, n = 1, \dots, N$. The inequality at the left hand side is the most strong for $n = N$ while the inequality at the right hand side is the most strong for $n = 1$. Thus, the system of inequalities (29.24) holds if $\nu \frac{1+e^{-\alpha\varepsilon}}{2} \leqslant u \leqslant \nu|\rho|^{-1} e^{-\alpha T} e^{\alpha\varepsilon \frac{1+e^{-\alpha\varepsilon}}{2}}$. Consequently, this inequality holds if the following stronger inequality $\nu \leqslant u \leqslant \nu|\rho|^{-1} e^{-\alpha T}$ holds. The remarks above lead to the following condition:

$\mathbf{E_1}$: $|\rho| < e^{-\alpha T}$.

If condition $\mathbf{E_1}$ holds then interval $[\nu,\ \nu|\rho|^{-1}e^{-\alpha T}]$ has non-zero length and one can choose any value $u \in [\nu,\ \nu|\rho|^{-1}e^{-\alpha T}]$. Moreover, one can always choose a rational value of u in this interval that is useful for numerical calculations. If the value of u chosen as described above we have the following values for the corresponding probabilities,

$$\begin{cases} p^{(\varepsilon)}_{n,++} = p^{(\varepsilon)}_{n,--} &= \frac{\sigma^2_{n,\varepsilon}}{4u^2\varepsilon} + \frac{\varrho_{n,\varepsilon}}{4\sigma u\varepsilon} \\ &= \frac{\nu^2 e^{-2\alpha T}}{4u^2} e^{2\alpha n\varepsilon} \frac{1-e^{-2\alpha\varepsilon}}{2\alpha\varepsilon} + \frac{\rho\nu e^{-\alpha T}}{4u} e^{\alpha n\varepsilon} \frac{1-e^{-\alpha\varepsilon}}{\alpha\varepsilon}, \\ p^{(\varepsilon)}_{n,+-} = p^{(\varepsilon)}_{n,-+} &= \frac{\sigma^2_{n,\varepsilon}}{4u^2\varepsilon} - \frac{\varrho_{n,\varepsilon}}{4\sigma u\varepsilon} \\ &= \frac{\nu^2 e^{-2\alpha T}}{4u^2} e^{2\alpha n\varepsilon} \frac{1-e^{-2\alpha\varepsilon}}{2\alpha\varepsilon} - \frac{\rho\nu e^{-\alpha T}}{4u} e^{\alpha n\varepsilon} \frac{1-e^{-\alpha\varepsilon}}{\alpha\varepsilon}, \\ p^{(\varepsilon)}_{n,+\cdot} = p^{(\varepsilon)}_{n,-\cdot} &= \frac{1}{2} - \frac{\sigma^2_{n,\varepsilon}}{2u^2\varepsilon} \\ &= \frac{1}{2} - \frac{\nu^2 e^{-2\alpha T}}{2u^2} e^{2\alpha n\varepsilon} \frac{1-e^{-2\alpha\varepsilon}}{2\alpha\varepsilon}, \\ n = 1, \dots, N. \end{cases} \tag{29.34}$$

In this case, the Markov chain $(n, \mathbf{Y}^{(\varepsilon)}(n\varepsilon)), n = 0, 1, \dots$ is a bivariate binomial–trinomial tree model with the initial node $(0, (0, 0))$ and $(n+1)(2n+1)$ nodes of the form $(n, \mathbf{y}_{n,l_1,l_2})$, where $\mathbf{y}_{n,l_1,l_2} = (2l_1 - n)\sigma\sqrt{\varepsilon}, l_2 u\sqrt{\varepsilon})$, $l_1 = 0, 1, \dots, n, l_2 = 0, \pm 1, \dots, \pm n$, after n steps. The corresponding tree has $(n + 1)(2n + 1)$ nodes after n steps. The number of nodes is a quadratic function of n.

The standard backward procedure can be applied in order to find the optimal expected reward at moment 0 for the discrete time exponential trinomial price process $\mathbf{S}^{(\varepsilon)}(t_n) = e^{\mathbf{Y}^{(\varepsilon)}(t_n)}, t_n = n\varepsilon, n = 0, 1, \dots, N$. This optimal expected reward coincides, in this case, with the reward functional $\Phi(\mathcal{M}^{(\varepsilon)}_{\Pi_\varepsilon,T})$ for the continuous time exponential price processes $\mathbf{S}^{(\varepsilon)}(t) = e^{\mathbf{Y}^{(\varepsilon)}(t)}, t \in [0, T]$.

To estimate the difference $\Phi(\mathcal{M}^{(\varepsilon)}_{\max,T}) - \Phi(\mathcal{M}^{(\varepsilon)}_{\Pi_\varepsilon,T})$, we can use Theorem 2. In this case, $d(\Pi_\varepsilon) = \varepsilon$, while $\Delta_\beta(Y^{(\varepsilon)}_1(\cdot), \varepsilon, T) = \mathsf{E}e^{\beta|Y^{(\varepsilon)}_{1,1}|} - 1 \leqslant e^{\beta\sigma\sqrt{\varepsilon}} - 1$ and $\Delta_\beta(Y^{(\varepsilon)}_2(\cdot), \varepsilon, T) = \max_{1\leqslant n\leqslant N}(\mathsf{E}e^{\beta|Y^{(\varepsilon)}_{n,2}|} - 1) \leqslant e^{\beta u\sqrt{\varepsilon}} - 1$, for $\beta > 0$. Theorem 2 yields, in this case, the following relation,

$$\begin{aligned} &\Phi(\mathcal{M}^{(\varepsilon)}_{\max,T}) - \Phi(\mathcal{M}^{(\varepsilon)}_{\Pi_\varepsilon,T}) \\ &\quad\leqslant L_2\varepsilon + L_3\left(\left(e^{\beta\sigma\sqrt{\varepsilon}} - 1\right)^{\frac{\beta-\gamma}{\beta}} + \left(e^{\beta u\sqrt{\varepsilon}} - 1\right)^{\frac{\beta-\gamma}{\beta}}\right) \to 0 \text{ as } \varepsilon \to 0. \end{aligned} \tag{29.35}$$

The reward functional $\Phi(\mathcal{M}^{(\varepsilon)}_{\Pi_\varepsilon,T})$ is the optimal expected reward for American type option in discrete time that corresponds to the discrete time Markov log-price process $\mathbf{Y}^{(\varepsilon)}(t_n), t_n = n\varepsilon, n = 0,1,\dots,N$ with parameter $\varepsilon = T/N$ and the payoff function $g(t, \mathrm{e}^{\mathbf{y}}) = \mathrm{e}^{-rt} C(t, s_0 \mathrm{e}^{\mu t} \mathrm{e}^{y_1}, \sigma \mathrm{e}^{y_2 \mathrm{e}^{\alpha(T-t)}}), t \in [0,t], \mathbf{y} = (y_1, y_2) \in \mathbb{R}^2$ defined according to relation (29.13). Let us introduce the corresponding reward functions,

$$w^{(\varepsilon)}(t_n, \mathbf{y}_{n,l_1,l_2}) = \sup_{\tau \in \mathcal{M}_{\Pi_\varepsilon,T},\, t_n \leqslant \tau \leqslant t_N = T} \mathsf{E}_{t_n, \mathbf{y}_{n,l_1,l_2}} g(\tau, \mathrm{e}^{\mathbf{Y}^{(\varepsilon)}(\tau)}),$$

where Markov moments τ depend only on trajectories $\mathbf{Y}^{(\varepsilon)}(t_k), k \geqslant n$, and the vector points $\mathbf{y}_{n,l_1,l_2}$, $l_1 = 0,1,\dots,n, l_2 = 0,\pm1,\dots,\pm n, n = 0,\dots,N$ are defined as above, and, in particular, $\mathbf{y}_{0,0,0} = (0,(0,0))$. Then, by the definition,

$$\Phi(\mathcal{M}^{(\varepsilon)}_{\Pi_\varepsilon,T}) = w^{(\varepsilon)}(0,(0,0)). \tag{29.36}$$

The reward functions $w^{(\varepsilon)}(t_n, \mathbf{y}_{n,l_1,l_2})$ can be found using the following recurrence relations, for $n = 0,1,\dots,N-1$,

$$\begin{aligned} w^{(\varepsilon)}(t_n, \mathbf{y}_{n,l_1,l_2}) &= g(t_n, \mathrm{e}^{\mathbf{y}_{n,l_1,l_2}}) \vee \big(w^{(\varepsilon)}(t_{n+1}, \mathbf{y}_{n+1,l_1+1,l_2+1}) p^{(\varepsilon)}_{n,++} \\ &+ w^{(\varepsilon)}(t_{n+1}, \mathbf{y}_{n+1,l_1+1,l_2}) p^{(\varepsilon)}_{n,+\cdot} + w^{(\varepsilon)}(t_{n+1}, \mathbf{y}_{n+1,l_1+1,l_2-1}) p^{(\varepsilon)}_{n,+-} \\ &+ w^{(\varepsilon)}(t_{n+1}, \mathbf{y}_{n+1,l_1,l_2+1}) p^{(\varepsilon)}_{n,-+} + w^{(\varepsilon)}(t_{n+1}, \mathbf{y}_{n+1,l_1,l_2}) p^{(\varepsilon)}_{n,-\cdot} \\ &+ w^{(\varepsilon)}(t_{n+1}, \mathbf{y}_{n+1,l_1,l_2-1}) p^{(\varepsilon)}_{n,--}\big), \\ & l_1 = 0,1,\dots,n,\ l_2 = 0,\pm1,\dots,\pm n, \end{aligned} \tag{29.37}$$

with the boundary conditions,

$$w^{(\varepsilon)}(t_N, \mathbf{y}_{N,l_1,l_2}) = g(t_N, \mathrm{e}^{\mathbf{y}_{N,l_1,l_2}}),\ l_1 = 0,1,\dots,N,\ l_2 = 0,\pm1,\dots,\pm N. \tag{29.38}$$

The corresponding approximation result for the bivariate binomial tree algorithm described above can be summarized in the following theorem.

Theorem 5. *Let condition* $\mathbf{E_1}$ *holds. Then the optimal reselling rewards,*

$$w^{(\varepsilon)}(0,(0,0)) = \Phi(\mathcal{M}^{(\varepsilon)}_{\Pi_\varepsilon,T}) \to \Phi(\mathcal{M}^{(0)}_{\max,T}) = \Phi(\mathcal{M}_T) \text{ as } \varepsilon \to 0. \tag{29.39}$$

In conclusion, let us comment condition $\mathbf{E_1}$. It is, in fact, a condition of weak correlation between the noise terms of the price and stochastic volatility processes. The restriction imposed by this condition can be lightened by dividing the time interval into smaller parts and constructing a tree that has k different values of jumps on different time intervals under assumption that the following condition holds for some $k \geqslant 1$:

$\mathbf{E_k}$: $|\rho| < \mathrm{e}^{-\frac{\alpha T}{k}}$.

Let us, for example, shortly describe the case where $k = 2$. This condition guarantees the existence of two intervals with non-zero length, $[\nu \mathrm{e}^{-\frac{\alpha T}{2}},\ \nu|\rho|^{-1}\mathrm{e}^{-\alpha T}]$ and $[\nu,\ \nu|\rho|^{-1}\mathrm{e}^{-\frac{\alpha T}{2}}]$ such that the constant values $u_{(1)} = u_n, n = 1,\dots,[\frac{N}{2}]$ and $u_{(2)} = u_n, n = [\frac{N}{2}]+1,\dots,N$ can be chosen, respectively, from the first and the second interval, when constructing a solution for system (29.22) satisfying the system of inequalities (29.23). Moreover, these values can be chosen to be positive rational numbers. This makes it possible to represent them in the following form

$u_{(1)} = m_1 u$ and $u_{(2)} = m_2 u$, where u is a positive rational number and m_1 and m_2 are positive integers. Note that the recombining condition hold for each subinterval. The corresponding tree for the trinomial component has, in this case, at most $N_n = m_1 \min(n, [\frac{N}{2}]) + m_2(\max(n, [\frac{N}{2}]) - [\frac{N}{2}]) + 1$ nodes located on the grid $-N_n u\sqrt{\varepsilon}, \ldots, -u\sqrt{\varepsilon}, 0, u\sqrt{\varepsilon}, \ldots, N_n u\sqrt{\varepsilon}$, after n steps. The corresponding bivariate binomial–trinomial tree has at most $(n+1)N_n$ nodes, after n steps. This function also has not more than the quadratic rate of growth as a function of n. A backward algorithm for finding the corresponding reward functions and convergence results are analogous to those presented above.

It is also worth to note that algorithm described above let one also to derive the corresponding optimal stopping domains for the approximation binomial–trinomial processes. Indeed, these domains are defined by the relations,

$$\Gamma^{(\varepsilon)} = \{(t_n, \mathbf{y}_{n,l_1,l_2}) : w^{(\varepsilon)}(t_n, \mathbf{y}_{n,l_1,l_2}) = g(t_n, e^{\mathbf{y}_{n,l_1,l_2}}), n = 0, \ldots, N-1\}. \quad (29.40)$$

These domains, according Theorem 5, well approximate the optimal stopping domains for the corresponding continuous time bivariate price processes with respect to the natural measure of closeness based on the deviation of the corresponding optimal expected rewards.

29.5 Numerical Examples

In this section, we illustrate some basic numerical aspects connected with the approximation tree algorithm described above.

We consider the model introduced in Sect. 29.2 with the following parameters. The risk free interest rate $r = 0.04$. The price process $S(t)$ has the initial value $S(0) = 10$, the drift parameter $\mu = 0.02$ and the initial volatility $\sigma = 0.2$. Note that these values correspond to the risk neutral setting for the price process. We also assume the parameters for the mean reverse volatility process $\sigma(t)$ to be $\alpha = 1$ for the mean reverting coefficient and $\nu = 0.2$ for the volatility of volatility. The correlation coefficient connecting the noise terms for the price and stochastic volatility processes is $\rho = 0.3$. We consider an European call option with the strike price $K = 10$ and the time to maturity $T = 0.5$ of a year. Note that condition $\mathbf{E_1}$ holds for the chosen values of parameters ρ, α and T.

In this case, we can choose the jump value for the trinomial tree to be $u = \nu$.

Numerical studies show that the optimal expected reselling reward values $w^{(\varepsilon)}(0, (0,0))$ stabilize good enough for $N \geqslant 15$. For example, they take values 0.9961 and 0.9955 if, respectively, $N = 15$ and $N = 50$, for the model with parameters pointed above. Note that the time needed for calculation of the approximate optimal expected reward value in the case of $N = 15$ is about 2.5 s on an 1.73 GHz Intel® Pentium-M processor, 1GB internal memory using Matlab®.

The above reward values should be compared with the expected reward corresponding to the reselling at maturity that is equivalent to the execution of the option at maturity. In this case, the expected reward is 0.7228 for the above model. Thus, optimal reselling of the option before maturity increases the expected reward (for the model with parameters pointed above) by about 25%.

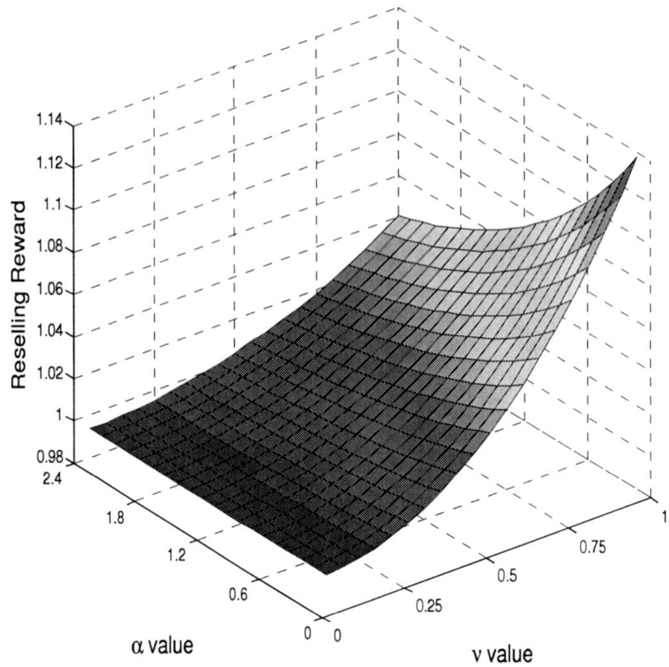

Figure 29.1. The optimal expected reselling rewards for the models with parameters $r = 0.04$; $S(0) = 10$, $\mu = 0.02$, $\sigma = 0.2$, $0.12 < \alpha < 2.4$, $0.05 < \nu < 1$, $\rho = 0.3$; and $K = 10$, $T = 0.5$

We also show in the figures below how the reselling reward depend upon parameters of price and stochastic volatility processes. In Fig. 29.1, we show how the optimal expected reselling reward depends upon parameters α and ν. We let α vary in the interval $(0.12, 2.4)$ and ν vary in the interval $(0.05, 1)$. Other parameters take the same values as in the above initial example. In this case, condition $\mathbf{E_1}$ is not violated for all values of α in the above interval. We see that the optimal reselling reward is a decreasing function of α and an increasing function of ν. In Fig. 29.2, we show how the reselling reward depend upon parameters μ and σ. We let μ vary in the interval $(-0.5, 0.5)$ and σ vary in the interval $(0.05, 1)$. Other parameters take the same values as in the above initial example. We see that the optimal reselling reward is an increasing function of both parameters μ and σ.

Let us also shortly comment alternative approaches to the problem. The complexity of the model makes it problematic to get close analytic solutions. As far as always existing alternative of Monte Carlo based algorithms, they require, in the case of American type options, to use discrete time-space approximations for the price processes. These algorithms have a slow performance even for univariate price processes due to large number of points at the corresponding time-space grids, as show the experimental studies realized, for example, in Jönsson [Jon01], [Jon05], Lundgren [Lun07], and Silvestrov et al. [SGS99]. This problem is very much aggravated for bivariate price processes. Also, the corresponding convergence results provided by strict proofs should be given. These problems do require additional theoretical and experimental studies.

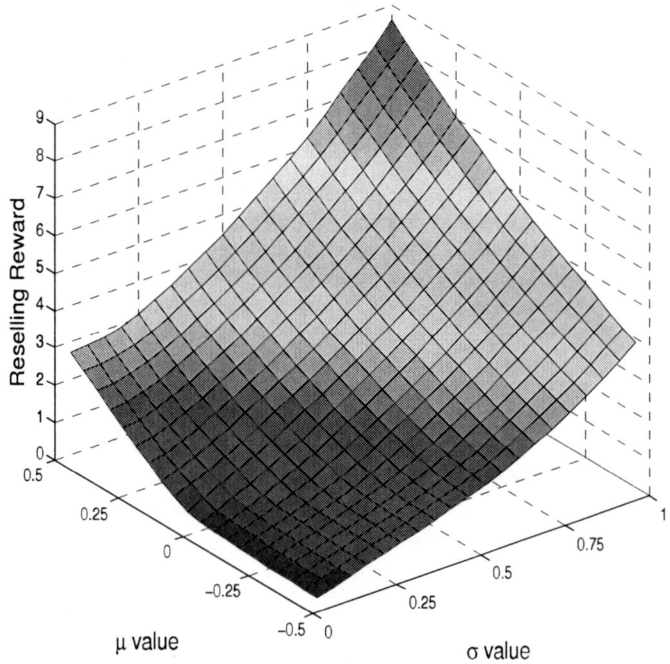

Figure 29.2. The optimal expected reselling rewards for the models with parameters $r = 0.04$; $S(0) = 10$, $-0.5 < \mu < 0.5$, $0.05 < \sigma < 1$, $\alpha = 1$, $\nu = 0.2$, $\rho = 0.3$; and $K = 10$, $T = 0.5$

In conclusion, we would like to note that the reselling of options is a new and complex problem. We hope that the promising results presented in the paper will stimulate further research studies in this area.

References

[AK94] Amin, K., Khanna, A.: Convergence of American option values from discrete – to continuous – time financial models. Math. Finance, **4**, no. 4, 289–304 (1994)

[CT07] Coquet, F., Toldo, S.: Convergence of values in optimal stopping and convergence in optimal stopping times. Electr. J. Probab., **12**, no. 8, 207–228 (2007)

[DK03] Dayanik, S., Karatzas, I.: On the optimal stopping problem for one-dimensional diffusions. Stoch. Proc. Appl., **107**, 173–212 (2003)

[DW05] Dupuis, P., Wang, H.: On the convergence from discrete to continuous time in an optimal stopping problem. Ann. Appl. Probab., **15**, 1339–1366 (2005)

[ELTW09] Ekström, E., Lindberg, C., Tysk, J., Wanntorp, H.: Optimal liquidation of a call spread. Preprint Uppsala University, Sweden (2009)

[GHS00] Gau, B., J.Z. Huang, Subrahmanyam, M.: The valuation of American barrier options using the decomposition technique. J. Econom. Dynam. Control, **24**, 1783–1827 (2000)

[HH08] Henderson, V., Hobson, D.: An explicit solution for an optimal stopping/optimal control problem which models an asset sale. Ann. Appl. Probab., **18**, 1681–1705 (2008)

[Hes93] Heston, S.L.: A closed form solution for options with stochastic volatility with applications to bond and currency options. Rev. Fin. Stud., **6**, no. 2, 327–343 (1993)

[Jac91] Jacka, S.D.: Optimal stopping and the American put. Math. Finance, **1**, 1–14 (1991)

[Jon01] Jönsson, H.: Monte Carlo studies of American type options with discrete time. Theory Stoch. Process., **7(23)**, no. 1-2, 163–188 (2001)

[Jon05] Jönsson, H.: Optimal Stopping Domains and Reward Functions for Discrete Time American Type Options. Doctoral Dissertation **22**, Mälardalen University, Sweden (2005)

[JKS04] Jönsson, H., Kukush, A.G., Silvestrov, D.S.: Threshold structure of optimal stopping strategies for American type options. I. Theor. Ĭmovirn. Mat. Stat., **71**, 113–123 (2004) (English translation in Theory Probab. Math. Statist., **71**, 93–103).

[JKS05] Jönsson, H., Kukush, A.G., Silvestrov, D.S.: Threshold structure of optimal stopping strategies for American type options. II. Theor. Ĭmovirn. Mat. Stat., **72**, 42–53 (2005) (English translation in Theory Probab. Math. Statist., **72**, 47–58).

[Kim90] Kim, I.: The analytic valuation of American options. Rev. Fin. Stud., **3**, 547–572 (1990)

[KMS06] Kukush, A.G., Mishura, Y.S. Shevchenko, G.M.: On reselling of European option. Theory Stoch. Process., **12(28)**, no. 3-4, 75–87 (2006)

[KS00] Kukush, A.G., Silvestrov, D.S.: Structure of optimal stopping strategies for American type options. In: Uryasev, S. (ed) Probabilistic Constrained Optimization: Methodology and Applications, Kluwer, 173–185 (2000)

[KS01] Kukush, A.G., Silvestrov, D.S.: Skeleton approximations of optimal stopping strategies for American type options with continuous time. Theory Stoch. Process., **7(23)**, no. 1-2, 215–230 (2001)

[KS04] Kukush, A.G., Silvestrov, D.S.: Optimal pricing for American type options with discrete time. Theory Stoch. Process., **10(26)**, no. 1-2, 72–96 (2004)

[Lun07] Lundgren, R.: Structure of optimal stopping domains for American options with knock out domains. Theory Stoch. Process., **13(29)**, no. 4, 98–129 (2007)

[LS09] Lundgren, R., Silvestrov, D.: Convergence of option rewards for multivariate price processes, Research Report 2009-10, Department of Mathematics, Stockholm University, 53 pages (`http://www.math.su.se`) (2009)

[LSK08] Lundgren, R., Silvestrov, D., Kukush A.: Reselling of options and convergence of option rewards. J. Num. Appl. Math., **96(1)**, 149–172 (2008)

[PS06] Peskir, G, Shiryaev, A.: Optimal Stopping and Free-Boundary Problems. Birkhäuser, Basel (2006)

[Pri03] Prigent, J.L.: Weak Convergence of Financial Markets. Springer, New York (2003)

[SZ99] Schöbel, R., Zhu, J.: Stochastic volatility with Ornstein -Uhlenbeck process: an extension. Europ. Fin. Stud., **6**, 327–343 (1999)

[SS93] Shepp, L.A., Shiryaev A.N.: The Russian option: reduced regret. Ann. Appl. Probab., **3**, no. 3, 631–640 (1993)

[SGM01] Silvestrov, D., Galochkin, V., Malyarenko, A.: OPTAN - a pilot program system for analysis of optaions. Theory Stoch. Process., **7(23)**, no. 1-2, 291–300 (2001)

[SGS99] Silvestrov, D.S., Galochkin, V.G., Sibirtsev, V.G.: Algorithms and Programs for optimal Monte Carlo pricing of American type options. Theory Stoch. Process., **5(21)**, no. 1-2, 175–187 (1999)

[SJS07] Silvestrov, D., Jönsson, H., Stenberg, F.: Convergence of option rewards for Markov type price processes. Theory Stoch. Process., **13(29)**, no. 4, 174–185 (2007)

[SJS08a] Silvestrov, D.S., Jönsson, H., Stenberg F.: Convergence of option rewards for Markov type price processes modulated by stochastic indices. Eurandom Research Reports, 2008-041, Eurandom, 34 pages (`http://www.eurandom.nl`) (2008)

[SJS08b] Silvestrov, D., Jönsson, H., Stenberg, F.: Convergence of option rewards for Markov type price processes modulated by stochastic indices I, II. Theor. Ĭmovirn. Mat. Stat., **79**, 138–154 (2008), **80** (2009) (English translation in Theory Probab. Math. Statist., **79**, 153–170 (2009), **80**, (2010), *to appear.*).

[Sko64] Skorokhod, A.V.: Random Processes with Independent Increments. Probability Theory and Mathematical Statistics, Nauka, Moscow (1964) (English edition: Nat. Lending Library for Sci. and Tech., Boston Spa, 1971)

[StSt91] Stein, E.M., Stein, J.C.: Stock price distribution with stochastic volatility: An analytical approach. Rev. Fin. Stud., **4**, no 4, 727–752 (1991)

[XZ07] Xia, J., Zhou, X.: Stock loans. Math. Finance, **17**, no. 2, 307–317 (2007)

[ZL06] Zhang, Z., Lim, K-G.: A non-lattice pricing model of American options under stochastic volatility. J. Futures Markets, **26**, no. 5, 417–448 (2006)

30

Bayesian Modeling of Health State Preferences

R. Muzaffer Musal[1*] and Refik Soyer[2]

[1] Texas State University, San Marcos, TX
[2] George Washington University, Washington, DC, soyer@gwu.edu

Abstract: In this paper, we present a Bayesian framework for modeling uncertainty about a population's health state preferences. Such a framework is motivated by the need to analyze preference-based measurement data that arise from evaluation of health states by a sample of individuals. The Bayesian framework leads to population utility estimation and health policy evaluation by introducing a probabilistic interpretation of the multiattribute utility theory (MAUT) models used in health economics. In doing so, our approach combines ideas from the MAUT approach of Keeney and Raiffa ("Decisions with Multiple Objectives-Preferences and Value Tradeoffs", Wiley, New York, 1976) and Bayesian view point to provide an alternate method of modeling preferences.

Keywords and phrases: Bayesian analysis, Health economics, Markov chain, Monte Carlo, Multiattribute utility

30.1 Introduction

Preference based measurement of health (PBMH) methods have been developed to be used in economic evaluation of health policies. Use of preference based measures requires quantification of health state preferences by a group of individuals. This preference data is used as a sample to develop an aggregate measure for the population preferences. A recent overview of PBMH methods can be found in Brazier and Roberts [BRA06]. The methods that quantify preference based measurement of health (PBMH) are referred to as the health related quality of life measures (HRQoL). The methods include

- Health Utilities Index (HUI) of Torrance et al. [TOR96]
- Quality of Well-Being (QWB) scale of Kaplan et al. [KAP88]
- Short Form (SF-6D) survey of Brazier, Roberts and Deverill [BRA02].

These measures are used to quantify a population's preferences over health states that have multiple dimensions. Thus, the measures are based on multiattribute evaluation of states using preference weights and scores. They provide a single index number

V.V. Rykov et al. (eds.), *Mathematical and Statistical Models and Methods in Reliability: Applications to Medicine, Finance, and Quality Control*, Statistics for Industry and Technology, DOI 10.1007/978-0-8176-4971-5_30,

for each health state. Typically, an index value "1" denotes perfect health and "0" denotes death. In the health economics literature, these index values are referred to as utility. Elicitation of utility requires sophisticated procedures based on standard gambles as discussed in Brazier and Deverill [BRA99]. A more simplified approach for obtaining preference measures is to ask respondents to assign values to health states directly and have the analyst convert these to utilities.

Preference based measures such as QWB and SF-6D use what is referred to as the *composite approach* for estimation of the multiattribute utility function for the health states. The composite approach involves direct elicitation of utility of multidimensional health states and requires more health states than that can be evaluated by a single respondent; see Brazier [BRA05]. Regression type models are used to extrapolate the values of health states that are not included in the survey. An alternative for estimation of the utilities for the health states is the *decomposed approach* employed in the HUI. This method uses the MAUT framework of Keeney and Raiffa [KR76] and determines a functional form for the multiattribute utility function of health states. The decomposed approach is based on simplifying assumptions such as *preferential independence* and *utility independence*; see for example, Keeney [K76]. The approach yields simpler forms of utility functions and substantially reduces the valuation effort by decomposing the problem into one-dimensional elicitation problems. In addition to providing evaluations of all possible health states, the decomposed approach is also flexible in modeling interactions using multiplicative utility functions. As noted by [BRA05], this is unlike the composite approach where there is no standard method for determining the states required to estimate a model with interaction terms. Hazen [HAZ04] describes how the additive or multiplicative decomposition within QALYs can be constructed using the independence concepts and discusses how they relate to HUI.

Both the composite and decomposed approaches provide us with a sample of health state valuation data, that is, with health state utilities from a sample of individuals. The objective is to estimate the health state utilities of the population based on this sample and use the estimated population utility function to evaluate different health policy alternatives. Statistical methods have been considered by earlier researchers such as Dolan [DOL97] and Brazier et al. [BRA02]. In general, these approaches employed linear models with normally distributed error terms. As pointed out by Brazier [BRA05], these models, that used data from the composite approach, "have estimated crude summary terms for interactions" and have required range of transformations to deal with highly skewed data.

Most of the earlier models and associated statistical methods have used classical sampling theoretic approaches. More recently, Bayesian approaches have been considered in the health economics literature. For example, nonparametric Bayesian models are proposed by Kharroubi, O'Hagan and Brazier [KH05] and by Kharroubi et al. [KH07] for HRQoL estimation of a population. Standard gambles are used by the authors to elicit utilities of health states from the sample of individuals. Although a sampling method is used by authors to avoid an individual going through large number of gambles, the proposed composite approach still requires large number of evaluations of health states. In view of this, following Musal et al. [OURS], we consider a parametric Bayesian approach and use a decomposed model in our framework. Our framework consists of modeling attribute utilities, modeling attribute weights and using a multi-attribute model for aggregation.

In Sect. 30.2, we introduce a model for single attribute utilities. Using a decomposed approach and an additive utility function, modeling of attribute weights is presented in Sect. 30.3. We discuss Bayesian aggregation of attribute utilities and weights as well as probabilistic evaluation of health state preferences in Sect. 3.4. Concluding remarks are given in Sect. 3.5.

30.2 Bayesian Modeling of Attribute Utilities

Assume that preference ordering of the $K_c + 1$ levels with respect to each attribute $c = 1, \ldots, C$,

$$X_{c,1} \prec X_{c,2} \prec \cdots \prec X_{c,K_c} \prec X_{c,K_c+1} \tag{30.1}$$

is identical for the population, where $X_{c,j}$ denotes the jth level of a single attribute c. We are interested in making inference about the unknown population utilities

$$u(X_{c,2}) < \cdots < u(X_{c,K_c}), \tag{30.2}$$

where $u(X_{c,1}) = 0$ and $u(X_{c,K_c+1}) = 1$. We may have a prior opinion on these values and we are interested in updating this prior opinion based on the sample utility measurements on the N individuals. In general, $u(X^i_{c,j}) = u^i_{c,j}$ is the utility declared by the i-th individual for attribute c at level j.

We focus on a single criterion and to reflect the ordering that applies to the population we assume that for all individuals

$$0 < u_2 < u_3 < \cdots < u_K < 1. \tag{30.3}$$

It is desirable to have a probability model for utility vectors $\mathbf{u}^i = (u^i_2, \ldots, u^i_K)$ which is consistent with the ordering and flexible enough to reflect the diminishing utility scenario encountered in many applications. As suggested in Musal et al. [OURS], the *ordered Dirichlet* model can provide that kind of flexibility in describing different utility scenarios. The ordered Dirichlet model has been previously used in reliability growth modeling by Mazzuchi and Soyer [IEEE93] and Erkanli, Mazzuchi and Soyer [TECH98] who pointed out such properties with respect to behavior of reliabilities over time. The ordered Dirichlet model for the utility vector $\mathbf{u} = (u_2, u_3, \ldots, u_K)$ is given by

$$p(\mathbf{u} \mid \beta, \alpha) = \frac{\Gamma(\beta)}{\prod_{j=2}^{K+1} \Gamma(\beta\alpha_j)} \prod_{j=2}^{K+1} (u_j - u_{j-1})^{\beta\alpha_j - 1}, \tag{30.4}$$

where $u_1 = 0$ and $u_{K+1} = 1$ and the distribution is defined over the simplex (30.3). The model parameters are β and α such that $\beta > 0, \alpha_j > 0$ and $\sum_{j=2}^{K+1} \alpha_j = 1$.

In the ordered Dirichlet model (30.4), the marginals are beta densities denoted as

$$(u_j \mid \beta, \alpha) \sim Beta\Big(\beta\alpha^*_j, \beta(1 - \alpha^*_j)\Big) \tag{30.5}$$

for $j = 2, \ldots K$, where $\alpha^*_j = \sum_{k=2}^{j} \alpha_k$, $E[u_j \mid \beta, \alpha] = \alpha^*_j$ and β is the precision parameter where lower values are associated with a more diffused distribution of the utility level. As pointed out by van Dorpe, Mazzuchi and Soyer [JSPI97], for $i < j$, the model implies that

$$(u_j - u_i) \mid \beta, \alpha \sim Beta\Big(\beta(\alpha_j^* - \alpha_i^*), \beta(1 - \alpha_j^* + \alpha_i^*)\Big). \tag{30.6}$$

Thus, the changes in the adjacent utilities $(u_j - u_{j-1})$, for $j = 2, \ldots, K+1$, follow a beta distribution where the mean is given by

$$E[u_j - u_{j-1} \mid \beta, \alpha] = (\alpha_j^* - \alpha_{j-1}^*) = \alpha_j. \tag{30.7}$$

It follows from the above that, α_j can be interpreted as the expected increase in utility as a result of going from attribute level X_{j-1} to attribute level X_j whereas α_j^* is the expected utility at attribute level X_j. We note that α_j^* is increasing with j, implying that for the population we expect utility is an increasing function of the attribute when high values of the attribute are desirable. If $E[u_j - u_{j-1} \mid \beta, \alpha] = \alpha_j$ is a decreasing sequence in j, then we expect that the marginal utility is diminishing as the attribute level gets larger. In this case, we will have $E[u_j \mid \beta, \alpha] = \alpha_j^*$ is discrete concave in j. Thus, the model is attractive in that it allows for incorporation of different prior information about expected behavior of utilities into the analysis. Such prior beliefs can be used in specification of the prior distribution of the parameters α and β.

In the Bayesian paradigm, uncertainty about all unknown quantities are described probabilistically. Thus, completion of Bayesian modeling of attribute utilities requires us to specify the prior distribution of α and β. We will denote the prior by $p(\beta, \alpha)$. It is not unreasonable to assume that α and β are independent a priori. Since β is the precision parameter, a reasonable prior distribution for β is the gamma distribution. As pointed out by Musal et al. [OURS], in specifying the prior distribution of α, one can use the properties of the ordered Dirichlet model discussed above. More specifically, if we have prior beliefs about the monotonicity of α_j's, then a distribution which reflects that belief will be the appropriate choice. However, in modeling health state preferences, one of the objectives is to infer about such monotonic behavior and thus, a prior which does not force an ordering of α_j's is more desirable. Since the ordered Dirichlet model (30.4) requires that $\sum_{j=2}^{K+1} \alpha_j = 1$, a natural prior for α is the Dirichlet distribution

$$p(\alpha \mid b, \mathbf{a}) = \frac{\Gamma(b)}{\prod_{j=2}^{K+1} \Gamma(b a_j)} \prod_{j=2}^{K+1} (\alpha_j)^{a b_j - 1}, \tag{30.8}$$

where $a_j > 0$ and $b > 0$ are specified parameters such that $\sum_{j=2}^{K+1} a_j = 1$. Note that a_j's are the expected values of α_j's and b is the precision parameter of the model.

30.2.1 Bayesian Analysis of Utilities

Our objective is to describe uncertainty about the population utility $\mathbf{u} = (u_2, \ldots, u_K)$ based on the information provided by the sample of N utility vectors

$$\mathbf{u}^i = (u_2^i, u_3^i, \ldots, u_K^i), i = 1, \ldots, N,$$

where $\mathbf{u}^i$'s are assumed to follow the ordered Dirichlet model (30.4). Thus, given sample utilities $\mathbf{u}^N = (\mathbf{u}^1, \mathbf{u}^2, \ldots, \mathbf{u}^N)$ from the N individuals, we are interested in the posterior predictive distribution $p(\mathbf{u}|\mathbf{u}^N)$. The distribution $p(\mathbf{u}|\mathbf{u}^N)$ describes our inferences about population utilities for a given attribute.

The posterior predictive distribution is obtained via the calculus of probability as

$$p(\mathbf{u}|\mathbf{u}^N) = \int_{\beta,\alpha} p(\mathbf{u}|\mathbf{u}^N,\beta,\alpha)p(\beta,\alpha|\mathbf{u}^N)\mathrm{d}\beta\mathrm{d}\alpha. \tag{30.9}$$

Using the conditional independence of $\mathbf{u}$ and $\mathbf{u}^N$, given β and α, (30.9) reduces to

$$p(\mathbf{u}|\mathbf{u}^N) = \int_{\beta,\alpha} p(\mathbf{u}|\beta,\alpha)p(\beta,\alpha|\mathbf{u}^N)\mathrm{d}\beta\mathrm{d}\alpha, \tag{30.10}$$

where $p(\mathbf{u}|\beta,\alpha)$ is given by (30.4) and $p(\beta,\alpha|\mathbf{u}^N)$ is the posterior distribution of the parameters of the ordered Dirichlet model. Once sample utilities $\mathbf{u}^N$ is available, uncertainty about α and β is revised via Bayes' rule

$$p(\beta,\alpha|\mathbf{u}^N) \propto L(\beta,\alpha;\mathbf{u}^N)p(\beta,\alpha), \tag{30.11}$$

where $L(\beta,\alpha;\mathbf{u}^N)$ is the likelihood function based on the ordered Dirichlet distribution (30.4). More specifically, we have

$$L(\beta,\alpha;\mathbf{u}^N) = \prod_{i=1}^{N}\left[\frac{\Gamma(\beta)}{\prod_{j=2}^{K+1}\Gamma(\beta\alpha_j)}(u_j^i - u_{j-1}^i)^{\beta\alpha_j-1}\right]. \tag{30.12}$$

The posterior distribution $p(\beta,\alpha|\mathbf{u}^N)$ can not be obtained analytically for any choice of the prior distributions discussed earlier. However, we can draw samples from the posterior distribution using Markov chain Monte Carlo (MCMC) methods; see Musal et al. [OURS] for details. Since $p(\beta,\alpha|\mathbf{u}^N)$ is not analytically available, evaluation of the posterior predictive distribution (30.10) requires use of Monte Carlo methods. Given samples $\left(\beta^{(s)},\alpha^{(s)}\right)_{s=1}^{S}$ from the posterior distribution $p(\beta,\alpha|\mathbf{u}^N)$, we can approximate (30.10) via the Monte Carlo average

$$p(\mathbf{u}|\mathbf{u}^N) \approx \frac{1}{S}\sum_{s=1}^{S} p(\mathbf{u}|\beta^{(s)},\alpha^{(s)}). \tag{30.13}$$

By using the predictive distribution we can make probability statements about utilities at each of the attribute levels and approximate the population's *expected utility function* by plotting the $E(u_j|\mathbf{u}^N)$ versus the attribute level X_j's. We can also provide posterior probability bounds for the utilities.

30.3 Modeling Attribute Weights

The development in Sect. 30.2 is presented for a single attribute, that is, for attribute c, with K_c+1 levels and observed utility vectors $\mathbf{u}_c^i = (u_{c,2}^i, \ldots, u_{c,K_c}^i)$ for individuals $i = 1,\ldots,N$. In general, for specifying the ordered Dirichlet model for the multiattribute problem, all model parameters are indexed by c, that is, we actually have (β_c, α_c) with prior $p(\beta_c,\alpha_c)$.

If we have mutual utility independence of the attributes (see for example, [KR76]), then the above development can be easily extended to C attributes. In this case, the parameters (β_c, α_c) are assumed to be independent for $c = 1, \ldots, C$ and thus the approach presented in Sect. 30.2 yields independent posterior predictive distributions $p(\mathbf{u}_c|\mathbf{u}_c^N)$ for $c = 1, \ldots, C$. The mutual utility independence of the attributes justifies the use of multiplicative multiattribute utility model as considered by Torrance, Boyle and Hardwood [TOR82] for describing society's preference for health states. Furthermore, if *additive utility independence* (see Keeney [K76]) can be justified, then the multiattribute utility function can be written as

$$u(X_1, X_2, \ldots, X_C) = \sum_{c=1}^{C} w_c u(X_c), \tag{30.14}$$

where w_c's are weights representing the relative importance of the attributes such that $0 < w_c < 1$ and $\sum_{c=1}^{C} w_c = 1$.

In the multiattribute utility (MAU) function (30.14), $u(X_c)$'s are unknown utilities whose posterior distributions are available to us via using Monte Carlo based methods. The weight vector $\mathbf{w} = (w_1, , w_2, \ldots, w_C)$ is also an unknown quantity. Thus, from a Bayesian perspective for given levels of $X_1, X_2, \ldots, X_C$, uncertainty about the MAU needs to be described probabilistically. Such a development requires us to consider a probability model for the attribute weight vector $\mathbf{w}$. The model needs to be consistent with the requirement that $\sum_{c=1}^{C} w_c = 1$ and that the weights are negatively correlated. An appropriate probability model for this case is the Dirichlet distribution

$$p(\mathbf{w} \mid \kappa, \gamma) = \frac{\Gamma(\kappa)}{\prod_{c=1}^{C} \Gamma(\kappa\gamma_c)} \prod_{c=1}^{C} (w_c)^{\kappa\gamma_c - 1}, \tag{30.15}$$

where $\gamma = (\gamma_1, \ldots, \gamma_C)$. It is well known that all the marginal distributions are Beta densities, that is,

$$(w_c \mid \kappa, \gamma) \sim Beta\Big(\kappa\gamma_c, \kappa(\gamma_0 - \gamma_c)\Big), \tag{30.16}$$

where $\gamma_0 = \sum_{c=1}^{C} \gamma_c = 1$. Prior distributions of the unknown parameters κ and γ can be specified to reflect these properties. For example, the prior for γ can be specified as a Dirichlet distribution. Since γ_c's represent expected attribute weights in the model, the prior parameters can be chosen to reflect our best guesses about attribute weights in the population. As before, a gamma density is a reasonable prior for the precision parameter κ.

Health state preference data typically include attribute weights $\mathbf{w}^i = (w_1^i, w_2^i, \ldots, w_C^i)$, $i = 1, \ldots, N$, elicited from the sample of N individuals (see for example, [OURS]). As before, we treat the N weight vectors as the samples from the Dirichlet distribution in (30.15). Thus, it is possible to develop Bayesian machinery to revise our uncertainty based on such data.

If we specify, $p(\kappa, \gamma)$, as the prior distribution for κ and γ then given sample weights, $\mathbf{w}^N$, from N individuals, the posterior distribution is obtained via

$$p(\kappa, \gamma|\mathbf{w}^N) \propto L(\kappa, \gamma; \mathbf{w}^N) p(\kappa, \gamma), \tag{30.17}$$

where the likelihood function, $L(\kappa, \gamma; \mathbf{w}^N)$ is based on the Dirichlet model (30.15). The posterior distribution (30.17) can not be analytically obtained for any reasonable choice of the prior $p(\kappa, \gamma)$. As before, MCMC methods can be used to draw samples from (30.17). As in the case of attribute utilities of Sect. 30.2, the objective is to make inference about the attribute weight vector of the population. Thus, we are interested in obtaining the posterior predictive distribution

$$p(\mathbf{w}|\mathbf{w}^N) = \int p(\mathbf{w}|\kappa, \gamma)p(\kappa, \gamma|\mathbf{w}^N)\mathrm{d}\kappa\mathrm{d}\gamma. \tag{30.18}$$

The integral in (30.18) can not be obtained analytically, but given samples $\left(\kappa^{(s)}, \gamma^{(s)}\right)_{s=1}^{S}$ from the posterior distribution (30.17) we can approximate it via the Monte Carlo average

$$p(\mathbf{w}|\mathbf{w}^N) \approx \frac{1}{S}\sum_{s=1}^{S} p(\mathbf{w}|\kappa^{(s)}, \gamma^{(s)}), \tag{30.19}$$

where $p(\mathbf{w}|\kappa^{(s)}, \gamma^{(s)})$ is the Dirichlet density. Note that once we simulate the weight vectors $\mathbf{w}^{(s)}$, for $s = 1, \ldots, S$, from the Dirichlet distribution, then we can make probability statements about attribute weights such as $Pr(w_i > w_j|\mathbf{w}^N)$. In other words, we can infer probabilistically if certain attributes are more important than the others for the population.

30.4 Bayesian Evaluation of Health States

In Sects. 30.2 and 30.3, we have presented a Bayesian framework for modeling attribute utilities and weights. In so doing, we have discussed how to obtain two sets of posterior samples, that is, sample of utilities for given attributes and attribute weights. Using these, we can make probability statements using the MAU function (30.14). Given samples from the posterior predictive distributions $p(\mathbf{u}_c|\mathbf{u}_c^N)$ for $c = 1, \ldots, C$ and $p(\mathbf{w}|\mathbf{w}^N)$, we can evaluate the population utility distribution for a specific health state.

For a specific health state, A_i with the attribute levels, $(X_{1,A_i}, ..., X_{C,A_i})$, we can obtain the probability distribution of the corresponding MAU via the Monte Carlo evaluation of $u\,(X_{1,A_i}, ..., X_{C,A_i})$ using (30.14). For A_i, we can write

$$u\,(X_{1,A_i}, ..., X_{C,A_i}) = \sum_{c=1}^{C} w_c u(X_{c,A_i}), \tag{30.20}$$

and using the posterior samples we can obtain a histogram estimate of the posterior distribution of $u\,(X_{1,A_i}, ..., X_{C,A_i})$. In a similar manner, we can make probability statements on whether health state A_i is preferred to state A_j in the population, that is,

$$Pr(A_i \succ A_j|\mathbf{u}_1^N, \ldots, \mathbf{u}_C^N, \mathbf{w}^N).$$

This probability is equivalent to

$$Pr\Big\{u\,(X_{1,A_i}, \ldots, X_{C,A_i}) > u\,\big(X_{1,A_j}, \ldots, X_{C,A_j}\big)\,|\mathbf{u}_1^N, \ldots, \mathbf{u}_C^N, \mathbf{w}^N\Big\} \tag{30.21}$$

which can be approximated using the posterior samples.

The literature on the preference based measures of health generally has considered other alternatives to the additive model. A common method of decomposition that is used to account for potential interactions in attribute utilities is the multiplicative utility model. Musal et al. [OURS] considered a multiplicative model and developed a Bayesian framework similar to what is presented here.

30.5 Concluding Remarks

In this paper, we have presented a Bayesian framework for modeling uncertainty about a population's health state preferences. Our development is based on the composite approach as in Torrance, Boyle and Hardwood [TOR82]. The Bayesian framework involves modeling both the attribute utilities and attribute weights and provides probabilistic evaluation of health state preferences. Since our focus has been on the Bayesian perspective here, computational issues involving implementation of MCMC methods for developing posterior inferences are not discussed in the paper. For this we refer the interested reader to Musal et al. [OURS] where such methods are applied in analysis of actual health state preference data.

It is possible to consider extensions of the models proposed here. For example, the precision parameters β_c's, in modeling utilities, can be assumed to be constant across the attributes. This can be one way to impose a dependence structure for different attribute utilities. Other possible extensions include incorporation of covariate information in the utility and attribute weight models and taking into consideration heterogeneity of individuals. Such issues have been considered in [OURS].

References

[BRA99] Brazier, J., Deverill, M.: Obtaining the Q in QALYs: A Comparison of Five Multi-attribute Utility Scales. School of Health and Related Research, The University of Sheffield, (1999)

[BRA02] Brazier, J. E., Roberts, J., Deverill, M.: The estimation of a preference-based measure of health from the SF-36. *Journal of Health Economics*, 21, 271–292 (2002)

[BRA05] Brazier, J.: Current State of the Art in Preference Based Measures of Health and Avenues for Further Research. Discussion Paper. School of Health and Related Research, The University of Sheffield, (2005)

[BRA06] Brazier, J., Roberts, J.: Methods for developing preference-based measures of Health: In Jones, A. (ed) *The Elgar Companion to Health Economics.* Edward Elgar, Cheltenham, UK, (2006)

[DOL97] Dolan, P.: Modeling Valuations for EuroQol Health States. *Medical Care*, 35, 1095–1108 (1997)

[JSPI97] van Dorp, J. R., Mazzuchi, T. A., Soyer, R.: Sequential inference and decision making for single mission systems development. *Journal of Statistical Planning and Inference*, 62, 207–218 (1997)

[TECH98] Erkanli, A., Mazzuchi, T. A., Soyer, R.: Bayesian Computations for a Class of Reliability Growth Models. *Technometrics*, 40, 14–23 (1998)

[HAZ04] Hazen, G.: Multiattribute Structure for QALYs. *Decision Analysis*, 1, 205–216 (2004)

[KAP88] Kaplan, R. M., Ganiats, G. M., Sieber, W. J., Anderson, J. P.: The Quality of Well-Being Scale: critical similarities and differences with SF-36. *International Journal of Quality of Health Care*, 10, 509–520 (1988)

[KR76] Keeney, R. L., Raiffa, H.: *Decisions with Multiple Objectives-Preferences and Value Tradeoffs*. Wiley, New York (1976)

[K76] Keeney, R. L.: A Group Preference Axiomatization with Cardinal Utility. *Management Science*, 23, 140–145 (1976)

[KH05] Kharroubi, S. A., O'Hagan, A., Brazier, J. E.: Estimating utilities from individual health preference data: a nonparametric Bayesian approach. *Applied Statistics*, 54, 879–895 (2005)

[KH07] Kharroubi, S. A., Brazier, J. E., O'Hagan, A., Roberts, J.: Modelling SF-6D health state preference data using a nonparametric Bayesian method. *Journal of Health Economics*, 26, 597–612 (2007)

[IEEE93] Mazzuchi, T. A., Soyer, R.: A Bayes methodology for assessing product reliability during development testing. *IEEE Transactions on Reliability*, 42, 503–510 (1993)

[OURS] Musal, R. M., Soyer, R., McCabe, C., Kharroubi, S. A.: Estimating the Population Utility Function: A Parametric Bayesian Approach. *European Journal of Operational Research*, under second review, (2009)

[TOR82] Torrance, G. W., Boyle, M. H., Horwood, S. P.: Application of Multiattribute Utility Theory to Measure Social Utility for Health States. *Operations Research*, 30, 1043–1069 (1982)

[TOR96] Torrance, G. W., Feeny, D. H., Furlong, W. J., Ronald, B., Yueming, Z., Qinan, W.: Multiattribute utility function for a comprehensive health status classification: Health Utilities Index Mark 2. *Medical Care*, 34, 702–722 (1996)

31

Information Measures in Biostatistics and Reliability Engineering

Filia Vonta[1] and Alex Karagrigoriou[2]

[1] Department of Mathematics, National Technical University of Athens, Greece
[2] Department of Mathematics and Statistics, University of Cyprus, Cyprus

Abstract: In this paper, we discuss the basic tools for modelling in Biomedicine and Reliability. In particular, we present the divergence measures and the tests of fit while optimal modelling issues are also addressed. The last section is devoted to various applications in Reliability, Biomedicine, Hydrology, and Insurance and Actuarial Science.

Keywords and phrases: Divergence measures, Tests of fit, Residual lifetimes, Past lifetimes, Optimal modelling

31.1 Introduction

Statistical theory and applications are developing an increasingly close relationship. Statistics cannot flourish without data whilst data cannot be handled without appropriate methodological techniques.

Statistical modelling or model building is an activity aimed at identifying the generating mechanisms or probability distributions that produce a set of observed data. Although, in some very special cases, the physical data generating mechanism suggests a good probability model, in general, we do not have enough knowledge of the machinery that generates the data to convert it into a probability distribution. Practitioners know that one can never find the 'true' data generating distribution, so that its identification must be regarded as an unreachable task. This leads us to estimate the true model by approaches based on information or divergence measures. Starting with the reasonable idea that in general no model in a collection of models can capture all the features in the data we should look for a model within the collection selected that does it best. In order to formalize this concept, Akaike [Aka73] assumed that the 'true model' exists but is not in the collection selected. Then, one searches within the collection of candidate models for that model which is "closest" to the true. An information theoretic approach needs to be incorporated into the analysis to measure the closeness or the distance between each candidate model and the true but unknown model. Statistical modelling is associated with the Model selection theory and the companion concept of tests of fit, both of which are associated with the information and divergence measures.

V.V. Rykov et al. (eds.), *Mathematical and Statistical Models and Methods in Reliability: Applications to Medicine, Finance, and Quality Control*, Statistics for Industry and Technology, DOI 10.1007/978-0-8176-4971-5_31, © Springer Science+Business Media, LLC 2010

The information or divergence (or distance) measures are used as indices of similarity or dissimilarity between populations. They are also used to measure the distance or the discrepancy between two functions or two populations. Finally, they are used either to measure mutual information concerning two variables or to construct model selection criteria.

In this paper, we discuss the basic tools for modelling in Biomedicine and Reliability. In particular, in Sect. 31.2.1 we present the divergence measures and in Sect. 31.2.2 the Tests of Fit. Optimal modelling issues are addressed in Sect. 31.2.3. The last section is devoted to various applications in Reliability, Biomedicine, Hydrology, and Insurance and Actuarial Science.

31.2 Modelling in Biomedicine and Reliability

31.2.1 Modelling Tools: Divergence Measures

Measures of divergence between two probability distributions have a very long history. One could consider as pioneers in this field the famous scientists of the 20th century, Pearson, Mahalanobis, Lévy and Kolmogorov. In our days the most popular measure of divergence is considered to be the Kullback–Leibler measure of divergence introduced in the 50's [KL51]. A well known family of measures is the φ-divergence known also as Csiszar's measure of information which was introduced and investigated independently by Csiszár [Csz63] and Ali and Silvey [AS66] and is defined by

$$I_X^C(f,g) = \int g\varphi(f/g)\mathrm{d}\mu,$$

where μ is the Lebesgue measure or a counting measure and φ is a convex function on $[0,\infty)$ such that $\varphi(1) = 0$ and $\varphi''(1) \neq 0$ with the conventions $0\varphi(0/0) = 0$ and $0\varphi(u/0) = \lim_{u\to\infty}[\varphi(u)/u], u > 0$. For various functions φ the measure takes different forms. Members of this family are among others, the Kullback–Leibler measure as well as Pearson's X^2 divergence measure.

Measures of divergence can be used in statistical inference for the construction of test statistics for tests of fit [ZFP90, Zha02, CLY04] or in statistical modelling for the construction of model selection criteria like the Kullback–Leibler measure which has been used for the development of various criteria [Aka73, Cav04, YTM06].

A unified analysis has been provided by Cressie and Read [CR84], who introduced for both the continuous and the discrete case a family of measures of divergence known as power divergence family of statistics that depends on a parameter λ and is used for goodness-of-fit tests for multinomial distributions. The family is a member of the Csiszar family with

$$\varphi(\lambda) = \frac{x^{\lambda+1} - x - \lambda(x-1)}{\lambda(\lambda+1)}, \ \lambda \neq 0, -1.$$

The Cressie and Read family includes for different values of λ among others, the well known Pearson's X^2 divergence measure and for multinomial models the loglikelihood ratio statistic. It should be noted that for the appropriate limit of λ to 0 the above measure becomes the Kullback–Leibler measure.

A new measure of divergence known as the BHHJ divergence measure, was recently introduced by Basu et al. [BHHJ98]. The measure which is given by

$$I_X^a(f,g) = \int \left\{ g^{1+a}(z) - \left(1+\frac{1}{a}\right) f(z)\, g^a(z) + \frac{1}{a} f^{1+a}(z) \right\} \mathrm{d}z, \; a>0 \quad (31.1)$$

and depends on a positive index a which controls the trade-off between robustness and efficiency when the measure is used as an estimating criterion for robust parameter estimation. Basu *et al.* [BHHJ98] showed that values of a close to zero provide parameter estimators with good robust features without significant loss in terms of efficiency. Note that for the appropriate limit of a to 0 the measure reduces to the Kullback-Leibler measure. Note also that the BHHJ measure has been recently generalized to the (φ,α)-family by Mattheou, Lee & Karagrigoriou [MLK09] as follows

$$I_X^a(g,f) = E_g\Big(g^a(X)\varphi\Big(\frac{f(X)}{g(X)}\Big)\Big) = \int g^{1+a}(z)\,\varphi\Big(\frac{f(z)}{g(z)}\Big)\mathrm{d}z, \quad a>0, \quad (31.2)$$

where μ represents the Lebesgue measure and φ belongs to a class Φ of convex functions such that $\varphi(1)=0$, $\varphi'(1)=0$, and $\varphi''(1)\neq 0$ with the conventions $0\varphi(u/0)=\lim_{u\to\infty}[\varphi(u)/u], u>0$ and $0\varphi(0/0)=0$.

As it was mentioned earlier measures of divergence can also be used in model selection. Since some measures of divergence have been proposed as distinguishability indices between two distributions which are far from each other or from two distributions which are close, they can be used for the construction of model selection criteria. A model selection criterion can be considered as an approximately unbiased estimator of the expected overall discrepancy, a nonnegative quantity which measures the *distance* between the true unknown model and a fitted approximating model belonging to a collection of candidate models. If the value of the criterion is small for a specific member of the candidate class, then the corresponding approximated model is good. The Kullback–Leibler divergence was the measure used by Akaike [Aka73] to develop the Akaike Information Criterion (AIC).

Similar measures of entropy and divergence are useful in reliability and survival analysis models. Specific measures have been introduced by Ebrahimi and Kirmani [EK96] between the lifetimes X and Y of two items at time t. In survival analysis or in reliability, we might know the current age t of a biomedical or technical system which we need to take into consideration when we compare two systems or populations. Ebrahimi and Kirmani [EK96] achieved this by replacing the distribution functions of the random variables X and Y in the Kullback–Leibler divergence of X and Y, by the distributions of their residual lifetimes. Di Crescenzo and Longobardi [DCL04] define a dual measure of divergence which constitutes a distance between past life distributions.

Let $f(x)$, $F(x)$ and $\overline{F}(x)=1-F(x)$ be the density function, the cumulative distribution function and the survival function of X, respectively. Let also $g(x)$, $G(x)$ and $\overline{G}(x)=1-G(x)$ be the density function, the cumulative distribution function and the survival function of Y respectively. Recently, Vonta and Karagrigoriou [VK10] have proposed a new family of divergence measures for lifetime distributions. More specifically they proposed two new measures of discrepancy which are based on the Csiszar's φ-divergence family, namely, the *φ-distance between residual lifetimes*

$$I_{X,Y}^{\varphi}(t) = \int_t^{\infty} \frac{g(x)}{\overline{G}(t)} \, \varphi\left(\frac{f(x)/\overline{F}(t)}{g(x)/\overline{G}(t)}\right) \mathrm{d}x, \qquad t > 0 \tag{31.3}$$

and the *φ-distance between past lifetimes*

$$\overline{I}_{X,Y}^{\varphi}(t) = \int_0^{t} \frac{g(x)}{G(t)} \varphi\left(\frac{f(x)/F(t)}{g(x)/G(t)}\right) \mathrm{d}x, \qquad t > 0 \tag{31.4}$$

where the function φ belongs to a class of functions Φ with properties

(1) $\varphi(x)$ is continuous, differentiable and convex for $x \geqslant 0$
(2) $\varphi(1) = 0$
(3) $\varphi'(1) = 0$.

From the above assumptions we deduce that $\varphi(x) \geqslant 0$ for all $x > 0$ and $\varphi'(x) > 0$ for $x > 1$ and $\varphi'(x) < 0$ for $x < 1$. The following theorem provides an upper bound for I_X^C, $I_{X,Y}^{\varphi}(t)$ and $\overline{I}_{X,Y}^{\varphi}(t)$ measures.

Theorem 1. *For the measures $I_X^C(f,g)$, $I_{X,Y}^{\varphi}(t)$ and $\overline{I}_{X,Y}^{\varphi}(t)$ we have that*

$$\text{(i)} \quad I_X^C(f,g) < \varphi(0) + \lim_{r\to\infty} \frac{\varphi(r)}{r}, \tag{31.5}$$

$$\text{(ii)} \quad I_{X,Y}^{\varphi}(t) < \varphi(0) + \lim_{r\to\infty} \frac{\varphi(r)}{r}, \tag{31.6}$$

and

$$\text{(iii)} \quad \overline{I}_{X,Y}^{\varphi}(t) < \varphi(0) + \lim_{r\to\infty} \frac{\varphi(r)}{r}. \tag{31.7}$$

The following theorems provide the ordering of measures between r.v's X_1, Y and X_2, Y. Similar results can be established for the ordering between Y_1, X and Y_2, X.

Theorem 2. *Let three random variables X_1, X_2 and Y have p.d.f's f_1, f_2 and g respectively. If $X_2^t \geqslant_{st} X_1^t$ and $f_1(x)/g(x)$ increasing in x then*

$$I_{X_1,Y}^{\varphi}(t) \leqslant I_{X_2,Y}^{\varphi}(t), \quad t > 0. \tag{31.8}$$

If $X_1^t \geqslant_{st} X_2^t$ and $f_2(x)/g(x)$ increasing in x, the inequality is reversed.

Theorem 3. *Let three random variables X_1, X_2 and Y have p.d.f's f_1, f_2 and g respectively. If $X_{2,t} \geqslant_{st} X_{1,t}$ and $f_1(x)/g(x)$ increasing in x then*

$$\overline{I}_{X_1,Y}^{\varphi}(t) \leqslant \overline{I}_{X_2,Y}^{\varphi}(t), \quad t > 0. \tag{31.9}$$

If $X_{1,t} \geqslant_{st} X_{2,t}$ and $f_2(x)/g(x)$ increasing in x, the inequality is reversed.

31.2.2 Modelling Tools: Tests of Fit

For goodness of fit tests the focus is on multinomial distributions. More specifically, for two multinomial distributions $P = (p_1, \ldots, p_m)$ and $Q = (q_1, \ldots, q_m)$ with sample space $\Omega = \{x : p(x) \cdot q(x) > 0\}$ and $p(x)$ and $q(x)$ the probability mass functions of the two distributions, the discrete version of the (φ, α)-family of divergences is given by

$$d_a \equiv d_a(Q, P) = E_q\Big(q^a(X)\varphi\Big(\frac{p(X)}{q(X)}\Big)\Big) \equiv \sum_{i=1}^{m} q_i^{1+a}\varphi\left(\frac{p_i}{q_i}\right), \tag{31.10}$$

with $\varphi(\cdot) \in \Phi$ defined in (31.2). Observe that for

$$\varphi(u) = \Phi_1(u) = u^{1+a} - \left(1 + \frac{1}{a}\right)u^a + \frac{1}{a}$$

measure (31.10) reduces to the discrete version of the BHHJ measure corresponding to (31.1) given by

$$d_a(Q, P) = \sum_{i=1}^{m} p_i^{1+a} - \left(1 + \frac{1}{a}\right)\sum_{i=1}^{m} q_i p_i^a + \frac{1}{a}\sum_{i=1}^{m} q_i^{1+a}. \tag{31.11}$$

Tests of fit have a very long history with recent contributions based on likelihood ratio [Zha02, Mar06], the multivariate settings [SF84, ZFP90], the survival data [VA07, HV04, CLY04] and the various modelling settings [AN94, BN02]. If we have to examine whether the data $(n_1, n_2, ..., n_m)$ come from a known multinomial distribution $M(N, P_0)$, where $P_0 = (p_{10}, p_{20}, ..., p_{m0})$ and $N = \sum_{i=1}^{m} n_i$, a well known test statistic is the chi-square goodness of fit test statistic. Mattheou and Karagrigoriou [MK10], for any function $\varphi \in \Phi$ such that $\varphi(1) = \varphi'(1) = 0$ and $\varphi''(1) \neq 0$, proposed the following statistic for the above goodness of fit test:

$$\mathrm{X}_a^2 \equiv \frac{2N\hat{d}_a}{\varphi''(1)}, \quad \hat{d}_a = d_f(P_0, \hat{P}) = \sum_{i=1}^{m} {p_{i0}}^{1+a}\varphi\left(\frac{\hat{p}_i}{p_{i0}}\right) \tag{31.12}$$

where $\hat{p}_i$ the MLE of p_i. The above test statistic for $\varphi \in \Phi$ as in (31.11), constitutes the test statistic associated with the BHHJ measure. Mattheou and Karagrigoriou [MK10] have showed that the above test statistic is superior to the well known tests including the popular X^2 Pearson's test.

The theorem below provides the asymptotic distribution of the test statistic (31.12).

Theorem 4. *Let* $(n_1, ..., n_m) \sim M(N, P)$ *with* $P = (p_1, ..., p_m)$ *and* p_i, $i = 1, ..., m$ *unknown parameters. Let also* $W = \sum_{i=1}^{m} \frac{N}{p_{i0}}\left(\frac{n_i}{N} - p_{i0}\right)^2$. *Under the null hypothesis* H_0 : $p_i = p_{i0}$, $i = 1, ..., m$ *we have:*

- $\left(\min_i p_{i0}^a\right) W \prec_{\mathrm{st}} \sum_{i=1}^{m} \frac{Np_{i0}^a}{p_{i0}}\left(\frac{n_i}{N} - p_{i0}\right)^2 \prec_{\mathrm{st}} \left(\max_i p_{i0}^a\right) W$
- $\mathrm{X}_a^2 - \sum_{i=1}^{m} \frac{Np_{i0}^a}{p_{i0}}\left(\frac{n_i}{N} - p_{i0}\right)^2 \xrightarrow{P} 0$ *and*
- *the distribution of (31.12) is estimated to be approximately* $c\mathcal{X}_{m-1}^2$, *where*

$$c = 0.5\big(\min_i p_{i0}^a + \max_i p_{i0}^a\big)$$

where $\mathcal{X}_{m-1}^2$ is the chi-squared distribution with $m-1$ degrees of freedom and $\prec_{\text{st}}$ the symbol for stochastic ordering.

The power of the test is given below.

Proposition 1. *Under the assumptions of Theorem 4 and for the alternative hypothesis $H_a : p_i = p_{ib}, i = 1, \ldots, m$ the power of the test is asymptotically equal to*

$$\gamma_a = P\left(Z \geq (2\sqrt{N}\sigma_a)^{-1}\left(\varphi''(1)\, c\mathcal{X}_{m-1,\alpha}^2 + 2N\varphi(1)\sum_{i=1}^{m} p_{i0}^{1+a} - 2Nd_a\right)\right) \quad (31.13)$$

where Z is a standard Normal random variable, $\mathcal{X}_{m-1,\alpha}^2$ the αth percentile of the $\mathcal{X}_{m-1}^2$ distribution and

$$\sigma_a^2 = \sum_{i=1}^{m} p_{ib}\left[p_{i0}^a \varphi'\left(\frac{p_{ib}}{p_{i0}}\right)\right]^2 - \left[\sum_{j=1}^{m} p_{ib} p_{i0}^a \varphi'\left(\frac{p_{ib}}{p_{i0}}\right)\right]^2.$$

It is important to point out that any type of data can be viewed as multinomial data by dividing the range of data into m categories. In that sense, data related to biomedicine, engineering and economics and finance that usually come from continuous distributions can be transformed into multinomial data and goodness of fit based on the above (discrete) measures can be applied. The proper modelling of such data plays a key role in determining the optimum goodness of fit test. In that sense the search for the ideal value of m, namely the optimal number of categories is vital so that the resulting test of fit for various null hypotheses given by

$$\mathrm{X}_a^2 \equiv \frac{2N\sum_{i=1}^{m} {p_{i0}}^{1+a}\varphi\left(\frac{\hat{p}_i}{p_{i0}}\right)}{\varphi''(1)} \quad (31.14)$$

will result in achieving

(a) The nominal size.
(b) The maximum power for a wide range of possible alternatives.

Examples of such continuous distributions are the exponential, lognormal, Gamma, Inverse Gaussian, Weibull, Pareto, and Positive Stable distributions. For instance, the family of two-parameter inverse Gaussian distribution (IG2) is one of the basic models for describing positively skewed data which arise in a variety of fields of applied research as cardiology, hydrology, demography, linguistics, employment service, etc. Such examples include the repair times of an airborne communication transceiver [CF77] and quality characteristics ([Sim03, ME06]). Recently, Huberman et al. [HPPL98] have argued and demonstrated the appropriateness of the inverse Gaussian family for studying the internet traffic and in particular the number of visited pages per user within an internet site. Most applications of IG2 are justified on the fact that the IG2 is the distribution of the first passage time in Brownian motion with positive drift. Furthermore, distributions like the Weibull, the Positive Stable and the Pareto are frequently encountered in survival modelling. The main problem of determining the appropriate distribution is extremely important for reducing the possibility of erroneous inference. In addition, the existence of censoring schemes in survival modelling makes the determination of the proper distribution an extremely challenging problem. Finally, distributions like the exponential, the Gamma, the lognormal and others are very common in lifetime problems.

31.2.3 Optimal Modelling

The asymptotic distribution of any goodness of fit test statistic based on a divergence measure, say F_T is given by (see e.g. [ZFP90, MK10])

$$F_T = F_{\chi^2_{m-1}} + o(1).$$

It is possible to provide techniques to improve the accuracy of the proposed test statistics which are indexed by a positive parameter a. More specifically, in addition to the asymptotic distribution other approximations of the exact distribution of the new family of test statistics can be provided. The methods that have been introduced and widely used in the last three decades, provide:

(a) The test statistic with the first three moments as close as possible to the first three moments of the asymptotic distribution.
(b) The corrected test statistic which has mean and variance equal to the mean and variance of the asymptotic distribution.
(c) The second order approximation to the exact distribution where the second order component from the Remainder term of the convergence to the asymptotic distribution is extracted. Note that this second order component was obtained by Yarnold [Yar72] for the chi-square test statistic under the null hypothesis (the approximation consists of a term of multivariate Edgeworth expansion for a continuous distribution and a discontinuous term), by Siotani and Fujikoshi [SF84] for the likelihood ratio test statistic and Freeman–Tukey test statistic, by Read [Rea84] for the power-divergence test statistic and by Menéndez et al. [MPPP97] for the Csiszar-family.

These approximations are useful in an attempt to improve the accuracy of the proposed test statistics especially when the sample size is not large. Using Taylor expansions one can identify the optimum values of the index a so that the convergence rate to the corresponding moments of the asymptotic distribution will be the fastest possible. Hence, it is expected that the resulting optimum test will produce sufficiently accurate results when the asymptotic distribution is used. Here we focus on the first method only and for the test statistic

$$I_n^{\phi}\left(\widehat{\mathbf{p}}, \mathbf{p}_0\right) = \frac{2N \sum\limits_{i=1}^{m} p_{i0} \phi_a \left(\frac{\hat{p}_i}{p_{i0}}\right)}{\phi_a''(1)} \tag{31.15}$$

with $\phi_a \in \Phi$ which is the discrete analogue of I_X^C or equivalently it is a special case of (31.15) with $a = 0$.

Pardo [Par06] has established that

$$E\left(I_n^{\phi_a}\left(\widehat{\mathbf{p}}, \mathbf{p}_0\right)\right)^{\beta} = E\left(\chi^2_{m-1}\right)^{\beta} + \frac{1}{n} f^{\beta}_{\phi_a} + O\left(n^{-3/2}\right), \quad \beta = 1, 2, 3. \tag{31.16}$$

According to (31.16), $f^{\beta}_{\phi_a}$ controls the speed at which the first exact three central moments of $I_n^{\phi}\left(\widehat{\mathbf{p}}, \mathbf{p}_0\right)$ converge to the corresponding moments of the χ^2_{m-1} distribution. The function ϕ_a for which $f^{\beta}_{\phi_a} = 0$, $\beta = 1, 2, 3$, will be the best. So, we need to find under what conditions these quantities converge faster to 0, taking into consideration the value of m.

If now we consider the equiprobable null hypothesis

$$H_0 : \mathbf{P} = \mathbf{P}_0 = (1/m, \ldots, 1/m)',$$

we have $m^2 = \sum_{j=1}^{m} (p_{j0})^{-1}$. However for m increasing the roots of the equation $f^1_{\phi_a} = 0$ converge to the roots of the equation

$$4\phi_a'''(1) + 3\phi_a^{(4)}(1) = 0, \quad \text{for} \quad \phi_a''(1) \neq 0, \tag{31.17}$$

since, in this case, $f^1_{\phi_a}$ can be written as

$$4\phi_a'''(1)\left(\frac{2 - 3m + m^2}{1 - 2m + m^2}\right) + 3\phi_a^{(4)}(1) = 0.$$

Then the roots of this last equation converge to the roots of (31.17) as $m \to \infty$. Similar arguments can be applied to $f^i_{\phi_a} = 0$, $i = 2, 3$. If we consider the Cressie and Read family of the power-divergence test statistics, $\phi_a = \phi(\lambda)$, we have that the roots of the equation (31.17) are $\lambda = 1$ and $\lambda = 2/3$. These values were found directly by Read and Cressie [RC88].

Let

$$\varphi_a(u) \equiv \Phi_{1,a}(u) = \frac{1}{1+a}\left(u^{1+a} - \left(1 + \frac{1}{a}\right)u^a + \frac{1}{a}\right) = \frac{1}{(1+a)}\Phi_1(u), \quad a \in \mathbb{R}. \tag{31.18}$$

The associated Csiszar's family of tests statistics has the expression

$$I_n^{\Phi_{1,a}}(\widehat{\mathbf{p}}, \mathbf{p}_0) = \frac{2n}{a(a+1)} \sum_{j=1}^{m} p_{j0}\left\{a\left(\frac{\widehat{p}_j}{p_{j0}}\right)^{1+a} - (a+1)\left(\frac{\widehat{p}_j}{p_{j0}}\right)^{a} + 1\right\}. \tag{31.19}$$

It is observed immediately that $I_n^{\Phi_{1,1}}(\widehat{\mathbf{p}}, \mathbf{p}_0)$ is the chi-square test statistic.

Proposition 2. *The rate of convergence of the moments of the test statistic $I_n^{\Phi_{1,a}}(\widehat{\mathbf{p}}, \mathbf{p}_0)$ to the moments of the χ^2_{m-1} is optimum for $a = 1$ or $a = 10/9$.*

It is important to know how large m has to be for using the roots $a = 1$ and $a = 10/9$. The solutions for the $f^1_{\Phi_{1,a}}$ are given in Table 31.1 as the number of classes m increases and $\sum_{j=1}^{m} (p_{j0})^{-1}$ changes. In particular, we have considered, for $\sum_{j=1}^{m} (p_{j0})^{-1}$, the values m^2, m^3, m^4 and m^5.

For the function $\Phi_{1,a}(\cdot)$ we conclude the following: *For $m \geqslant 20$ all roots are within ± 0.05 of the limiting roots $a = 1$ and $a = 10/9$. Therefore, for $m > 20$ and for the choices $a = 1$ or $a = 10/9$ the convergence of the first moment of the test statistic to the corresponding moment of a chi-square random variable with $m - 1$ degrees of freedom is faster. The range [1, 1.5] is optimal for all values of m which are not too small. For small m one should use the values given in Table 31.1.*

Table 31.1. Roots ($a_1 < a_2$) for $f^1_{\Phi_{1,a}} = 0$

S	$S = m^2$		$S = m^3$		$S = m^4$		$S = m^5$	
m	a_1	a_2	a_1	a_2	a_1	a_2	a_1	a_2
2	1	2.000000	1	1.288889	1	1.179487	1	1.141762
3	1	1.555556	1	1.191919	1	1.134503	1	1.118581
4	1	1.407407	1	1.157895	1	1.121821	1	1.113733
5	1	1.333333	1	1.141762	1	1.116883	1	1.112252
10	1	1.209877	1	1.119266	1	1.111913	1	1.111191
20	1	1.157895	1	1.113233	1	1.111217	1	1.111116
40	1	1.133903	1	1.111653	1	1.111125	1	1.111111
50	1	1.129252	1	1.111460	1	1.111118	1	1.111111
100	1	1.120090	1	1.111199	1	1.111112	1	1.111111
200	1	1.115578	1	1.111133	1	1.111111	1	1.111111
500	1	1.112892	1	1.111115	1	1.111111	1	1.111111
700	1	1.112383	1	1.111113	1	1.111111	1	1.111111
1000	1	1.112001	1	1.111112	1	1.111111	1	1.111111

31.3 Applications

Applications of the above proposed models and methods can be found *in Insurance and Actuarial Science, in Reliability and Engineering Systems, in Accelerated Lifetime Models, in Biomedicine and Biostatistics, in Hydrology*, etc. Some types of such applications are briefly presented below:

Insurance and Actuarial Science

Let us assume that the expected frequency of claims varies within the portfolio. Let us further assume that any particular risk in the portfolio has a Poisson distribution of claim frequencies with mean θ which is itself a random variable with distribution representing the expected risks inherent in the given portfolio. The unconditional distribution of claim frequencies of an individual drawn from the portfolio is mixed Poisson. One interesting distribution for the mean θ is the Inverse Gaussian which has thick tails and provides the advantage of having closed form expression for the moment generating function. It is considered to be a reasonable distribution for modelling in many insurance situations. Such Poisson Inverse Gaussian models are appropriate in bonus-malus systems in the insurance industry [Tre92].

Similar models like the Exponential Inverse Gaussian model are also frequently used in describing the amount paid in a contract. The mean of the exponential distributions assumed follows the Inverse Gaussian distribution although the Gamma distribution may also be considered as an adequate candidate.

Accelerated Life Time Models

(a) Shelf Life Failures. The distribution of the shelf life of a product by sensory evaluation is a key to the improvement of the method of estimation of shelf life. The definition of time to failure in sensory testing is purely subjective. The mean time to failure based on simple averages is biased by the inclusion of unfailed data. This is often overcome by defining failure time as the time required for a sample to reach a median or an average panel score of 3.5 on a seven-point rating scale. In a shelf life test, a sample of n items of a product is evaluated for failures at predesignated periods, and the age to failure is recorded. When the allotted experimental time elapses, the ages of the samples that did not fail are also recorded. The total experimental time is chosen by the examiner. Thus, life testing produces two sets of data: the times to failure of the flawed items and the running times of the unflawed items. Data sets of shelf lives of several products have been examined by Gacula and Kubala [GK75], Folks and Chhikara [FC78] and O'Reilly and Rueda [OR92].

In such cases, the recording failures are based on the average panel score of 3.5 on a seven-point off-flavor rating scale. The contribution from the presence of yeasts and high bacterial count, rendering the product unsuitable for taste, was also a factor. Either the lack of samples or the expiration of the total time from the date of production was a factor in deciding the running time, with the samples obtained from the manufacturer and sent to the laboratory for sensory testing. Such data are usually analyzed using the normal, log-normal, exponential, Weibull, and extreme-value distributions with limiting success. On the other hand some statistical techniques identify the IG2 as a good fit while others fail to do so. New advanced tests of fit are required to show beyond any doubt the appropriateness of the IG2 distribution.

(b) Endurance of Ball Bearings. Ball bearings can be used in a life test study by measuring in millions the revolutions to failure. Similar examples include the hours of operation of light bulbs.

Reliability and Engineering

(a) Traffic Data. The length of intervals between the times at which vehicles pass a point on a road in seconds are recorded and used for a life test study.

(b) Machine Operation Data. The operating hours between successive failures of aircondition units in aircrafts or in hospital operating rooms.

(c) Occupational Exposure Concentrations. In manufacturing, workers are exposed to fluids (often called metalworking fluids, MWF) which are used during machining and grinding to prolong the life of the tool, carry away debris, and protect the surfaces of work pieces. These fluids reduce friction between the cutting tool and the work surface, reduce wear and galling, protect surface characteristics, reduce surface adhesion or welding and carry away generated heat. Workers can be exposed to MWFs by inhaling aerosols (mists) and by skin contact with the fluid. Skin contact occurs by dipping the hands into the fluid, splashes, or handling workpieces coated with the fluids. The amount of mist generated (and the resulting level of exposure) depends on many factors: the type of MWF and its application process; the MWF temperature; the specific machining or grinding operation; the presence of splash guarding; and the effectiveness of the ventilation system in capturing and removing the mist. Substantial scientific evidence indicates that workers currently exposed to MWF aerosols have an increased risk of respiratory (lung) and skin diseases. These health effects vary based on the type of MWF, route of exposure, concentration, and length of exposure. Occupational exposure concentrations are generally assumed to vary in an Inverse Gaussian or log-normal manner.

Hydrology

Precipitation and Storage. Precipitation data from a certain geographic area and the amount of water maintained in a dam and its discharge provide interesting and important examples which can be analysed by distributions such as the ones discussed here. The water in a dam is a well known storage problem where the dam collects irregularly amounts of water and discharges water (usually) at a steady rate except when it is empty.

Biomedicine

(a) Survival Modelling. Many failure time data in epidemiological studies are simultaneously truncated and interval-censored. Interval-censored data occur in grouped data or when the event of interest is assessed on repeated visits. Right and left censored data are particular cases of interval-censored data. Right truncated data occur in registers. For instance, an acquired immune deficiency syndrome (AIDS) register only contains AIDS cases, which have been reported. This generates right-truncated samples of induction times. Recently, Huber-Carol and Vonta [HV04] introduced frailty models in the case of arbitrarily censored and truncated data and focused on estimation of the parameter of interest as well as the nuisance parameter of their model.

The concept of frailty models was introduced by Vaupel et al. [VMS79] who studied models with Gamma distributed frailties. There are many frailty distributions one could consider, like the Gamma which corresponds to the well-known Clayton–Cuzick model [CC85, CC86], the Inverse Gaussian and the Positive Stable. The choice of Gamma is the most popular; however, due to its mathematical convenience. The determination of appropriate regression parameters on frailty models plays a key role in describing as accurately as possible the AIDS data. As a result, hypothesis testing about the regression parameter of the frailty model is of particular interest and should be investigated in different situations such as the case of independent and dependent covariates, the misspecification of the truncated proportion of the population and the misspecification of the frailty distribution producing the data. The performance of such tests of fit is based on the determination of the asymptotic distribution of the proposed test statistic under both the null and the alternative distributions.

(b) Length of Hospital Stay. Description of length-of-hospital stay for certain types of illness and, particularly, for psychiatric illnesses can be examined under the present setting. Some preliminary research on schizophrenic patients released from hospital show that although there is a significant departure from the IG2 distribution, the fit is still quite good considering the difficulty which is experienced generally in finding acceptable theoretical models for psychiatric hospital-stay data.

References

[AN94] Aguirre, N., Nikulin, M.: Chi-squared goodness-of-fit test for the family of logistic distributions. Kybernetika, **30**, 214–222 (1994)

[Aka73] Akaike, H.: Information theory and an extension of the maximum likelihood principle. In: Petrov B. N. and Csaki F. (eds) Proceedings of the 2nd International Symposium on Information Theory, Akademiai Kaido, Budapest (1973)

[AS66] Ali, S. M., Silvey, S. D.: A general class of coefficients of divergence of one distribution from another. J. Roy. Statist. Soc. B, **28**, 131–142 (1966)

[BN02] Bagdonavičius, V., Nikulin, M. S.: Goodness-of-fit tests for accelerated life models. In: Huber-Carol, C., Balakrishnan, N., Nikulin, M. S., and Mesbah, M. (eds) Goodness-of-fit tests and Model Validity, Birkhäuser, Boston, 281–297 (2002)

[BHHJ98] Basu, A., Harris, I. R., Hjort, N. L., Jones, M. C.: Robust and efficient estimation by minimising a density power divergence. Biometrika, **85**, 549-559 (1998)

[Cav04] Cavanaugh, J. E.: Criteria for linear model selection based on Kullback's symmetric divergence. Aust. N. Z. J. Stat., **46**, 257–274 (2004)

[CLY04] Chen, H. S., Lai, K., Ying, Z.: Goodness-of-fit tests and minimum power divergence estimators for survival data. Stat. Sin., **14**, 231–248 (2004)

[CF77] Chhikara, R. S., Folks, J. L.: The inverse Gaussian distribution as a lifetime model. Technometrics, **19**, 461–468 (1977)

[CC85] Clayton, D., Cuzick, J.: Multivariate generalizations of the proportional hazards model (with discussion). J. Roy. Stat. Soc., **A 148**, 82–117 (1985)

[CC86] Clayton, D., Cuzick, J.: The semiparametric Pareto model for regression analysis of survival times. Papers on Semiparametric Models MS-R8614, Centrum voor Wiskunde en Informatica, Amsterdam, 19–31 (1986)

[CR84] Cressie, N., Read, T. R. C.: Multinomial goodness-of-fit tests, JRSSB, **5**, 440–454 (1984)

[Csz63] Csiszar, I.: Eine Informationstheoretische Ungleichung und ihre Anwendung auf den Bewis der Ergodizitat on Markhoffschen Ketten. Publication of the Mathematical Institute of the Hungarian Academy of Sciences, **8**, 84–108 (1963)

[DCL04] Di Crescenzo, A., Longobardi, M.: A measure of discrimination between past lifetime distributions. Stat. Probab. Lett., **67**, 173–182 (2004)

[EK96] Ebrahimi, N., Kirmani, S. N. U. A.: A measure of discrimination between two residual life-time distributions and its applications. Ann. Inst. Stat. Math., **48**, 257–265 (1996)

[FC78] Folks, J. L., Chhikara, R. S.: The inverse Gaussian distribution and its statistical application – a review. J. Roy. Stat. Soc. B, **40**, 263–289 (1978)

[GK75] Gacula Jr., M. C., Kubala, J. J.: Statistical models for shelf life failures. J. Food Sci., **40**, 404–409 (1975)

[HV04] Huber-Carol, C., Vonta, F.: Frailty models for arbitrarily censored and truncated data. Lifetime Data Anal., **10(4)**, 369–388 (2004)

[HPPL98] Huberman, B. A., Pirolli, P. L. T., Pitkow, J. E., Lukose, R. M.: Strong regularities in world wide web surfing. Science, **280**, 95–97 (1998)

[KL51] Kullback, S., Leibler, R.: On information and sufficiency. Ann. Math. Stat., **22**, 79–86 (1951)

[Mar06] Marsh, P.: Data driven likelihood ratio tests for goodness-to-fit with estimated parameters. Discussion Papers in Economics, Department of Economics and Related Studies, The University of York, **2006/20** (2006)

[MK10] Mattheou, K., Karagrigoriou, A.: A new family of divergence measures for tests of fit. Aust. N. Z. J. Stat. **52**, 187–200 (2010)

[MLK09] Mattheou, K., Lee, S., Karagrigoriou, A.: A model selection criterion based on the BHHJ measure of divergence. J. Stat. Plan. Infer., **139**, 128–135 (2009)

[MPPP97] Menéndez, M. L., Pardo, J. A., Pardo, L., Pardo, M. C.: Asymptotic approximations for the distributions of the (h, ϕ)-divergence goodness-of-fit statistics: applications to Rényi's statistic. Kybernetes, **26**, 442–452 (1997)

[ME06] Moghadam, M., Eskandari, F.: Quality improvement by using inverse Gaussian model in robust engineering. Qual. Quantity, **40**, 157–174 (2006)

[OR92] O'Reilly, F.J., Rueda, R.: Goodness of fit for the inverse Gaussian distribution. Can. J. Stat., **20**, 387–397 (1992)

[Par06] Pardo, L.: Statistical Inference Based on Divergence Measures. Chapman and Hall/CRC, London (2006)

[Rea84] Read, T. R. C.: Closer asymptotic approximations for the distributions of the power divergence goodness-of-fit statistics. Ann. Inst. Stat. Math., **36**, 59–69 (1984)

[RC88] Read, T. R. C., Cressie, N.: Goodness-of-Fit Statistics for Discrete Multivariate Data. Springer Verlag, New York (1988)

[Sim03] Sim, C. H.: Inverse Gaussian control charts for monitoring process variability. Commun. Stat., Simul. Comput., **32**, 223–239 (2003)

[SF84] Siotani, M., Fujikoshi, Y.: Asymptotic approximations for the distributions of multinomial goodness-of-fit statistics. Hiroshima Math. J. **14**, 115–124 (1984)

[Tre92] Tremblay, L.: Using the Poisson inverse Gaussian in Bonus-Malus systems. ASTIN Bulletin International Actuarial Association, **22:1**, 97–106, http://www.casact.org/library/astin/vol22no1/97.pdf (1992)

[VMS79] Vaupel, J. W., Manton, K. G., Stallard, E.: The impact of heterogeneity in individual frailty on the dynamics of mortality. Demography, **16**, 439–454 (1979)

[VA07] Vonta, F., Artemiou, A.: Hypothesis testing in frailty models for arbitrarily censored and truncated data. Comm. Dependability Qual. Man. **10(1)**, 110–121 (2007)

[VK10] Vonta, F., Karagrigoriou, A.: Generalized measures of divergence in survival analysis and reliability. J. Appl. Probab., **47(1)**, 216–234 (2010)

[YTM06] Yanagihara, H., Tonda, T., Matsumoto, C.: Bias correction of cross-validation criterion based on the Kullback-Leibler information under a general condition. J. Multivariate Anal. **97**, 1965–1975 (2006)

[Yar72] Yarnold, J. K.: Asymptotic approximations for the probability that a sum of lattice random vectors lies in a convex set. Ann. Math. Stat., **43**, 1566–1580 (1972)

[Zha02] Zhang, J.: Powerful goddness-of-fit tests based on likelihood ratio. J. R. Stat. Soc. Ser. B, **64(2)**, 281–294 (2002)

[ZFP90] Zografos, K., Ferentinos, K., Papaioannou, T.: Φ-divergence statistics: sampling properties, multinomial goodness of fit and divergence tests. Comm. Statist. Theor. Meth., **19(5)**, 1785–1802 (1990)

Part IV

Reliability Computer Tools

32

Software System for Simulation and Research of Probabilistic Regularities and Statistical Data Analysis in Reliability and Quality Control

Ekaterina V. Chimitova, Boris Yu. Lemeshko, Stanislav B. Lemeshko, Sergey N. Postovalov, and Andrey P. Rogozhnikov

Novosibirsk State Technical University, Novosibirsk, Russia

Abstract: The computer approach to the investigation of estimation methods and statistical tests is considered as an effective technique for developing apparatus of applied mathematical statistics. It has been shown that basing on the considered approach and software system one can investigate statistical properties of estimates for distribution parameters including estimates by grouped and censored samples. The statistic distributions of nonparametric goodness-of-fit tests in testing composite hypotheses have been investigated. The statistic distributions and the power of χ^2 goodness-of-fit tests have been investigated depending on the number of intervals and the grouping method. A number of tests for deviation from the normal distribution law have been investigated. Homogeneity tests (for testing hypotheses about equality of means, equality of variances and homogeneity of distributions) have been studied. Various classical tests have been investigated in case of non-normal distributions of observations.

Keywords and phrases: Computer simulation, Nonparametric goodness-of-fit tests, χ^2 goodness-of-fit tests, Normality tests, Tests for homogeneity of distributions, Tests for homogeneity of means, Tests for homogeneity of variances, The test power

32.1 Introduction

The practice of using statistical analysis methods in applications is full of various problems whose statements are not described within the framework of classical assumptions. A wide range of statistical methods are based on the assumption of measurement error normality. Under real conditions normality and often some other assumptions are not satisfied. The use of classical methods of mathematical statistics in such situations can turn out to be incorrect.

Many classical results have an asymptotical nature. At the same time in practice one usually works with samples of a limited size. The application of asymptotical results is not always valid for limited sample sizes.

The form of data (measurements) registration doesn't often conform to complete samples considered in mathematical statistics textbooks. Actually, samples of

V.V. Rykov et al. (eds.), *Mathematical and Statistical Models and Methods in Reliability: Applications to Medicine, Finance, and Quality Control*, Statistics for Industry and Technology, DOI 10.1007/978-0-8176-4971-5_32,

observations can be grouped, censored, partially grouped or interval. Mathematical techniques must give an ability to analyze data in any form and must take into account this form and not to neglect it.

As a rule revealing fundamental statistic regularities in nonstandard conditions is a complicated problem for researchers. And the analytical methods for investigating properties of statistical estimates and test statistic distributions are very difficult and as a result of their complexity don't allow researchers to solve a great number of problems. The best way out is to use the numerical approach that is computer modeling of statistical regularities under conditions which simulate some real situations of measurement taking. Then mathematical models approximating the regularities obtained are constructed. Such an approach allows us to obtain good results in dealing with problems which are difficult to solve by analytical methods only. That is why computer simulation methods for statistical regularity analysis are becoming more and more popular.

This paper is devoted to the consideration of results obtained in various chapters of applied mathematical statistics with usage of the developed computer approach and software system meant for research of statistical regularities and statistical data analysis.

32.2 The Investigation of Parameter Estimates Properties

It has been shown in [Lem97a, Lem97b] that the usage of data grouping in tasks of distribution model identification enables to obtain robust estimates, eliminating an influence of gross measurement errors existed in samples. And the usage of asymptotically optimal grouping, for which losses of the Fisher information are minimized, enables to obtain estimates with good asymptotical properties.

The Fisher information losses caused by sample censoring have been considered in [LGP01, Lem01]. It has been shown that in some cases even for the considerable censoring degree the losses of the Fisher information induced with censoring of samples are not large. This enables to obtain rather good estimates of distribution parameters. The distributions of maximum likelihood estimates (MLE) of distribution parameters by censored samples have been investigated by computer simulation methods for various censoring degrees and various sample sizes. It has been shown that for the limited sample sizes the distribution of MLE turns out to be asymmetric and MLE is biased. The distribution laws frequently used in "life time" data analysis, such as lognormal, exponential, gamma, Rayleigh, Weibull, and other distributions have been considered.

32.3 The Investigation of Nonparametric Goodness-of-Fit Test Statistic Distributions

In composite hypotheses testing of the form $H_0 : F(x) \in \{F(x, \theta), \quad \theta \in \Theta\}$, when the estimate $\hat{\theta}$ of the scalar or vector distribution parameter $F(x, \theta)$ is calculated by

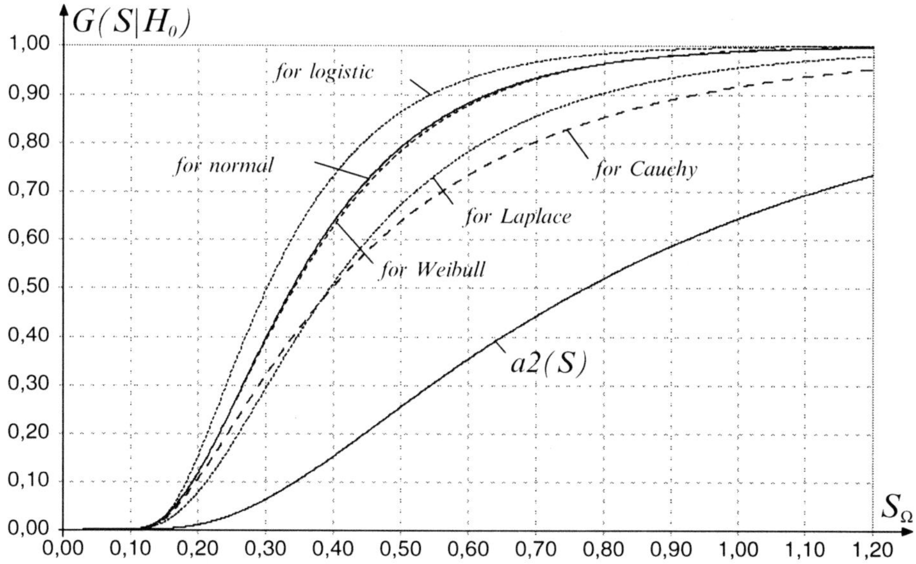

Figure 32.1. The Anderson–Darling statistic distributions for testing composite hypotheses with calculating MLE of two law parameters

the same sample, the nonparametric goodness-of-fit Kolmogorov, ω^2 Cramer–Mises–Smirnov, Ω^2 Anderson–Darling tests lose the free distribution property [KKW55]. In this case the conditional distribution law of the statistic $G(S\,|H_0)$ is affected by a number of factors: the form of the observed law $F(x, \theta)$ corresponding to the true hypothesis H_0; the type of the parameter estimated and the number of parameters to be estimated; sometimes, it is a specific value of the parameter (e.g., in the case of gamma-distribution and beta-distribution families and others); the method of parameter estimation [LP99]. The distinctions in the limiting distributions of the same statistics in testing simple and composite hypotheses are so significant that we cannot neglect them. For example, Fig. 32.1 shows distributions of the Anderson–Darling statistic while testing the composite hypotheses subject to various laws using maximum likelihood estimates (MLE) of two parameters.

Figure 32.2 illustrates the dependence of Kolmogorov test statistic distribution upon the type and the number of estimated parameters by the example of *Su*-Jonson law.

In our research [LP99, LP98a, LP01, Rec02ii, LM04a, Lem04, DELT04, LL09a, LNS09, LL09b, LL09c] statistic distributions of the nonparametric goodness-of-fit tests have been investigated by the methods of statistical simulating. Then basing on the obtained empirical statistic distributions we have constructed approximate analytical models of the statistic distribution laws. Table 32.1 contains a list of distributions relative to which we can test composite goodness-of-fit hypotheses using the constructed approximations of the limiting nonparametric statistic distributions.

One can use the models presented in [DELT04, LL09a, LNS09, LL09b, LL09c] in tasks of statistical data analysis, beginning from the sample size $n > 25$ and using the maximum likelihood estimates of unknown parameters.

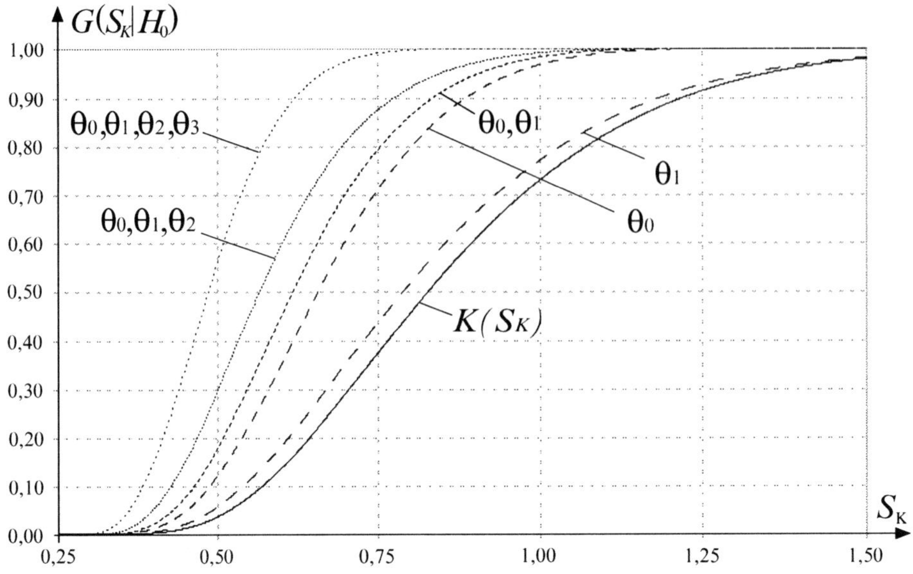

Figure 32.2. The Kolmogorov statistic distributions for testing composite hypotheses with calculating MLE of *Su*-Jonson distribution law parameters

---

32.4 The Investigation of Statistic Distributions and the Power of χ^2 Tests

It has been shown in [DL79] that the less information losses caused by grouping are, the higher power of χ^2 tests (the Pearson χ^2 test and the likelihood ratio test) for close competing hypotheses.

Information losses can be decreased by selecting boundary points so, that $\mathbf{J}_G(\theta)$ tends to the information matrix for nongrouped data $\mathbf{J}(\theta)$, i.e. by solving asymptotically optimal grouping problem.

In case of scalar parameter, the problem reduces to the maximization of Fisher information quantity for grouped sample

$$\max_{x_{(1)}<x_{(2)}<\cdots<x_{(k-1)}} \sum_{i=1}^{k} \left(\frac{\partial \ln P_i(\theta)}{\partial \theta} \right)^2 P_i(\theta) = \max_{x_{(1)}<x_{(2)}<\cdots<x_{(k-1)}} J_G(\theta) .$$

And in case of vector parameter various functionals of the Fisher information matrix can be chosen.

D-optim: the determinant of information matrix is maximized with respect to the boundary points

$$\max_{x_0<x_1<\ldots<x_{k-1}<x_k} \det \mathbf{J}_G(\theta)$$

(asymptotically D-optimal grouping problem).

Table 32.1. Random variable distribution laws

Random variable distribution	Density function $f(x,\theta)$	Random variable distribution	Density function $f(x,\theta)$
Exponential	$\frac{1}{\theta_0}\mathrm{e}^{-x/\theta_0}$	Laplace	$\frac{1}{2\theta_0}\mathrm{e}^{-\lvert x-\theta_1\rvert/\theta_0}$
Seminormal	$\frac{2}{\theta_0\sqrt{2\pi}}\mathrm{e}^{-x^2/2\theta_0^2}$	Normal	$\frac{1}{\theta_0\sqrt{2\pi}}\mathrm{e}^{-\frac{(x-\theta_1)^2}{2\theta_0^2}}$
Rayleigh	$\frac{x}{\theta_0^2}\mathrm{e}^{-x^2/2\theta_0^2}$	Log-normal	$\frac{1}{x\theta_0\sqrt{2\pi}}\mathrm{e}^{-(\ln x-\theta_1)^2/2\theta_0^2}$
Maxwell	$\frac{2x^2}{\theta_0^3\sqrt{2\pi}}\mathrm{e}^{-x^2/2\theta_0^2}$	Cauchy	$\frac{\theta_0}{\pi[\theta_0^2+(x-\theta_1)^2]}$

Random variable distribution	Density function $f(x,\theta)$
Logistic	$\frac{\pi}{\theta_0\sqrt{3}}\exp\left\{-\frac{\pi(x-\theta_1)}{\theta_0\sqrt{3}}\right\}\Big/\left[1+\exp\left\{-\frac{\pi(x-\theta_1)}{\theta_0\sqrt{3}}\right\}\right]^2$
Extreme-value (maximum)	$\frac{1}{\theta_0}\exp\left\{-\frac{x-\theta_1}{\theta_0}-\exp\left(-\frac{x-\theta_1}{\theta_0}\right)\right\}$
Extreme-value (minimum)	$\frac{1}{\theta_0}\exp\left\{\frac{x-\theta_1}{\theta_0}-\exp\left(\frac{x-\theta_1}{\theta_0}\right)\right\}$
Weibull	$\frac{\theta_0 x^{\theta_0-1}}{\theta_1^{\theta_0}}\exp\left\{-\left(\frac{x}{\theta_1}\right)^{\theta_0}\right\}$
Sb- Johnson $Sb(\theta_0,\theta_1,\theta_2,\theta_3)$	$\frac{\theta_1\theta_2}{(x-\theta_3)(\theta_2+\theta_3-x)}\exp\left\{-\frac{1}{2}\left[\theta_0-\theta_1\ln\frac{x-\theta_3}{\theta_2+\theta_3-x}\right]^2\right\}$
Sl-Johnson $Sl(\theta_0,\theta_1,\theta_2,\theta_3)$	$\frac{\theta_1}{(x-\theta_3)\sqrt{2\pi}}\exp\left\{-\frac{1}{2}\left[\theta_0+\theta_1\ln\frac{x-\theta_3}{\theta_2}\right]^2\right\}$
Su-Johnson $Su(\theta_0,\theta_1,\theta_2,\theta_3)$	$\frac{\theta_1}{\sqrt{2\pi}\sqrt{(x-\theta_3)^2+\theta_2^2}}$ $\times\exp\left\{-\frac{1}{2}\left[\theta_0+\theta_1\ln\left\{\frac{x-\theta_3}{\theta_2}+\sqrt{\left(\frac{x-\theta_3}{\theta_2}\right)^2+1}\right\}\right]^2\right\}$
Gamma-distribution $\gamma(\theta_0,\theta_1,\theta_2)$	$\frac{1}{\theta_1^{\theta_0}\Gamma(\theta_0)}(x-\theta_2)^{\theta_0-1}\mathrm{e}^{-(x-\theta_2)/\theta_1}$
Double-exponential	$\frac{\theta_0}{2\theta_1\Gamma(1/\theta_0)}\exp\left\{-\left(\frac{\lvert x-\theta_2\rvert}{\theta_1}\right)^{\theta_0}\right\}$
Beta-distribution of the I type	$\frac{1}{\theta_2\mathrm{B}(\theta_0,\theta_1)}\left(\frac{x}{\theta_2}\right)^{\theta_0-1}\left(1-\frac{x}{\theta_2}\right)^{\theta_1-1}$
Beta-distribution of the II type	$\frac{1}{\theta_2\mathrm{B}(\theta_0,\theta_1)}\frac{[x/\theta_2]^{\theta_0-1}}{[1+x/\theta_2]^{\theta_0+\theta_1}}$
Generalized Weibull	$\frac{\theta_0}{\theta_1}\theta_2^{\theta_0}x^{\theta_0-1}\left(1+\left(\frac{x}{\theta_2}\right)^{\theta_0}\right)^{\frac{1}{\theta_1}-1}\mathrm{e}^{1-\left(1+\left(\frac{x}{\theta_2}\right)^{\theta_0}\right)^{\frac{1}{\theta_1}}}$
Inverse Gaussian	$\left(\frac{\lambda}{2\pi x^3}\right)^{1/2}\exp\left(-\frac{\lambda(x-\mu)^2}{2\mu^2 x}\right)$

A-optim: the trace of information matrix is maximized with respect to the boundary points

$$\max_{x_0<x_1<\cdots<x_{k-1}<x_k} Sp\mathbf{J}_G(\theta)$$

(asymptotically A-optimal grouping problem).

E-optim: the minimal eigenvalue of information matrix is maximized with respect to the boundary points

$$\max_{x_0<x_1<\cdots<x_{k-1}<x_k}\ \min_{i=\overline{1,r}}\lambda_i\left(\mathbf{J}_G(\theta)\right)$$

(asymptotically E-optimal grouping problem).

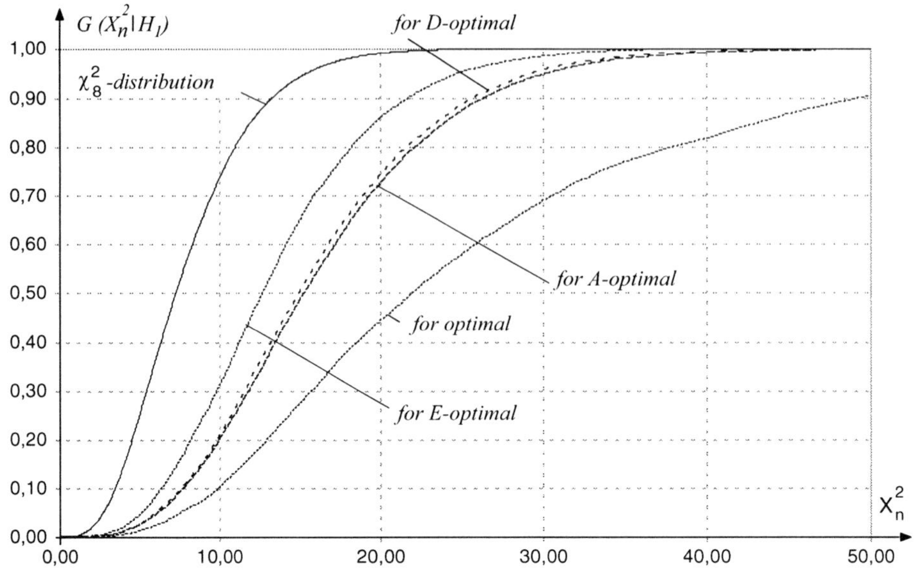

Figure 32.3. Distributions of Pearson's statistic X_n^2 in testing simple hypothesis H_0 if the hypothesis H_1 is true depending on the grouping method with $k = 9$ and $n = 500$

Optimum: the ν is maximized with respect to the boundary points

$$\max_{x_{(1)}<x_{(2)}<\cdots<x_{(k-1)}} s = \max_{x_{(1)}<x_{(2)}<\cdots<x_{(k-1)}} \left(n \sum_{i=1}^{k} \frac{\left(P_i(\theta_1) - P_i(\theta)\right)^2}{P_i(\theta)} \right).$$

The tables of asymptotically D-optimal grouping for rather wide range of distribution laws which are most frequently used in applications were constructed previously [DLT93]. At the present time we have solved the problems of A- and E-optimal grouping for a number of distributions. The tables of asymptotically optimal grouping which can be used in estimating distribution parameters by grouped samples and in testing goodness-of-fit have been constructed. The use of asymptotically optimal grouping tables ensures the maximal power of χ^2 tests for close competing hypotheses. In Fig. 32.3, there are the Pearson χ^2 test statistic distributions in testing simple hypothesis of goodness-of-fit to the normal distribution $H_0 : f(x) = \frac{1}{\theta_0\sqrt{2\pi}} \exp\left\{-\frac{(x-\theta_1)^2}{2\theta_0^2}\right\}, \theta_0 = 1, \theta_1 = 0$, in case of the true competing hypothesis $H_1 : f(x) = \frac{\pi}{\theta_0\sqrt{3}} \exp\left\{-\frac{\pi(x-\theta_1)}{\theta_0\sqrt{3}}\right\} \Big/ \left[1 + \exp\left\{-\frac{\pi(x-\theta_1)}{\theta_0\sqrt{3}}\right\}\right]^2, \theta_0 = 1, \theta_1 = 0$ (logistic distribution), in dependence on the grouping method.

It has been shown for the first time that there is an optimal number of intervals k depending on sample size, concrete alternatives and a way of grouping. The optimal number of intervals k depends on the sample size n and on the concrete pair of competing hypotheses H_0 and H_1. As a rule, the optimal k turns out to be significantly less than values recommended by a number of empirical formulas for the choice of k. In Fig. 32.4 the power functions of the Pearson χ^2 tests are represented depending on the interval number k in case of D-, A-, and E-optimal grouping in simple hypothesis testing, for $n = 200$ (H_0 : normal distribution; against H_1 : logistic distribution).

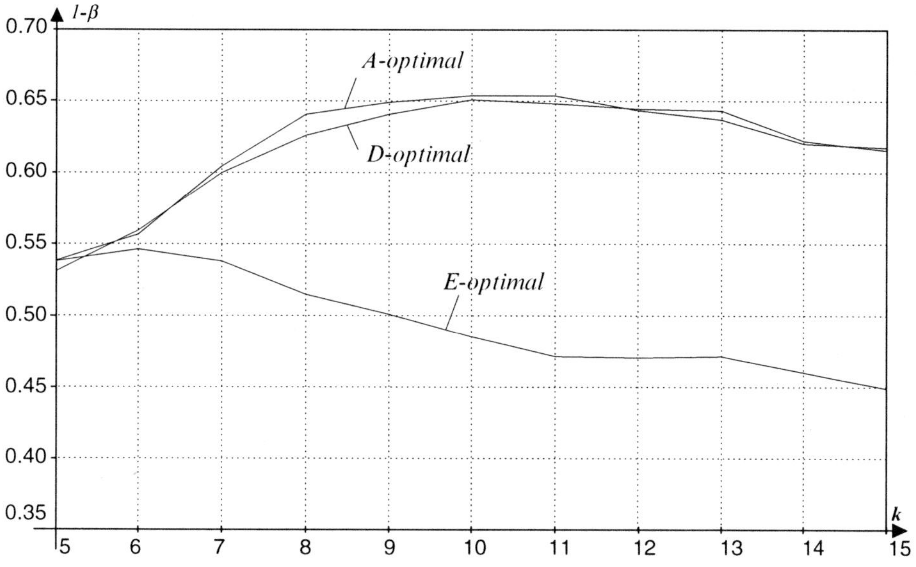

Figure 32.4. The dependence of the Pearson χ^2 test power on the number of intervals k for various grouping methods, $n = 200$, in case of testing H_0: normal distribution against H_1: logistic distribution

The results of investigating [Lem97c, Lem98, LP98b, LC00, LPC01, LC03, LC02] properties of the χ^2 goodness-of-fit tests (Pearson, Rao-Robson-Nikulin [Nik73, NC73, RR74, GN96]) were included to the developed recommendations [Rec02i].

The power of the χ^2 Dzhaparidze–Nikulin test has been investigated depending on the grouping method and the number of intervals. The problem of power maximization for the χ^2 Pearson and Rao–Robson–Nikulin tests has been investigated for specified pairs of competing hypotheses. Moreover, we have considered the use of the so called Neumann–Pearson intervals [GN96], for which the boundary points coincide with cross points of density functions of competing hypotheses. It has been shown that such intervals are reasonable to be used. But at the same time, the use of these intervals doesn't ensure the maximal power of the test for given pair of competing hypotheses.

32.5 The Comparative Analysis of the Power of Goodness-of-Fit Tests

The power of a number of nonparametric and parametric goodness-of-fit tests with respect to a series of pairs of competing hypotheses has been studied [LLP07, LLP09, LLP08] in case of testing simple and composite hypotheses by statistical simulation methods. We can rank the tests by power for simple hypothesis testing as follows:

$$\chi^2 \text{ Pearson (AOG)} \succ \Omega^2 \text{ Anderson–Darling} \succ \omega^2 \text{ Mises} \succeq \text{Kolmogorov}.$$

This scale holds while using asymptotically optimal grouping in the Pearson χ^2 test, which minimizes the losses in the Fisher information. For quite close hypotheses, the advantage by power of the Pearson χ^2 test can be essential.

In testing composite hypotheses the preference sequence turns out to be quite different:

$$\Omega^2 \text{ Anderson–Darling} \succ \omega^2 \text{ Mises} \succ Y_n^2 \text{ Rao–Robson–Nikulin (AOG)} \succ \\ \succ \chi^2 \text{ Pearson (AOG)} \succ \text{Kolmogorov.}$$

For very close competing hypotheses the following sequence can take place:

$$\Omega^2 \text{ Anderson–Darling} \succ Y_n^2 \text{ Rao–Robson–Nikulin (AOG)} \succ \omega^2 \text{ Mises} \succ \\ \succ \chi^2 \text{ Pearson (AOG)} \succ \text{Kolmogorov.}$$

The conclusions stated have an integrated nature.

32.6 The Investigation of Statistic Distributions and the Power of Normality Tests

Statistic distributions and the power of a number of criteria for testing deviation from the normal law (Shapiro-Wilk, Epps-Pulley, D'Agostino, Frosini, Hegazy-Green, Spiegelhalter, Geary, David-Hartley-Pearson and some others) have been investigated in [LL05a, LR09].

The considered tests can be ranked by power as follows:

$$\text{Geary} \succ \text{Spiegelhalter} \succ \text{Hegazy-Green } (T_2) \succ \text{Hegazy-Green } (T_1) \succ \\ \succ \text{Epps-Pulley} \succ \text{David-Hartley-Pearson} \succ \text{Shapiro-Wilk} \succ \text{Frosini.}$$

The advantages and disadvantages of various tests have been shown. It has been shown for the first time that for small sample sizes a number of tests are biased, including the Spiegelhalter, Shapiro-Wilk, Epps-Pulley and Hegazy-Green tests relative to symmetrical alternatives with the kurtosis value less than three. Some of the considered tests are not reasonable to be applied at all because of their fundamental disadvantages. The normality tests have been compared by power with the goodness-of-fit tests.

32.7 The Investigation of Homogeneity Test Statistic Distributions

The homogeneity tests are intended for checking whether two random samples represented by the variation series

$$x_1 < x_2 < \cdots < x_m \text{ and } y_1 < y_2 < \cdots < y_n$$

belong to the same distribution, i.e. $H_0 : F(x) = G(x)$ for any x.

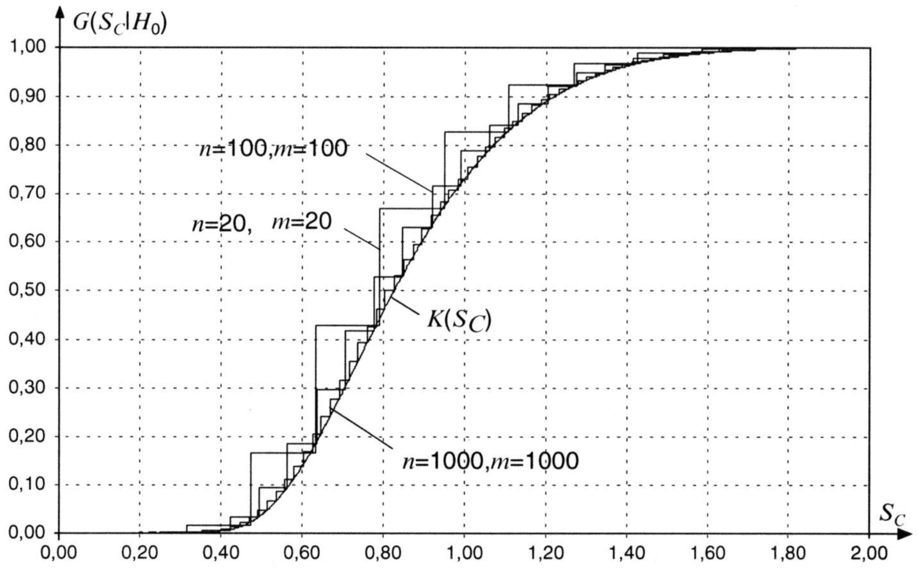

Figure 32.5. The Smirnov statistic distributions when the null hypothesis is true depending on the sample sizes m and n

The statistic distributions and the power of the Smirnov and Lehmann–Rosenblatt tests for homogeneity of two samples were investigated in [LL05b]. The Smirnov test statistic is a discrete random variable and its distribution converges slowly (from the left!) to the limiting Kolmogorov distribution (see Fig. 32.5). Hereupon, the use of the Kolmogorov distribution $K(s)$ as the limiting law when sample sizes are limited lead to the overrated values of significance level achieved and, hence, to increasing the number of beta errors. The recommendations of choosing sample sizes m and n are given in this paper. The empirical correction for the Smirnov statistic which improves convergence of the statistic distribution to the limiting Kolmogorov law has been obtained.

The power of the Lehmann–Rosenblatt test, as a rule, turns out to be higher than the power of the Smirnov test.

32.8 The Investigation of Statistic Distributions and the Power of Tests for Homogeneity of Means

The comparative analysis of the power of parametric and nonparametric criteria used for testing homogeneity of means has been carried out in [LL08]. In the general case, the hypothesis of mathematical expectations equality corresponding to k samples has the form

$$H_0 : \mu_1 = \mu_2 = \cdots = \mu_k$$

under the competing hypothesis

$$H_1 : \mu_{i_1} \neq \mu_{i_2}$$

for at least some pair of indices i_1, i_2.

There are a number of parametric tests that may be used to compare two sample means to check some hypothesis H_0: with known variances; with unknown, but equal variances (Student's test); with unknown and unequal variances; and with the F-test. There also exists a number of nonparametric tests that may be used for this purpose, e.g., the Wilcoxon, Mann–Whitney, and Kruskal–Wallis tests. Membership of the particular sample being analyzed to a normal law is the basic assumption determining whether parametric tests should be used. Nonparametric tests are free of this requirement.

It has been shown that parametric tests associated with testing a hypothesis of mathematical expectations are robust with respect to deviations of the observed laws from the normal distribution. If the distribution law (laws) of analyzed samples is different from the normal law but doesn't have the "heavy tails" than the use of parametric tests is correct, at least it doesn't result in considerable errors.

Some conclusions can be arrived on the basis of the test power investigation results. Firstly, the parametric tests have the greater power than do nonparametric tests. Secondly, it may be stated that nonparametric tests are absolutely slightly inferior in terms of power to parametric tests, thus, the Mann–Whitney test is inferior to the Student's test, and the Kruskal–Wallis test to the Fisher test, respectively.

32.9 The Investigation of Statistic Distributions and the Power of Tests for Homogeneity of Variances

One of the main assumptions which should be taken into account while constructing the classical tests for homogeneity of variances is the normality of observed random variables (measurement errors). Therefore, the application of classical tests is always associated with the question whether obtained conclusions are correct in a certain situation. The conditional test statistic distributions relative to a true hypothesis under test, as a rule, change significantly if the assumption of the normal distribution of analyzed random variables is disturbed.

The tested hypothesis about equality of variances in k samples has the form

$$H_0 : \sigma_1^2 = \sigma_2^2 = \cdots = \sigma_k^2,$$

And the competing hypothesis is

$$H_1 : \sigma_{i_1}^2 \neq \sigma_{i_2}^2,$$

where inequality holds at least for one pair of indices i_1, i_2.

The distributions of the classical test statistics have been investigated in case when distributions of observed random variables differ from the normal law. The possibility of the classical tests application under conditions of non-normal distributions has been studied. The comparative analysis by power of the classical variance homogeneity tests (Fisher's, Bartlett's, Cochran's, Hartley's and Levene's tests) and the nonparametric (rank) tests Ansari–Bradley, Mood's, Siegel–Tukey tests) have been carried out by statistical simulation methods in [LM04b, LP06] and further works.

We have investigated the power of Fisher's, Bartlett's, Cochran's, Hartley's and Levene's tests relative to the competing hypothesis of the kind $H_1 : \sigma_2 = d\sigma_1, d \neq 1$, for the number of samples $k = 2$ in case of normal distribution of random variables. It has been shown that in this situation the Fisher, Bartlett, Cochran and Hartley tests are equal by power. The Levene test considerably yields to them.

The Fisher, Bartlett, Cochran and Hartley tests remain to be equal by power if the random variable distribution is different from the normal law, e.g. in case of belonging of two analyzed samples to the double-exponential distribution law, and the Levene test yields to them. But in case of distributions with "heavy tails" the Levene test has an advantage in power.

The tests of Bartlett, Cochran, Hartley and Levene may be applied for number of samples $k > 2$. In such situations, the power of these tests turns out to be different. When the assumption of the normal distribution holds for $k > 2$ these tests may be ranked by power decrease as follows:

$$\text{Cochran} \succ \text{Bartlett} \succ \text{Hartley} \succ \text{Levene}.$$

This preference order also holds in case when the normality assumption is disturbed. An exception concerns the situations when samples belong to some distributions which have more heavy tails than the normal law. For example, in case of belonging samples to the Laplace distribution the Levene test turns out to be slightly more powerful than three others.

The results of investigating the nonparametric tests have shown an evident advantage of the Mood test and practical equivalence of the Ansari–Bradley and Siegel–Tukey tests. The nonparametric tests are obviously inferior by power to the Bartlett, Cochran, Hartley and Levene tests. In Fig. 32.6, there are the graphs of the test power relative to the competing hypotheses $H_1^1 : \sigma_2 = 1.1\sigma_1$ and $H_1^2 : \sigma_2 = 1.5\sigma_1$ depending on the sample sizes n_i for $\alpha = 0.1$ in case of the normal law. As it is seen from the figure the power of Cochran's test comparing with the most powerful nonparametric Mood's test is rather considerable. Let us remind that in case of two samples the power of the Fisher, Bartlett, Cochran and Hartley tests coincide.

The distributions of nonparametric tests don't depend on the observed distribution law if both samples belong to one and the same family of distributions. But if samples have different distributions than for the true tested hypothesis H_0 of variance homogeneity the distributions of nonparametric test statistics change: they depend on these distribution laws.

The classical tests have a considerable advantage by power over the nonparametric tests. This advantage remains even when analyzed samples belong to the distribution which is considerably different from the normal law. So there are all reasons for investigation of the classical test statistic distributions (construction of the distribution models or the tables of percentage points) for the non-normal laws frequently used in practice. The Cochran test is the most appropriate for this role among the considered tests.

The table of upper percentage points (1%, 5%, 10%) of the Cochran test (for the numbers of samples $k = 2/5$) have been constructed in case of some certain families

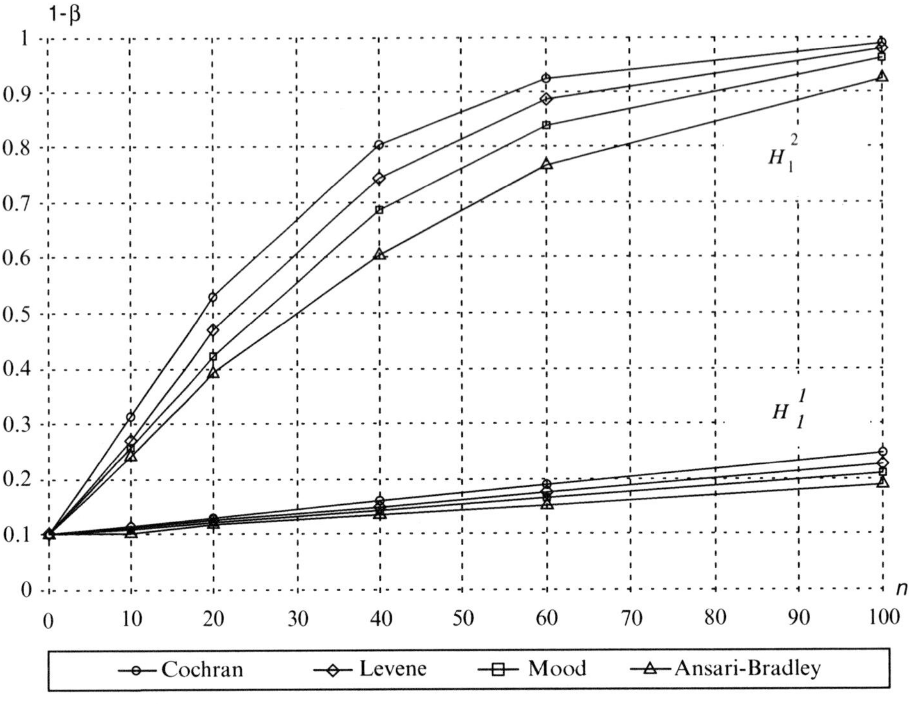

Figure 32.6. The power of tests relative to the competing hypotheses H_1^1 and H_1^2 depending on the sample size n for $\alpha = 0.1$ in case of the normal law

of observed random variable distributions for a number of sample size values n. The developed software system enables to solve this problem for any random variable distribution law and for any classical test of variance homogeneity, as well as it enables to construct models of statistic distributions for these tests when necessary.

32.10 Conclusion

The computer simulation technique of data analysis and investigation of probabilistic regularities have been used for solving other problems of applied mathematical statistics. In particular, we have investigated the robustness and power of the Abbe test used for testing hypothesis about the trend absence [Lem06]. The distributions of the Grabbs test statistic used in tasks of rejecting anomalous measurements have been investigated by statistical simulation methods in case when an observed law is different from the normal distribution [LL05c]. The statistic distributions of classical criteria of testing hypotheses about variances have been investigated when a random variable distribution differs from the normal law [LP04a]. We have developed the facilities of modeling and investigating the distribution laws of arbitrary functions of random variables and functions of random variable systems as well as the facilities of constructing approximate models for these distribution laws [LC07]. We have also developed the technique for simulation and investigation of distributions of multivariate random variables statistics [LP02].

So we can state that the computer technologies of data analysis and investigation of probabilistic and statistical regularities present the powerful tool for the development and improvement of the applied mathematical statistics apparatus including solving problems of reliability and survival analysis.

At the same time there are some own problems on the way of developing the computer technologies of data analysis and statistical regularities research [LP04b]. First of all, the construction of sufficiently precise models; for example, the models of test statistic distributions, basing on statistical simulation results frequently requires large amount of simulations (tens and hundreds hours of processor time). Secondly, the classical results involve mainly the most elementary situations. In more complicated cases, the decision can turn out to be ambiguous. For example, the distribution of some test statistic may depend on the value of a certain parameter of observed distribution law and cannot be expressed in the form of regularity, someway depending on this parameter. This means that the statistic distribution changes in dependence on solved task conditions. There exists the infinite number of combinations of the conditions and so it's impossible to construct an infinite number of models for all the affairs. Consequently, it turns out to be reasonable to construct a probabilistic model "in real time" when there occurs the necessity of decision making under conditions of available assumptions. It means that when testing a statistical hypothesis we have to specify the test statistic distribution, corresponding to the true hypothesis under test, in process of statistical analysis itself. And then basing on this distribution one will reject or will not reject the hypothesis under test. For the present condition of computing facilities and encouraging perspectives of their development the achievement of the purpose is feasible by organizing distributed computations using free facilities of computers and computer clusters in the networks. Our computing experiments have confirmed the possibility and efficiency of such approach because of comparatively simple paralleling simulation operations.

Acknowledgements. These researches have been performed with partial support from the Russian Foundation for Basic Research (Grant No. 09-01-00056-a) and the Federal Agency on Education of the Ministry of Education and Science of the Russian Federation within the framework of the analytical departmental targeted program "Development of higher school scientific potential" (Project No. 2.1.2/3970) and federal targeted program of the Ministry of Education and Science of the Russian Federation "Scientific and scientific-educational personnel of innovation Russia" (Projects NK-15P/15, NK-178P/1, NK-437P/6).

References

[Lem97a] Lemeshko, B.Yu.: Robust methods for the estimation and rejection of anomalous measurements. Ind. Lab., **V.63**, N 5, 297–302 (1997)

[Lem97b] Lemeshko, B.Yu.: Grouping observations as the way to obtain robust estimates. Reliab. Qual. Contrl., **V.5**, 26–35 (1997) [In Russian]

[LGP01] Lemeshko, B.Yu., Gildebrant, S.Ya., Postovalov, S.N.: On the estimation of reliability parameters by censored samples. Ind. Lab. Mat. Diag., **V.67**, N 1, 52–64 (2001) [In Russian]

[Lem01] Lemeshko, B.Yu.: On the estimation of distribution parameters and testing hypotheses by censored samples. Methods Qual. Manag., **V.4**, 32–38 (2001) [In Russian]

[KKW55] Kac, M., Kiefer, J., Wolfowitz, J.: On tests of normality and other tests of goodness of fit based on distance methods. Ann. Math. Stat., **V.26**, 189–211 (1955)

[LP99] Lemeshko, B.Yu., Postovalov, S.N.: On the rules of testing goodness-of-fit of empirical distribution with a theoretical law. Methods of Quality Management. Reliab. Qual. Contrl., **V.11**, 34–43 (1999) [In Russian]

[LP98a] Lemeshko, B.Yu., Postovalov, S.N.: On the distributions of nonparametric goodness-of-fit test statistics when estimating distribution parameters by samples. Ind. Lab. Mat. Diag., **V.64**, N 3, 61–72 (1998) [In Russian]

[LP01] Lemeshko, B.Yu., Postovalov, S.N.: Application of the nonparametric goodness-of-fit Tests in testing composite hypotheses. Optoelectronics, Instrument. Data Process., **V.2**, 76–88 (2001)

[Rec02ii] R 50.1.037-2002. Recommendations for Standardization. Applied Statistics. Rules of Check of Experimental and Theoretical Distribution of the Consent. Part II. Nonparametric Goodness-of-Fit Tests. Publishing House of the Standards, Moscow (2002) [In Russian]

[LM04a] Lemeshko, B.Yu., Maklakov, A.A.: Nonparametric test in testing composite hypotheses on goodness of fit exponential family distributions. Optoelectron. Instrument. Data Process., **V.40**, N 3, 3–18 (2004)

[Lem04] Lemeshko, B.Yu.: Errors when using nonparametric fitting criteria. Meas. Tech., **V.47**, N 2, 134–142 (2004)

[DELT04] Denisov, V.I., Eger, K.-H., Lemeshko, B.Yu., Tsoy, E.B.: Design of Experiments and Statistical Analysis for Grouped Observations. NSTU Publishing House, Novosibirsk (2004)

[LL09a] Lemeshko, B.Yu., Lemeshko, S.B.: Models of statistic distributions of nonparametric goodness-of-fit tests in composite hypotheses testing in case of double exponential law. In: Sakalauskas, L., Skiadas, C., Zavadskas, E.K. (eds) Proceedings of the 13th International Conference "Applied Stochastic Models and Data Analysis" (ASMDA-2009), Selected Papers. Technika, Vilnius, 153–157 (2009)

[LNS09] Lemeshko, S.B., Nikulin, M.S., Saaidia, N.: Simulation of the statistics distributions and power of the goodness-of-fit tests in composite hypotheses testing rather inverse Gaussian distribution. In: Ermakov, S.M., Melas, V.B., Pepelyshev, A.N. (eds) Proceedings of the 6th St. Petersburg Workshop on Simulation. VVM comm. Ltd., St. Petersburg, V.1, 323–328 (2009)

[LL09b] Lemeshko, B.Yu., Lemeshko, S.B.: Distribution models for nonparametric tests for fit in verifying complicated hypotheses and maximum-likelihood estimators. Part I. Meas. Tech., **V.52**, N 6, 555–565 (2009)

[LL09c] Lemeshko, B.Yu., Lemeshko, S.B. Models for statistical distributions in nonparametric fitting tests on composite hypotheses based on maximum-likelihood estimators. Part II. Meas. Tech., **V.52**, N 8, 799–812 (2009)

[DL79] Denisov, V.I., Lemeshko, B.Yu.: Optimal grouping in experimental data processing. In: Measuring Information Systems. Novosibirsk, 5–14 (1979) [In Russian]

[DLT93] Denisov, V.I., Lemeshko, B.Yu., Tsoy, E.B.: Optimal Grouping, Parameter Estimation and Design of Regression Experiments. In 2 parts. NSTU Publishing House, Novosibirsk (1993) [In Russian]

[Lem97c] Lemeshko, B.Yu.: Asymptotically optimal grouping of observations is ensuring of maximal test power. Reliab. Qual. Contrl., **V.8**, 3–14 (1997) [In Russian]

[Lem98] Lemeshko, B.Yu.: Asymptotically optimal grouping of observations in goodness-of-fit tests. Ind. Lab. Mat. Diag., **V.64**, N 1, 56–64 (1998) [In Russian]

[LP98b] Lemeshko, B.Yu., Postovalov, S.N. On the dependence of the limiting statistic distributions of the Pearson chi-squire and the likelihood ratio tests upon the grouping method. Ind. Lab. Mat. Diag., **V.64**, N 5, 56–63 (1998) [In Russian]

[LC00] Lemeshko, B.Yu., Chimitova, E.V.: Maximization of the power of chi-squire tests. Papers of Siberian branch of the Academy of Sciences of higher school, **V.2**, 53–61 (2000) [In Russian]

[LPC01] Lemeshko, B.Yu., Postovalov, S.N., Chimitova, E.V.: On statistic distributions and the power of the Nikulin chi-squire test. Ind. Lab., **V.67**, N 3, 52–58 (2001) [In Russian]

[LC03] Lemeshko, B.Yu., Chimitova, E.V.: On the choice of the number of intervals in chi-squire goodness-of-fit tests. Ind. Lab. Mat. Diag., **V.69**, N 1, 61–67 (2003) [In Russian]

[LC02] Lemeshko, B.Yu., Chimitova, E.V.: Errors and incorrect procedures when utilizing χ^2 fitting criteria. Meas. Tech., **V.45**, N 6, 572–581 (2002)

[Nik73] Nikulin, M.S.: On a chi-squared test for continuous distributions. Theory Probab. Appl., **V.18**, N 3, 638–639 (1973) [In Russian]

[NC73] Nikulin, M.S.: Chi-squared test for continuous distributions with shift and scale parameters. Theory Probab. Appl., **V.18**, N 3, 559–568 (1973) [In Russian]

[RR74] Rao, K.C., Robson, D.S.: A chi-squared statistic for goodness-of-fit tests within the exponential family. Commun. Statist., **V.3**, 1139–1153 (1974)

[GN96] Greenwood, P.E., Nikulin, M.S.: A Guide to Chi-Squared Testing. Wiley, New York (1996)

[Rec02i] R 50.1.033-2001. Recommendations for Standardization. Applied Statistics. Rules of Check of Experimental and Theoretical Distribution of the Consent. Part I. Goodness-of-Fit Tests of Chi-Square Type. Publishing House of the Standards, Moscow (2002) [In Russian]

[LLP07] Lemeshko, B.Yu., Lemeshko, S.B., Postovalov, S.N.: The power of goodness of fit tests for close alternatives. Meas. Tech., **V.50**, N 2, 132–141 (2007)

[LLP09] Lemeshko, B.Yu., Lemeshko, S.B., Postovalov, S.N.: Comparative analysis of the power of goodness-of-fit tests for near competing hypotheses. I. The verification of simple hypotheses. J. Appl. Ind. Math., **V.4**, N 4, 462–475 (2009)

[LLP08] Lemeshko, B.Yu., Lemeshko, S.B., Postovalov, S.N.: Comparative analysis of the power of goodness-of-fit tests for near competing hypotheses. II. The verification of composite hypotheses. Siberian J. Ind. Math., **V.11**, N 4(36), 78–93 (2008)

[LL05a] Lemeshko, B.Yu., Lemeshko, S.B.: Comparative analysis of criteria of testing deviation from the normal law. Metrology, **V.2**, 3–24 (2005) [In Russian]

[LR09] Lemeshko, B.Yu., Rogozhnikov, A.P.: The investigation of features and the power of some tests for normality. Metrology, **V.4**, 3–24 (2009) [In Russian]

[LL05b] Lemeshko, B.Yu., Lemeshko, S.B.: Statistical distribution convergence and homogeneity test power for the Smirnov and Lehmann-Rosenblatt tests. Meas. Tech., **V.48**, N 12, 1159–1166 (2005)

[LL08] Lemeshko, B.Yu., Lemeshko, S.B.: Power and robustness of criteria used to verify the homogeneity of means. Meas. Tech., **V.51**, N 9, 950–959 (2008)

[LM04b] Lemeshko, B.Yu., Mirkin, E.P.: Bartlett and Cochran tests in measurements with probability laws different from normal. Meas. Tech., **V.47**, N 10, 960–968 (2004)

[LP06] Lemeshko, B.Yu., Ponomarenko, V.M.: The investigation of distributions of statistic used for testing hypotheses of variances homogeneity for non-normal error distributions. Sci. Herald of NSTU, **V.2**, N 23, 21–33 (2006) [In Russian]

[Lem06] Lemeshko, S.B.: The Abbe independence test with deviations from normality. Meas. Tech., **V.49**, N 10, 962–969 (2006)

[LL05c] Lemeshko, B.Yu., Lemeshko, S.B.: Extending the application of Grubbs-type tests in rejecting anomalous measurements. Meas. Tech., **V.48**, N 6, 536–547 (2005)

[LP04a] Lemeshko, B.Yu., Pomadin, S.S.: Testing hypotheses about mathematical expectations and variances in problems of metrology and quality control when probabilistic laws are different from the normal law. Metrology, **V.3**, 3–15 (2004) [In Russian]

[LC07] Lemeshko, B.Yu., Ogurtsov, D.V.: Statistical modeling as an effective instrument for investigating the distribution laws of functions of random quantities. Meas. Tech., **V.50**, N 6, 593–600 (2007)

[LP02] Lemeshko, B.Yu., Pomadin, S.S.: Correlation analysis of observations of multidimensional random variables with deviations from normality. Siberian J. Ind. Math., **V.5**, N 3, 115–130 (2002) [In Russian]

[LP04b] Lemeshko, B.Yu., Postovalov, S.N.: Computer Technologies of Data Analysis and Investigation of Statistical Regularities: Textbook. NSTU Publishing House, Novosibirsk (2004) [In Russian]

33

Inverse Gaussian Model and Its Applications in Reliability and Survival Analysis

Boris Yu. Lemeshko,[1] Stanislav B. Lemeshko,[1] Kseniya A. Akushkina,[1] Mikhail S. Nikulin,[2] and Noureddine Saaidia[2,3]

[1] Novosibirsk State Technical University, Novosibirsk, Russia
[2] IMB, Victor Segalen University, Bordeaux, France
[3] IMB, Victor Segalen University, Bordeaux, France and Université Badji Mokhtar, Annaba, Algérie

Abstract: Statistical properties of the parametric family of inverse Gaussian distributions are studied. Different goodness-of-fit tests for this family are compared. Some applications of the inverse Gaussian model in survival analysis and reliability are considered.

Keywords and phrases: Inverse Gaussian distribution, Generalized Weibull distribution, Hazard rate function, Goodness-of-fit test, Composite hypotheses, RRN statistic, Nonparametric goodness-of-fit tests, Kolmogorov test, Cramér–Mises–Smirnov test, Anderson-Darling test, Computer simulation, Power of the test, Reliability, Survival analysis

33.1 Introduction

Over a century the family of inverse Gaussian distributions (IGD) had attracted the attention of many researchers in several fields. The origin of this distribution goes back to the famous botanist Robert Brown (1773–1858). He interested in the study of particles motion (which now is well-known Brownian motion). In 1905, Albert Einstein derived the normal distribution as the model for Brownian motion, also in 1915 Schrödinger has obtained the distribution of first passage time as inverse Gaussian, for more details, we can see ([CF89, Ses93, Ses99]). Use of the IGD as a lifetime model is particularly appealing [GDAM97, Sin06]. The hazard rate function of the IGD has ∩-shape like Log-normal, generalized Weibull and Log-logistic distributions, i.e. the hazard rate of IGD is unimodal which increases from 0 to its maximum value and then decreases asymptotically to a constant. For these reasons the family of IGD is used often in reliability and survival analysis.

V.V. Rykov et al. (eds.), *Mathematical and Statistical Models and Methods in Reliability: Applications to Medicine, Finance, and Quality Control*, Statistics for Industry and Technology, DOI 10.1007/978-0-8176-4971-5_33,

33.2 The Family of the Inverse Gaussian Distributions

Let consider $X_1, X_2, \ldots, X_n$ be n independent and identically distributed random variables. We say that X_i follows the IGD and we note $X_i \sim IG(\mu, \lambda)$ if the density function is defined by

$$f(x,\theta) = \left(\frac{\lambda}{2\pi x^3}\right)^{\frac{1}{2}} \exp\left\{-\frac{\lambda(x-\mu)^2}{2\mu^2 x}\right\}, \quad x \geqslant 0, \quad \theta = (\mu, \lambda)^T \in \mathbb{R}^1_+ \times \mathbb{R}^1_+ \subset \mathbb{R}^2, \tag{33.1}$$

where μ is the mean and λ is the shape parameter.

The density is unimodal with mode equal to

$$M_o = \mu\left\{\left(1+\frac{9}{4\phi^2}\right)^{\frac{1}{2}} - \frac{3}{2\phi}\right\}, \qquad \phi = \frac{\lambda}{\mu},$$

and it is easy to verify that

$$\mathbf{E}(X_i) = \mu, \quad \mathbf{Var}(X_i) = \frac{\mu^3}{\lambda}.$$

All the positive and negative moments of the IGD exist with

$$\mathbf{E}(X_i^k) = \mu^k \sum_{i=0}^{k-1} \frac{(k-1+i)!}{i!(k-1-i)!}(2\phi)^{-i}, \quad k \geqslant 1 \quad \text{and} \quad \mathbf{E}(X_i^{-k}) = \frac{E(X^{k+1})}{\mu^{2k+1}}.$$

The distribution function is

$$F(x,\theta) = \Phi\left(\sqrt{\frac{\lambda}{x}}\left(\frac{x}{\mu}-1\right)\right) + \exp\left(\frac{2\lambda}{\mu}\right)\Phi\left(-\sqrt{\frac{\lambda}{x}}\left(\frac{x}{\mu}+1\right)\right), \quad x \geqslant 0, \quad \mu, \lambda > 0. \tag{33.2}$$

The hazard rate function of IGD is

$$h(x,\theta) = \frac{\left(\frac{\lambda}{2\pi x^3}\right)^{\frac{1}{2}} \exp\{-\frac{\lambda(x-\mu)^2}{2\mu^2 x}\}}{\Phi\left(-\sqrt{\frac{\lambda}{x}}\left(\frac{x}{\mu}-1\right)\right) - \exp(\frac{2\lambda}{\mu})\Phi\left(-\sqrt{\frac{\lambda}{x}}\left(\frac{x}{\mu}+1\right)\right)}, \quad x \geqslant 0.$$

Since h has $\cap$-shape we may say that the family of IGD is the natural competitor of the family of Log-normal distributions (LND), the family of generalized Weibull distributions (GWD) and the family of Log-logistic distributions (LLD). We can note for example, that if we choose two densities (one from IGD and another from LND) such that the first moments are equals, then we may see that these two distributions are close to each other (Fig. 33.1).

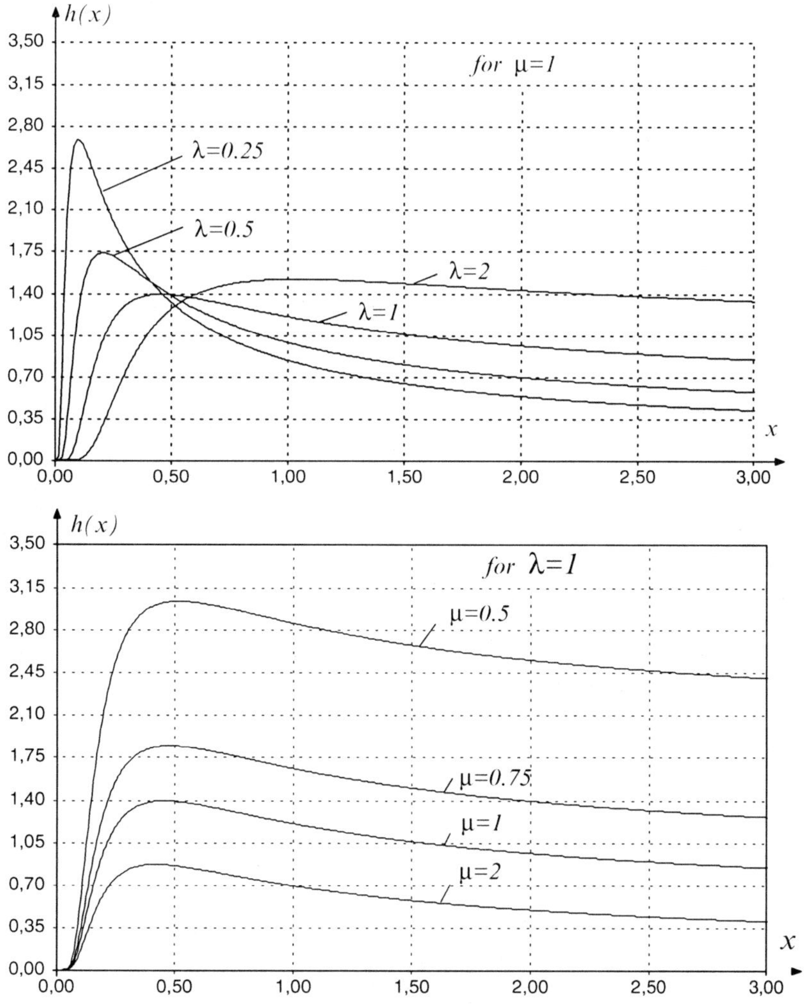

Figure 33.1. Hazard rate function of IGD

The loglikelihood function $\ell_n(\theta)$ of the sample $X_1, X_2, \ldots, X_n$ is

$$\ell_n(\theta) = \frac{n}{2}\ln\lambda - \frac{n}{2}\ln(2\pi) - \frac{3}{2}\sum_{i=1}^{n}\ln X_i - \sum_{i=1}^{n}\frac{\lambda(X_i-\mu)^2}{2\mu^2 X_i},$$

from where it follows that the bivariate statistic $T = \left(\overline{X}, V\right)^T$ is the complete minimal sufficient statistic for θ, where

$$\overline{X} = \frac{1}{n}\sum_{i=1}^{n} X_i, \quad V = \sum_{i=1}^{n}(X_i^{-1} - \frac{1}{\overline{X}}).$$

We may note here that the components $\overline{X}$ and V of the sufficient statistic are independent. It is easy to show that the maximum likelihood estimators (MLE) of μ and λ are

$$\hat{\mu}_n = \overline{X} \quad \text{and} \quad \hat{\lambda}_n = \frac{n}{\sum_{i=1}^{n}(X_i^{-1} - \frac{1}{\overline{X}})} = \frac{n}{V}$$

respectively.

The Fisher's information matrix of X_i is

$$I(\theta) = \begin{pmatrix} \frac{\lambda}{\mu^3} & 0 \\ 0 & \frac{1}{2\lambda^2} \end{pmatrix}.$$

Remark 1. The minimum variance unbiased estimators (MVUE) of μ and λ [VN93] are respectively:

$$\hat{\mu} = \overline{X} \quad \text{and} \quad \hat{\lambda} = \frac{n-3}{V}, \quad n > 3.$$

Remark 2. The MVUE of the density of IGD with unknown parameters μ and λ [VN93] is

$$\hat{f}(x, \mu, \lambda) = \begin{cases} 0, & 0 < V < C \\ \dfrac{n(n-1)\sqrt{\overline{X}^3} \quad \Gamma(\frac{n-1}{2})(V-C)^{\frac{n-4}{2}}}{\sqrt{\pi} \quad V^{\frac{n-3}{2}}\Gamma(\frac{n-2}{2})\sqrt{nx^3(n\overline{X}-x)^3}}, & V > C \end{cases}, \tag{33.3}$$

where

$$C = \frac{n(x-\overline{X})^2}{\overline{X}x(n\overline{X}-x)}, \quad n > 2.$$

Remark 3. The MVUE of the distribution function of IGD with unknown parameters μ and λ [VN93] is

$$\hat{F}(x) = \mathbf{P}(X_i \leqslant x)$$
$$= \begin{cases} 0, & x < x_1 \\ 1 - F_{n-2}(-\omega_1) + \frac{n-2}{n}\left\{1 + \frac{4(n-1)}{nV\overline{X}}\right\}^{\frac{n-3}{2}} F_{n-2}(-\omega_2), & x_1 \leqslant x \leqslant x_2 \\ 1, & x > x_2 \end{cases}, \tag{33.4}$$

where

$$\omega_1 = \frac{\sqrt{n(n-2)}(x-\overline{X})}{\sqrt{V\overline{X}(n\overline{X}-x)x - n(n-\overline{X})^2}},$$

$$\omega_2 = \frac{\sqrt{n(n-2)}\left[\overline{X} + \frac{n-2}{n}x\right]}{\sqrt{V\overline{X}(n\overline{X}-x)x - n(n-\overline{X})^2}},$$

$$F_{n-2}(x) = \frac{\Gamma(\frac{n-1}{2})}{\sqrt{\pi(n-2)}\Gamma(\frac{n-2}{2})}\int_{-\infty}^{x}\left(1 + \frac{u^2}{n-2}\right)^{-\frac{n-1}{2}} du,$$

$$x_{1,2} = \frac{\overline{X}}{2(n+V\overline{X})}\left\{n(2+V\overline{X}) \pm \sqrt{4n(n-1)V\overline{X} + n^2V^2\overline{X}^2}\right\}.$$

33.3 Goodness-of-Fit Tests for the Family of IGD

Let $X_1, X_2, \ldots, X_n$ be n independent and identically distributed random variables. We consider the problem of testing the composite hypothesis H_0:

$$H_0 : \mathbf{P}(X_i \leqslant x) = F(x, \theta), \quad x \geqslant 0, \quad \theta = (\mu, \lambda)^T.$$

Goodness-of-fit tests measure the degree of agreement between the distribution of an observed data sample and a theoretical probability distribution. In all cases, a test statistic is compared with a known critical value to accept or reject the hypothesis H_0. Many statisticians have developed numerous nonparametric methods including the Chi-squared test and various empirical distribution function tests for testing H_0. The best known tests include the following one.

33.3.1 The RRN Statistic

We divide the real line into r intervals $I_1, I_2, \ldots, I_r$ by the points

$$0 = a_0 < a_1 < \cdots < a_{r-1} < a_r = +\infty,$$

$$I_i = [a_{i-1}, a_i[, \quad I_i \cap I_j = \emptyset, i \neq j, \quad \cup_{i=1}^{r} I_i = \mathbf{R}^1,$$

and we group the sample over these intervals, we obtain the vector of frequencies $\nu = (\nu_1, \nu_2, \ldots, \nu_r)^T$ and the probability vector

$$p(\theta) = (p_1(\theta), p_2(\theta), \ldots, p_r(\theta))^T, \quad \text{where} \quad p_j(\theta) = \mathbf{P}(X_1 \in I_j | H_0), \quad j = 1, 2, \ldots, r.$$

The Fisher's information matrix of the vector of frequencies ν is

$$nJ(\theta) = nB^T(\theta)B(\theta),$$

where

$$B(\theta) = \left[\frac{1}{\sqrt{p_l(\theta)}} \frac{\partial p_l(\theta)}{\partial \theta_j} \right]_{r \times 2}.$$

Let

$$q(\theta) = (\sqrt{p_1(\theta)}, \sqrt{p_2(\theta)}, \ldots, \sqrt{p_r(\theta)})^T,$$

and consider the vector

$$X_n(\theta) = \left(\frac{\nu_1 - np_1(\theta)}{\sqrt{np_1(\theta)}}, \frac{\nu_2 - np_2(\theta)}{\sqrt{np_2(\theta)}}, \ldots, \frac{\nu_r - np_r(\theta)}{\sqrt{np_r(\theta)}} \right)^T.$$

From the structure of the vector ν it follows by the multivariate Lindeberg-Levy central theorem that under H_0 and under the Cramer's regularity conditions the vector $X_n(\hat{\theta}_n)$ is $AN(\mathbf{0}_r, W(\theta))$, where $\mathbf{0}_r = (0, ..., 0)^T \in \mathbf{R}^r$, and W is the limit covariance matrix:

$$W(\theta) = I_r - q(\theta)q^T(\theta) - B(\theta)I^{-1}(\theta)B^T(\theta), \quad \text{rank}W(\theta) = r - 1.$$

For testing H_0 one may use the Chi-squared test based on the RRN statistic Y_n^2, (see, for example, [Nik73a, Nik73b, RR74, HR76, Dro88, AN94, GN96, Van98]). The RRN statistic is defined as the next quadratic form

$$Y_n^2(\hat{\theta}_n) = X_n^T(\hat{\theta}_n)W^-(\hat{\theta}_n)X_n(\hat{\theta}_n),$$

where $W^-(\theta)$ is the generalized inverse matrix of $W(\theta)$. The asymptotic behavior of the statistic $Y_n^2(\hat{\theta}_n)$ is given by the next

Theorem 1.

$$\lim_{n\to\infty} \mathbf{P}(Y_n^2(\hat{\theta}_n) \leqslant x|H_0) = \mathbf{P}(\chi_{r-1}^2 \leqslant x).$$

According to this the hypothesis H_0 must be rejected at a significance level α, if $Y_n^2(\hat{\theta}_n) > C_\alpha$, where C_α is the critical value of the test, and $C_\alpha = \chi_{r-1,\alpha}^2$ is the upper $\alpha-$ quantile of the χ^2 distribution with $r-1$ degrees of freedom.

Remark 4. For the RRN statistic, one can use the MVUE instead of the MLE.

33.3.2 The Kolmogorov, Cramér–Mises–Smirnov and Anderson–Darling Statistics

An extension of the Kolmogorov goodness-of-fit test for testing H_0 is based on the application of the random variable

$$D_n = \sup_{|n|<\infty} |F_n(x) - F(x,\theta)|\,, \quad \theta \in \Theta, \tag{33.5}$$

where $F_n(x)$ is the empirical distribution function. In practice it is better to use the test based on D_n with Bolshev correction [Bol87] in the form [BS83]

$$S_K = \frac{6nD_n + 1}{6\sqrt{n}}, \tag{33.6}$$

where $D_n = \max(D_n^+, D_n^-)$,

$$D_n^+ = \max_{1\le i\le n}\left\{\frac{i}{n} - F(x_i,\theta)\right\}, D_n^- = \max_{1\le i\le n}\left\{F(x_i,\theta) - \frac{i-1}{n}\right\},$$

$x_1, x_2, \ldots, x_n$ are sample values in increasing order. The distribution of statistic (33.6) obeys the Kolmogorov distribution law $K(S)$ [BS83] in the testing simple hypotheses.

The statistic of ω^2 Cramér–Mises–Smirnov goodness-of-fit test can be written as

$$S_\omega = \frac{1}{12n} + \sum_{i=1}^{n}\left\{F(x_i,\theta) - \frac{2i-1}{2n}\right\}^2, \tag{33.7}$$

and the statistic of Ω^2 Anderson–Darling test can be written in the form

$$S_\Omega = -n - 2\sum_{i=1}^{n}\left\{\frac{2i-1}{2n}\ln F(x_i,\theta) + \left(1 - \frac{2i-1}{2n}\right)\ln(1 - F(x_i,\theta))\right\}. \tag{33.8}$$

In simple hypothesis testing, the statistic (1.7) has the $a_1(S)$ distribution and statistic (1.8) has the $a_2(S)$ distribution (see [BS83]).

When testing composite hypotheses the conditional distribution law of the statistic $G(S|H_0)$ is affected by a number of factors: the form of the law $F(x,\theta)$ corresponding to the true hypothesis H_0; the method of parameter estimation and the number of estimated parameters; sometimes it is a specific value of a parameter (e.g., in the case of gamma-distribution, beta-distribution, IGD or GWD). The distinctions in the marginal statistic distributions in testing simple and composite hypotheses are so significant that we cannot neglect them.

Distribution statistic models and tables of percentage points for nonparametric goodness-of-fit tests for testing hypotheses relative to the IGD are given in [LNS09].

33.4 About Testing Hypotheses for the Inverse Gaussian Distribution

The IGD is used in reliability and survival studies along with the LND and the GWD. When constructing models of laws for really observed random variables it is sometimes difficult to choose one of the distributions mentioned above because it is complicated to distinguish these families of distributions using parametric and nonparametric goodness-of-fit tests.

Let consider an example about problems to distinguish these laws.

The sample presented (set out) below with size $n = 200$ was simulated in accordance with the IGD with parameters $\mu = 2$ and $\lambda = 2$. Pseudorandom values are given with 3 decimal digits in increasing order (by columns) in the table below (Table 33.1).

Table 33.1. Pseudorandom values

0.183	0.501	0.689	0.894	1.075	1.386	1.690	2.393	2.952	4.450
0.185	0.505	0.701	0.896	1.081	1.397	1.694	2.410	2.973	4.521
0.266	0.509	0.713	0.903	1.151	1.405	1.698	2.419	3.010	4.763
0.298	0.537	0.716	0.912	1.171	1.416	1.776	2.421	3.100	4.955
0.309	0.538	0.722	0.913	1.180	1.421	1.780	2.431	3.477	5.011
0.315	0.542	0.758	0.913	1.202	1.448	1.814	2.446	3.538	5.158
0.320	0.563	0.761	0.917	1.210	1.539	1.874	2.452	3.557	5.392
0.324	0.568	0.768	0.919	1.221	1.543	1.901	2.454	3.663	5.460
0.343	0.578	0.787	0.925	1.223	1.547	1.903	2.482	3.674	5.578
0.367	0.593	0.835	0.929	1.247	1.563	1.904	2.483	3.749	5.625
0.386	0.593	0.839	0.955	1.258	1.585	2.007	2.522	3.777	6.295
0.390	0.600	0.852	0.960	1.280	1.586	2.029	2.559	3.886	6.376
0.416	0.605	0.854	0.984	1.316	1.599	2.067	2.598	3.900	6.717
0.421	0.623	0.855	0.992	1.326	1.626	2.087	2.609	3.901	7.185
0.427	0.624	0.865	1.029	1.335	1.634	2.099	2.626	3.992	7.772
0.438	0.628	0.866	1.030	1.346	1.645	2.159	2.640	4.006	8.265
0.457	0.628	0.873	1.041	1.351	1.657	2.171	2.701	4.105	10.100
0.464	0.636	0.873	1.045	1.364	1.674	2.175	2.706	4.108	13.896
0.466	0.637	0.880	1.051	1.368	1.675	2.199	2.730	4.109	14.844
0.470	0.688	0.889	1.074	1.374	1.688	2.352	2.846	4.297	15.503

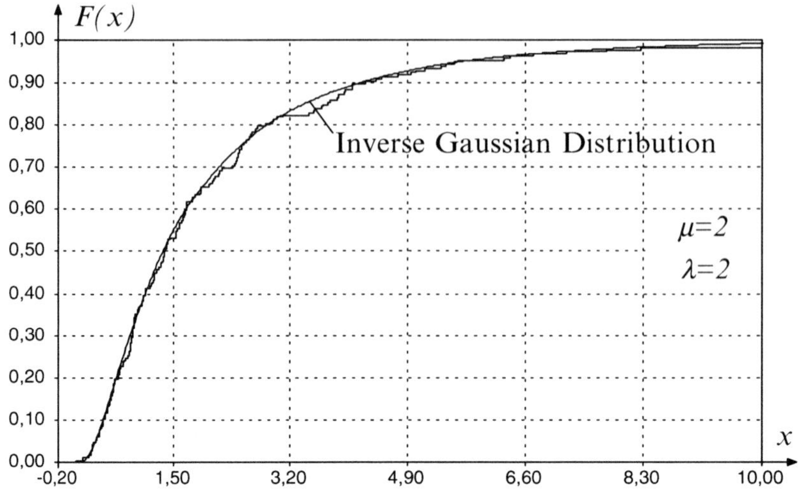

Figure 33.2. Empirical distribution and the IGD with parameters $\mu = 2$ and $\lambda = 2$

Testing a hypothesis about goodness-of-fit of the empirical distribution to the theoretical IGD has been carried out by following four criteria: RRN test [Nik73a, Nik73b, RR74] (Pearson test for simple hypothesis); Kolmogorov test; Cramér–Mises–Smirnov test; Anderson–Darling test. In case of the RRN test the number of intervals is $k = 10$. And domain of random variable is divided into equiprobable intervals (equiprobable grouping). The Neyman–Pearson intervals [GN96] are also considered in investigation of the test power. In this case, the interval bounds are chosen at the points in which the density functions corresponding to competing hypotheses intersect. The Table 1.1 contains the results of testing simple hypothesis about belonging of the sample to the IGD with parameters $\mu = 2$ and $\lambda = 2$. In the third column, there are achieved significance levels (p-values) for each test. The empirical and theoretical laws are presented in the Fig. 1.2 (Fig. 33.2).

The results of testing composite hypothesis about belonging of the sample to the IGD are presented in Table 33.3. Parameters are estimated by the maximum likelihood method ($\mu = 1.9848$ and $\lambda = 2.1119$) (by non-grouped data). In the Fig. 33.3 corresponding results are shown.

If composite hypotheses are tested the distributions of the nonparametric Kolmogorov, Cramér–Mises–Smirnov, Anderson–Darling goodness-of-fit tests depend on certain μ and λ parameter values of the IGD. In the paper [LNS09], the models of statistic distributions (and percentage points) of the nonparametric Kolmogorov, Cramér–Mises–Smirnov, Anderson–Darling goodness-of-fit tests were obtained for integer parameter values of the IGD. That is why in this case (in this situation) to obtain p-values it is necessary to model statistic distributions of the those three tests for the values $\mu = 1.9848$ and $\lambda = 2.1119$. In the Table 33.2 there are the p-values obtained basing on such modelling.

The same sample is tested for goodness-of-fit to the LND with density function

$$f(x, \mu, \sigma) = \frac{1}{x\sigma\sqrt{2\pi}} \exp\left(-(lnx - \mu)^2/2\sigma^2\right), \quad x > 0, \quad \mu \in R^1, \quad \sigma > 0,$$

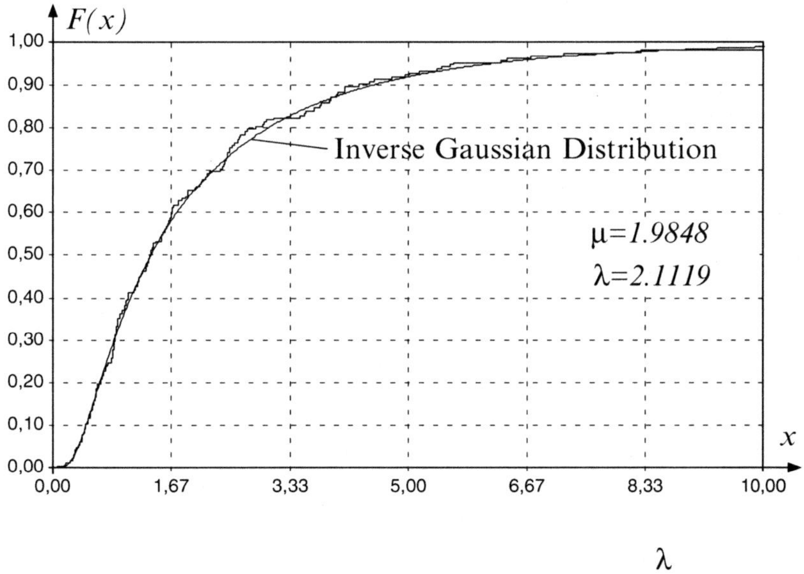

Figure 33.3. Empirical distribution and the IGD with the maximum likelihood estimates $\mu = 1.9848$ and $\lambda = 2.1119$

Table 33.2. The results of testing a simple hypothesis about belonging of the sample to the IGD with parameters $\mu = 2$ and $\lambda = 2$

Test	Statistics value	p-value
RRN test (Pearson test)	9.7000	0.3753
Kolmogorov test	0.6204	0.8362
Cramér-Mises-Smirnov test	0.0397	0.9346
Anderson-Darling test	0.2712	0.9581

and to the GWD with density function has the form

$$f(x;\theta_0,\theta_1,\theta_2) = \frac{\theta_0}{\theta_1}\theta_2{}^{\theta_0}x^{\theta_0-1}\left(1+\left(\frac{x}{\theta_2}\right)^{\theta_0}\right)^{\frac{1}{\theta_1}-1}\exp\left\{1-\left(1+\left(\frac{x}{\theta_2}\right)^{\theta_0}\right)^{\frac{1}{\theta_1}}\right\}$$

where $x > 0, \quad \theta_0,\theta_1,\theta_2 > 0$.

As you can see in the figure below three distributions (the LND, the IGD, the GWD) are close to each other. In the Fig. 33.4, there are distribution functions, in the Fig. 33.5 there are density functions and in the Fig. 33.6 there are hazard rate functions. Obviously, in practice it is difficult to make decision what law is the most appropriate. At the same time this issue plays an important role in reliability and survival studies because for close distribution functions (Fig. 33.4) we have considerable distinctions in hazard rate functions (Fig. 33.6).

In the Table 33.3, the results of testing composite hypotheses about belonging of the sample to the LND are presented. The models of statistic distributions of nonparametric goodness-of-fit tests for testing composite hypotheses included to the developed software system. These models are presented in the papers [LL09a, LL09b, LLP10].

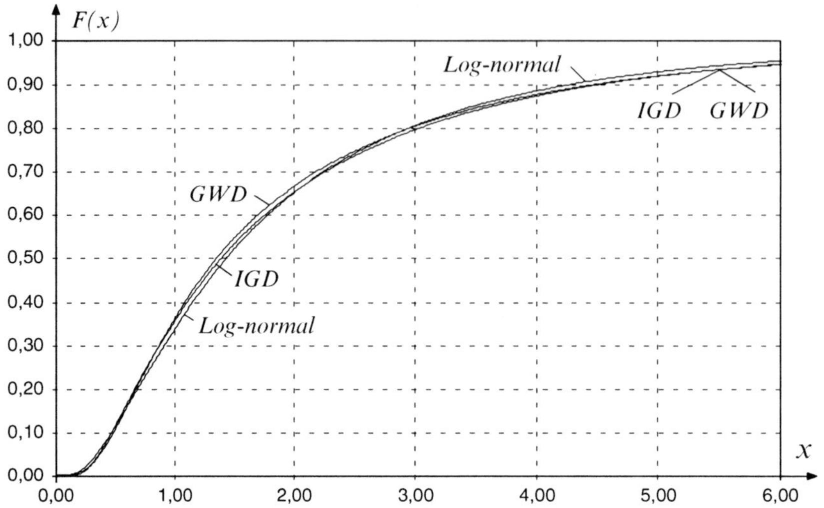

Figure 33.4. IGD ($\mu = 1.9848$, $\lambda = 2.1119$), LND ($\mu = 0.3663$, $\sigma = 0.8558$) and GWD ($\theta_0 = 3.1955$, $\theta_1 = 5.6772$, $\theta_2 = 0.5423$)

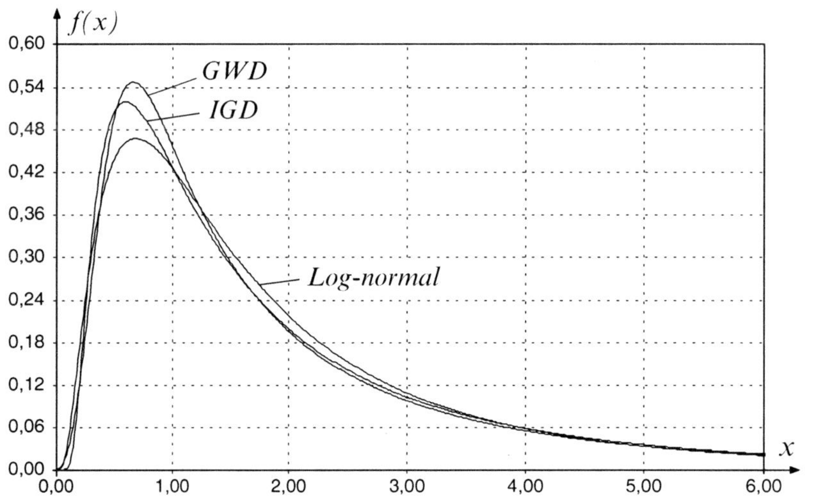

Figure 33.5. Density of the IGD ($\mu = 1.9848$, $\lambda = 2.1119$), density of the LND ($\mu = 0.3663$, $\sigma = 0.8558$) and density of the GWD ($\theta_0 = 3.1955$, $\theta_1 = 5.6772$, $\theta_2 = 0.5423$)

The Table 33.5 contains the results of testing composite hypothesis about belonging of the sample to the GWD. To obtain p-value it is necessary to model statistic distributions of the Kolmogorov, Cramér–Mises–Smirnov and Anderson–Darling test statistics with the values $\theta_0 = 3.1955$, $\theta_1 = 5.6772$ and $\theta_2 = 0.5423$. The p-value (which is presented in the Table 33.5) are obtained on the basis of modeled statistic distributions of the nonparametric goodness-of-fit tests.

The p-values obtained in testing hypothesis about the GWD practically the same as in the case of the IGD. Notice that there are no reasons to rejected hypothesis about belonging the sample to the LND.

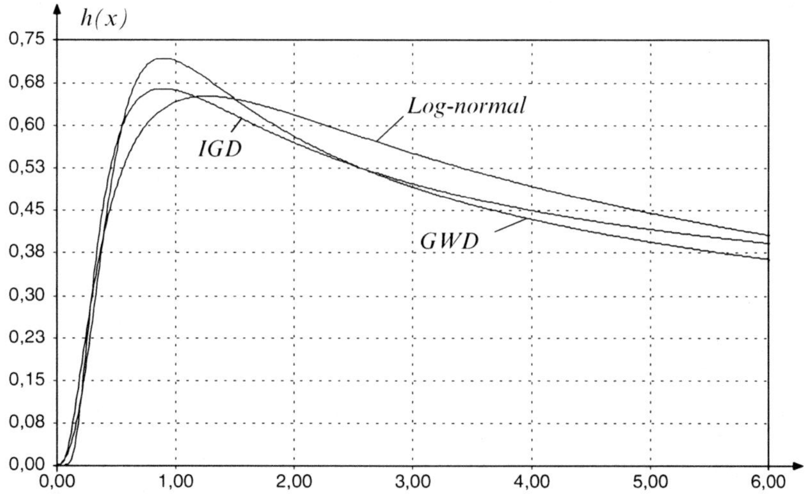

Figure 33.6. Hazard rate function of the IGD ($\mu = 1.9848$, $\lambda = 2.1119$), the LND ($\mu = 0.3663$, $\sigma = 0.8558$) and the GWD ($\theta_0 = 3.1955$, $\theta_1 = 5.6772$, $\theta_2 = 0.5423$)

Table 33.3. The results of testing composite hypothesis of goodness-of-fit to the IGD with the parameters $\mu = 1.9848$ and $\lambda = 2.1119$ estimated by the maximum likelihood method (use ungrouped observations)

Test	Statistics value	p-value
RRN test	2.6469	0.9545
Kolmogorov test	0.4875	0.9006
Cramér–Mises–Smirnov test	0.0263	0.9351
Anderson–Darling test	0.1754	0.9549

Table 33.4. The results of testing composite hypothesis of goodness-of-fit to the LND with maximum likelihood estimates (by ungrouped observations) $\mu = 0.3636$ and $\sigma = 0.8558$

Test	Statistics value	p-value
RRN test	9.1500	0.4235
Kolmogorov test	0.6524	0.4084
Cramér–Mises–Smirnov test	0.0500	0.5136
Anderson–Darling test	0.3055	0.5914

According to results in the Tables 33.3–33.5 the IGD model is more appropriate than the GWD or the LND for the sample.It is logical because the sample was modeled from the IGD. However, an isolated case does not give an opportunity (make it possible) to recognize given laws by goodness-of-fit tests. We can assess the capacity of tests to differ laws by the power of the test in testing hypothesis about belonging of the sample to the IGD, considering the LND and the GWD as the competing distributions.

The obtained power estimates of tests are presented in the Tables 33.6 and 33.7. Presented power values allow to suggest about sample sizes n due to which we can differ corresponding laws. It is evident that the LND is easier to be differed from the IGD than the GWD. At the same time it is obvious that a sure distinction requires a large value of a sample size. In particular, to get the probability of the second type error

Table 33.5. The results of testing composite hypothesis of goodness-of-fit to the GWD with maximum likelihood estimates (by ungrouped observations) $\theta_0 = 3.1955$, $\theta_1 = 5.6772$ and $\theta_2 = 0.5423$

Test	Statistics value	p-value
RRN test	6.1602	0.7238
Kolmogorov test	0.4853	0.8047
Cramér–Mises–Smirnov test	0.0257	0.8594
Anderson–Darling test	0.1636	0.9131

Table 33.6. The power of the test in testing composite hypothesis concerning the IGD ($\mu = 1.9848$, $\lambda = 2.1119$) against the LND ($\mu = 0.3636$, $\sigma = 0.8558$) as a competing hypothesis

Test	n						
	50	100	200	300	400	500	1,000
RRN test (k = 10, equiprobable grouping)	0.147	0.212	0.314	0.415	0.506	0.587	0.834
RRN test ($k = 5$, Neyman–Pearson classes)	0.133	0.194	0.291	0.377	0.468	0.552	0.821
Kolmogorov test	0.161	0.221	0.322	0.412	0.500	0.568	0.798
Cramér–Mises–Smirnov test	0.173	0.247	0.366	0.476	0.577	0.650	0.868
Anderson–Darling test	0.174	0.249	0.377	0.501	0.610	0.688	0.923

The probability of the first type error $\alpha = 0.1$

Table 33.7. The power of the test in testing composite hypothesis concerning the IGD ($\mu = 1.9848$, $\lambda = 2.1119$) against the GWD ($\theta_0 = 3.1955$, $\theta_1 = 5.6772$ and $\theta_2 = 0.5423$) as a competing hypothesis

Test	n						
	50	100	200	300	400	500	1,000
RRN test (k = 10, equiprobable grouping)	0.122	0.147	0.200	0.248	0.288	0.337	0.524
RRN test ($k = 5$, Neyman–Pearson classes)	0.132	0.145	0.179	0.211	0.226	0.242	0.325
Kolmogorov test	0.116	0.133	0.166	0.200	0.228	0.258	0.400
Cramér–Mises–Smirnov test	0.125	0.157	0.198	0.238	0.282	0.318	0.499
Anderson–Darling test	0.128	0.161	0.216	0.266	0.316	0.366	0.588

The probability of the first type error $\alpha = 0.1$

with the condition $\beta \leqslant 0.1$ for the given probability of the first type error $\alpha = 0.1$ the sample sizes $n > 2000$ are required. In this case, we test the hypothesis concerning the IGD against the competing hypothesis concerning the GWD.The power of Anderson–Darling test (Table 33.7) is 0.588 for n = 1,000 ($\beta = 0.412$). If n = 2,000 the power is 0.853 ($\beta = 0.147$), if n = 2,500 the power is 0.916 ($\beta = 0.084$). Under the same conditions to differ the IGD from the LND n should be about 1,000 (the Anderson–Darling test, Table 33.6).

33.5 Chi-Squared Goodness-of-fit Test for the Family of IGD in Case of Censored Data

In Reliability and survival analysis, we often encounter incomplete observations, and in this situation the usual methods are no longer valid. In the case of random censorship, one can use the RRN statistic $\hat{Y}_n^2$ which is well adapted for right censored data, (see [HT86]), where the Kaplan–Meier estimator $\hat{S}_n(x)$ is compared with the parametric estimator $S(x, \hat{\theta}_n)$, where $\hat{\theta}_n$ is the MLE of θ (see also [NS99]). Consider now the problem of testing the hypothesis H_0 that the data are coming from the IGD.

Under the random censorship model, we assume that the failure times $T_1, T_2, \ldots, T_n$ are nonnegative and independent. The censoring variables $C_1, C_2, \ldots, C_n$ are also nonnegative and assumed to be random sample. We observe only $X_i = min(T_i, C_i)$ and the indicator functions δ_i defined as:

$$\delta_i = \begin{cases} 1 & , \; if \;\; T_i \leqslant C_i \\ 0 & , \; \text{otherwise} \end{cases} .$$

Let $S(t, \theta) = 1 - F(t, \theta)$, $\theta = (\mu, \lambda)^T$ is the survival function (or reliability function) of IGD, $f(t, \theta)$ is the density function corresponding to $F(t, \theta)$, $H(t)$ the unknown survival function of consortship and $h(t)$ the density function corresponding to $H(t)$.

The loglikelihood function is

$$\ell_n(\mu, \lambda) = \sum_{i=1}^{n} \delta_i \left\{ \frac{1}{2} \ln \lambda - \frac{1}{2} \ln 2\pi - \frac{3}{2} \ln T_i - \lambda \frac{(T_i - \mu)^2}{2\mu^2 T_i} \right\}$$

$$+ \sum_{i=1}^{n} (1 - \delta_i) \ln \left\{ \Phi(A_i) - \exp\left(\frac{2\lambda}{\mu}\right) \Phi(B_i) \right\}, \tag{33.9}$$

where $A_i = -\sqrt{\frac{\lambda}{T_i}}(\frac{T_i}{\mu} - 1)$ and $B_i = -\sqrt{\frac{\lambda}{T_i}}(\frac{T_i}{\mu} + 1)$.

The score functions $U_l(\mu, \lambda)$, $l = 1, 2$ are

$$U_1(\mu, \lambda) = \frac{\partial \ell_n(\mu, \lambda)}{\partial \mu} = \frac{\lambda}{\mu^3} \sum_{i=1}^{n} \delta_i (T_i - \mu)$$

$$+ \frac{1}{\mu^2} \sum_{i=1}^{n} (1 - \delta_i) \frac{\sqrt{\lambda T_i} \varphi(A_i) + \exp(\frac{2\lambda}{\mu}) \left(2\lambda \Phi(B_i) - \sqrt{\lambda T_i} \varphi(B_i)\right)}{S(T_i, \mu, \lambda)},$$

$$U_2(\mu, \lambda) = \frac{\partial \ell_n(\mu, \lambda)}{\partial \lambda} = \sum_{i=1}^{n} \delta_i \left(\frac{1}{2\lambda} - \frac{(T_i - \mu)^2}{2\mu^2 T_i} \right)$$

$$+ \sum_{i=1}^{n} (1 - \delta_i) \frac{\frac{1}{2\lambda} A_i \varphi(A_i) - \exp(\frac{2\lambda}{\mu}) \left(\frac{2}{\mu} \Phi(B_i) + \frac{1}{2\lambda} B_i \varphi(B_i) \right)}{S(T_i, \mu, \lambda)},$$

where $\varphi(t)$ is the density function of the standard normal distribution. To have the MLE $\hat{\theta}_n$ of θ one can solve the system of equations of the score functions.

Habib and Thomas [HT86,Hjo90] have shown that $\sqrt{n}\left(\hat{S}_n(t) - S(t,\hat{\theta}_n)\right)$ converges to the Gaussian process under the hypothesis H_0.

We divide the real line into r intervals: $I_1, I_2, \ldots, I_r$ mutually disjoint by the points:

$$0 = t_0 < t_1 < \cdots < t_{r-1} < t_r = +\infty.$$

Let consider the vector

$$\hat{Z}_n = \sqrt{n}\left(\hat{S}_n - S_{\hat{\theta}_n}\right),$$

where

$$\hat{S}_n = (\hat{S}_n(t_1), \hat{S}_n(t_2), \cdots, \hat{S}_n(t_{r-1}))^T$$

and

$$S_{\hat{\theta}_n} = (S(t_1,\hat{\theta}_1), S(t_2,\hat{\theta}_2), \cdots, S(t_{r-1},\hat{\theta}_n))^T.$$

Under those assumptions:

1. $f(t,\theta)$ and $F(t,\theta)$ are twice differentiable in θ with continuous derivatives.

2. The Fisher's information matrix $I(\theta)$ is positive definite , and continuous in θ, where

$$I_{ij} = -\int \frac{\partial^2 \ln f(t,\theta)}{\partial\theta_i\theta_j} H(t) f(t,\theta)\mathrm{d}t - \int \frac{\partial^2 \ln S(t,\theta)}{\partial\theta_i\theta_j} h(t) S(t,\theta)\mathrm{d}t, \quad i,j = 1,2.$$

3. The MLE $\hat{\theta}_n$ exist and is $\sqrt{n}$- consistent estimator of θ with

$$\sqrt{n}(\hat{\theta}_n - \theta) = I^{-1} W_n + o_p(1),$$

where

$$W_n = n^{-1/2} \sum_{i=1}^{n} \frac{\partial \ln g(X_i, \delta_i, \theta)}{\partial\theta},$$

and g is the density of joint distribution of (X, δ).

Let

$$B = B(\theta) = \left[\frac{\partial F(t_i,\theta)}{\partial\theta_j}\right]_{(r-1)\times 2},$$

and

$$V = V(\theta) = \left[\mathrm{Cov}\left(Z(t_i), Z(t_j)\right)\right]_{(r-1)\times(r-1)},$$

where

$$\mathrm{Cov}\left(Z(t_i), Z(t_j)\right) = S(t_i,\theta) S(t_j,\theta) \int_0^{t_i \wedge t_j} \frac{f(t,\theta)}{H(t) F^2(t,\theta)} \mathrm{d}t.$$

To test H_0 we construct the general modified Chi-squared type of Pearson for random censored data which has the quadratic form

$$\hat{Y}_n^2(\hat{\theta}_n) = \hat{Z}_n^T \hat{\Sigma}^{-}(\hat{\theta}_n) \hat{Z}_n,$$

where the matrix $\hat{\Sigma}$ is the estimator of the covariance matrix Σ and Σ^- its general inverse, such that

$$\Sigma(\theta) = V(\theta) - B(\theta) I^{-1}(\theta) B^T(\theta), \quad \mathrm{rank}\Sigma = r - 1.$$

The asymptotic behavior of the statistic $\hat{Y}_n^2(\hat{\theta}_n)$ is given by the following

Theorem 2 ([HT86]).

$$\lim_{n\to\infty} \mathbf{P}(\hat{Y}_n^2(\hat{\theta}_n)) \leqslant x|H_0) = \mathbf{P}(\chi^2_{r-1} \leqslant x).$$

Note that for uncensored data, the statistic $\hat{Y}_n^2$ reduces to the considered before RRN statistic. We can also consider the case of doubly censored data (see [IL99]).

33.6 Models with Covariates Based on the Family of IGD

Failure time regression models relating the lifetime distribution to possibly time dependent external explanatory variables are considered in this section. Failure time regression models relating failure time distribution not only with external but also with internal explanatory variables will be discussed in the next section. Such models are used now not only in reliability but also in demography, dynamics of populations, gerontology, biology, survival analysis, genetics, radiobiology, biophysics, everywhere people study the problems of longevity, aging and degradation using the stochastic modeling.

In reliability, in accelerated life testing (ALT) in particular, the choice of a good regression model often is more important than in survival analysis. For example, in ALT units are tested under accelerated stresses which shorten the lifetime. Using such experiments the lifetime under the usual stress is estimated using some regression model. The values of the usual stress are often not in the range of the values of accelerated stresses, since the wide separation between experimental and usual stresses is possible, so if the model is misspecified, the estimators of survival under the usual stress may be very bad.

Let E be a set of all admission possible time-depending stresses (covariables)

$$E = \{x(\cdot) = (x_1(\cdot), x_2(\cdot), \ldots, x_m(\cdot)) : [0, +\infty) \to \mathbf{R}^m\}.$$

We denote by $T_{x(\cdot)}$ the failure time under $x(\cdot)$ and by $f_{x(\cdot)}(t)$, $S_{x(\cdot)}(t)$, $F_{x(\cdot)}(t)$ the density function , survival function and the distribution function, respectively, where

$$S_{x(\cdot)}(t) = \mathbf{P}\left(T_{x(\cdot)} \geqslant t\right) = 1 - F_{x(\cdot)}(t), \quad x(\cdot) \in E.$$

The hazard rate function of $T_{x(\cdot)}$ under $x(\cdot)$is:

$$\lambda_{x(\cdot)}(t) = \lim_{h\to 0} \frac{1}{h}\mathbf{P}\left(t \leqslant T_{x(\cdot)} < t+h \mid T_{x(\cdot)} \geqslant t\right) = -\frac{S'_{x(\cdot)}(t)}{S_{x(\cdot)}(t)}, \quad x(\cdot) \in E,$$

and we denote by:

$$\Lambda_{x(\cdot)}(t) = \int_0^t \lambda_{x(\cdot)}(u)\mathrm{d}u = -\ln(S_{x(\cdot)}(t)), \quad x(\cdot) \in E,$$

the cummulative hazard rate function of $T_{x(\cdot)}$.

33.6.1 Cox Model

The famous Cox model on E is the most popular in survival analysis. It is given in terms of the hazard rate function:

$$\lambda_{x(\cdot)} = r(x(t))\lambda_0(t), \quad x(\cdot) \in E,$$

where $\lambda_0(t)$ is the baseline hazard rate function(generally unknown) and r is a positive function often parameterized as

$$r(x) = \mathrm{e}^{\beta^T x}, \quad \beta = (\beta_1, \ldots, \beta_m)^T.$$

Let us suppose that

$$\lambda_0(t) \in \left\{ \frac{\left(\frac{\lambda}{2\pi t^3}\right)^{\frac{1}{2}} \exp\{-\frac{\lambda(t-\mu)^2}{2\mu^2 t}\}}{\Phi\left(-\sqrt{\frac{\lambda}{t}}\left(\frac{t}{\mu}-1\right)\right) - \exp(\frac{2\lambda}{\mu})\Phi\left(-\sqrt{\frac{\lambda}{t}}\left(\frac{t}{\mu}+1\right)\right)}, \ t \geqslant 0, \ \mu > 0, \ \lambda > 0 \right\}.$$

In such, we have the so-called parametric Cox Inverse Gaussian model, where the vector β expresses the effect of covariate $x(\cdot)$ on the distribution of the lifetime $T_{x(\cdot)}$. Let consider the hypothesis

$$H_0 : \beta = 0.$$

Under H_0 there is no influence of covariates on distribution of the lifetime $T_{x(\cdot)} = T$, and in this case (under H_0) the lifetime T follows an inverse Gaussian distribution. So to test H_0 one can use, for example, the Chi-squared test exposed before.

33.6.2 AFT Model

The term accelerated life testing applies that the type of study where failure times can be accelerated by applying higher "stress" to the component or system reliability, and higher stress may bring quicker failure. For example, some component may fail quicker at a higher temperature; however, it may have a long lifetime at normal temperatures. At normal stress conditions, the time required may be too large for its reliability estimation which may be tested under higher stress factors terminating the experiment. We look at case where the hazard rate function has the ∩-shape.

We consider some applications of the family of IGD as the baseline survivals in the construction of the accelerated failure time (*AFT*) model which is very natural competitor of lognormal and generalized Weibull distributions.

The *AFT* model holds on E, (see [BN02]), if there exists on E a positive function r and a survival function S_0 such that:

$$S_{x(\cdot)}(t) = S_0\left(\int_0^t r(x(u))\mathrm{d}u\right), \quad t \geqslant 0, \quad x(\cdot) \in E, \tag{33.10}$$

where S_0 is the baseline survival function. In term of hazard rate function, the expression (1.10) holds if and only if there exists on E a positive function r and on $[0, +\infty)$ a positive function q such that

$$\alpha_{x(\cdot)}(t) = r(x(t)q(S_{x(\cdot)}(t)), \quad x(\cdot) \in E.$$

In parametric case S_0 belongs to a parametric family, and $r(x)$ is often parameterized as in the Cox model: $r(x) = \mathrm{e}^{\beta^T x}$.

If we consider the inverse Gaussian distribution as models for the baseline survival function S_0 such that

$$S_0(t) = \Phi\left(-\sqrt{\frac{\lambda}{t}}\left(\frac{t}{\mu} - 1\right)\right) - \exp\left(\frac{2\lambda}{\mu}\right)\Phi\left(-\sqrt{\frac{\lambda}{t}}\left(\frac{t}{\mu} + 1\right)\right), \quad t \geqslant 0, \quad \mu, \lambda > 0,$$

we obtain the AFT_{IG} model.

Let consider the hypothesis

$$H_0 : \beta = 0.$$

Under H_0 there is no influence of covariates on distribution of the lifetime $T_{x(\cdot)} = T$, and under H_0 the distribution of the lifetime T is inverse Gaussian, and one can test it using the chi-square test, for example.

Remark 5. We say that the distribution of $T_{x(\cdot)}$ belongs to the class GPH (Generalized Proportional Hazards) on E (see [BN02]) if its hazard rate function is given by formula

$$\lambda_{x(\cdot)} = r(x(t))q(\Lambda_{x(\cdot)})\lambda_0(t), \quad x(\cdot) \in E,$$

where $q(\cdot)$ is a positive function on R^1_+. Model GPH is very interesting since it generalizes the famous Cox and AFT models on E. It is evident that in some situations we need to test the Cox model against the AFT model, and in this situation it is very interesting to apply the GPH model.

33.6.3 Inverse Gaussian Family and Analysis of Redundant System

Now we shall consider, following the papers of Bagdonavičius, Masiulaityle and Nikulin [BMN08a, BMN08b, BMN09, BMN10], one example of the parametric estimation of redundant system reliability when the distribution of the failure time in "hot" and "warm" conditions belongs to the IGD. Let consider redundant system $S(1, m-1)$ with one principal main unit operating in "hot" and $(m-1)$ stand-by units operating in "warm" conditions. The problem is to estimate the parameters of the redundant system, using failure data of two groups of units, when we suppose that switching from warm to hot does not cause shock or damage to units.

Denote by T_1, F_1 and f_1 the failure time, the cumulative distribution function and the probability density function of the main unit. The failure times of the stand-by units denote by $T_2, \ldots, T_m$. In "hot" conditions their distribution functions are also F_1. In "warm" conditions the distribution function of T_i is F_2 and the density function is f_2, $i = 2, \ldots, m$. If a stand-by unit is switched to "hot" conditions, its cumulative distribution function is different from F_1 and F_2.

The failure time of the system $S(1, m-1)$ is

$$T^{(m)} = T_1 \vee T_2 \vee \cdots \vee T_m.$$

Denote by K_j and k_j the distribution function and the density function of $T^{(j)}$, respectively, $(j = 2, \ldots, m)$, $K_1 = F_1$, $k_1 = f_1$. The distribution function K_j can be written in terms of the distribution function K_{j-1} and F_1:

$$K_j(t) = \mathbf{P}(T^{(j)} \leqslant t) = \int_0^t \mathbf{P}(T_j \leqslant t | T^{(j-1)} = y) \mathrm{d}K_{j-1}(y).$$

The "fluent switch on" hypothesis H_0 formulated by Bagdonavičius et al. [BMN08a, BMN08b] states that

$$f_{T_j|T^{(j-1)}=y}(t) = \begin{cases} f_2(t) & \text{if} \quad t \leqslant y, \\ f_1(t + g(y) - y) & \text{if} \quad t > y; \end{cases},$$

where $g(y) = F_1^{-1}(F_2(y))$. This model implies that

$$K_j(t) = \int_0^t F_1(t + g(y) - y) \mathrm{d}K_{j-1}(y), \quad j = 2, ..., m.$$

So the distribution function K_m of the system with $m - 1$ stand-by units is defined recurrently by the last formula. We consider here the situation when the distribution of units functioning in "warm" and "hot" conditions differ only in scale, i.e. we suppose that $g(y) = ry$ and hence $F_2(t) = F_1(rt)$ for all $t \geqslant 0$ and some $r > 0$. In such a case the cumulative distribution function of units functioning in "hot" and "warm" conditions mostly belong to the same parametric classes of distributions, for example, to the family of IGD. If the cumulative distribution function of units belongs to the family of IGD then

$$S_1(t, \mu, \lambda) = 1 - F_1(t, \mu, \lambda) = \Phi\left(-\sqrt{\frac{\lambda}{t}}\left(\frac{t}{\mu} - 1\right)\right) - \exp\left(\frac{2\lambda}{\mu}\right)\Phi\left(-\sqrt{\frac{\lambda}{t}}\left(\frac{t}{\mu} + 1\right)\right).$$

Parametric Estimators of the Parameter $\gamma = (r, \mu, \lambda)^T$

Suppose that the following data are available:

(a) Complete ordered sample $T_{11}, \ldots, T_{1n_1}$ of size n_1, T_{1i} is the failure time of units tested in 'hot' condition;
(b) Complete ordered sample $T_{21}, \ldots, T_{2n_2}$ of size n_2, T_{2i} is the failure time of units tested in 'warm' condition.

Let $\gamma = (r, \mu, \lambda)^T$, the MLE $\hat{\gamma} = (\hat{r}, \hat{\mu}, \hat{\lambda})^T$ of the parameter γ maximizes the loglikelihood function

$$\ell(r, \mu, \lambda) = \frac{n}{2} \ln \lambda - \frac{n}{2} \ln(2\pi) - \frac{n_2}{2} \ln r + \frac{\lambda n}{\mu} - \frac{3}{2} \sum_{i=1}^{n_1} \ln(T_{1i}) - \frac{3}{2} \sum_{i=1}^{n_2} \ln(T_{2i})$$

$$- \frac{\lambda}{2\mu^2} \sum_{i=1}^{n_1} T_{1i} - \frac{\lambda}{2} \sum_{i=1}^{n_1} T_{1i}^{-1} - \frac{\lambda r}{2\mu^2} \sum_{i=1}^{n_2} T_{2i} - \frac{\lambda}{2r} \sum_{i=1}^{n_2} T_{2i}^{-1},$$

where $n = n_1 + n_2$.

The score functions are

$$\frac{\partial \ell}{\partial r} = -\frac{n_2}{2r} - \frac{\lambda}{2\mu^2}\sum_{i=1}^{n_2} T_{2i} + \frac{\lambda}{2r^2}\sum_{i=1}^{n_2} T_{2i}^{-1},$$

$$\frac{\partial \ell}{\partial \mu} = -\frac{\lambda n}{\mu^2} + \frac{\lambda}{\mu^3}\sum_{i=1}^{n_1} T_{1i} + \frac{\lambda r}{\mu^3}\sum_{i=1}^{n_2} T_{2i},$$

$$\frac{\partial \ell}{\partial \lambda} = n\left(\frac{1}{2\lambda} + \frac{1}{\mu}\right) - \frac{1}{2\mu^2}\sum_{i=1}^{n_1} T_{1i} - \frac{1}{2}\sum_{i=1}^{n_1} T_{1i}^{-1} - \frac{r}{2\mu^2}\sum_{i=1}^{n_2} T_{2i} - \frac{1}{2r}\sum_{i=1}^{n_2} T_{2i}^{-1}.$$

To find the estimator $\hat{\gamma}$ one can solve the system formed by equalizing the score functions to zero.

The Fisher information matrix is

$$I(\gamma) = \begin{pmatrix} \frac{(\mu+2\lambda)n_2}{2\mu r^2} & -\frac{\lambda n_2}{r\mu^2} & -\frac{n_2}{2r\lambda} \\ -\frac{\lambda n_2}{r\mu^2} & \frac{\lambda n}{\mu^3} & 0 \\ -\frac{n_2}{2r\lambda} & 0 & \frac{n}{2\lambda^2} \end{pmatrix},$$

The inverse of this matrix is

$$I^{-1}(\gamma) = \begin{pmatrix} \frac{2nr^2\mu}{n_1 n_2(\mu+2\lambda)} & \frac{2r\mu^2}{n_1(\mu+2\lambda)} & \frac{2r\mu\lambda}{n_1(\mu+2\lambda)} \\ \frac{2r\mu^2}{n_1(\mu+2\lambda)} & \frac{(n_1\mu+2n\lambda)\mu^3}{n_1 n(\mu+2\lambda)\lambda} & \frac{2n_2\mu^2\lambda}{n_1 n(\mu+2\lambda)} \\ \frac{2r\mu\lambda}{n_1(\mu+2\lambda)} & \frac{2n_2\mu^2\lambda}{n_1 n(\mu+2\lambda)} & \frac{2(n\mu+2n_1\lambda)\lambda^2}{n_1 n(\mu+2\lambda)} \end{pmatrix}.$$

Taking $j = 2$, the cumulative distribution function $K_2(t)$ is estimated by

$$\hat{K}_2(t) = \int_0^t \sqrt{\frac{\hat{\lambda}}{2\pi y^3}} \exp\left\{-\frac{\hat{\lambda}(y-\hat{\mu})^2}{2\hat{\mu}^2 y}\right\} \Phi\left(\sqrt{\frac{\hat{\lambda}}{t+\hat{r}y-y}}\left(\frac{t+\hat{r}y-y}{\hat{\mu}} - 1\right)\right) dy$$

$$+ \int_0^t \sqrt{\frac{\hat{\lambda}}{2\pi y^3}} \exp\left\{-\frac{\hat{\lambda}(y-\hat{\mu})^2}{2\hat{\mu}^2 y} + \frac{2\hat{\lambda}}{\hat{\mu}}\right\} \Phi\left(-\sqrt{\frac{\hat{\lambda}}{t+\hat{r}y-y}}\left(\frac{t+\hat{r}y-y}{\hat{\mu}} + 1\right)\right) dy.$$

Now using the results of Bagdonavičius et al. [BMN08a, BMN08b, BMN09, BMN10] we can construct the asymptotic $1-\alpha$ confidence interval $(\underline{K}_2(t), \overline{K}_2(t))$ for $K_2(t)$, with

$$\underline{K}_2(t) = \left(1 + \frac{1-\hat{K}_2(t)}{\hat{K}_2(t)} \exp\left\{\frac{\hat{\sigma}_{\hat{K}_2} z_{1-\alpha/2}}{\sqrt{\hat{K}_2(t)(1-\hat{K}_2(t))}}\right\}\right)^{-1},$$

$$\overline{K}_2(t) = \left(1 + \frac{1-\hat{K}_2(t)}{\hat{K}_2(t)} \exp\left\{-\frac{\hat{\sigma}_{\hat{K}_2} z_{1-\alpha/2}}{\sqrt{\hat{K}_2(t)(1-\hat{K}_2(t))}}\right\}\right)^{-1},$$

here

$$\hat{\sigma}^2_{\hat{K}_2(t)} = C_2^T(t,\hat{\gamma}) I^{-1}(\hat{\gamma}) C_2(t,\hat{\gamma}),$$

where

$$C_2(t,\gamma) = (C_{21}(t,\gamma), C_{22}(t,\gamma), C_{23}(t,\gamma))^T,$$

$$C_{21}(t,\gamma) = \int_0^t \frac{\partial F_1}{\partial r}(t+ry-y,\mu,\lambda)\mathrm{d}F_1(y,\mu,\lambda),$$

$$C_{22}(t,\gamma) = \int_0^t \frac{\partial F_1}{\partial \mu}(t+ry-y,\mu,\lambda)\mathrm{d}F_1(y,\mu,\lambda) + F_1(y,\mu,\lambda)\mathrm{d}\left(\frac{\partial F_1}{\partial \mu}(y,\mu,\lambda)\right),$$

$$C_{23}(t,\gamma) = \int_0^t \frac{\partial F_1}{\partial \lambda}(t+ry-y,\mu,\lambda)\mathrm{d}F_1(y,\mu,\lambda) + F_1(y,\mu,\lambda)\mathrm{d}\left(\frac{\partial F_1}{\partial \lambda}(y,\mu,\lambda)\right),$$

and $z_{1-\alpha/2}$ is the $(1-\alpha/2)$-quantile of the standard normal distribution.

At the end of this section we note that using the results of [BMN10] it is easy to estimate the parameter γ when the data are censored.

Acknowledgements. These researches have been performed with partial support from the Russian Foundation for Basic Research (Grant No. 09-01-00056-a) and the Federal Agency on Education of the Ministry of Education and Science of the Russian Federation within the framework of the analytical departmental targeted program "Development of higher school scientific potential" (Project No. 2.1.2/3970) and federal targeted program of the Ministry of Education and Science of the Russian Federation "Scientific and scientific-educational personnel of innovation Russia" and the support of the Ministry of Higher Education and Scientific Research of Algeria.

References

[AN94] Aguirre, N., Nikulin, M.: Chi-squared goodness-of-fit test for the family of logistic distributions. Kybernetika, **V.30**, 214–222, (1994).

[BMN08a] Bagdonavičius, V., Masiulaityle, I., Nikulin, M.: Statistical analysis of redundant system with "warm" stand-by units. Stochastics: An International Journal of Probability and Stochastical Processes, **80**, N 2–3, 115–128, (2008a).

[BMN08b] Bagdonavičius, V., Masiulaityle, I., Nikulin, M.: Statistical analysis of redundant system with one stand-by unit. In: Mathematical Methods in Survival Analysis, Reliability and Quality of Life. (Huber, C., Limnios, N., Balakrishnan, N., Messbah, M., Nikulin, M. (Eds)), ISTE & Wiley: London, 189–202, (2008b).

[BMN09] Bagdonavičius, V., Masiulaityle, I., Nikulin, M.: Advances in Degradation Modeling: Applications to Reliability, Survival Analysis, and Finance. (Nikulin, M., Limnios, N., Balakrishnan, N., Kahle, W., Huber, C. (Eds)), Springer Verlag: Boston, 293–310, (2009).

[BMN10] Bagdonavičius, V., Masiulaityle, I., Nikulin, M.: Parametric estimation of redundant system reliability from censored data. In: Mathematical and Statistical Methods in Reliability. (Balakrishnan, N., Nikulin, M., Rykov, V. (Eds)), Springer Verlag: Boston, (2010).

[BS83] Bolshev, L.N., Smirnov, N.V.: Tables of Mathematical Statistics. Science: Moscow, (1983). [in Russian].

[Bol87] Bolshev, L.N.: On the question on testing some composite statistical hypotheses. In: Theory of Probability and Mathematical Statistics. Selected Works. Nauka: Moscow, 5–63, (1987).

[CF89] Chhikara, R.S., Folks, J.L.: The Inverse Gaussian Distribution. Marcel Dekker: New York, (1989).

[Dro88] Drost, F.: Asymptotics for generalized Chi-squared goodness-of-fit tests. CWI Tracs, Amsterdam: Centre for Mathematics and Computer Sciences, **V.48**, (1988).

[GN96] Greenwood, P.E., Nikulin, M.S.: A Guide to Chi-Squared Testing. Wiley: New York, (1996).

[GDAM97] Gunes, H., Dietz, D.C., Auclai, P.F., Moor, A.H.: Modified goodness-of-fit tests for the inverse Gaussian distribution. Computational Statistics and Data analysis, **V.24**, 63–77, (1997).

[HT86] Habib, M.G., Thomas, D.R.: Chi-squared goodness-of-fit tests for randomly censored data. Annals of Statistics, **V.14**, 759–765, (1986).

[Hjo90] Hjort, N.L.: Goodness-of-fit tests in models for life history data based on cumulative hazard rates. Annals of Statistics, **V.18**, 1221–1258, (1990).

[HR76] Hsuan, T.A., Robson, D.S.: The χ^2-goodness-of-fit tests with moment type estimator. Communications in Statistics – Theory and Methods, **V.16**, 1509–1519, (1976).

[IL99] Ionescu, D.C., Lmmnios, N. (Eds) Statistical and Probabilistic Models in Reliability, Birkhauser: Boston, (1999).

[LNS09] Lemeshko, S.B., Nikulin, M.S., Saaidia, N.: Simulation of the statistics distributions and power of the goodness-of-fit tests in composite hypotheses testing rather inverse Gaussian distribution. Proceedings of the 6th St. Petersburg Workshop on Simulation, St. Petersburg, June 28–July 4. VVM comm. Ltd., St. Petersburg, **V.1**, 323–328, (2009).

[LL09a] Lemeshko, B.Yu., Lemeshko, S.B.: Distribution models for nonparametric tests for fit in verifying complicated hypotheses and maximum-likelihood estimators. Part 1. Measurement Techniques, **V.52**, N 6, 555–565, (2009).

[LL09b] Lemeshko, B.Yu., Lemeshko, S.B.: Models for statistical distributions in nonparametric fitting tests on composite hypotheses based on maximum-likelihood estimators. Part II. Measurement Techniques, **V.52**, N 8, 799–812, (2009).

[LLP10] Lemeshko, B.Yu., Lemeshko, S.B., Postovalov, S.N.: Statistic distribution models for some nonparametric goodness-of-fit tests in testing composite hypotheses. Communications in Statistics – Theory and Methods, **V.39**, Issue 3, 460-471, (2010).

[Nik73a] Nikulin, M.S.: On a Chi-squared test for continuous distributions. Theory Probability and its Applications. **V.18**, N 3, 638–639, (1973).

[Nik73b] Nikulin, M.S.: Chi-squared test for continuous distributions with shift and scale parameters. Theory Probability and its Applications, **V.18**, N 3, 559–568, (1973).

[NS99] Nikulin, M.S., Solev, V.N.: Chi-squares goodness-of-fit test for doubly censored data with applications in survival analysis and reliability. In: Statistical and Probabilistic Models in Reliability. (Ionescu, D.C., Limnios, N. (Eds)). Birkhauser: Boston, 101–112, (1999).

[RR74] Rao, K.C., Robson, D.S.: A Chi-squared statistic for goodness-of-fit tests within the exponential family. Communication Statistics. **V.3**, 1139–1153, (1974).

[Ses93] Seshadri, V.: The Inverse Gaussian Distribution: A Case Study in Exponential Families. Clarendon Press: Oxford, (1993).

[Ses99] Seshadri V.: The Inverse Gaussian Distribution: Statistical Theory and Applications. Springer: New York, (1999).

[Sin06] Singpurwalla, N.D.: Reliability and Risk. Wiley: Chichester, (2006).

[Van98] Van Der Vaart, A.W.: Asymptotic Statistics. Cambridge Series in Statistics and probabilistic Mathematics. Cambridge University Press: Cambridge, (1998).

[VN93] Voinov, V., Nikulin, M.: Unbiased Estimators and their Applications, V.1 Univariate case. Kluwer: Dordrecht, (1993).

Index

V.V. Rykov et al. (eds.), *Mathematical and Statistical Models and Methods in Reliability: Applications to Medicine, Finance, and Quality Control*, Statistics for Industry and Technology, DOI 10.1007/978-0-8176-4971-5,

Breinigsville, PA USA
16 November 2010

249150BV00009BA/1/P